Peptidomimetics in Drug Discovery

Peptidomimetics in Drug Discovery

ANDREA TRABOCCHI
University of Florence, Department of Chemistry "Ugo Schiff",
Florence, Italy

WILEY

This edition first published 2027

Registered Offices
John Wiley & Sons, Inc., 111 River Street, Hoboken, NJ 07030, USA
John Wiley & Sons Ltd. New Era House, 8 Oldlands Way, Bognor Regis, West Sussex, PO22 9NQ, UK

For details of our global editorial offices, customer services, and more information about Wiley products visit us at www.wiley.com.

The manufacturer's authorized representative according to the EU General Product Safety Regulation is Wiley-VCH GmbH, Boschstr. 12, 69469 Weinheim, Germany, e-mail: Product_Safety@wiley.com.

Wiley also publishes its books in a variety of electronic formats and by print-on-demand. Some content that appears in standard print versions of this book may not be available in other formats.

Library of Congress Cataloging-in-Publication Data Applied for:

Hardback ISBN: 9781394326228

Cover Design: Wiley
Cover Image: marabird/Getty Images

Set in 10/12pt Times LT Std by Straive, Chennai, India
Printed and bound by CPI Group (UK) Ltd, Croydon, CR0 4YY
C9781394326228_030926

Contents

PART II Peptidomimetic Compounds 123

Preface

Peptidomimetics in Drug Discovery follows the previous title *Peptidomimetics in Organic and Medicinal Chemistry* published in 2014, which outlines the concepts and chemistry toolbox underlying the building of bioactive molecules of peptidomimetic nature. A peptidomimetic is designed to mimic a peptide with a druggable profile such as enhanced stability and retained biological activity. Over the years, the peptidomimetic approach continues to increase a successful track record for the generation of small-molecule drugs, mostly as enzyme inhibitors and receptor ligands. The topics included in this follow-up book cover the concepts in an updated view that extends to more complex peptides, as well as the chemistry toolbox taking advantage of modern C—H activation strategies and late-stage functionalization to synthesize and modify unnatural amino acids and peptides, respectively. Modern peptidomimetic scaffolds herein discussed take advantage of novel concepts in peptide bioisosterism and the successful application of peptide stapling and foldamer design for developing protein-protein interaction (PPI) inhibitors and antimicrobials. The application section with case studies in drug discovery follows the structure of the previous title, reporting the recent approval of peptidomimetic SARS-CoV-2 antiviral drug nirmatrelvir, along with other significant applications in CNS and cancer therapy to illustrate the successful application of peptidomimetics in drug discovery.

Part I outlines the chemistry toolbox to develop peptidomimetics, and this section aims to update and expand what was previously described in *Peptidomimetics in Organic and Medicinal Chemistry* book, giving room to multicomponent reactions (MCR) and modern C—H functionalization of both amino acids and peptide sequences in late-stage functionalization, and outlining advances in the development of cyclic amino acid scaffolds and their applications.

Part II is dedicated to the description of various peptidomimetic molecules that have emerged over the last two decades as key molecular entities in drug discovery in the fields of cancer and infectious diseases, spanning from macrocyclic peptidomimetics to stapled peptides and foldamers, and reporting successful examples in drug discovery for each class of peptidomimetics.

Part III is organized in five chapters discussing the applications and case studies in drug discovery. The first two chapters of this section report the peptidomimetic approach toward antiviral SARS-CoV-2 drugs and antimicrobial compounds, respectively. Chapter 10 outlines the application of peptidomimetics in the context of PPIs as a relevant class of therapeutic targets, discussing successful entries in the fields of viral diseases and cancer. Chapter 11 is dedicated to drug discovery in the context of Alzheimer's disease, particularly describing the development of peptidomimetics addressing BACE1 as a relevant protein target. The last chapter of the book is organized into 11 fact sheets reporting approved or investigational new drugs spanning various biomedical fields, highlighting the different approaches utilizing the peptidomimetic toolbox toward the development of successful drug structures.

Overall, this title aims to expand upon the basics of peptidomimetics described in the former book title, giving an updated book with new contents and reporting significant advances in the peptidomimetic toolbox and applications in drug discovery, with the aim to deliver a reference title for the entire drug discovery community interested in peptides and peptidomimetics, including scientists engaged in academia and industry.

Abbreviations

3CLpro	3-chymotrypsin-like protease
8-AQ	8-aminoquinoline
ABC	ATP-binding cassette
ACBC	aminocyclobutanecarboxylic acid
ACE2	angiotensin converting enzyme 2
ACPC	aminocyclopentanecarboxylic acid
AD	Alzheimer's disease
ADME	absorption, distribution, metabolism, and excretion
Aib	α-aminoisobutyric acid
AICD	APP intracellular domain
Alloc	allyloxycarbonyl
AML	acute myeloid leukemia
AMCA	*trans*-4-(aminomethyl)cyclohexanecarboxylic acid
AMP	antimicrobial peptide
AMR	antimicrobial resistance
APOE	apoliprotein E
APP	amyloid precursor protein
AR	androgen receptor
ART	antiretrovital therapy
BACE1	beta-site APP cleaving enzyme 1; β-secretase 1
BBB	blood–brain barrier
CAL-B	Candida antarctica lipase B
CCR	Castagnoli–Cushman reaction
CD	circular dichroism
CLL	chronic lymphocytic leukemia
CNS	central nervous system
COVID-19	coronavirus disease 2019
CPP	cell-penetrating peptide
CRAB	carbapenem-resistant *A. baumannii*
CRBN	cereblon
Cryo-EM	cryogenic electron microscopy
CuAAC	copper-catalyzed azide-alkyne cycloaddition
Da	Dalton
DAA	direct-acting antiviral
DAST	diethylaminosulfur trifluoride
DCE	dichloroethane

DDQ	2,3-dichloro-5,6-dicyano-1,4-benzoquinone
DG	directing group
DKP	diketopiperazine
DMF	dimethylformamide
DMSO	dimethylsulfoxide
DNA	deoxyribonucleic acid
DOS	diversity-oriented synthesis
DPP-4	dipeptidyl peptidase 4
DTT	dithiothreitol
EMA	European Medicines Agency
EOAD	early onset AD
ESKAPE	*Enterococcus faecium, Staphylococcus aureus, Klebsiella pneumoniae, Acinetobacter baumannii, Pseudomonas aeruginosa*, and *Enterobacter* spp.
FAH	fluoro-aminohydantoin
FBDD	fragment-based drug design
FDA	Food and Drug Administration
FITC	fluorescein isothiocyanate
GBB	Groebke–Blackburn–Bienaymé reaction
GDP	guanosine diphosphate
GH	growth hormone
GHRP-6	growth hormone-releasing hormone-6
GHS	growth hormone secretagogue
GLP-1	glucagon-like peptide-1
GPCR	G protein-coupled receptor
GPE	glycine-proline-glutamate
GTP	guanosine triphosphate
HAT	hydrogen atom transfer
HBD	hydrogen-bond donor
HCV	hepatitis C virus
HD	Huntington's disease
HDAC	histone deacetylase
hIAPP	human islet amyloid polypeptide
HDP	host defense peptide
HIV	human immunodeficiency virus
HIF-1α	hypoxia-inducible factor 1α
HR1	heptad repeat 1
HOMO	highest occupied molecular orbital
IAM1	inhibitor of amyloid 1
IAP	inhibitor of apoptosis proteins
IAPP	islet amyloid polypeptide
IEG	iterative exponential growth
IGF-1	insulin-like growth factor 1
IMCR	isocyanide-based MCR
ITC	isothermal titration calorimetry
KRAS	Kirsten rat sarcoma viral oncogene homolog

LHRH	luteinizing hormone-releasing hormone
LLP3	lentiviral lytic peptide-3
LOAD	late onset AD
LPS	lipopolysaccharide
LptD	lipopolysaccharide transport protein D
LSD1	lysine-specific demethylase 1
LUMO	lowest unoccupied molecular orbital
mAb	monoclonal antibody
MC4R	melanocortin-4 receptor
MCI	mild cognitive impairments
MCR	multicomponent reaction
mCRPC	metastatic castration-resistant prostate cancer
MDM2	mouse double minute 2 homolog
MDMX	double minute X human homolog
MDR	multidrug-resistant
MECP2	methyl CpG binding protein 2
MERS-CoV	Middle East Respiratory Syndrome coronavirus
MIC	minimum inhibitory concentration
Mpro	main protease
mRNA	messenger ribonucleic acid
β-MSH	β-melanocyte-stimulating hormone
Mtt	4-methyltrityl
NFT	neurofibrillary tangles
NGF	nerve growth factor
NMDA	*N*-methyl-D-aspartate
NMR	nuclear magnetic resonance
NMPA	National Medical Products Administration
NOX	NADPH oxidase
NRG1	neuregulin-1
NRPS	non-ribosomal peptide synthetase
NSP	non-structural protein
ORFs	open reading frames
PADAM	Passerini-Amine Deprotection-Acyl Migration process
PARP	poly(ADP-ribose) polymerase
PD	Parkinson's disease
PDB	Protein Data Bank
Pgp	P-glycoprotein
Phth	phthaloyl group
PIP	2-(pyridin-2-yl)isopropyl
PLpro	papain-like protease
PMI	principal moments of inertia
PPI	protein-protein interaction
PROTAC	Proteolysis Targeting Chimera
PSEN	presenilin
PSMA	prostate-specific membrane antigen
RBD	receptor-binding domain

RCM	ring-closing metathesis
RdRp	RNA-dependent RNA polymerase
RGD	arginine–glycine–aspartate
RIPPs	ribosomally synthesized peptides
ROCS	rapid overlay of chemical structures
RNS	reactive nitrogen species
ROS	reactive oxygen species
RTV	ritonavir
SAH	stabilized alpha-helix
SAR	structure–activity relationship
SARS-CoV-2	severe acute respiratory syndrome coronavirus 2
SLL	small lymphocytic lymphoma
SMILES	simplified molecular input line entry system
SnAP	saturated *N*-heterocycle building blocks
SPAAC	strain-promoted azide-alkyne cycloaddition
SPPS	solid-phase peptide synthesis
SPR	surface plasmon resonance
STEMtides	sulfonate-linked macrocyclic peptides
SuFEx	sulfur(VI) fluoride exchange
TAM	triazolyldimethylmethyl
TAR	trans-activation response element
TAT	trans-activator of transcription
TBAI	tetrabutylammonium iodide
TBI	traumatic brain injury
TCEP	*tris*(2-carboxyethyl)phosphine
TCICA	trichloroisocyanuric acid
TDG	transient directing group
TEM	transmission electron microscopy
TEMPO	2,2,6,6-tetramethylpiperidin-1-oxyl
TMPRSS2	transmembrane serine protease 2
TS	transition state
VEGF	vascular endothelial growth factor
VEGFR	vascular endothelial growth factor receptor
WHO	World Health Organization

Part I

Chemistry Toolbox

1

Overview of Peptidomimetic Concepts and Design

1.1 Introduction

Peptides represent a valuable class of bioactive compounds that are of considerable interest and significance as therapeutic agents across diverse areas of biomedical intervention [1]. Peptides and protein-based drugs have become increasingly prominent in the treatment of numerous disorders due to their high specificity, strong target affinity, and favorable degradation properties in some cases. Also, peptides possessing well-defined structures have proved to show capabilities for cellular targeting, molecular recognition, and internalization [2]. Specifically, the field of peptide drug discovery has progressed significantly from the development of molecules that are able to target a limited range of receptors and proteases to the generation of more elaborated compounds, encompassing a broad spectrum of biological targets including protein–protein interactions (PPIs). In particular, peptides have begun to overcome the long-standing barrier of cellular penetration, thus enabling the modulation of intracellular targets and broadening the landscape of druggable biological pathways. In this context, a renaissance in peptide-based drug research has emerged over the last two decades due to this class of molecules being particularly efficient for targeting molecular pathways by interfering with complex interactions at the molecular level [3, 4], thus addressing diseases of high impact and concern such as cancer, cardiovascular diseases, and diabetes [5]. From an industrial perspective, advances in scalable peptide synthesis taking advantage of modern solid-phase techniques and peptide delivery solutions, including nanoparticle-based formulation, enabled the optimization of bioavailability, stability, and drug delivery, thus improving their clinical applicability. Also, the high specificity of peptide therapeutics toward specific protein targets, such as enzymes or receptors, enabled their application in precision medicine studies, particularly in the fields of cancer immunotherapy, endocrine functions, and infectious diseases [6]. Finally, great efforts have been made to study peptide drug delivery strategies as an application beyond therapeutic use, particularly in cancer research for targeted therapy and molecular imaging applications, as well as in the fields of infectious diseases and material science [7].

Several thousand naturally occurring peptides have been characterized for their biological functions and roles as hormones, neurotransmitters, growth factors, ion channel ligands, or anti-infectives [8]. Also, more than a hundred peptide-based drugs have been approved as therapeutics and diagnostic agents by the FDA [9], and several hundred are under clinical trials and preclinical development [10, 11]. The knowledge of post-translational modifications leading to bioactive peptide natural products is a source of inspiration toward synthetic approaches to improve the druggability of peptides [12]. Major post-translational modifications can be found at the backbone level, including masking of amide bonds to improve resistance toward proteases and modulating the overall conformation of the peptide. Specific natural modifications include L- to D-amino acid conversion, N-methylation at amide bonds, and more extensive modifications such as heterocycle formation by side chain to backbone cyclization as natural amide bond isosteres, such as succinimide, thiazole, and oxazole involving aspartic acid, cysteine, and serine amino acids, respectively. More generally, when conceived as drugs, peptides may experience some important well-known drawbacks, including low bioavailability, stability in the bloodstream, penetration through biological membranes, including the blood–brain barrier, and difficult targeting of intracellular proteins, although recent studies concerning the application of peptides in drug delivery are improving their druggability, as anticipated (Table 1.1).

The structural modification of bioactive peptides has been pursued to improve their druggability, resulting in the development of peptidomimetics as a class of small molecules showing peptide similarity [13, 14]. The stabilization of peptides can be achieved in multiple approaches, and similar to what happens in nature, the easiest approach is the structural modification of amide bonds that can be replaced by isosteric groups, leading to increased resistance toward proteolytic cleavage. Other structural modifications can be introduced to add conformational restrictions and to lock the structure to the bioactive three-dimensional conformation, resulting in the improvement of both selectivity and affinity to the molecular target [15]. The peptide chemical space has been expanded thanks to innovations in synthetic chemistry, particularly in developing novel non-coded amino acids and other peptide-based building blocks, such as secondary-structure mimetics (Table 1.2). Similarly, advances in tagged combinatorial techniques such as DNA-encoded libraries using peptides as tag molecules [16] and biotechnological tools like phage- and mRNA-display proved to expand the interest in peptides in the drug discovery field. Finally, target identification, particularly in elucidating key PPIs at the molecular level [17], has benefited from innovations in the fields of molecular genetics, structural biology, and computational chemistry.

Table 1.1 *Strengths and limitations of peptides as drugs.*

Advantages	Disadvantages
Direct structural relationship between the bioactive peptide and the corresponding engineered peptide molecule	Poor delivery technologies
Good safety and tolerability profile	Low metabolic stability
Tissue penetration as compared to antibodies	Limited potency as compared to antibodies; generally applied to targets having bioactive peptides as ligands
Less immunogenic than recombinant proteins and antibodies	Short half-life (rapid renal clearance) and lack of *in vivo* stability (protease degradation)
Reduced size with respect to antibodies	Limited access to intracellular targets

Table 1.2 *Technological advancements in drug discovery encompassing peptides and peptidomimetics.*

Technique	Role in peptide and peptidomimetic drug discovery
X-ray crystallography	Identification and characterization of bioactive secondary structures
Technologies for peptide-based libraries (solid-phase peptide synthesis, phage-display, and mRNA-display)	Library screening for the identification of bioactive peptide epitopes
Technologies for peptidomimetic development from bioactive epitopes (stapled helical peptides, macrocyclic peptides, *N*-methyl amino acids, D-amino acids, and non-coded amino acid building blocks)	Novel peptide-derived lead molecules and extension of druggable target space
Chemical modification of clinically investigated peptides (lipidation, pegylation, and macrocyclization)	Improvement of pharmacokinetic properties

1.2 Definition and Classification

The term *peptidomimetic* [18, 19] refers to organic molecules that have similar structural motifs to those of native peptides that allow them to interact with the amino acid residues of the molecular target binding site. Historically, they have been classified into three categories, as a function of their structural elements and their functional characteristics [20] (Figure 1.1):

- *Structural mimetics*: they possess the same backbone topology as the native peptide, and they contain all the functional groups that exert the bioactivity. Some examples of them are peptides in which non-natural amino acids are inserted (i.e. β- and γ-amino acids and modified amino acids).
- *Functional mimetics*: they have an analogy with the native peptide only concerning the interactions with the molecular target, but generally they do not possess any peptide backbone element.
- *Structural–functional mimetics*: they usually possess a non-peptide backbone, but they contain all the functional groups that exert the bioactivity, properly arranged in the 3D space.

During the last decade, the classification of mimetics has evolved, together with the progress made in this field. Specifically, peptidomimetics can be classified into four major groups (Classes A–D) based on their peptide character and level of similarity to the native substrate (Table 1.3) [22].

Class A mimetics are the most similar to the corresponding bioactive peptide epitope, involving only minor structural modification at the local level to stabilize the conformation and to reduce enzymatic degradation. This group can be associated with type I mimetics of initial classification, where both the backbone and side chains closely match those of the native peptide.

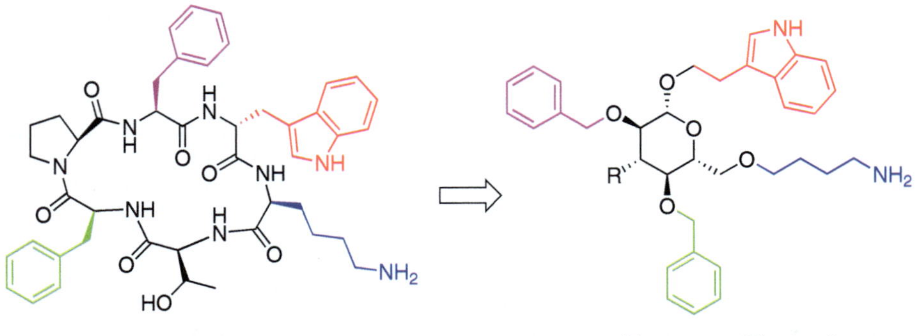

Figure 1.1 Example of a peptidomimetic compound of the cyclopeptide L-363,301, binding the somatostatin receptor and consisting of a D-glucose scaffold. Source: Adapted from [21].

Table 1.3 Classification of peptidomimetics based on peptide character.

Classification	Type	Description
Class A	Modified peptides	Minor local modifications and peptide stapling
Class B		Various backbone and side-chain modifications/peptide foldamers
Class C	Small molecules	Structural mimetics employing small-molecule scaffolds
Class D		Mechanistic/functional mimetics without direct link to side chain of parent peptide epitope

Class B mimetics, still possessing a peptide-like character, incorporate more substantial changes both at the backbone and side chains, as they may include non-natural amino acids or synthetic building blocks, such as foldamers and peptoids.

Class C mimetics comprise molecules that deviate further from the natural peptide structure, using a fully synthetic, non-peptide framework that mimics the three-dimensional arrangement of key functional groups. These compounds are closer to type III mimetics of traditional classification.

Class D mimetics are the most structurally distant to peptides, not resembling the parent peptide directly although mimicking its biological effect. These compounds are generally discovered through high-throughput screening or by further processing of Class C scaffolds and may refer to either type II or III categories of conventional classification depending on how much they vary from the original structure.

The development of a peptidomimetic compound is based on the information of the electronic and conformational elements of the reference bioactive peptide and of the structural data of the interaction with the molecular target. Moreover, some basic principles must be considered:

- the peptide backbone must be replaced by a non-peptide framework, provided that such an amide bond does not have a role in the interaction with the molecular target; otherwise, an isostere group must be used;
- the maintenance of the functional groups forming the pharmacophore of the compound, which can also be modified in a way that increases the bioactivity;

- the introduction of rigidity elements, such as cycles, double bonds, or modified amino acids, can lock the structure into the bioactive three-dimensional conformation to improve the biological activity;
- a study of the molecular target and of the pharmacophoric groups must be carried out to achieve information on designing a good peptidomimetic compound. In this view, structure–activity relationship (SAR) and structural biology studies of the binding site of the target are key tools to rational design;
- D-amino acids can be introduced to stabilize the metabolic profile of the peptide.

Following these principles, numerous synthetic strategies following hierarchical approaches and peptidomimetic scaffolds have been proposed to improve the druggability of the parent peptide with local and global modifications [19]. *In silico* approaches have been proposed and are being developed and applied more often as a powerful tool to achieve optimal peptidomimetic candidates, particularly for addressing complex protein targets.

1.3 *In Silico* Tools for Peptidomimetic Design

As stated before, a challenging aspect in designing peptidomimetics is related to PPIs. In fact, proteins interact with each other in a variety of recognition modalities with complex and diverse conformational shapes [23], displaying the highest degree of structural diversity when interacting with another protein or a peptide and producing binding pockets and recognition sites with varying degrees of specificity of interaction. Thus, although of great interest in target drug discovery for their variability and specificity [24], PPIs are difficult to study and to target [25]. The large size of the interacting area, the modulation of the interaction upon pH change, cellular functions, and cellular environment in general are the main factors contributing to the complexity of studying PPI [26]. The first is particularly relevant in the challenge of identifying small peptidomimetics targeting large PPI sites; nevertheless, PPIs are characterized by key recognition regions defined as "hot spots," which are small surface areas that determine the binding [27]. Indeed, the identification of hot spots and their incorporation into virtual screening are an established *in silico* strategy of the peptidomimetic discovery process in this context. Thus, the knowledge of the structure of the protein–protein complex is considered of key importance for the identification of peptidomimetics. In this view, the growing number of 3D structures available from X-ray diffraction, cryo-EM, and NMR methods has improved the knowledge of the structural features of protein–protein recognition and binding processes, giving detailed information on the structure of proteins and of both protein–protein and peptide–protein complexes that are of great value in peptidomimetic design. In such a context, AlphaFold is revolutionizing the structural biology field by predicting accurate structures of proteins and their complexes with interacting peptides and other proteins [28]. However, the identification of hot spots is often complicated, for example, in the study with structural biology approaches like X-ray crystallography, due to the intrinsic flexibility of proteins.

Peptides and peptidomimetics are valuable molecules to address PPIs by addressing the large binding surfaces involved in the molecular recognition of the interacting proteins [29]. Peptide-derived compounds are particularly suited to address such interactions, contrary to traditional small molecular platforms, and peptides and the peptidomimetics as their structural analogues have been considered as privileged molecules [30]. In the context of peptidomimetic design based on the structural features of PPIs, many computational approaches have been proposed during the last two decades, taking advantage of the improvement of methodologies, as well as of the high-performance computing facilities. Such approaches include those based on sequence, fragments, geometry similarity, peptide-derived pharmacophore, hierarchical and machine learning approaches, and *de novo* design. Although the field of peptide and peptidomimetic design for targeting PPIs is complicated by the wide number of prediction and refinement protocols, the combination of bioinformatics approaches with molecular simulations proved a valuable strategy to achieve drug candidates of this type. This is also due to increased calculation capabilities and accuracy in determining binding free energy and characterizing the energy levels of the binding interaction events. The information of peptide–protein complexes is generally obtained with molecular docking procedures that are successively validated by simulation techniques [17]. These are also applied to predict the binding affinity of the designed peptides for their targets. Finally, concerning the study and prediction of peptide conformation, bioinformatics tools are useful to address this issue with reasonable accuracy. Nevertheless, docking methods often oversimplify the calculation by treating backbones as rigid. While experimental methods can be applied in simple cases, complex peptides require atomistic simulations and advanced conformational search methods to estimate energetically equivalent structures to support peptide design with reliable energy estimation. Efficient conformational searching methods have been developed for this purpose, including simulated annealing, distance geometry, Monte Carlo random search, and molecular dynamics-based algorithms.

1.3.1 Classification of Methods

The numerous bioinformatics tools for peptide and peptidomimetic design can be broadly classified as ligand-based and structure-based methods [31], with the two approaches often being combined in the design process [32]. Also, methods can be categorized based on the availability of the information to be exploited in peptidomimetic proteins, whether the information about PPI, protein target, or only the bioactive peptide is available.

Ligand-based methods focus on known ligands for a target to establish a relationship between the structure and the activity and to find applications for drug optimization or the design of new drug candidates [33]. These approaches can be further classified into sequence-based and conformation-based.

Structure-based methods explore the 3D structure of protein targets to identify key sites and interactions that are important for their respective biological functions. These strategies can be further classified into pharmacophore-based and *de novo* design approaches.

In silico peptidomimetic design methods can be grouped based on the information available as the starting point [34]. The choice of the computational tool depends on the structural information available for addressing the selected PPI or protein target in general. Also, the application of different computational approaches is commonly pursued by a consensus strategy approach.

(a) If only the structure of the target protein is known, mainly from X-ray crystallography, NMR, or cryo-EM, and no known peptide ligand is available, *in silico* peptidomimetic design can still proceed using several computational strategies. The main approaches that can be followed are the following:
   - *Binding site identification and characterization*: This approach aims to identify pockets or surfaces on the protein that are likely to bind peptides or small molecules. There are a number of algorithms that are used for this purpose, including *FTMap*, *SiteMap* (Schrödinger), *DoGSiteScorer*, *CASTp*, *PockDrug*, and *fpocket*.
   - *De novo ligand design*: This strategy allows for obtaining peptide and peptidomimetic structures often characterized by specific secondary structural features, taking advantage of structural information of the target. The following algorithms have been employed for this purpose:
     - *GrowMol*: This is the first computational tool used for this approach, an algorithm capable of exploring potential ligands for an enzyme binding site with a combinatorial *de novo* growth algorithm [35].
     - *Pepsec tool of Rosetta* [36]: Pepsec is another algorithm used for this purpose, working with an "anchor and grow" docking algorithm capable of assembling a peptide from a starting residue positioned on the target protein surface [37]. With this approach, the generation of a library of peptoid foldamers was achieved using non-coded amino acids [38].
     - *PEP-FOLD* [39]: This is a conformation-based algorithm for building peptide structures and conformational pools refined by investigation of SAR studies.
   - *Molecular docking*: Computational approach used for docking small peptide libraries consisting of 2–6 amino acids to the identified binding site. Generally, flexible docking tools are used for this purpose, for example *Rosetta FlexPepDock*, *AutoDock*, *MOE*, and *Glide*.
   - *Fragment-based drug design (FBDD)*: Methods for screening small chemical fragments able to bind to different subsites in the pocket, such as Schrödinger's Fragment-Based Drug Discovery, *FTMap,* and *SeeSAR,* are generally employed for this purpose, allowing the identification of candidate fragments for subsequent linking or growing strategies to achieve a mimetic structure.
   - *Pharmacophore modeling*: This approach is carried out once peptide-like or fragment hits are docked, allowing for extracting common interacting elements such as hydrogen-bond donors/acceptors, hydrophobic sites, and so on, to build a pharmacophore model useful as a tool for subsequent virtual screening or design of peptidomimetics.

(b) If the structure of the protein–protein complex is available, pharmacophore-based methods can exploit such information. In particular, a popular method is the peptide-driven pharmacophoric approach [40]. This computational tool facilitates the search for peptidomimetics taking advantages of protein–protein and peptide–protein structure and the corresponding pharmacophore model, characterized by the type and spatial orientation of backbone and side-chain epitopes involved in peptide–target interactions, including hydrogen-bond network, hydrophobic, and electrostatic interactions [41]. Major computational tools include the following:

- *pepMMsMIMIC* [42], a web-based virtual screening tool of peptidomimetic compounds based on a 3D-similarity search strategy. This tool has been parametrized using a multimillion-member library of conformers generated from about four million commercially available chemicals implemented in the *MMsINC* database [43]. Based on pharmacophore and shape similarity, *pep-MMsMIMIC* is able to deliver compounds similar to a peptide structure bound to a protein as the input for the calculation and responsible for a PPI as a key epitope.
- *PiPred* is used to find anchor residues from the structure of protein complexes to be applied for identifying and ranking the best peptides matching the target surface from databases of fragments [44].
- *PepComposer* explores an array of protein surfaces and selects a pool of scaffolds capable of targeting them, followed by Monte Carlo simulation to refine the conformation of the newly designed peptides [45].
- *PepCrawler* and *PinaColada* select starting peptide epitopes from the analysis of protein complexes and improve their affinity through the generation of peptide derivatives ranked for the predicted affinity based on the evaluation of a library of peptide analogues randomly mutated.

(c) If the structure of the binding peptide is the only available information, *in silico* peptidomimetic design can proceed using the following computational strategies:

- *Geometry similarity*: This approach establishes a geometric similarity among peptide epitopes and non-peptide templates. The program *SuperMimic* [46] generates peptide mimetics starting from a known protein structure and conformations of a peptide mimetic, which are superimposed with the protein backbone.
- *Sequence-based approach*: This strategy is based on the identification of conserved functional motifs through arrays of sequence alignment that are modified to obtain a ranking of different candidates potentially able to interact with a specific target protein. The *PeptideMine* web server is a tool that allows for the identification of potential interacting epitopes for a protein and design criteria for peptides based on desired biochemical properties [47].
- *Fragment-based approach* [48]: Using this approach, non-peptide mimetics for specific elements of known peptide ligands are computationally identified.

Once a binding mode is known *via* docking or pharmacophore modeling, the scaffold design proceeds *in silico* through a number of approaches, including:

- Backbone/side-chain mimicry, using computational algorithms such as ROCS (shape-based comparison), *LigBuilder*, *PepMMsMIMIC*, and *PeptoGrid* (β-turn and helix mimetics, etc.).
- Macrocyclic or constrained design to mimic peptide conformations like α-helix and β-turns, using computational tools like *MacroModel*, *CycloPs*, *RDKit,* and *RosettaRemodel.*

- Virtual screening and machine learning using docking or machine-learning models to screen peptidomimetic libraries from large datasets such as *ZINC, ChemDiv,* and *Enamine* and applying filters like Lipinski's rule of 5, synthetic accessibility, and ADME properties.

In silico design is taking advantage of fast-forward progress in computational methods and algorithms, together with advances in the field of structural biology that are rapidly implementing database of protein structures and PPI information, thanks to cryo-EM technology. Machine learning is delivering a new set of tools to the field of structural biology. In particular, the AlphaFold technology paved the way for the possibility of structure prediction of unknown proteins and to move forward to protein design and protein–protein and protein–peptide interactions, thus enabling new opportunities in the *in silico* design.

1.4 Overview of Synthetic Approaches Toward Peptidomimetics

Traditionally, the modification of a bioactive peptide is addressed through a hierarchical approach, consisting of sequential modifications of the native peptide sequence toward druggable peptidomimetic structures to achieve a combination of high affinity to target and proteolytic stability (Table 1.4) [49]. Such modifications are carried out consecutively based on the information retrieved by molecular binding predictions, biophysical or cell-based screening assays, and validation data [50]. The various types of peptide modification can be classified into local and global modifications [18, 19], integrating chemical modifications of the native peptide epitope, including amino acid substitutions (e.g. linear or cyclic α-/β-amino acids), backbone amide N-alkylation (e.g. N-methylation) [51], replacement backbone amide bonds with non-hydrolyzable isosteres, peptide scaffold modification, introduction of secondary-structure mimetics (α-helix, β-strand/β-sheet or β-/γ-turns) and non-peptide templates, and macrocyclization to stabilize conformationally restricted peptide chemotypes.

The generation of peptidomimetics follows a hierarchical approach that follows some basic principles to maintain the features of the parent peptide for the interaction with the target:

- Replacement of peptide backbone with non-peptide moieties to improve bioavailability; this is applied to those amide bonds not participating directly in the interaction with the target.
- Maintenance of side chain responsible for the biological activity of the reference peptide and constituting the pharmacophoric elements for the interaction. Modifications at this level include chain homologation and use of functional-group isosteres.
- Maintenance of flexibility at the initial stage if bioactivity is observed for a flexible mimetic; the application of conformational constraints allows for improving the activity and stability of the first-generation mimetic.
- Pharmacophore hypothesis based on SAR data and structural information of either the target or the bioactive ligand, or both.

Table 1.4 *Representative local and global modifications in the peptide-to-peptidomimetic process.*

Modification type	Description	Effect	Example
		Local modifications	
Amino acid substitution	Replaces natural amino acids with non-natural analogs	Changes side-chain properties (e.g. hydrophobicity and charge)	D-amino acids, *N*-Me-amino acids, and β-amino acids
Backbone modification	Alters peptide bond geometry or connectivity	Alters secondary structure and stability	α/β-Peptides and cyclic peptides
Peptoid substitution	Shifts side chains to amide nitrogen atoms in the backbone	Reduces susceptibility to proteolysis	*N*-substituted glycine
Aryl/alkyl substitution	Replaces side chains with aromatic or alkyl groups	Modulates peptide binding and increases hydrophobicity	Aromatic substitutions (e.g. phenylalanine to biphenyl)
Cyclic modification	Introduces cyclic structures	Increases conformational rigidity and resistance to proteases	Disulfide bonds, lactam formation
Fluorination	Introduces fluorine atoms into side chains	Modifies lipophilicity and receptor binding	E.g. 4-fluoro-phenylalanine
Proline isomerization	Modifies proline residues	Affects peptide conformation and folding	Proline mimetics
		Global modifications	
Peptide cyclization	Forms cyclic structures (head-to-tail and head-to-side chain)	Enhances binding affinity and protease resistance	Cyclization using amide bond formation, click chemistry, and RCM
Stapling	Uses synthetic staples (e.g. hydrocarbon staples) to stabilize folded peptides	Enhances structural rigidity and stability	Stapled peptides (e.g. α-helix stapling)
Macrocyclization	Creates large ring structures to increase rigidity	Increases receptor selectivity, stability and cell permeability	Cyclization *via* peptide bonds, click chemistry, and RCM

In this view, the hierarchical approach toward peptidomimetic design follows a gradual modification of the starting bioactive peptide to improve druggability in terms of potency and stability and to understand the role of peptide residues in molecular recognition as a guide to rational design.

Several steps are then considered in the hierarchical approach toward peptidomimetic design, including:

- *Size reduction*: this is carried out by systematic peptide truncation at both *N*- and *C*-termini to generate a library of small peptides that are assayed for their potency in view of identifying the minimum peptide sequence responsible for bioactivity.
- *Alanine scan*: it is a common approach to study the role of each amino acid in the parent peptide toward bioactivity that consists of the generation of a library of peptides differing by the presence of an alanine residue replacing a specific amino acid in the sequence. If substitution reduces or abolishes activity, then the residue that has been replaced by alanine is essential for binding to the target and is a key element of the pharmacophore.
- D-*Amino acid scan*: like alanine scanning, this is a powerful approach that involves systematically replacing individual L-amino acids in the reference peptide with their D-enantiomers to study how stereochemistry affects the bioactivity. This study also furnishes information on the role of stereochemistry in peptide stability and conformational preferences.
- *N-methyl scan*: similar to alanine- and D-amino acid scan, by replacing each amino acid systematically with the corresponding *N*-methyl analogue, allows for achieving information on stability toward peptide degradation by proteases, potency by modulating hydrogen-bonding interactions, and conformational preferences.
- *Local and global modifications*: following the identification of the key peptide epitope and information on the properties of each residue, the application of structural modifications to modulate the conformation and the rigidity of the molecule are applied locally and over the entire peptide sequence (global approach).

During peptide modification, sequential generation of peptidomimetics can be envisaged, based on the peptide features and target structural characteristics, starting from minimal modification of the peptide structure to the introduction of molecular scaffolds to achieve a non-peptide peptidomimetic:

- *First-generation peptidomimetics*: hit peptide epitopes are the basis for the synthesis of pools of modified peptides possessing local modifications at the peptide backbone with corresponding isosteres and at the side chain with suitable non-coded amino acids bearing isosteric moieties. Overall, the aim of the modification is to improve resistance to peptide degradation without losing the inhibitory activity peptide and induce conformational restriction.
- *Second-generation peptidomimetics*: both selected peptide epitopes and hits resulting from first-generation peptidomimetics are subjected to macrocycle arrays as a global modification using diverse chemistries (i.e. peptide stapling through CuAAC, RCM, and disulfide bonds) and local modification with amino acid homologues; operationally, this is carried out by solid-phase synthesis, taking advantage of macrocyclization on the solid phase prior to peptide cleavage from the resin.

- *Third-generation peptidomimetics*: this stage is devised taking into account the SAR analysis resulting from previous hit compounds and conceived by applying dipeptide isosteres for small epitopes or developing peptoids and foldamers for hit peptide epitopes of >10 amino acids or requiring the conformation to mimic helix or sheet as secondary structures. The synthesis of non-peptide peptidomimetics is achieved by developing suitable dipeptide isosteres such as delta-amino acids and using these scaffolds in the synthesis of the parent first-generation peptidomimetic. For example, the synthesis of foldamers is achieved by solid-phase peptide synthesis (SPPS) using suitable non-coded amino acids, whereas peptoid fragments are developed on solid phase using the sub-monomer approach [52].

1.5 Amino Acid and Peptide Bioisosterism

Peptide bioisosteres follow the general concept of bioisosterism that comprises those groups possessing size, shape, and electronic properties capable to mimic the function of another group in a molecule to preserve or improve the biological activity of the parent chemotype. In the peptide field, such structural modifications can replace peptide bonds or amino acid side chains with groups possessing similar biological activity and improving properties like stability, bioavailability, and selectivity. This is a crucial approach in peptide drug discovery, as these modifications can address issues like proteolytic degradation and poor membrane permeability. The class of peptide bioisosteres can be grouped based on the structure being addressed:

- *Peptide bond replacements*, particularly involving amide bioisosteres with groups like ethers (—O—), ketones (—CO—), thioamides (—CS—NH—), or constrained rings (e.g. triazoles and oxadiazoles).
- *Side-chain replacements*, involving amino acid substitution with non-natural amino acids or amino acid mimetics. This is a broad definition that may envision a further breakdown based on single-atom replacement, addition of conformational constraints, and modulation of hydrophobic content.
- *Dipeptide mimetics*, involving a dipeptide sequence in the bioisosteric replacement, often applied to mimic secondary-structure portions of the parent peptide epitope.

1.5.1 Peptide Bond Isosteres

Over the last 30 years, extensive chemical modifications have been directed toward the peptide backbone to introduce a range of advantageous features, such as:

- Greater resistance to enzymatic proteolysis and thereby improved metabolic stability.
- Optimized molecular recognition to facilitate active-site binding in the development of peptidomimetic inhibitors targeting proteases.
- Enhanced pharmacokinetic properties, including bioavailability.

As described in the previous paragraph, the modification of amide bonds for improving the *in vivo* proteolytic stability of peptides is viable for those moieties that are not directly engaged in the biological recognition of the target. Such substitutions, however,

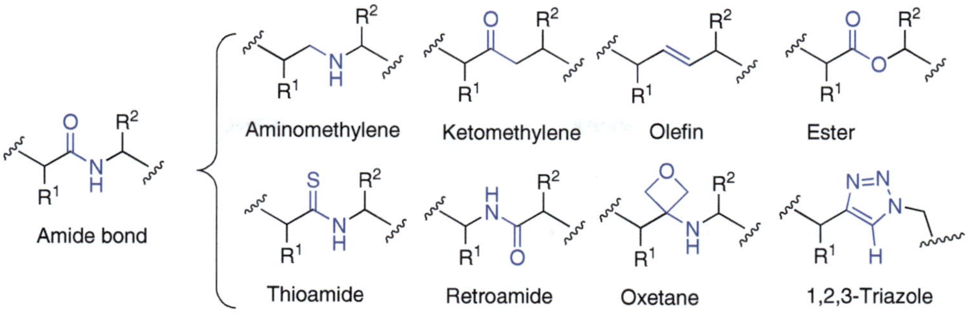

Figure 1.2 *Representative amide bond isosteres used in peptidomimetic design.*

can alter the intrinsic chemical properties of the peptide bond, particularly its conformational preferences, and consequently modulate the binding affinity toward the target protein. Therefore, these strategies must be applied with caution. Over the years, numerous amide bond isosteres have been applied in peptidomimetic design, including aminomethylene, ketomethylene, ethylene, olefin (which preserve the planarity and *cis*/*trans* isomerism of the amide bond while displaying a reduced dipole moment and loss of hydrogen-bonding capacity), ether, thioether, retroamide, thioamide, ester, triazole, thiazole, and oxazole as non-classic heterocyclic bioisosteres of the amide bond (Figure 1.2 for representative amide bond isosteres). Specifically, the triazole ring proved to be a versatile amide bond surrogate [19, 53, 54], thanks to its stereoelectronic features, being planar, and displaying similar electronic content, dipole moment, and hydrogen-bonding capability as the nitrogen and oxygen atoms of the amide bond [55]. Of particular value is also its direct accessibility *via* copper-catalyzed azide–alkyne cycloaddition (CuAAC) as the exemplary "click chemistry" reaction. Sulfonamide and phosphonamide have also been proposed as amide surrogates, being able to promote distortions in the peptide backbone due to the ~90° dihedral angle these groups adopt to delocalize π electrons. Similarly, the substitution of the carbonyl moiety of an amide bond with an oxetane ring has proven highly effective in leading to analogs of lactams and ketones with improved physicochemical properties and resistance toward proteolysis, demonstrating compatibility with modern practices in peptide synthesis [56]. The oxetane ring demonstrated utility in promoting peptide macrocyclization [57], as the four-membered heterocycle brings the *C*- and *N*-termini into close proximity more efficiently than most local modifications.

A distinct class of isosteres is derived from hydroxyl-containing fragments, which function as stable, non-hydrolyzable transition-state analogues of the tetrahedral intermediate formed during peptide bond cleavage. Peptidomimetic compounds developed as protease inhibitors have long been studied considering the inclusion of transition-state isosteres, where a hydroxyl group displaces the water molecule at the cleaving site and coordinates with key residues, forming strong hydrogen bonds, thus precluding enzymatic activity. The most used transition-state isosteres are reported in Figure 1.3, that enabled the generation of several classes of peptidomimetics, for example, in the development of inhibitors of HIV protease and BACE1. The hydroxyethylamine group, which conserves the hydrogen-bonding features of the transition state, is present in several approved drugs as HIV protease inhibitors. Other notable transition-state mimics include the

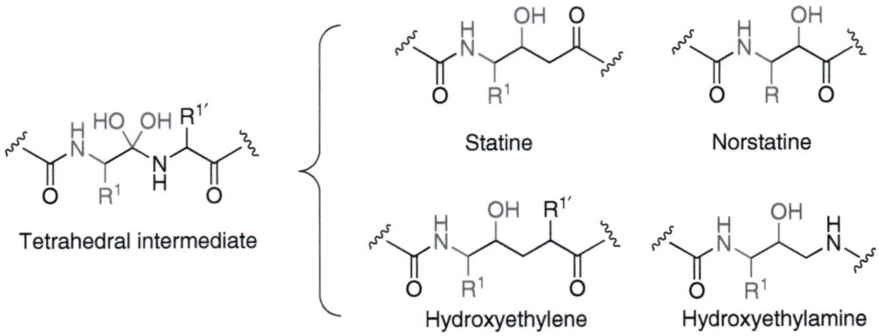

Figure 1.3 Transition-state isosteres commonly used for peptidomimetic protease inhibitors.

hydroxyethylene group (with nitrogen replaced by a methylene, as in the antihypertensive renin inhibitor aliskiren) and β-hydroxy-γ-amino acids (i.e. statine, as exemplified by the natural aspartic protease inhibitor pepstatin).

1.5.2 Side-Chain Isosteres

Side-chain modifications have been employed to probe steric and electronic contributions in pharmacophoric regions, such as enhancing hydrophobic interactions through the incorporation of aromatic substituents or introducing polar groups to engage in hydrogen bonding and other polar interactions with the target receptor [19]. Modification of amino acid side chains represents a complementary strategy to introduce local variations into the parent peptide. Initially limited to the incorporation of D-, β-, or γ-amino acids, peptidomimetic chemistry has advanced significantly with the development of conformationally constrained amino acid analogues.

A particularly important area of side-chain modification involves the incorporation of non-natural amino acids bearing substituents designed to strengthen the interactions with "hot spots" on target protein surfaces. Considerable effort has focused on substituting aromatic side chains with heterocycles or extended aromatic moieties to increase size and hydrophobicity, as for phenylalanine where naphthyl, biphenyl, and thienyl groups have been studied as isosteres of the phenyl ring (Figure 1.4, top). Similarly, surrogates of basic and acidic residues have been engineered to modulate their ionization properties and, consequently, their ionic interactions. For example, efforts have targeted the carboxyl groups of aspartic and glutamic acids, leading to isosteres that modulate the acidity and lipophilicity [58], such as phosphonates and acylsulfonamides, with higher and lower acidity as compared to carboxylic acid, hydroxamic acids, and the non-classical isosteric tetrazole, possessing similar electrostatic potential as of carboxylic acid (Figure 1.4, middle). Concerning basic residues, the guanidinium group of arginine plays a critical role in several natural and synthetic inhibitors of serine proteases and integrins. However, its high basicity often contributes to low selectivity and poor oral bioavailability; thus, various guanidinium isosteres have been developed to modulate the basicity (i.e. amino-imidazoline, amino-pyridine, and acyl-arginine, Figure 1.4, bottom).

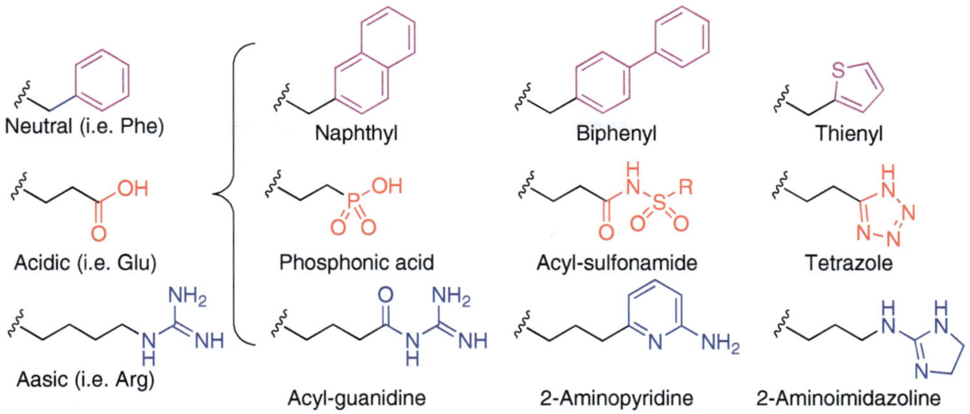

Figure 1.4 Representative side-chain isosteres in acidic, neutral, and basic amino acids.

1.5.3 Dipeptide Isosteres

The design of new amino acid derivatives and molecular frameworks that function as dipeptide isosteres has become a central focus in peptidomimetic research [19]. This strategy is particularly valuable because it enhances resistance to enzymatic proteolysis, typically by substituting the conventional amide linkage with alternative structural motifs, many of which are incorporated into cyclic systems. Also, dipeptide isosteres are capable of restricting backbone conformational flexibility by locking ϕ, ψ, and ω torsional angles. A pivotal advance in this field was introduced by Freidinger, who proposed the use of dipeptide lactams as conformational restraints [59]. These structures enforce a *trans* configuration at the peptide bond and simultaneously restrict backbone dihedral flexibility. These structures featured δ-amino acid moieties given atom-by-atom structural similarity to dipeptide sequences have been exploited as a versatile platform to develop dipeptide isosteres possessing linear or more often constrained structures. Subsequent investigations from Freidinger's group and others demonstrated that such lactams not only improve metabolic stability but also increase receptor selectivity and often lead to stronger biological activity [60]. Based on this framework, a variety of heterocyclic scaffolds, including lactams, oxopiperazines, and spirocyclic scaffolds, have been employed to modulate the geometry of dipeptide units (Figure 1.5). The ring size and architecture of these cyclic surrogates were found to exert significant influence on backbone geometry. Among the most conformationally constrained scaffolds are spirocyclic derivatives, in which two fused rings simultaneously restrict ϕ, ψ, and ω angles, thereby enforcing a rigid architecture. For example, spiro-γ-lactam-based δ-amino acids (Figure 1.5) have been successfully employed in the design of constrained mimetics of the neurotensin *C*-terminal hexapeptide, resulting in more than 1,000-fold increase of binding affinity when compared to the more flexible analogue [61]. Another widely explored structural motif is the piperazine-2-one ring, which has emerged as a versatile scaffold in peptidomimetic chemistry. Tian and co-workers developed highly potent and selective melanocortin-4 receptor (MC4R) agonists based on 1,3,4-trisubstituted 2-oxo-piperazine derivatives, demonstrating the pharmacological utility of this scaffold [62]. Other δ-amino acid structures have been

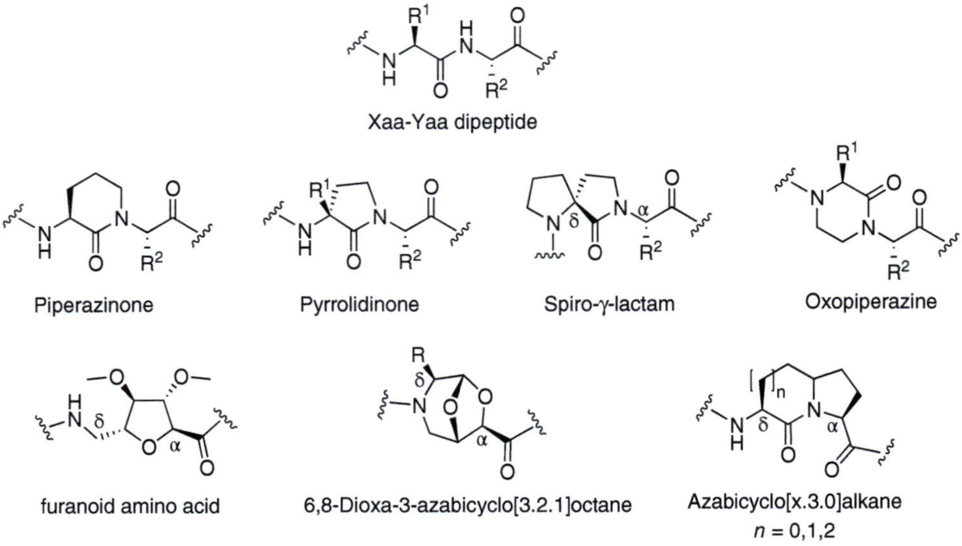

Figure 1.5 *Representative examples of constrained dipeptide isosteres.*

exploited as a versatile platform to develop dipeptide isosteres possessing both linear and constrained structures. Among cyclic δ-amino acids as dipeptide isosteres, furanoid and pyranoid δ-sugar amino acids have been found valuable for constraining the peptide backbone, introducing stereochemical diversity and enhancing hydrophilicity due to the hydroxyl-rich framework [19]. For example, based on the biological activity of some natural lipopeptides, a library of linear analogues incorporating a furanoid sugar amino acid was recently synthesized and evaluated for their *anti*-leishmanial activity [63]. Representative examples of bicyclic δ-amino acids include molecular scaffolds containing pyrrolidine, such as azabicyclo[x.3.0]alkanes as robust mimetics of Xaa-Pro units, and bridged systems like 6,8-dioxa-3-azabicyclo[3.2.1]octane-7-carboxylic acids (Figure 1.5). These conformationally biased scaffolds, providing atom-by-atom mimics of dipeptide units, have been extensively used to design β-turn mimetics enabling precise substitution of *i*+1 and *i*+2 residues in β-turn systems. Collectively, dipeptide isosteres, ranging from simple lactams to spirocyclic and δ-amino acids, constitute a versatile toolkit for enhancing proteolytic stability, receptor selectivity, and conformational control in peptidomimetic design.

1.6 Local Modifications: The Tethering Approach

The "tethering" approach is a key strategy for side-chain modification to target dihedral torsion angles (ϕ, ψ, χ; Figure 1.6) through the introduction of conformational constraints to control the flexibility of rotatable bonds within amino acid residues, thereby shaping the overall peptide conformational landscape. Local modifications of amino acid side chains primarily influence the conformational landscape of peptides by regulating the flexibility

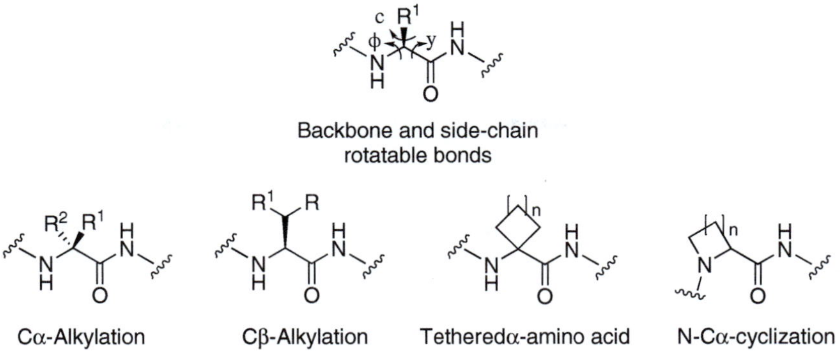

Figure 1.6 *Strategies for tethering the rotational space.*

of the rotatable bonds within each residue. To achieve this, various tethering strategies have been developed to restrict specific dihedral angles (Figure 1.6), including [64]:

- C_α*-alkylation* is one of the simplest approaches, producing quaternary carbon centers.
- *α-substituted amino acids* reduce rotation about the $N—C_\alpha$ and C_α—CO bonds, thus decreasing backbone flexibility by nearly 90%. The best-studied example, α-methylalanine, is known to restrict ϕ and ψ angles in α-helices and 3_{10}-helices.
- *N-alkylation* restricts backbone dihedral angles without significantly altering χ.
- *β-alkylation* selectively constrains χ^1 and χ^2, proving especially useful in favoring distinct rotamers of aromatic side chains.
- $N_\alpha—C_\alpha$ *cyclization* enables conformational locking by restricting both ϕ and χ torsions and limiting ψ flexibility due to steric and non-bonded interactions between the introduced ring and neighboring residues. Constrained amino acids were initially designed as proline analogues, with ring-size variations, heteroatom incorporation, and substitution at the 3-, 4-, and 5-positions [65].
- $C_\alpha—C_\beta$ *conformational restriction* can be accomplished by introducing double bonds or cyclopropane units, yielding α,β-unsaturated amino acids (dehydro-amino acids) or cyclopropyl analogues. Appropriate selection of *E*/*Z* stereochemistry enables precise positioning of side chains. This approach is relevant to controlling side-chain mobility, particularly when these side chains are directly involved in target recognition.

Tethering strategies have been applied to aromatic amino acids, too, with bicyclic α-amino acids offering rigidified phenylalanine and histidine analogues.

1.7 Heterocyclic Peptidomimetics

Heterocyclic scaffolds play a pivotal role in peptidomimetic chemistry due to their ability to impose conformational constraints and to improve metabolic stability and target selectivity. Their occurrence and significance in peptidomimetic design have already been shown in previous paragraphs, highlighting the role of such scaffolds in local modifications, both at the backbone and the side chain. The panorama of heterocyclic peptidomimetics

Table 1.5 Use of heterocycles in peptidomimetic design.

Approach	Heterocyclic scaffolds	Role	Representative examples
Backbone modification	Lactams, piperazinones, and imidazolinones	Replaces amide bond, rigidifies backbone, and improves proteolytic stability	Freidinger's lactams and 2-oxopiperazine MC4R agonists
Proline mimetics	Pyrrolidinones, imidazolidinones, and azabicycles	Reproduce proline's conformational constraint; lock bioactive geometries	Constrained pyrrolidinone-based peptides
Side-chain surrogates	Triazoles, oxazoles, and thiazoles	Mimic amino acid side chains; adjust lipophilicity and receptor interactions	1,2,3-Triazole Tyr/Phe analogues
β-Turn, secondary-structure stabilization	Diketopiperazines, benzodi-azepinones, and quinazolinones	Induces or stabilizes β-turns and helices; enhances receptor binding	Diketopiperazine-based integrin ligands

is wide, and over the years, they have found application in several peptidomimetic design approaches [19]. Recent reviews account for continued improvements and studies in this field, for example, reporting on solid-phase synthesis strategies for protein secondary-structure mimetics [66]. The application of heterocycles in peptidomimetic design can be summarized as follows (see also Table 1.5):

- *Backbone modification* employs heterocycles, such as lactams and 2-oxopiperazines, to mimic the peptide bond and to lock or constrain the rotatable bonds to restrict torsional flexibility.
- *Proline mimetics* represent another important approach, where heterocycles such as pyrrolidinones or imidazolidinones are used to replicate the conformational influence of proline residues, often locking peptides into bioactive geometries. Recent reviews have thoroughly reported the wide panorama of proline analogues and their relevance in drug design [67, 68], classifying all proline derivatives in several groups, including substituted prolines, unsaturated and fused structures, ring-size homologues, heterocyclic, and bridged proline-resembling structures.
- *Side-chain surrogates* are introduced through heterocyclic frameworks to mimic natural amino acid functionalities and for tuning lipophilicity or receptor interactions.
- *β-Turn and secondary-structure inducers* often rely on heterocycles to stabilize specific folding patterns, thereby enhancing binding affinity in peptide-receptor recognition.

1.7.1 Aromatic Scaffolds

Heterocyclic compounds have been applied in local modification of peptides in the context of peptidomimetic design both at side chain and backbone level (see Sections 1.5 and 1.7.3), particularly applying oxazole, thiazole, imidazole, pyrimidine, and 1,2,3-triazole as peptide bond replacements.

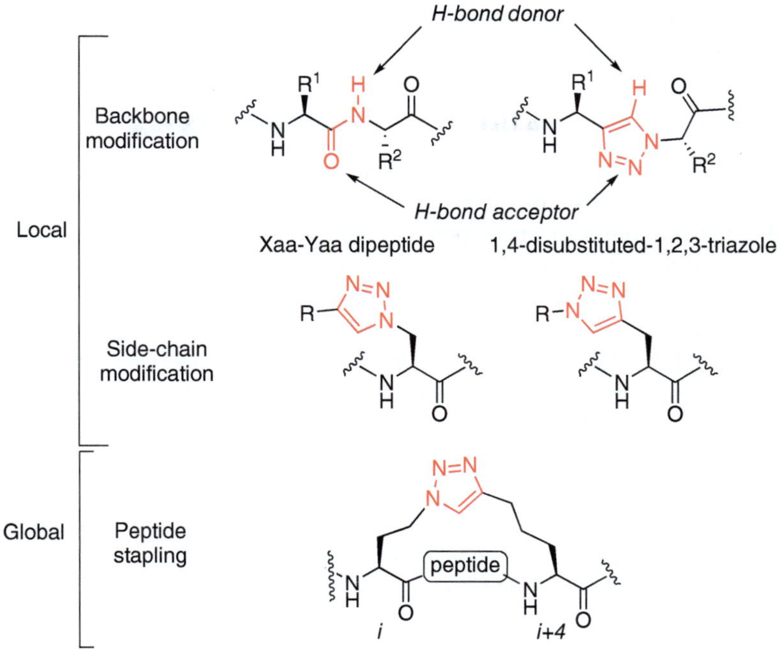

Figure 1.7 Use of 1,4-substituted-1,2,3-triazole as a peptide bond isostere and in side-chain modification (local modification) or as a peptide stapling element (global modification).

Of particular relevance with aromatic heterocycles in peptidomimetic design is the 1,2,3-triazole ring (Figure 1.7). Among the wide number of applications that such a molecule has found in chemistry and the biomedical field following the advent of the CuAAC as the most representative "click chemistry" reaction, the triazole ring proved to effectively work as a peptide bond isostere, by mimicking the *trans*-amide bonds because of their similar size, planarity, hydrogen-bonding capabilities, and dipole moment [69], and such substitution proved to afford protease-resistant peptidomimetics [70, 71]. The triazole ring has also been applied in the context of side-chain modification, particularly for bioconjugation issues. In addition, 1,5-disubstituted-1,2,3-triazoles were assessed as *cis*-peptide bond surrogates [72].

1.7.2 Nonaromatic Scaffolds

Among nonaromatic heterocyclic scaffolds applied to peptidomimetic design, saturated aza-heterocycles have been studied with particular interest, resulting in the development of aziridine, azetidine, pyrrolidine, piperidine, and piperazine ring-containing unnatural cyclic amino acids [73]. The structure of such molecules is characterized by having a C_α—N_α cyclized amino acid structure that mimics the structural constraint imposed by proline, such as the steric hindrance between the cyclic amino acid and neighboring residues, which affects the local conformation around the unnatural amino acid. The different ring size is instrumental for modulating the constraint at the rotatable ψ dihedral angle in peptidomimetic design. These heterocyclic structures have found wide applications as enzyme inhibitors and more recently to modulate PPIs efficiently when inserted in bioactive peptide epitopes.

1.7.3 Backbone Editing with Heterocyclic Peptidomimetics

While traditional approaches to peptide modification have largely focused on amide bonds, side chain, or dipeptide isosteres, a strategy that introduces heterocyclic units directly into the peptide chain is a complementary approach to backbone modification. Such heterocyclic motifs, including oxazole/oxazolidine and thiazole/thiazolidine rings, provide substantial rigidity to the peptide backbone and enhance molecular recognition with biological targets [74]. A representative example is provided by the synthesis of the isopentenylated cyclic peptide trunkamide A through heterocycle formation in the backbone [75]. In this case, treatment of a heptapeptide thioamide **1** with Burgess' reagent promoted cyclodehydration to afford a thiazoline moiety **2** (Figure 1.8).

Similarly, intramolecular cyclization of adjacent serine residues as in **3** has been achieved using diethylaminosulfur trifluoride (DAST), enabling oxazole ring formation (**5**), applied to the total synthesis of cyclopeptides, such as YM-216391 (Figure 1.9), and highlighting the utility of reactive polar residues, such as the hydroxyl group of serine and threonine, for heterocyclic backbone construction [76].

Carboxylic acid has been exploited as a polar reactive functional group for heterocycle formation too. This reactivity is of interest for direct backbone functionalization as this group is largely present in peptides. For instance, radical decarboxylative cyclization allows for the introduction of oxazole rings at defined positions without perturbing other functional groups. Specifically, Boto developed a one-pot process combining radical decarboxylation, oxidation, isomerization, and iodine-promoted cyclization to convert *N*-acyl amino acids into 2,5-disubstituted oxazoles under mild, metal-free conditions that was successfully applied to a range of peptide substrates. When the acyl group of the amino acid participated in the cyclization, the resulting oxazole derivative (**7**) preserved a terminal amino group, allowing for further extension of the peptidyl chain (Figure 1.10) [77]. The

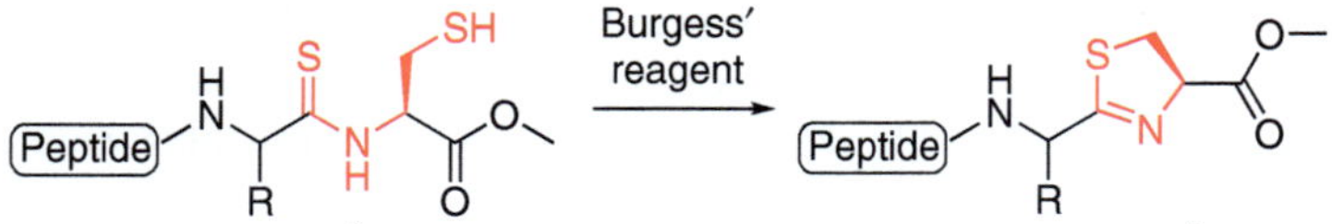

Figure 1.8 The formation of dihydrothiazole (thiazoline) as a heterocyclic unit for backbone editing using Burgess' reagent.

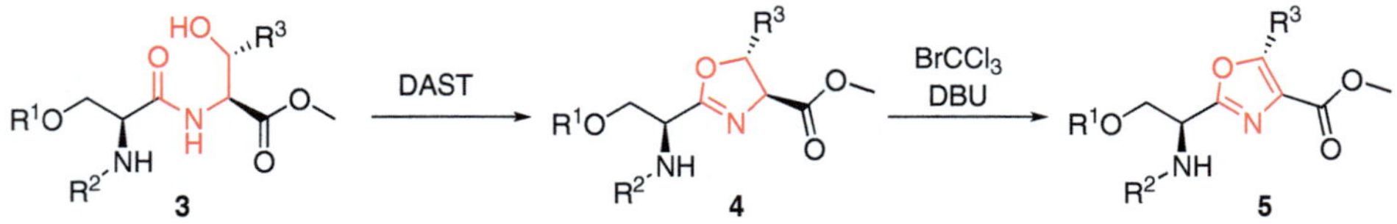

Figure 1.9 Backbone modification with dihydro-oxazole/oxazole as heterocyclic units by applying DAST reagent to Xaa-Thr or Xaa-Ser dipeptides.

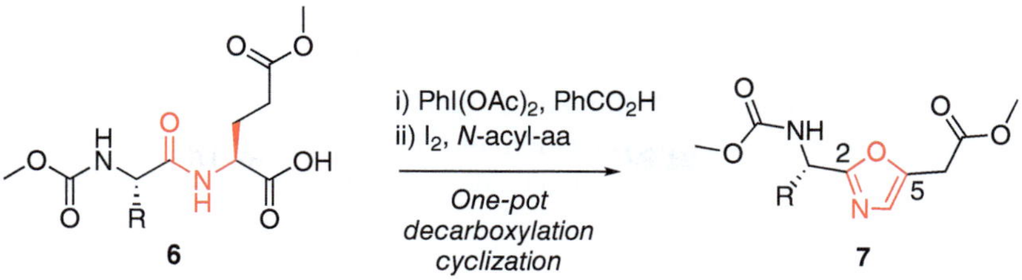

Figure 1.10 One-pot decarboxylation–cyclization to convert N-acyl amino acids into 2,5-disubstituted oxazole.

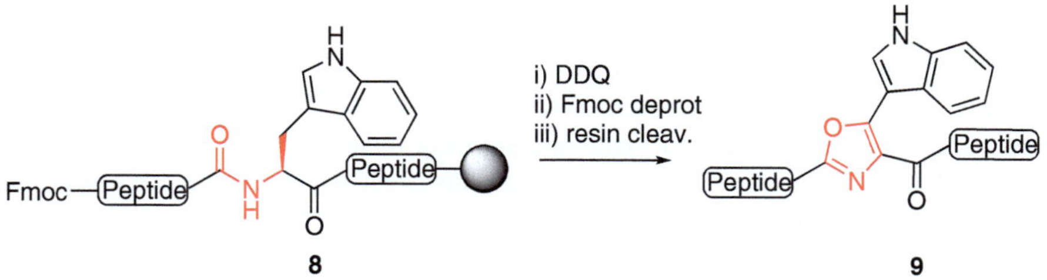

Figure 1.11 Chemoselective modification of tryptophan-containing peptides.

reaction proceeds through decarboxylation to generate an iminium ion, which undergoes isomerization to an enamide intermediate. Subsequent iodination yields iodonium species, which are trapped intramolecularly to form oxazoline intermediates, ultimately delivering oxazole scaffolds.

Additional innovation in backbone heterocycle formation has been reported by Qvortrup, who developed a chemoselective backbone modification method on solid phase through insertion of oxazole in tryptophan-containing peptides (Figure 1.11) [78]. In this approach, exposure to 2,3-dichloro-5,6-dicyano-1,4-benzoquinone (DDQ) promotes the selective cyclodehydration between the peptide backbone and the indole side chain of **8**, affording fully conjugated indolyl-oxazole motifs (**9**). This transformation not only rigidifies the peptide structure but also introduces intrinsic fluorescence, thus enabling this strategy as a potential peptide-labeling method. The reaction demonstrates broad sequence compatibility but is limited by intolerance to unprotected lysine residues, which lead to undesired side products.

1.8 Global Restrictions

A detailed understanding of the three-dimensional conformational features of peptides, and their correlation with binding affinity and modulation of therapeutic targets, has been central to advances in peptide drug design. Historically, regiospecific incorporation of D-amino acids, *N*-methyl, C_α-methyl, dehydro-amino acids, and cyclic amino acids into peptide leads has provided key insights into their ability to induce or stabilize β-turns, γ-turns, β-strands, α-helices, and 3_{10}-helices. Starting from these findings, the development of versatile synthetic peptidomimetics and carefully designed non-peptide scaffolds has emerged as a major strategy to reproduce key side-chain orientations and backbone geometries characteristic of such secondary structures. The interest to macrocyclic peptides in the context of targeting PPI stimulated significant efforts toward the study of new synthetic approaches toward macrocyclization, thus enabling the expansion and advances of chemical methodologies for peptidomimetic design in the context of global restrictions.

1.8.1 Macrocyclic Peptides

Macrocyclization of linear peptides is a highly effective strategy for generating peptidomimetics with restricted conformational flexibility [79]. Unlike methods that introduce local cyclic scaffolds, this approach alters the global conformational profile of the peptide, stabilizing specific secondary structures and reducing overall structural freedom. Such macrocyclic peptides gathered significant interest as therapeutic agents, particularly for modulating PPIs, which as described in Section 1.3 are often difficult to target with small molecules. They exhibit enhanced metabolic stability relative to linear peptides and can adopt well-defined three-dimensional conformations, improving both receptor selectivity and binding affinity. Structural features that contribute to these properties include a high prevalence of *cis* amide bonds, absence of free *N*- and *C*-termini, and restricted backbone mobility. Macrocyclic peptides are widely represented in nature, with relevant examples including somatostatin and its analogue octreotide, cyclosporine A, nisin, and colistin (Figure 1.12).

Strategies for peptide cyclization are commonly classified into four main categories, based on diverse chemical approaches to constrain the peptide conformation:

- *Head-to-tail*, occurring between *N*- and *C*-terminal ends.
- *Side chain-to-side chain*, achieved by exploiting reactive functional groups on the side chain of natural or non-natural amino acids.
- *Head-to-side chain,* occurring between *N*-terminal end and a complementary side-chain functional group of an amino acid.
- *Side chain-to-tail*, occurring between *C*-terminal end and a suitable functional group of an amino acid.

A variety of synthetic methodologies have been developed to convert linear peptides into macrocyclic frameworks. Among the most frequently employed strategies are lactamization, lactonization, disulfide bridge formation between cysteine residues, triazole generation *via* CuAAC click chemistry, and ring-closing metathesis (Figure 1.13).

Macrocyclization is typically conducted under high-dilution conditions to minimize undesirable intermolecular reactions such as oligomerization and polymerization, often achieved by immobilizing the linear peptide on an insoluble polymer support. The efficiency of cyclization is strongly influenced by ring size, as peptides comprising more than seven residues generally undergo cyclization with relative ease, whereas smaller sequences often present substantial challenges. In such cases, the inclusion of turn-inducing structural motifs as internal conformational auxiliaries can facilitate cyclization by promoting a backbone geometry that reduces transition-state strain. Efforts in the chemistry of peptide macrocyclization facilitated this approach being taken into account as a pivotal strategy in contemporary peptide drug discovery, enabling access to extensive chemical space and the design of peptides capable of modulating challenging intracellular targets. Overall,

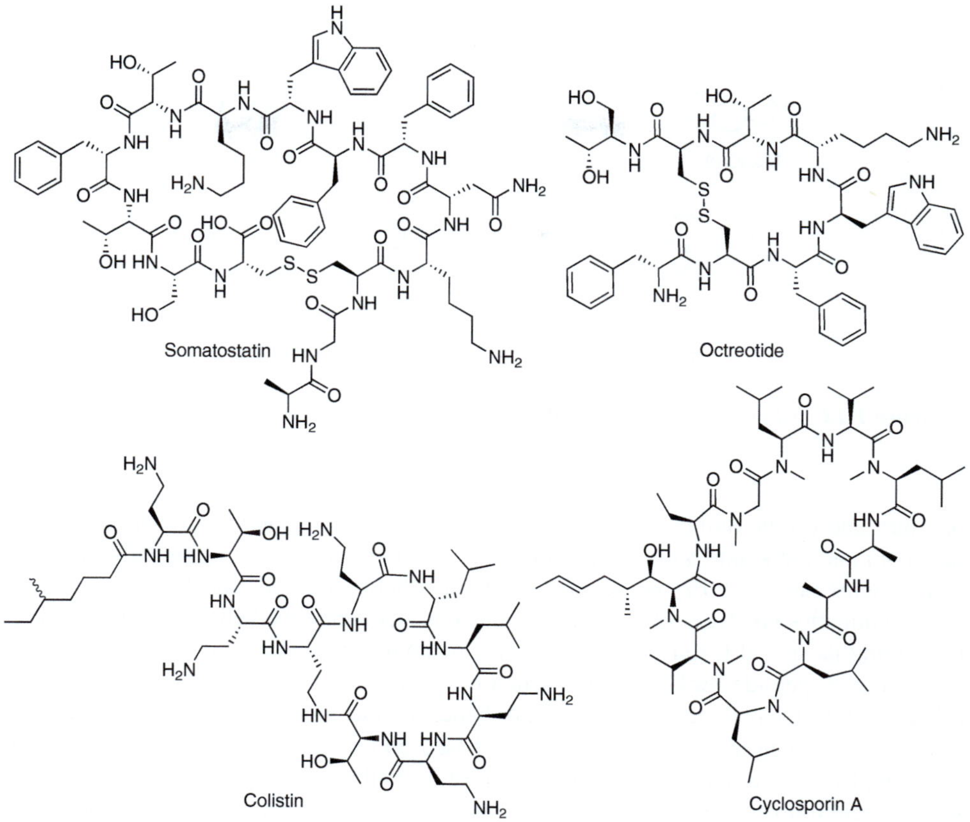

Figure 1.12 *Representative naturally occurring bioactive cyclopeptides.*

the utility of macrocyclization, whether through classical chemical linkages or advanced stapling methodologies proved as a robust strategy to constrain peptide conformations, to modulate biological activity and capable of generating drug-like macrocyclic compounds with significant translational potential.

1.8.2 Secondary-Structure Peptidomimetics

The stabilization of secondary structures, such as α-helices, β-strands, and β-turns, has been pursued to develop compounds capable of addressing PPIs, that are generally established among structured peptide fragments within large interacting surfaces. PPIs are central to numerous cellular processes and disease pathways, yet their large and flat interfaces make small-molecule targeting difficult. Over the years, several strategies and peptidomimetic scaffolds have been studied to stabilize a specific peptide secondary structure.

1.8.2.1 α-Helix Mimetics

Helices are formed by repetitive sequences stabilized by intramolecular hydrogen bonds between sequential residues. They are classified depending on the number of residues in

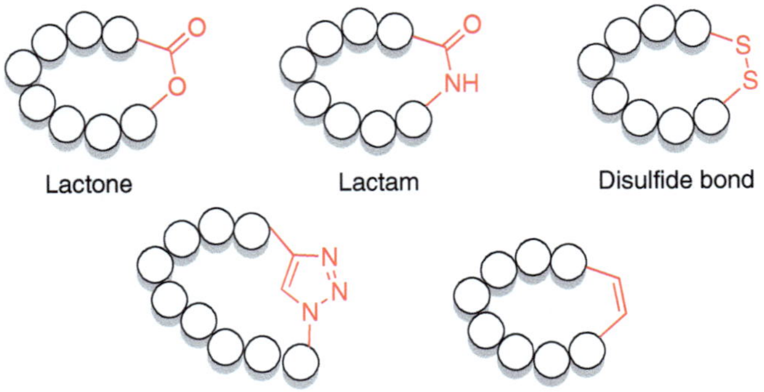

Figure 1.13 *Chemistry most frequently employed for the macrocyclization of peptides.*

one turn and on the number of atoms between the carbonyl group of the amino acid at position *i* and the amide proton at position *i*+n. The most widely occurring helix in globular proteins (around 90%) is the 3.6_{13}-helix, or α-helix. Methods for designing α-helix peptidomimetics involve several strategies aimed at stabilizing the helical conformation or mimicking its spatial features (Figure 1.14). One common approach is the application of constraints to peptide backbone, also called covalent staples such as lactam or disulfide bridges to lock the peptide into an α-helical shape. Another strategy uses non-peptide scaffolds in developing the so-called foldamers, based on terphenyl (**10**) [80] and terpyridyl (**11**) templates, as aromatic oligoamides [81], or 3-*O*-alkylated oligobenzamide (**12**) possessing rigid frameworks and a hydrogen-bonding network, which replicate the side-chain orientation of an α-helix while providing rigidity. β-Peptides and peptoids (**13**) have been exploited in other peptide-based foldamer approaches that mimic the helical structure and resist proteolysis, too. Finally, a widely exploited approach for helix stabilization is based on peptide stapling (**14**), capable of maintaining the folding of short peptide sequences in a helical fashion.

1.8.2.2 β-Sheet Mimetics

β-Sheets are defined by a regular network of intramolecular hydrogen bonds connecting adjacent β-strands, and they constitute critical structural elements influencing the three-dimensional architecture and biological function of proteins. Interest in β-sheet mimetics is particularly strong in the context of CNS disorders associated with the aggregation of oligomeric, flat-structured species [82]. Unlike α-helices, the structure and stability of β-sheets remain less fully understood, and β-sheet peptidomimetics are often employed as model systems to probe folding propensities. Design strategies exploit specific chemical motifs, such as urea linkages, or flat molecular scaffolds capable of mimicking β-strands and forming parallel hydrogen-bonding interactions as both donors and acceptors [83]. Notable examples include β-strand mimetics based on methoxypyrrole-derived amino acids (**16**) [84], and 5-amino-2-methoxybenzamides and hydrazides (**17**) for stabilizing antiparallel β-sheets [85], which serve as central strands orienting hydrogen-bonding functionality appropriately.

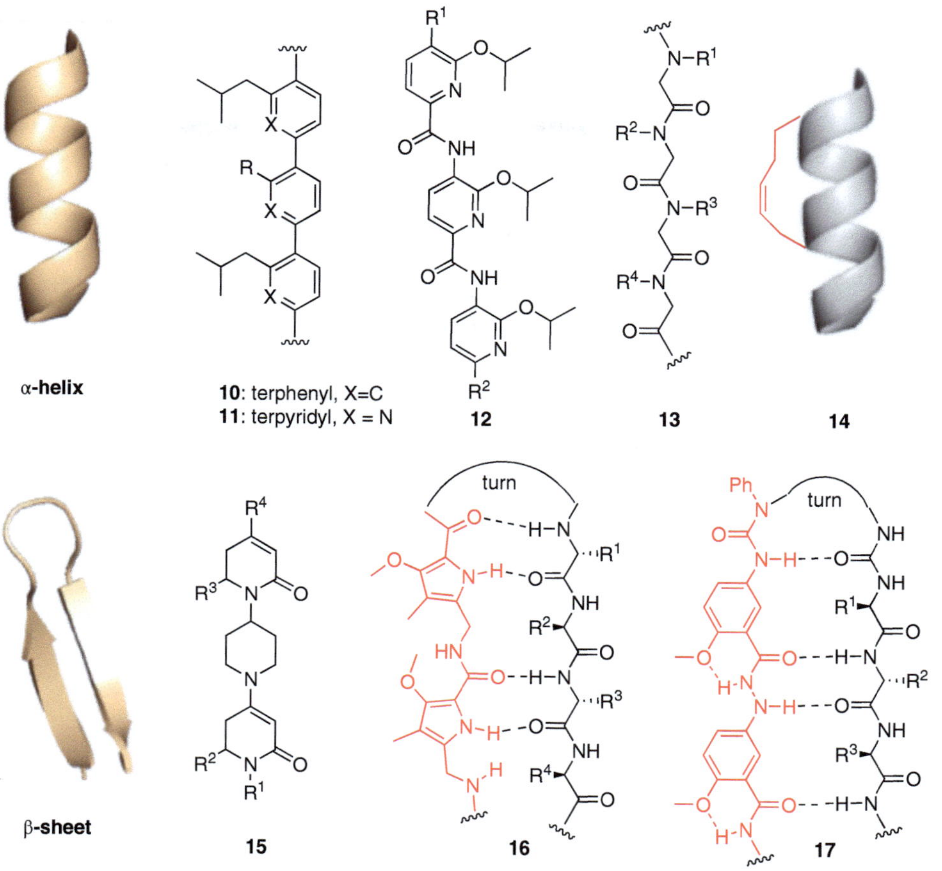

Figure 1.14 *Chemical approaches for mimicking α-helix (top) and β-sheet (bottom).*

1.8.2.3 β-Turn Mimetics

Turns are irregular protein secondary structures that let the polypeptide chain fold back on itself. They are common in loop regions and important for PPIs. Turns are classified by their length and hydrogen-bonding pattern [19]:

- *γ-turns*: three amino acids ($n = 2$), with a hydrogen bond forming a seven-membered ring.
- *β-turns*: four amino acids ($n = 3$), with a hydrogen bond forming a 10-membered ring.

β-Turns represent the most extensively investigated secondary structure for small-molecule mimetics [86]. A β-turn is defined as a tetrapeptide motif in which the distance between C_α i and C_α $i + 3$ is ≤ 7 Å, and its conformation can be stabilized through intramolecular hydrogen bonding. Designing effective turn mimetics requires identifying rigid scaffolds that orient side-chain residues in the same spatial arrangement as the natural peptide while enhancing solubility and resistance to enzymatic degradation. Various strategies have been employed (Figure 1.15), including dipeptide isosteres that induce turn

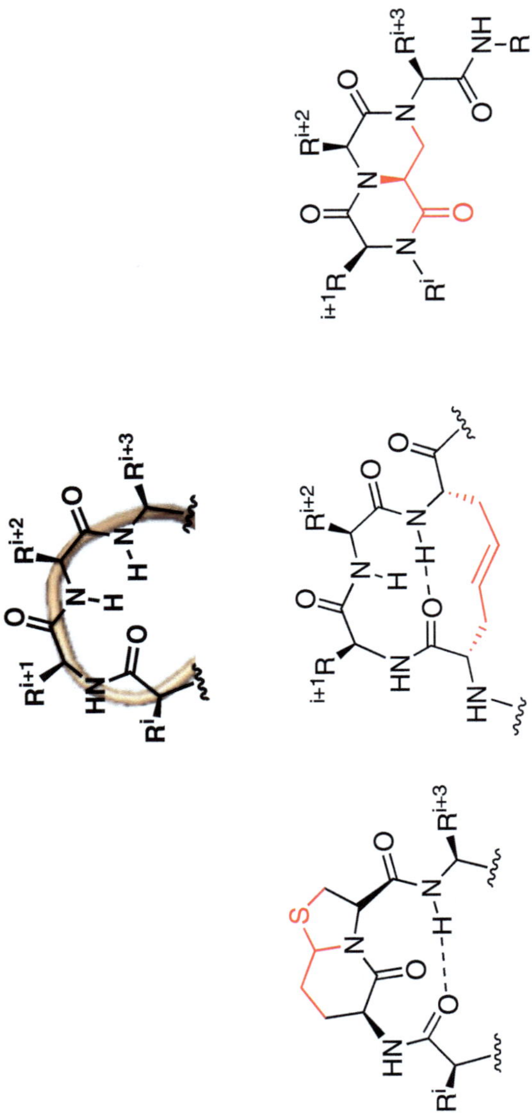

18, turn inducer

19, peptide tether

20, β-turn mimetic

Figure 1.15 *Chemical approaches for mimicking β-turns.*

formation within a peptide sequence (**18**), chemical tethers incorporated into macrocyclic constructs to constrain and stabilize β-turn conformations (**19**), and scaffolds that replicate the entire peptide turn motif (**20**).

1.9 Therapeutic and Diagnostic Applications of Peptidomimetics

Over the decades, peptidomimetics proved a straightforward tool to transform bioactive peptides into approved drugs with improved stability, oral bioavailability, or selectivity as enzyme inhibitors, receptor agonists or antagonists, antimicrobial agents, and modulators of intracellular signaling, as well as tools for drug delivery and diagnostic imaging. Such development of peptide-derived small-molecule therapeutics has resulted in the achievement of a wide number of drugs approved as antivirals, anticancer agents, cardiovascular drugs, metabolic regulators, and imaging agents (Table 1.6).

Also, the following are representative examples of peptidomimetics that are under clinical trials:

- *ALRN-6924*, a stapled peptide mimicking the p53 *N*-terminal domain, that targets MDM2/MDMX to induce tumor cell apoptosis.
- *Nemifitide (INN-00835)*, a pentapeptide antidepressant derived from MIF-1.
- *Brilacidin*, a non-peptide chemical mimic (an aryl-amide foldamer) that replicates the amphiphilic nature of antimicrobial peptides (AMPs), active against *Methicillin-Resistant Staphylococcus aureus* (MRSA) and a broad range of bacteria by disrupting microbial membranes.
- *Tavilermide (MIM-D3)*, a cyclic tripeptide acting as a TrkA partial agonist, mimicking nerve growth factor, in Phase III trials for dry eye disease, administered as eye drops.

Concerning the therapeutic applications of peptidomimetics, these can be summarized as follows:

- *Modulation of PPIs*: peptidomimetics can mimic α-helices, β-sheets, or β-turns to disrupt or stabilize PPIs that are otherwise "undruggable" by small molecules. Key examples include inhibitors of MDM2/p53, Bcl-2 family interactions, KRAS effectors, and viral protein assemblies.
- *Enzyme inhibition*: peptidomimetics are widely used as selective inhibitors of proteases, including viral proteases (HIV, hepatitis C virus (HCV)), matrix metalloproteases, caspases, kinases, and other enzymes. They are generally designed as conformationally constrained competitive inhibitors by mimicking substrate motifs.
- *Receptor targeting*: peptidomimetics can mimic endogenous peptide ligands to selectively activate or block receptors, such as G protein-coupled receptors (GPCRs) and ion channels. Examples include somatostatin analogs, calcitonin analogs, and opioid receptor modulators.
- *Antimicrobial and antiviral activity*: conformationally constrained peptidomimetics conceived as mimetics of AMPs improve stability and membrane-disrupting activity against bacteria and fungi and interfere with viral entry or replication, such as HIV fusion inhibitors.

Table 1.6 *Representative list of FDA-approved peptidomimetic drugs.*

Drug name	Approval	Indication(s)	Peptidomimetic approach	Target
			Enzyme inhibitors	
Saquinavir	1995	HIV/AIDS	Hydroxyethylamine isostere	HIV protease
Ritonavir	1996	HIV/AIDS; PK booster	Hydroxyethylene isostere	
Nelfinavir	1997	HIV/AIDS	Hydroxyethylamine PI	
Amprenavir	1999	HIV/AIDS	Hydroxyethylamine PI	
Atazanavir	2003	HIV/AIDS	Azapeptide protease inhibitor (PI)	
Fosamprenavir	2003	HIV/AIDS	Prodrug of amprenavir	
Darunavir	2006	HIV/AIDS	Advanced PI with *bis*-THF group	
Boceprevir	2011	HCV	Linear ketoamide PI	HCV protease (NS3/A4)
Grazoprevir	2016	HCV	Macrocyclic PI	
Voxilaprevir	2017	HCV	Macrocyclic PI	
Nirmatrelvir	2023	COVID-19 antiviral	3CLpro peptidomimetic	SARS-CoV-2 main protease
Bortezomib	2003	Multiple myeloma, mantle cell lymphoma	Boronic acid dipeptidomimetic	26S proteasome (chymotrypsin-like site)
Romidepsin	2009	Cutaneous and peripheral T-cell lymphoma	Cyclic depsipeptide	Histone deacetylases
			Receptor antagonist	
Icatibant	2011	Acute hereditary angioedema	Bradykinin analog	Bradykinin B_2 receptor
Tirofiban	1998	Acute coronary syndrome	Non-peptide RGD mimetic	Platelet GP IIb/IIIa receptor
			Receptor agonist	
Bremelanotide	2019	Female sexual dysfunction	Cyclic α-MSH analog	Melanocortin-4 receptor (MC4R)
Setmelanotide	2020	Genetic obesity disorders	Cyclic disulfide α-MSH analog	
			Other receptor ligands	
Piflufolastat-F18	2021	PET imaging (PSMA+)	Urea-based peptidomimetic ligand	PSMA (prostate-specific membrane antigen)

- *Immunomodulation and vaccines*: peptidomimetics can mimic B- or T-cell epitopes to generate targeted immune responses or modulate immune pathways. This opens the way to vaccine design, autoimmune disease modulation, and checkpoint inhibition.
- *Drug delivery and cell penetration*: macrocyclic or stapled peptidomimetics can function as mimetics of cell-penetrating peptides (CPPs) to enhance cell uptake of cargo molecules, such as therapeutic peptides, nucleic acids, or small-molecule drugs.
- *Diagnostic and imaging tools*: peptidomimetics designed as molecular probes can selectively bind biomarkers for imaging or diagnostic assays. Examples include radiolabeled cyclic peptides for tumor imaging and fluorescent β-sheet mimetics for amyloid detection.

1.9.1 Inhibitors of Protein–Protein Interactions

A major area of interest in peptidomimetic design for drug discovery applications is connected to the regulation of PPIs, which control at the molecular level important cellular functions including signal transduction, transcription, translation, and apoptosis [87]. As introduced before, PPIs are essential mediators of numerous biological processes and are implicated in the progression of many diseases [88]. Approximately 650,000 disease-associated PPIs have been reported within the human interactome, highlighting their potential as therapeutic targets. Generally, PPI interfaces present wide and flat interfaces of polar nature, rendering many PPIs "undruggable," as small-molecule therapeutics frequently fail to achieve effective binding. Despite these challenges, certain amino acid residues within PPI interfaces, known as "hot spots," contribute significantly to binding energy and provide opportunities for rational drug design [89, 90]. These hot spots often occur in regions of the protein secondary structure, including α-helices, β-sheets, and β-turns, with α-helices being particularly prominent. Over half of all secondary structures in protein complexes are α-helices, which interact with binding partners often using residues on a single helical face.

Thus, peptidomimetics can be rationally designed to mimic key structural motifs found in those PPIs, such as α-helices, β-strands, and β-turns, by stabilizing the bioactive conformation responsible for inhibition (antagonist) or activation (agonist). Indeed, peptidomimetics designed to mimic these helices have successfully modulated therapeutically relevant PPIs, sometimes replicating multiple faces of the helix. Overall, while the structural complexity of PPIs poses significant challenges for drug discovery, the identification of hot spots and the development of secondary-structure mimetics provide promising avenues for designing potent PPI inhibitors.

Stapled α-helical peptides have been particularly significant in this context [91]. These molecules are generated through covalent cyclization chemistries, including ring-closing metathesis, Cu-cat. azide–alkyne cycloaddition, and thioether formation, to stabilize helical conformations while enhancing proteolytic resistance, solubility, and cellular uptake. As a relevant example in this context, dual MDM2/MDMX inhibitor ATSP-7041 is a successful entry toward targeting PPI with peptidomimetics through the application of the stapling concept to the key peptide sequence (Figure 1.16) [92]. This result followed the clinical development of ALRN-6924, the first stapled α-helical peptide targeting an intracellular protein in humans [93]. The incorporation of α-methylated amino acids within macrocyclic

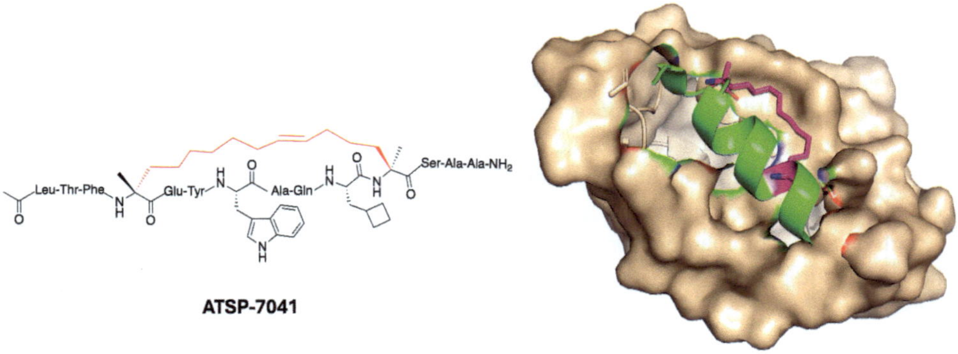

Figure 1.16 *Left: structure of ATSP-7041, the stapling moiety is highlighted in red; right: X-ray crystallographic structure of ATSP-7041 (peptide structure in green and the stapling moiety in magenta) binding to MDMX (PDB: 4N5T).*

frameworks has further refined the conformational and pharmacological properties of stapled peptides, extending their application to diverse PPI targets, including Bcl-2 family members, Notch, HIV proteins, HIF-1, KRAS, β-catenin, and aurora-A kinase.

1.9.2 Enzyme Inhibitors

The development of enzyme inhibitors is one of the most successful applications of peptidomimetic design in drug discovery [15]. This is because proteases are attractive therapeutic targets for their role in numerous physiological and pathological processes, including the activation of peptide hormones and neuromodulators from their inactive precursors [94]. The classification of proteases into different groups or families is based on the structural features of the active site and the catalytic mechanism.

The development of viral protease inhibitors represents a milestone in antiviral drug discovery, as viral proteases are essential for replication by cleaving polyprotein precursors into functional proteins [95]. Most inhibitors are peptidomimetics that block protease activity via peptide bond cleavage mechanisms [96]. The first example, saquinavir, was developed in 1990 and FDA approved in 1995 for HIV treatment [97], paving the way for other HIV and HCV protease inhibitors [98]. The proven success of this approach has since guided efforts to develop inhibitors against SARS-CoV-2.

HCV infection is a significant health issue, inducing chronic liver injury with an estimated >100 million individuals affected worldwide. The HCV protease, named NS3/4A, is a serine protease that plays a critical role in viral replication and maturation by cleaving the HCV polyprotein to generate functional nonstructural proteins necessary for the assembly of the viral replication complex. Thus, NS3/4A has become a key target for direct-acting antiviral agents, leading to the development of a range of covalent and non-covalent inhibitors [99]. Following first generation of covalent inhibitors, such as boceprevir and teleprevir, second-generation inhibitors no longer included the ketoamide group as a serine warhead but were designed to enhance the binding affinity through conformational preorganization, including the incorporation of bulky aromatic moieties connected to the P2 position and macrocyclization strategies at P1–P3 as in simeprevir

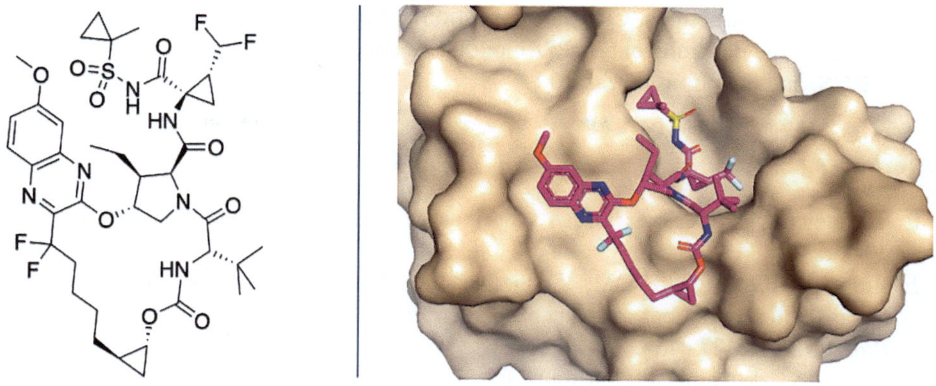

Figure 1.17 *Left: Structure of voxilaprevir; right: 3D close-up of the X-ray crystallographic structure of voxilaprevir in the active site of NS3/4A protease (PDB: 6NZT).*

or at P2–P4, as in grazoprevir and voxilaprevir inhibitors. Nowadays, five out of twelve NS3/4A protease inhibitors approved for treating HCV infections remain in clinical use primarily due to the emergence of drug resistance. Based on these successful results, the macrocyclization strategy has been increasingly applied to the development of inhibitors targeting other proteases. The design of macrocyclic inhibitors has proved particularly effective for addressing selective binding to protein surfaces involved in PPIs of clinical relevance and for augmenting inhibitor potency through the insertion of conformational bias to reduce entropic loss upon binding [100].

Voxilaprevir, as a third-generation NS3/4A protease inhibitor, was approved as a component of the pan-genotypic STR Vosevi® (sofosbuvir, velpatasvir, and voxilaprevir) [101], for treating patients who have failed prior therapies with other direct-acting antivirals. This molecule exerts its inhibitory activity through a series of key interactions within the active site of the enzyme, including binding to the catalytic triad through hydrogen-bonding, electrophilic interactions engaged by the difluoromethylene moiety with the backbone carbonyl oxygen atoms of the protease, and hydrophobic interactions by the cyclopropyl group at the P4 position with the complementary S4 pocket of the protease (Figure 1.17).

1.10 Conclusion

Peptide and peptidomimetic drug discovery has advanced from targeting limited receptors and proteases to addressing complex biological targets such as PPIs. This enabled the development of peptidomimetic design using *in silico* methods, taking advantage of significant progress in computational algorithms, bioinformatics, and structural biology, together with chemical methodologies for peptide modifications informed by molecular predictions and screening assays. Such modifications, classified as local or global, follow a hierarchical approach to enhance potency, stability, and druggability. Key strategies include peptide bioisosterism at backbone or side-chain levels, the use of heterocyclic scaffolds to modulate the peptide structure, and the design of more complex molecular architectures such as foldamers, macrocyclic peptides, and stapled peptides addressing key peptide secondary structures. Current applications of peptidomimetics in drug discovery are focused on the development of PPI modulators and enzyme inhibitors in the fields of cancer, CNS, and infectious diseases.

References

1. Sachdeva S. Peptides as 'Drugs': the journey so far. *Int J Pept Res Ther*. 2017;**23**: 49–60.
2. Branco MC, Sigano DM, Schneider JP. Materials from peptide assembly: towards the treatment of cancer and transmittable disease. *Curr Opin Chem Biol*. 2011;**15**:427–434.
3. Pereira AJ, de Campos LJ, Xing H, Conda-Sherindan M. Peptide-based therapeutics: challenges and solutions. *Med Chem Res*. 2024;**33**:1275–1280.
4. Barman P, Joshi S, Sharma S, Preet S, Sharma S, Saini A. Strategic approaches to improvise peptide drugs as next generation therapeutics. *Int J Pept Res Ther*. 2023;**29**:61.
5. Craik DJ, Fairlie DP, Liras S, Price D. The future of peptide-based drugs. *Chem Biol Drug Des*. 2013;**81**:136–147.
6. Ratnaparkhi MP, Chaudhari SP, Pandya VA. Peptides and proteins in pharmaceuticals. *Int J Curr Pharm Res*. 2011;**3**:1–9.
7. Merrifield RB. Solid phase peptide synthesis. I. The synthesis of a tetrapeptide. *J Am Chem Soc*. 1963;**85**:2149–2154.
8. Fosgerau K, Hoffmann T. Peptide therapeutics: current status and future directions. *Drug Discov Today*. 2015;**20**:122–128.
9. Jimmidi R. Synthesis and applications of peptides and peptidomimetics in drug discovery. *Eur J Org Chem*. 2023;**26**:e202300028.
10. Kaspar AA, Reichert JM. Future directions for peptide therapeutics development. *Drug Discov Today*. 2013;**18**:807–817.
11. Muttenthaler M, King GF, Adams DJ, Alewood PF. Trends in peptide drug discovery. *Nat Rev Drug Discov*. 2021;**20**:309–325.
12. Müller MM. Post-translational modifications of protein backbones: unique functions, mechanisms, and challenges. *Biochemistry*. 2018;**57**:177–185.
13. Grauer A, König B. Peptidomimetics – a versatile route to biologically active compounds. *Eur J Org Chem*. 2009;**2009**:5099–5111.
14. Ahn JM, Boyle NA, MacDonald MT, Janda KD. Peptidomimetics and peptide backbone modifications. *Mini Rev Med Chem*. 2002;**2**:463–473.
15. Giannis A, Kolter T. Peptidomimetics for receptor ligands – discovery, development, and medical perspectives. *Angew Chem Int Ed Engl*. 1993;**32**:1244–1267.
16. Heinis C, Winter G. Encoded libraries of chemically modified peptides. *Curr Opin Chem Biol*. 2015;**26**:89–98.
17. Perez JJ, Perez RA, Perez A. Computational modeling as a tool to investigate PPI: from drug design to tissue engineering. *Front Mol Biosci*. 2021;**8**:681617.
18. Lenci E, Trabocchi A. Peptidomimetic toolbox for drug discovery. *Chem Soc Rev*. 2020;**49**:3262–3277.
19. Trabocchi A, Guarna A. *Peptidomimetics in Organic and Medicinal Chemistry: The Art of Transforming Peptides in Drugs*. Chichester, West Sussex, United Kingdom: John Wiley & Sons Inc., 2014.

20. Ripka AS, Rich DH. Peptidomimetic design. *Curr Opin Chem Biol*. 1998;**2**:441–452.
21. Gante J. Peptidomimetics – tailored enzyme inhibitors. *Angew Chem Int Ed Engl*. 1994;**33**:1699–1720.
22. Pelay-Gimeno M, Glas A, Koch O, Grossmann TN. Structure-based design of inhibitors of protein-protein interactions: mimicking peptide binding epitopes. *Angew Chem Int Ed Engl*. 2015;**54**:8896–8927.
23. Kim WK, Henschel A, Winter C, Schroeder M. The many faces of protein-protein interactions: a compendium of interface geometry. *PLoS Comput Biol*. 2006;**2**:e124.
24. Vidal M, Cusick ME, Barabási AL. Interactome networks and human disease. *Cell*. 2011;**144**:986–998.
25. Arkin MR, Wells JA. Small-molecule inhibitors of protein-protein interactions: progressing towards the dream. *Nat Rev Drug Discov*. 2004;**3**:301–317.
26. Dumetz AC, Chockla AM, Kaler EW, Lenhoff AM. Effects of pH on protein-protein interactions and implications for protein phase behavior. *Biochim Biophys Acta*. 2008;**1784**:600–610.
27. Clackson T, Wells JA. A hot spot of binding energy in a hormone-receptor interface. *Science*. 1995;**267**:383–386.
28. Chang L, Perez A. Ranking peptide binders by affinity with AlphaFold. *Angew Chem Int Ed Engl*. 2023;**62**:e202213362.
29. Bakail M, Ochsenbein F. Targeting protein-protein interactions, a wide open field for drug design. *C R Chim*. 2016;**19**:19–27.
30. Henninot A, Collins JC, Nuss JM. The current state of peptide drug discovery: back to the future? *J Med Chem*. 2018;**61**:1382–1414.
31. Tomasella C, Floris M, Guccione S, Pappalardo M, Basile L. Peptidomimetics in silico. *Mol Inf*. 2021;**40**:2000087.
32. D'Annessa I, Di Leva FS, La Teana A, Novellino E, Limongelli V, Di Marino D. Bioinformatics and biosimulations as toolbox for peptides and peptidomimetics design: where are we? *Front Mol Biosci*. 2020;**7**:66.
33. Yu W, MacKerell AD Jr. Computer-aided drug design methods. *Methods Mol Biol*. 2017;**1520**:85–106.
34. Floris M, Moro S. Mimicking peptides in silico. *Mol Inf*. 2012;**31**:12–20.
35. Bohacek RS, McMartin C. Multiple highly diverse structures complementary to enzyme binding sites: results of extensive application of a de novo design method incorporating combinatorial growth. *J Am Chem Soc*. 1994;**116**:5560–5571.
36. Raman S, Vernon R, Thompson J, Tyka M, Sadreyev R, Pei J, Kim D, Kellogg E, DiMaio F, Lange O, Kinch L, Sheffler W, Kim BH, Das R, Grishin NV, Baker D. Structure prediction for CASP8 with all-atom refinement using Rosetta. *Proteins*. 2009;**77**(Suppl 9):89–99.
37. King CA, Bradley P. Structure-based prediction of protein-peptide specificity in Rosetta. *Proteins*. 2010;**78**:3437–3449.
38. Renfrew PD, Craven TW, Butterfoss GL, Kirshenbaum K, Bonneau R. A rotamer library to enable modeling and design of peptoid foldamers. *J Am Chem Soc*. 2014;**136**:8772–8782.
39. Thévenet P, Shen Y, Maupetit J, Guyon F, Derreumaux P, Tufféry P. PEP-FOLD: an updated de novo structure prediction server for both linear and disulfide bonded cyclic peptides. *Nucleic Acids Res*. 2012;**40**:W288–W293.

40. Kim KH, Kim ND, Seong BL. Pharmacophore-based virtual screening: a review of recent applications. *Expert Opin Drug Discov*. 2010;**5**:205–222.
41. Thomas Seidel T, Ibis G, Bendix F, Wolber G. Strategies for 3D pharmacophore-based virtual screening. *Drug Discov Today Technol*. 2010;**7**:e221–e228.
42. Floris M, Masciocchi J, Fanton M, Moro S. Swimming into peptidomimetic chemical space using pepMMsMIMIC. *Nucleic Acids Res*. 2011;**39**:W261–W269.
43. Masciocchi J, Frau G, Fanton M, Sturlese M, Floris M, Pireddu L, Palla P, Cedrati F, Rodriguez-Tomé P, Moro S. MMsINC: a large-scale chemoinformatics database. *Nucleic Acids Res*. 2009;**37**:D284–D290.
44. Oliva B, Fernandez-Fuentes N. Knowledge-based modeling of peptides at protein interfaces: PiPreD. *Bioinformatics*. 2015;**31**:1405–1410.
45. Obarska-Kosinska A, Iacoangeli A, Lepore R, Tramontano A. PepComposer: computational design of peptides binding to a given protein surface. *Nucleic Acids Res*. 2016;**44**:W522–W528.
46. Goede A, Michalsky E, Schmidt U, Preissner R. *BMC Bioinformatics*. 2006;**10**:7–11.
47. Shameer K, Madan LL, Veeranna S, Gopal B, Sowdhamini R. PeptideMine–a webserver for the design of peptides for protein-peptide binding studies derived from protein-protein interactomes. *BMC Bioinformatics*. 2010;**11**:473.
48. Andrews MJ, Kontopidis G, McInnes C, Plater A, Innes L, Cowan A, Jewsbury P, Fischer PM. REPLACE: a strategy for iterative design of cyclin-binding groove inhibitors. *Chembiochem*. 2006;**7**:1909–1915.
49. Hruby VJ. Designing peptide receptor agonists and antagonists. *Nat Rev Drug Discov*. 2002;**1**:847–858.
50. Hruby VJ, al-Obeidi F, Kazmierski W. . Emerging approaches in the molecular design of receptor-selective peptide ligands: conformational, topographical and dynamic considerations. *Biochem J*. 1990;**268**:249–262.
51. Chatterjee J, Gilon C, Hoffman A, Kessler H. N-methylation of peptides: a new perspective in medicinal chemistry. *Acc Chem Res*. 2008;**41**:1331–1342.
52. Connolly MD, Xuan S, Molchanova N, Zuckermann RN. Submonomer synthesis of sequence defined peptoids with diverse side-chains. *Methods Enzymol*. 2021;**656**:241–270.
53. Guan Q, Xing S, Wang L, Zhu J, Guo C, Xu C, Zhao Q, Wu Y, Chen Y, Sun H. Triazoles in medicinal chemistry: physicochemical properties, bioisosterism, and application. *J Med Chem*. 2024;**67**:7788–7824.
54. Angell YL, Burgess K. Peptidomimetics via copper-catalyzed azide-alkyne cycloadditions. *Chem Soc Rev*. 2007;**36**:1674–1689.
55. Kumari S, Carmona AV, Tiwari AK, Trippier PC. Amide bond bioisosteres: strategies, synthesis, and successes. *J Med Chem*. 2020;**63**:12290–12358.
56. McLaughlin M, Yazaki R, Fessard TC, Carreira EM. Oxetanyl peptides: novel peptidomimetic modules for medicinal chemistry. *Org Lett*. 2014;**16**:4070–4073.
57. Roesner S, Saunders GJ, Wilkening I, Jayawant E, Geden JV, Kerby P, Dixon AM, Notman R, Shipman M. Macrocyclisation of small peptides enabled by oxetane incorporation. *Chem Sci*. 2019;**10**:2465–2472.
58. Stefanic P, Dolenc MS. Aspartate and glutamate mimetic structures in biologically active compounds. *Curr Med Chem*. 2004;**11**:945–968.

59. Freidinger RM. Design and synthesis of novel bioactive peptides and peptidomimetics. *J Med Chem*. 2003;**46**:5553–5566.
60. Freidinger RM, Perlow DS, Randall WC, Saperstein R, Arison BH, Veber DF. Conformational modifications of cyclic hexapeptide somatostatin analogs. *Int J Pept Protein Res*. 1984;**23**:142–150.
61. Bittermann H, Einsiedel J, Hübner H, Gmeiner P. Evaluation of lactam-bridged neurotensin analogues adjusting psi(Pro10) close to the experimentally derived bioactive conformation of NT(8-13). *J Med Chem*. 2004;**47**:5587–5590.
62. Tian X, Mishra RK, Switzer AG, Hu XE, Kim N, Mazur AW, Ebetino FH, Wos JA, Crossdoersen D, Pinney BB, Farmer JA, Sheldon RJ. Design and synthesis of potent and selective 1,3,4-trisubstituted-2-oxopiperazine based melanocortin-4 receptor agonists. *Bioorg Med Chem Lett*. 2006;**16**:4668–4673.
63. Das D, Khan HP, Shivahare R, Gupta S, Sarkar J, Siddiqui MI, Ampapathi RS, Chakraborty TK. Synthesis, SAR and biological studies of sugar amino acid-based almiramide analogues: N-methylation leads the way. *Org Biomol Chem*. 2017;**15**:3337–3352.
64. Hruby VJ, Li G, Haskell-Luevano C, Shenderovich M. Design of peptides, proteins, and peptidomimetics in chi space. *Biopolymers*. 1997;**43**:219–266.
65. Van der Poorten O, Knuhtsen A, Sejer Pedersen D, Ballet S, Tourwé D. Side chain cyclized aromatic amino acids: great tools as local constraints in peptide and peptidomimetic design. *J Med Chem*. 2016;**59**:10865–10890.
66. Abdildinova A, Kurth MJ, Gong YD. Heterocycles as a peptidomimetic scaffold: solid-phase synthesis strategies. *Pharmaceuticals*. 2021;**14**:449.
67. Kubyshkin V, Rubini M. Proline analogues. *Chem Rev*. 2024;**124**:8130–8232.
68. Kubyshkin V, Mykhailiuk PK. Proline analogues in drug design: current trends and future prospects. *J Med Chem*. 2024;**67**:20022–20055.
69. Valverde IE, Bauman A, Kluba CA, Vomstein S, Walter MA, Mindt TL. 1,2,3-Triazoles as amide bond mimics: triazole scan yields protease-resistant peptidomimetics for tumor targeting. *Angew Chem Int Ed Engl*. 2013;**52**:8957–8960.
70. Pedersen DS, Abell A. 1,2,3-Triazoles in peptidomimetic chemistry. *Eur J Org Chem*. 2011;**2011**:2399–2411.
71. Kolb HC, Sharpless KB. The growing impact of click chemistry on drug discovery. *Drug Discov Today*. 2003;**8**:1128–1137.
72. Tam A, Arnold U, Soellner MB, Raines RT. Protein prosthesis: 1,5-disubstituted[1,2,3]triazoles as cis-peptide bond surrogates. *J Am Chem Soc*. 2007;**129**:12670–12671.
73. Singh MK, Lakshman MK. Recent developments in the utility of saturated azaheterocycles in peptidomimetics. *Org Biomol Chem*. 2022;**20**:963–979.
74. Shi J, Sun T, Yang M. Site-selective editing of peptides via backbone modification. *Org Chem Front*. 2024;**11**:1623–1640.
75. McKeever B, Pattenden B. Total synthesis of the prenylated cyclopeptide trunkamide A, a cytotoxic metabolite from lissoclinum sp. *Tetrahedron Lett*. 2001;**42**:2573–2577.
76. Deeley J, Bertram A, Pattenden G. Novel polyoxazole based cyclopeptides from streptomyces sp. total synthesis of the cyclopeptide YM-216391 and synthetic studies towards telomestatin. *Org Biomol Chem*. 2008;**6**:1994–2010.
77. Romero-Estudillo V, Batchu R, Boto A. One-pot conversion of amino acids into 2,5-disubstituted oxazoles: no metals needed. *Adv Synth Catal*. 2014;**356**:3742–3748.

78. Petersen J, Christensen KE, Nielsen MT, Mortensen KT, Komnatnyy VV, Nielsen TE, Qvortrup K. Oxidative modification of tryptophan-containing peptides. *ACS Comb Sci*. 2018;**20**:344–349.
79. Bechtler C, Lamers C. Macrocyclization strategies for cyclic peptides and peptidomimetics. *RSC Med Chem*. 2021;**12**:1325–1351.
80. Orner BP, Ernst JT, Hamilton AD. Toward proteomimetics: terphenyl derivatives as structural and functional mimics of extended regions of an alpha-helix. *J Am Chem Soc*. 2001;**123**:5382–5383.
81. Ernst JT, Becerril J, Park HS, Yin H, Hamilton AD. Design and application of an alpha-helix-mimetic scaffold based on an oligoamide-foldamer strategy: antagonism of the Bak BH3/Bcl-xL complex. *Angew Chem Int Ed Engl*. 2003;**42**:535–539.
82. Irvine GB, El-Agnaf OM, Shankar GM, Walsh DM. Protein aggregation in the brain: the molecular basis for Alzheimer's and Parkinson's diseases. *Mol Med*. 2008;**14**:451–464.
83. Moriuchi T, Hirao T. Highly ordered structures of peptides by using molecular scaffolds. *Chem Soc Rev*. 2004;**33**:294–301.
84. Khakshoor O, Nowick JS. Artificial beta-sheets: chemical models of beta-sheets. *Curr Opin Chem Biol*. 2008;**12**:722–729.
85. Bonauer C, Zabel M, König B. Synthesis and peptide binding properties of methoxypyrrole amino acids (MOPAS). *Org Lett*. 2004 Apr;**29**(6):1349–1352.
86. Nair RV, Baravkar SB, Ingole TS, Sanjayan GJ. Synthetic turn mimetics and hairpin nucleators: Quo Vadimus? *Chem Commun*. 2014;**50**:13874–13884.
87. Mabonga L, Kappo AP. Peptidomimetics: a synthetic tool for inhibiting protein–protein interactions in cancer. *Int J Pept Res Ther*. 2020;**26**:225–241.
88. Greenblatt JF, Alberts BM, Krogan NJ. Discovery and significance of protein-protein interactions in health and disease. *Cell*. 2024;**187**:6501–6517.
89. Geppert T, Hoy B, Wessler S, Schneider G. Context-based identification of protein-protein interfaces and "hot-spot" residues. *Chem Biol*. 2011;**18**:344–353.
90. Lu H, Zhou Q, He J et al. Recent advances in the development of protein-protein interactions modulators: mechanisms and clinical trials. *Signal Transduct Target Ther*. 2020;**5**:213.
91. Schafmeister CE, Po J, Verdine GL. An all-hydrocarbon cross-linking system for enhancing the helicity and metabolic stability of peptides. *J Am Chem Soc*. 2000;**122**:5891–5892.
92. Chang YS, Graves B, Guerlavais V, Tovar C, Packman K, To KH, Olson KA, Kesavan K, Gangurde P, Mukherjee A, Baker T, Darlak K, Elkin C, Filipovic Z, Qureshi FZ, Cai H, Berry P, Feyfant E, Shi XE, Horstick J, Annis DA, Manning AM, Fotouhi N, Nash H, Vassilev LT, Sawyer TK. Stapled α-helical peptide drug development: a potent dual inhibitor of MDM2 and MDMX for p53-dependent cancer therapy. *Proc Natl Acad Sci USA*. 2013;**110**:E3445–E3454.
93. Guerlavais V, Sawyer TK, Carvajal L, Chang YS, Graves B, Ren JG, Sutton D, Olson KA, Packman K, Darlak K, Elkin C, Feyfant E, Kesavan K, Gangurde P, Vassilev LT, Nash HM, Vukovic V, Aivado M, Annis DA. Discovery of sulanemadlin (ALRN-6924), the first cell-permeating, stabilized α-helical peptide in clinical development. *J Med Chem*. 2023;**66**:9401–9417.

94. Bond JS. Proteases: history, discovery, and roles in health and disease. *J Biol Chem.* 2019;**294**:1643–1651.
95. Tompa DR, Immanuel A, Srikanth S, Kadhirvel S. Trends and strategies to combat viral infections: a review on FDA approved antiviral drugs. *Int J Biol Macromol.* 2021;**172**:524–541.
96. Borges PHO, Ferreira SB, Silva FP Jr. Recent advances on targeting proteases for antiviral development. *Viruses*. 2024;**16**:366.
97. Roberts NA, Martin JA, Kinchington D, Broadhurst AV, Craig JC, Duncan IB, Galpin SA, Handa BK, Kay J, Kröhn A et al. Rational design of peptide-based HIV proteinase inhibitors. *Science*. 1990;**248**:358–361.
98. Weber IT, Wang YF, Harrison RW. HIV protease: historical perspective and current research. *Viruses*. 2021;**13**:839.
99. Martino SD, Petri GL, De Rosa M, Hepatitis C. The story of a long journey through first, second, and third generation NS3/4A peptidomimetic inhibitors. what did we learn? *J Med Chem*. 2024;**67**:885–921.
100. Vinogradov AA, Yin Y, Suga H. Macrocyclic peptides as drug candidates: recent progress and remaining challenges. *J Am Chem Soc*. 2019;**141**:4167–4181.
101. Taylor JG, Zipfel S, Ramey K, Vivian R, Schrier A, Karki KK, Katana A, Kato D, Kobayashi T, Martinez R, Sangi M, Siegel D, Tran CV, Yang ZY, Zablocki J, Yang CY, Wang Y, Wang K, Chan K, Barauskas O, Cheng G, Jin D, Schultz BE, Appleby T, Villaseñor AG, Link JO. Discovery of the pan-genotypic hepatitis C virus NS3/4A protease inhibitor voxilaprevir (GS-9857): a component of Vosevi®. *Bioorg Med Chem Lett*. 2019;**29**:2428–2436.

2

Multicomponent Reactions Toward Peptidomimetics

2.1 Introduction

Over the years, several approaches have been conceived for the design and synthesis of peptidomimetics, most of them including multiple reactions that include protection and deprotection steps. This is the traditional synthetic approach to achieve the desired compound with high selectivity. Nevertheless, this is not the optimal approach to introduce structural diversity in the context of peptidomimetics that are effective for fine-modulating the biological activity of parent peptides [1–3]. A well-established approach that allows for introducing high structural diversity while reducing the number of synthetic steps is the use of multicomponent reactions (MCRs) [4]. These are chemical processes in which three or more organic compounds react simultaneously through a well-defined order and form a single product with high selectivity. Most MCRs are old reactions that date back to the nineteenth and early twentieth centuries and are all characterized by high atom economy, as most of the atoms of the building blocks are retained in the structure of the product. Over the decades, the possibility to access a structurally complex, densely substituted precursor through a simple one-step procedure has found relevance for the drug discovery process [5], particularly during the advent of combinatorial chemistry and high-throughput chemistry [6, 7]. The study of MCRs in combination with transition-metal catalysis experienced an increased interest over last decades, especially in applications using palladium [8], and more recently using copper [9] as an inexpensive alternative to other precious metals [10–12].

The first example of MCRs was reported in 1850 by Strecker for the synthesis of α-amino acids, in which the combination of a carbonyl compound, ammonia, and hydrogen cyanide brings to the formation of α-amino cyanides as the final precursors of α-amino acids [13]. Nevertheless, in the same years between the end of the eighteenth century and early decades of the nineteenth century, other MCRs were reported and proved to show relevance in a wide range of applications. A representative example is the pyridine synthesis reported in 1881 by Hantzsch (Figure 2.1a), in which 1,4-dihydropyridine-3,5-dicarboxylate is obtained through a MCR between an aldehyde, ammonium acetate, and two equivalents of a β-keto ester and then reduced to the corresponding 2,6-dialkylpyridine-3,5-dicarboxylate [14]. Ten years later, Biginelli developed a MCR to generate 3,4-dihydropyrimidin-2-ones-5-carboxylates using a β-keto ester, an aldehyde, and urea (Figure 2.1b). Two other important examples are the Passerini three-component

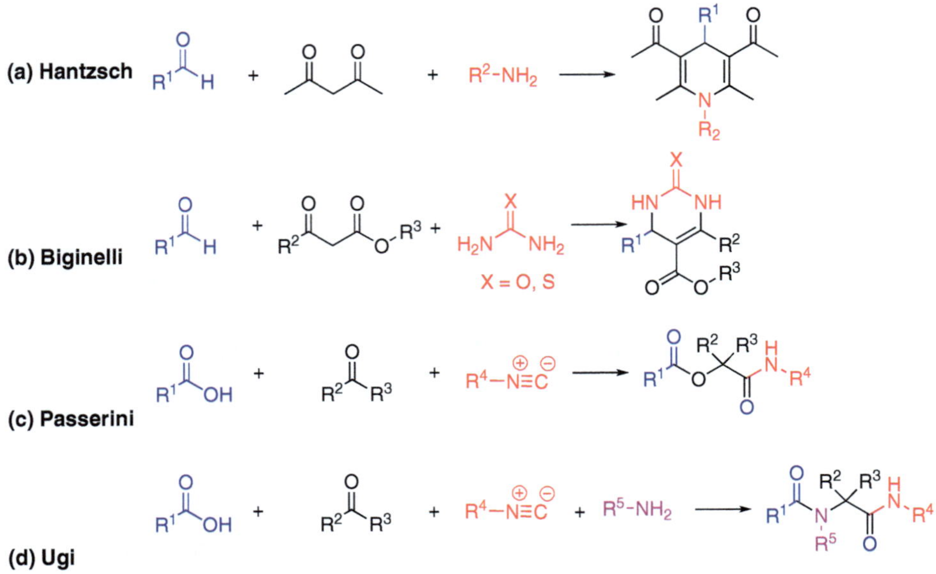

Figure 2.1 Early examples of MCRs, specifically of three- (a–c) and four- (d) component reactions.

reaction (1921), as the first entry to isocyanide-based MCR, and the Ugi four-component reaction (1959) that is widely recognized as the most famous isocyanide-based MCR. These two reactions allow achieving 2-(alkyl/arylamino)-2-oxo-ethyl acetates employing an isocyanide, a carboxylic acid, and a carbonyl compound (Figure 2.1c), or *N*-alkyl/aryl-2-(*N*-alkylacetamido)-acetamides employing an isocyanide, a carboxylic acid, a carbonyl compound, and a primary amine, respectively (Figure 2.1d).

MCRs can be classified according to the most characteristic building block or functional group present in the process, as for the isocyanide in Ugi and Passerini reactions, for example. The isocyanide-based MCRs are the most explored and studied reactions among all MCRs, due to the versatility of the isocyanide functional group, which displays both nucleophilic and electrophilic reactivity [15].

Another group of MCRs of significant impact is the imine-based MCRs for their relevance to the synthesis of nitrogen-containing compounds. Imines possess a double reactivity as isocyanides, namely the nitrogen atom acting as a nucleophile, and the masked carbonyl compound providing Mannich-type reactions with C-nucleophiles [16]. Imines can be easily generated employing several commercially available amines and aldehydes, thus promoting high structural diversity. In this view, Biginelli, Hantzsch, and Ugi reactions can also be considered as imine-based MCRs, as they involve the formation and reactivity of imine species. Other relevant examples belonging to such group of MCRs are the Mannich reaction (1912) [17], the Petasis–Borono–Mannich reaction (1993) [18], the Castagnoli–Cushman reaction (CCR) (1969–1977), and the A3 coupling reaction (1998) [19, 20], all employing a carbonyl compound and an amine to generate an imine (Figure 2.2).

Over the years, all these reactions have found applications in the synthesis of peptidomimetic scaffolds, taking advantage of suitable functional groups present in a wide range of amino acid precursors, such as amino and carboxylic groups and their derivatives

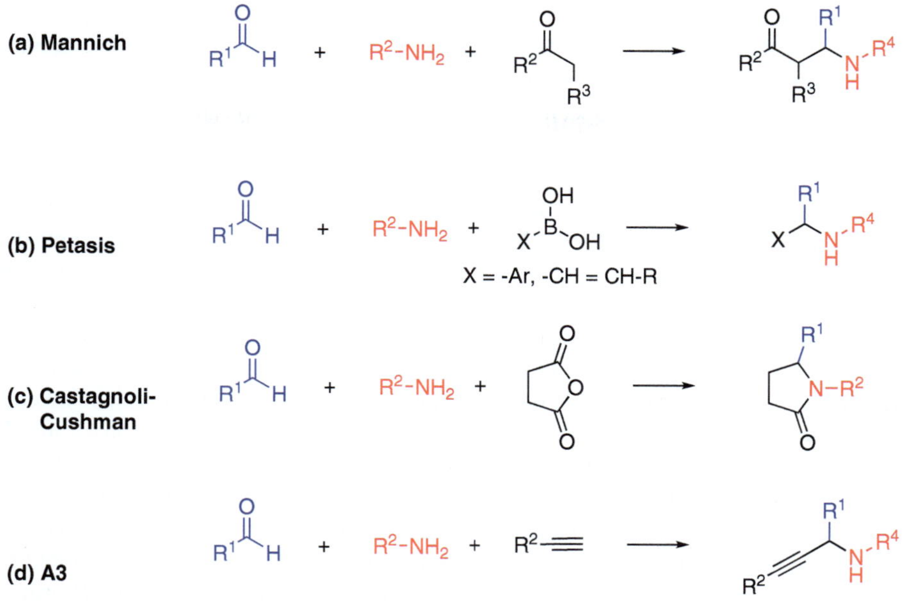

Figure 2.2 Examples of imine-based MCRs.

like aldehydes. Also, the chemical diversity around peptidomimetic products resulting from MCR processes has been expanded over the years, taking advantage of the combination with other reactions, particularly those enabling cyclization processes, giving access to more complex molecular scaffolds and macrocyclic peptidomimetics [21].

2.2 Isocyanide-Based MCRs

The development of peptidomimetic scaffolds taking advantage of isocyanide-based MCRs has been widely reported in the literature [22]. The isocyanide functionality has a dual reactivity at the carbon atom, being capable of displaying both nucleophile and electrophile reactivities, thus allowing for the generation of acyclic peptide-like compounds when using amino acid derivatives as MCR components [23]. Also, the use of bifunctional substrates or subsequent elaboration of initial MCR products enables the achievement of cyclic peptidomimetics [5]. Among isocyanide-based MCR (also abbreviated as IMCR), the Ugi and Passerini reactions have been studied most over the years for specific applicability of these transformations in the context of peptide and peptidomimetic chemistry. In both cases, several chemical approaches toward cyclization have been applied to obtain cyclic constructs [4, 24].

2.2.1 Ugi Reaction

The Ugi four-component condensation between an aldehyde, an amine, a carboxylic acid, and an isocyanide leads to the straightforward synthesis of α-aminoacyl amide derivatives. The Ugi reaction allows for exploiting high chemical diversity; thus, it has been considered of relevance for developing chemical libraries for screening applications. The hypothetical mechanism involves the initial formation of an imine species obtained by the corresponding amine and aldehyde components, followed by the attack of the carboxylic acid oxygen to give the corresponding iminium species that is attacked by the carbon atom of the isocyanide group. Then, an acyl transfer rearrangement of the resulting acylated isoamide occurs to achieve the final MCR product.

As Ugi adducts are peptide-like products, they have been frequently used for the synthesis of various peptidomimetics. Particular interest around the Ugi reaction in the context of peptidomimetic research has been focused on the possibility of applying amino acid-derived components useful for achieving cyclic peptidomimetic constructs following a subsequent cyclization step. The synthetic strategies toward cyclic peptidomimetics combining the Ugi reaction and follow-up cyclization are based on the presence of suitable functional groups in the component used for the MCR process. Specifically, functionalities such as alkenes, alkynes, and azides are considered orthogonal to the MCR reaction, thus providing the cyclic peptidomimetic via 1,3-dipolar cycloadditions or ring-closing metathesis (RCM) taking place after the multicomponent process. Alternatively, protecting group strategies are taken into account for any components, and an additional deprotection step is required before the cyclization step. Specifically, when a bifunctional compound is exploited for MCR and to enable subsequent cyclization, it requires protection during the MCR process when both groups may take part in the multicomponent process, as in the case of amino acids that in the Ugi reaction, require, for example, *N*-Boc protection at the amino group or an ester function in place of the carboxylic group.

An interesting general approach of combining the Ugi reaction and a cyclization step has been reported in a review by Cankařová and Krchňák [25], where several examples were reported on the use of solid phase as a convenient tool to apply a cyclative cleavage following the on-resin isocyanide-based Ugi MCR to develop cyclic peptidomimetic scaffolds released from the insoluble support. As a general approach, an internal oxygen or nitrogen nucleophile attacks an ester-type bond of the acyclic Ugi MCR product linked to the resin, resulting in the release of the cyclic peptidomimetics under mild conditions. These approaches have been reported for the achievement of several types of six- and seven-membered ring peptidomimetic scaffolds in high purity.

As a representative example of the application of the Ugi reaction on solid phase, the 3,4-dihydropyrazin-2(1*H*)-one **2** was achieved in only two steps (MCR and cyclative cleavage) using a solid-supported isocyanide component and an amino acetal as a bifunctional component (Figure 2.3) [26]. The acidic treatment allowed for releasing the product **1** from the resin and promoted cyclization of the deprotected aldehyde group.

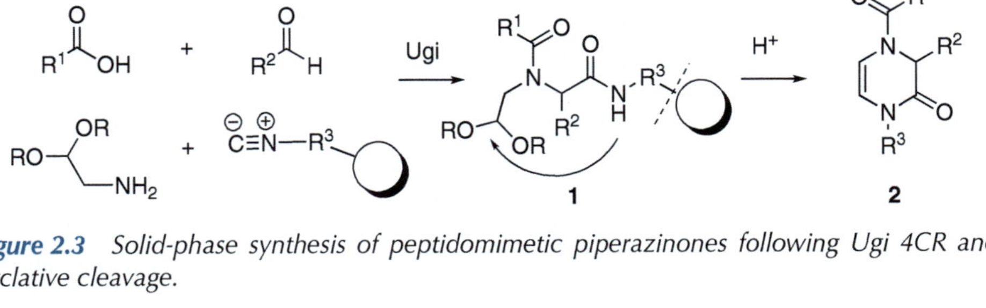

Figure 2.3 Solid-phase synthesis of peptidomimetic piperazinones following Ugi 4CR and cyclative cleavage.

Peptidomimetics of special interest that have been obtained using the Ugi MCR and follow-up cyclization steps are macrocyclic peptidomimetics and small-molecule scaffolds as dipeptide isosteres and reverse-turn mimetics.

Macrocyclic peptidomimetics. Over the last decades, the interest toward macrocyclic peptidomimetics has increased for their diverse range of sizes and shapes, biocompatible nature, and applications as inhibitors of protein–protein interactions. Macrocyclization as a reaction to achieve constrained peptides through a global restriction can be achieved by taking advantage of disulfide bonds that can be formed between cysteine residues within the same protein through oxidative folding. Thus, macrocyclic peptidomimetics have been achieved exploiting the disulfide bond [27], too, and through several cyclization methods, including intramolecular ester or amide bond formation, transition-metal-catalyzed coupling reaction, copper-catalyzed 1,3-dipolar cycloaddition between an azide and an alkyne (generally referred to as "click chemistry" or CuAAC), and ring-closing metathesis. The latter has been considered as the most versatile and straightforward cyclization approach following the Ugi reaction, taking advantage of the orthogonality of the alkyne and azide functional groups with the MCR process.

For example, the combination of the Ugi and click chemistry (using CuAAC reaction) was applied by Tahoori et al. for the synthesis of anti-cancer macrocyclic peptidomimetics using a propargyl group on the side chain of *N*-terminal serine as the alkyne moiety and C-terminal azido-lysine (Figure 2.4) [28].

In a similar approach, Zakharova et al. used a combination of Ugi reaction and click chemistry for the preparation of macrocyclic peptidomimetics using an alkynylcarboxylic acid and azido-isocyanide as bifunctional components for the MCR, resulting in the achievement of both monomeric and dimeric macrocycles [29].

Reverse-turn peptidomimetics. A reverse-turn can generally be defined as a structural motif in which a polypeptide chain reverses its overall direction. They play a central role in the structural arrangement of proteins and as recognition elements in protein–protein and peptide–protein interactions [30]. The most common structures belonging to the class of reverse turns are γ- and β-turns, which are established by three or four consecutive residues, respectively. They are generally stabilized by intramolecular hydrogen bonds, forming a 7- or 10-membered pseudocycle for γ-turns and β-turns, respectively [31].

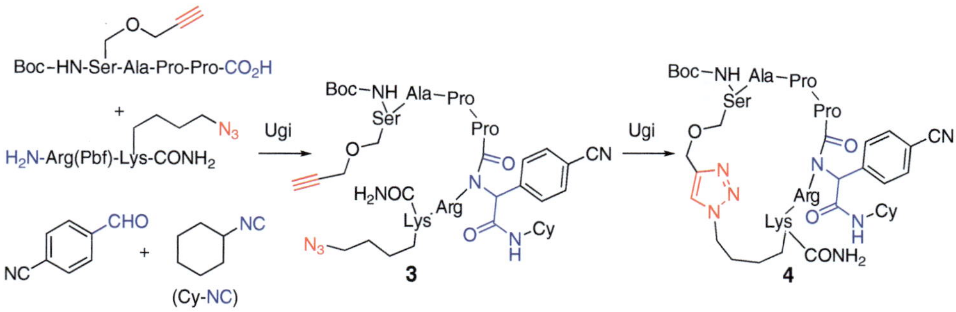

Figure 2.4 Combined Ugi 4CR and CuAAC to generate a macrocyclic peptidomimetic.

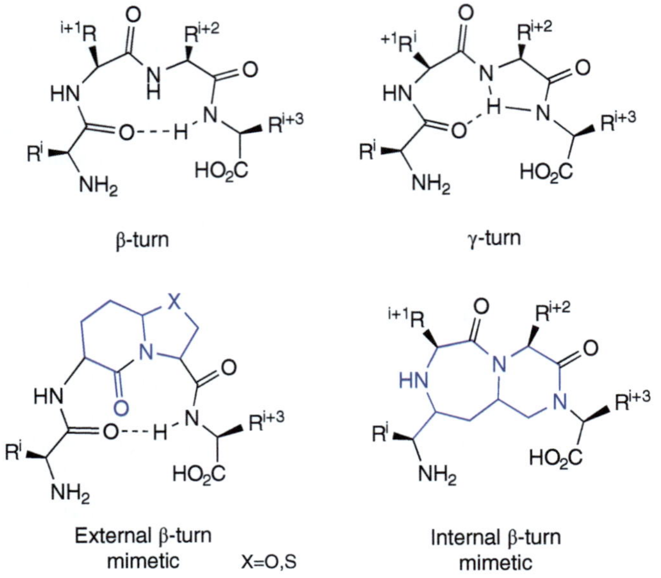

Figure 2.5 Top: structure of γ- and β-turns; bottom: external and internal β-turn mimetics.

β-Turn mimetics have been devised taking advantage of two different approaches and can be categorized into two distinct types: external and internal mimics (Figure 2.5) [32]. Internal mimics, characterized by their rigidifying component residing within the 10-membered pseudocycle of the parent peptide, include bicyclic compounds like diketopiperazines (DKPs). In particular, bicyclic DKPs have gathered significant attention due to their ability to constrain molecules into a conformation resembling the type I β-turn found in native peptides [33, 34]. Consequently, β-turn mimetics built upon this bicyclic core offer valuable insights into the conformation of the corresponding biologically active peptide.

DKPs are a class of naturally occurring six-membered ring dipeptides that have been exploited as peptidomimetic building blocks, and specifically studied for their capability to induce the structure of β-turns and β-hairpins, specifically acting as "internal" turn mimetics (Figure 2.6) [35]. Given the difficulty in achieving such a scaffold directly from the

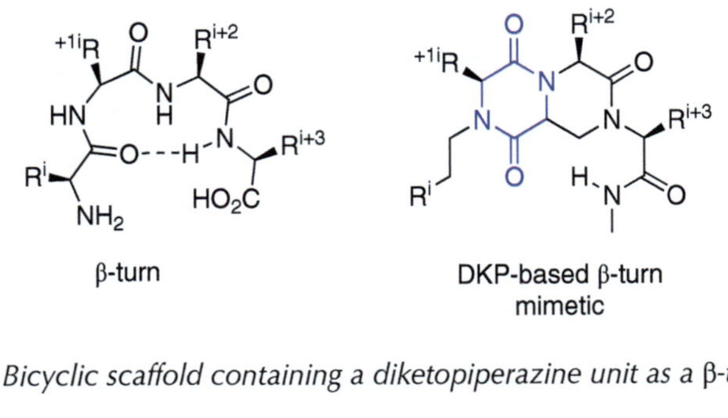

Figure 2.6 Bicyclic scaffold containing a diketopiperazine unit as a β-turn mimetic.

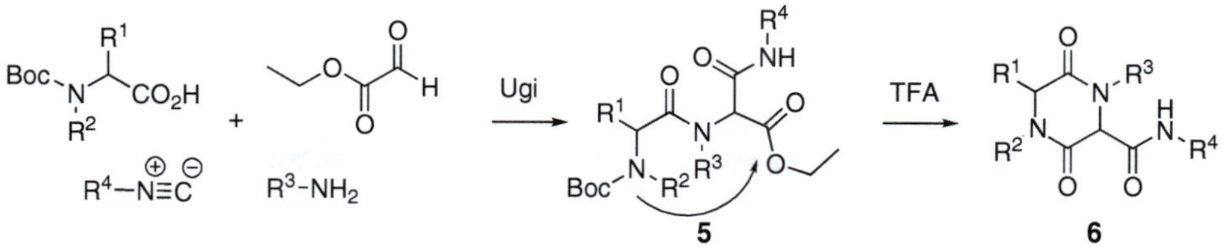

Figure 2.7 Combined Ugi–deprotection–cyclization to achieve a library of DKP scaffolds.

Ugi reaction, synthetic strategies to achieve DKPs have been conceived combining the Ugi MCR and subsequent post-condensation reactions such as Ugi–deprotection–cyclization, Ugi–activation–cyclization, or the combination of both deprotection and activation processes.

Hulme et al. reported for the first time a library of DKPs using a solution-phase Ugi-deprotection–cyclization synthesis [36, 37]. Specifically, *N*-Boc amino acid and a formyl-ester (2-oxoacetic acid) were used as bifunctional components in the MCR process, enabling the cyclization to the DKP scaffolds **6** by acidic *N*-Boc deprotection and subsequent lactam formation of adduct **5** (Figure 2.7).

A similar approach carried out on the solid phase was reported by Golebiowski et al., who constructed a library of bicyclic DKPs as potential β-turn mimics possessing high chemical diversity [38], following previous work by the same authors on β-turn mimetics [39]. A crucial step involved the Ugi 4-component reaction of a resin-bound β-*N*-Boc-diaminopropionic acid ester with β-bromoacid, aldehyde, and isocyanide (Figure 2.8). Although the Ugi reaction exhibited limited diastereoselectivity, providing diastereomeric mixtures (1:1 to 1:2) at the newly generated stereocenter of **7**, it allowed for the precise control of other stereogenic centers. After coupling of additional amino acid to

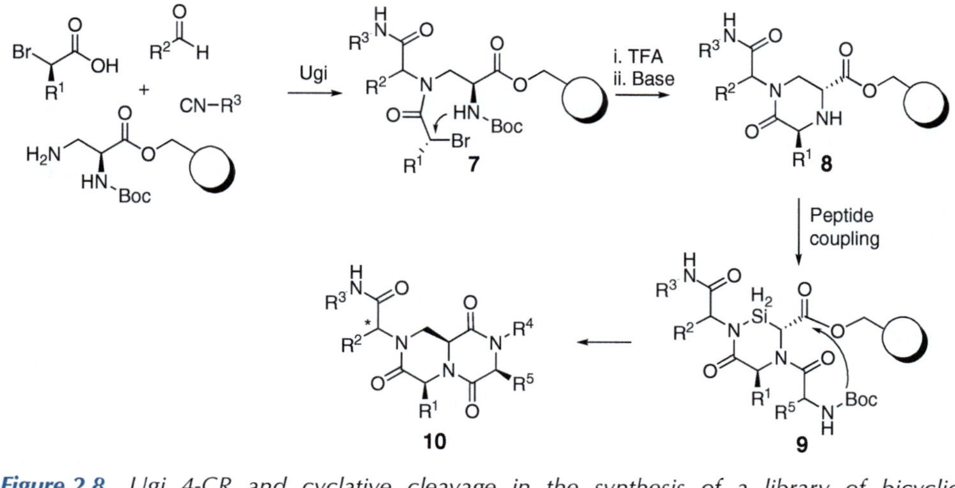

Figure 2.8 Ugi 4-CR and cyclative cleavage in the synthesis of a library of bicyclic diketopiperazines.

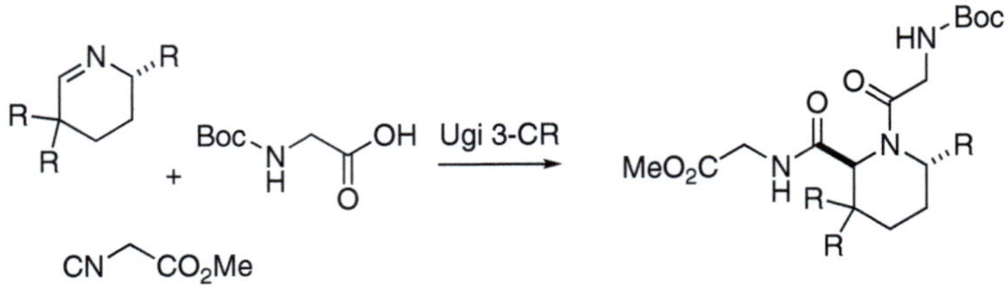

11

Figure 2.9 *Diastereoselective Ugi-3CR of pipecolic acid-based peptidomimetics.*

resin-bound ketopiperazines **8**, the cyclative cleavage of **9** was achieved upon treatment of the resin with acetic acid to give final DKPs **10**.

Cyclic imines have been exploited and reported in several papers for the synthesis of constrained peptidomimetic scaffolds as turn mimetics. The replacement of proline with pipecolic acid as its six-membered homolog proved to be a powerful tool for the study of peptide folding and bioactive conformations. Pipecolic acid derivatives are frequently used as β-turn mimetics [40], leading to their incorporation into various pharmaceutical agents. Maison et al. reported a compelling diastereoselective multicomponent synthesis of these six-membered pipecolic acid-based peptidomimetics [41]. Specifically, they explored the synthesis of pipecolic acid **11** substituted at 3- and 6-positions through a reaction involving the combination of cyclic imines, methyl-2-isocyanoacetate, and *N*-Boc-protected glycine (Figure 2.9).

Similarly, a three-component Ugi reaction was applied for the synthesis of peptidomimetic scaffolds based on the tetrahydroisoquinoline moiety [42]. The library of mimetics of reverse turns was achieved by applying chemical diversity at the isocyanide and carboxylic acid components and using the 3,4-dihydroisoquinoline component easily obtained by Bischler–Napieralski cyclization of the corresponding formamide precursor [43, 44].

2.2.2 Passerini Reaction

The Passerini reaction is a three-component chemical reaction that involves a carboxylic acid, a carbonyl compound, and an isocyanide, furnishing an α-acyloxy amide in a single step [45]. The established pathway for the reaction is conceived as a concerted mechanism, where the hydrogen-bonded carboxylic acid and the carbonyl components react with the isocyanide in a single step, involving an α-addition of the electrophilic carbonyl component and the nucleophilic carboxylate to the isocyanide, yielding the coupling intermediate that is subjected to a rearrangement to the final α-acyloxy amide product.

The synthesis of peptidomimetics has been carried out with the Passerini reaction by using protected amino acid-derived α-amino aldehydes and the three-step protocol MCR/Boc deprotection/cyclization [46], inspired by a previous similar approach developed for the Ugi reaction [39]. This approach proved quite efficient for the combinatorial synthesis of libraries of peptidomimetics for hit identification in drug discovery applications. Specifically, the application of the Passerini reaction to this protocol, the so-called

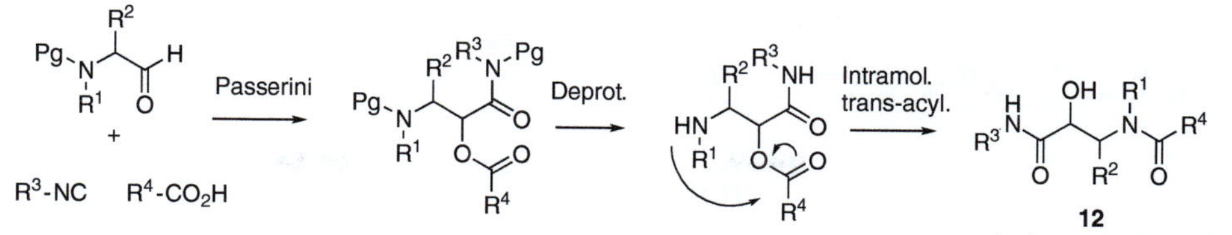

Figure 2.10 The Passerini-Amine Deprotection-Acyl Migration process (PADAM) for the synthesis of α-hydroxy-β-amino acid derivatives.

Passerini-Amine Deprotection-Acyl Migration process (PADAM), enabled the synthesis of α-hydroxy-β-acylaminoamides **12**, displaying the α-hydroxy-β-amino acid moiety typical of norstatine, or the corresponding oxidized derivative, that have been studied for their potential role as inhibitors of aspartic or serine proteases, respectively (Figure 2.10).

2.3 Imine-Based MCRs

Over the years, MCRs encompassing imines as substrates or intermediates have experienced continuous interest due to the reactivity that characterizes the chemical behavior of imines and the wide occurrence of amines and carbonyl compounds. As anticipated, the nitrogen atom of imines can function as a nucleophile, and the carbon atom is the electrophilic site to provide Mannich-type products in the presence of C-nucleophiles or those generated under transition-metal catalysis. This reactivity has proved to be useful in both target- and diversity-oriented synthesis [47] exploiting MCR approaches and for carrying out stereoselective reactions by using chiral catalysts. Concerning peptidomimetics, the application of imine-based MCR has been straightforward, taking advantage of the amino group present in amino acid and peptide precursors and the carbonyl function derived from the corresponding carboxylic acid functional groups. Additional chemical diversity of these approaches has been achieved based on suitable functional groups present on the side chain of peptidomimetic precursors.

2.3.1 A3 Coupling

The synthesis of small molecules possessing an alkyne has continued to gather interest in drug discovery, as the triple bond is a useful handle for a large number of chemical reactions [48] and has demonstrated its value as a functional group in the synthesis of several bioactive compounds and approved drugs, such as ladostigil, efavirenz, and ponatinib. The alkyne moiety allows for displaying functional groups on a linear arrangement due to the geometry of the triple bond, thus enabling a complementary entry in the structural topology of a bioactive compound with respect to the corresponding saturated and alkenyl hydrocarbon chains.

In this view, the A3 coupling received particular attention as a MCR comprising the alkyne as one of the components, to give access to corresponding propargylamine chemotypes [49]. The A3 coupling is generally referred to as a MCR process consisting of an imine or iminium ion formed *in situ* reacting under transition-metal catalysis with

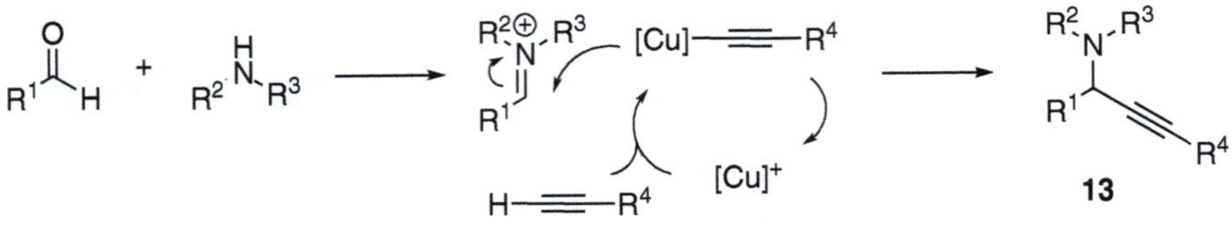

Figure 2.11 Tentative mechanism of Cu-catalyzed A3 coupling.

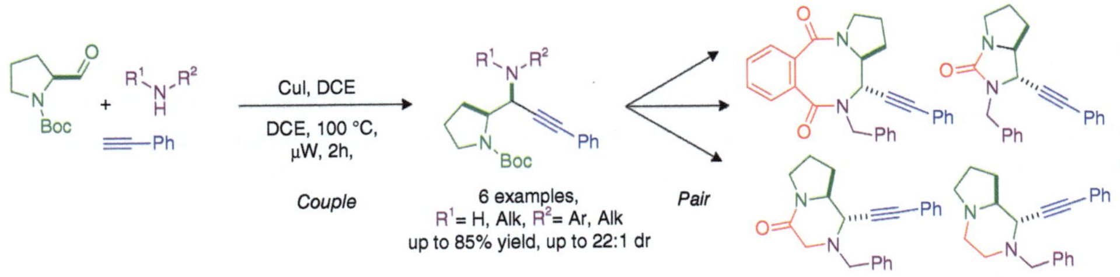

Figure 2.12 Application of the A3 coupling reaction in diversity-oriented synthesis of proline-derived peptidomimetics. Source: Reproduced with permission from Ref. 50; published by Elsevier, 2020.

an aldehyde and an alkyne. The mechanism is still poorly understood and consists of an intermediate π-metal–alkyne complex that is initially formed to increase the acidity of the alkyne proton at the terminal carbon atom for subsequent deprotonation (Figure 2.11). Then, the metal acetylide that is generated *in situ* reacts with the imine or iminium ion to achieve the corresponding propargylamine adduct **13**.

In 2019, Trabocchi et al. reported the application of Boc-prolinal as a cyclic α-amino aldehyde moiety for the Cu-catalyzed A3-coupling reaction using phenylacetylene and benzylamine under microwave irradiation, to give peptidomimetic propargylamines in good yield and diastereoselectivity [50] (Figure 2.12). Following the interest in expanding the chemical diversity around such coupling products in a couple-pair approach, the resulting products containing an orthogonally protected 1,2-diamine moiety and an alkyne functionality were subjected to intramolecular pairing reactions to achieve different peptidomimetic scaffolds according to diversity-oriented synthesis principles [47]. Also, chemoinformatics was applied to characterize the structural diversity of the compounds in terms of shape analysis in a 2D chemical space [51] using principal moments of inertia (PMI) [52] analysis and compared to a library of reference blockbuster drugs, indicating wide coverage of the chemical space as compared to brand-name blockbuster drugs.

In another example employing amino acids as building blocks for the amino component, a decarboxylative A3-coupling was reported. Specifically, a synthetic method based on α-amino acid and alkyne components for the synthesis of propargylamines was reported by Li et al. via a carbonyl-induced Cu(I)-catalyzed decarboxylative transformation [53]. In this approach, the intermediate in the MCR process is a key iminium ylide resulting from the condensation of a secondary α-amino acid with a carbonyl compound followed by decarboxylation.

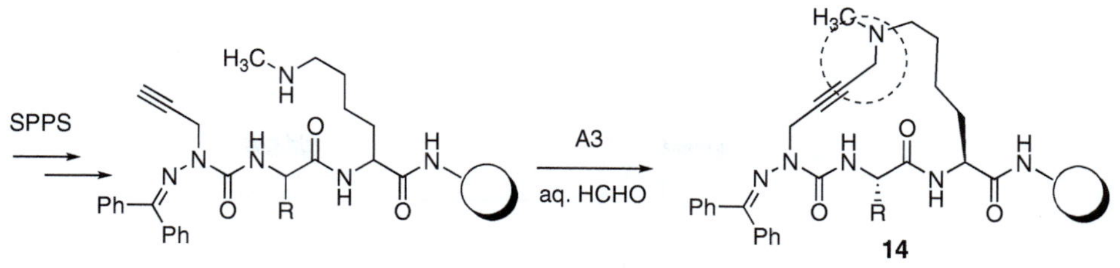

Figure 2.13 Cu-catalyzed macrocyclization reaction on solid phase using the A3 coupling between formaldehyde, azaPra, and ε-N-alkyllysine residues.

The group of Lubell reported the A3 coupling for the synthesis of macrocyclic peptidomimetics as growth hormone-releasing peptide-6 (GHRP-6) analogues possessing affinity for the CD36 receptor, resulting in their capacity to modulate Toll-like receptor agonist-induced overproduction of nitric oxide and reduce proinflammatory cytokine and chemokine production in macrophages [54]. Such peptidomimetic design took advantage of azapeptides as peptide mimics containing semicarbazides, in which the nitrogen atom replaces an amino acid α-carbon. Such aza-residues proved to reproducing the β-turn geometry and increase the metabolic stability of the peptide [55, 56]. The Cu-catalyzed macrocyclization reaction consisted of incorporating aza-propargylamine (azaPra) and ε-*N*-alkyl-lysine in the peptide sequence on solid phase and carrying out the A3 coupling using aqueous formaldehyde (Figure 2.13). The chemical diversity in this approach was achieved by keeping the ε-*N*-alkyl-lysine at the *C*-terminus, moving systematically the azaPra residue to the *N*-terminal of the sequence to achieve macrocycles of different ring sizes ranging from 16 to 25 atoms. Subsequent work on macrocyclic peptidomimetics **14** proved to expand the chemical diversity through modulation of the side-chain length and typology of the alkyne and amine components, and other peptide-bond isosteres beyond semicarbazide, resulting in an enlarged library of peptidomimetics for added insight into SAR analysis [57].

2.3.2 Castagnoli–Cushman Reaction

For half a century, the CCR, a condensation between imines and cyclic anhydrides, has been employed for the synthesis of lactams [58]. Although it combines an imine and an anhydride, the CCR is considered a multicomponent one because the imine is formed from amine and aldehyde. Castagnoli first reported this reaction in 1969 [59], demonstrating the synthesis of 4,5-disubstituted-2-pyrrolidinones **15** from imines and succinic anhydride (Figure 2.14). The scope of the reaction was later expanded by Cushman [60], who synthesized diastereomeric piperidones **16** from imines and glutaric anhydride, producing *trans* isoquinolones from homophthalic anhydride and imines [61]. These advancements illustrate the importance of CCR in generating complex structures and natural product analogues.

The CCR mechanism is still debated, and different hypotheses have been proposed accordingly. Castagnoli and Cushman suggested iminolysis of the anhydride, followed by cyclization and rearrangement [62]. This is in line with observations that electron-donating Schiff base substituents accelerate the reaction and electron-withdrawing succinic

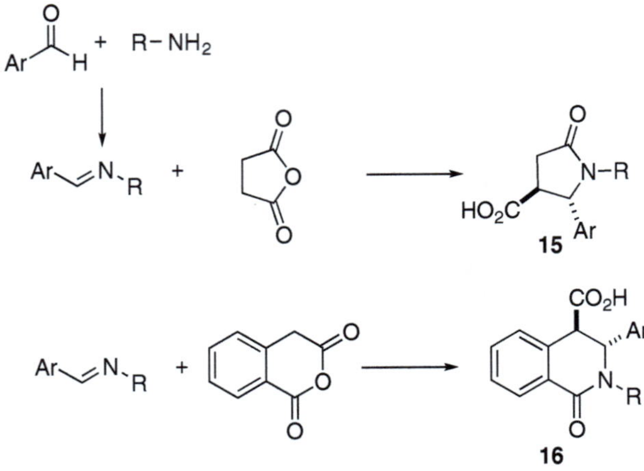

Figure 2.14 Representative examples of Castagnoli–Cushman reaction.

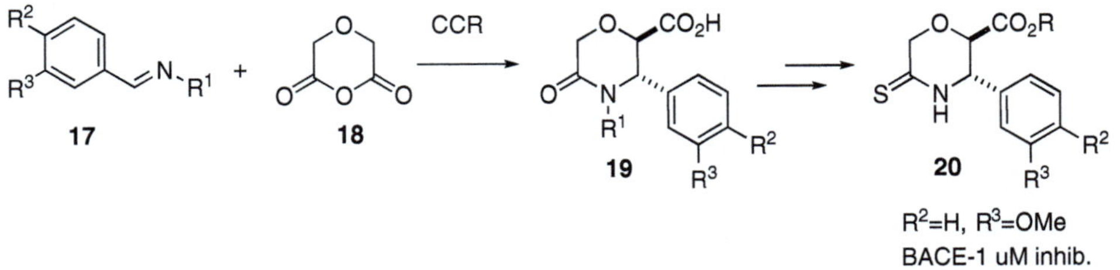

Figure 2.15 Synthesis of morpholine derivatives through CCR as BACE1 inhibitors in the micromolar range.

anhydride substituents enhance reactivity via enolate stabilization. The limitation to non-enolizable aldehyde-derived imines further supports this mechanism.

The condensation of diglycolic anhydride **18** and imines **17** was taken into account in the CCR for the straightforward one-step synthesis of peptidomimetics bearing a C-2 substituted morpholinone scaffold. Subsequent esterification of **19** provided the ester derivatives that were converted into the corresponding thioamide derivatives **20** (Figure 2.15), which showed inhibitory activity toward β-secretase 1 (BACE1) in the micromolar range [63]. Specifically, the cyclic scaffolds proved to expose the thiolactam moiety toward catalytic Asp228 and Arg235 for establishing hydrogen-bonding interactions, the carboxylate group toward Arg128 and Tyr198, and the phenyl ring toward Tyr71 on the flap region, establishing a π-stacking interaction.

In view of developing β-turn mimetics characterized by cyclic nonpeptide small-molecule scaffolds [64], bicyclic compounds consisting of fused DKP and ketopiperazine heterocycles [39, 65, 66] have been conceived as suitable constrained and stable molecules mimicking the 10-membered cyclic arrangement found in β-turn peptide motifs (Figure 2.16). Specifically, bicyclic piperazine-based β-turn mimetics are readily accessible through post-condensation transformations of the piperazines **23** obtained from

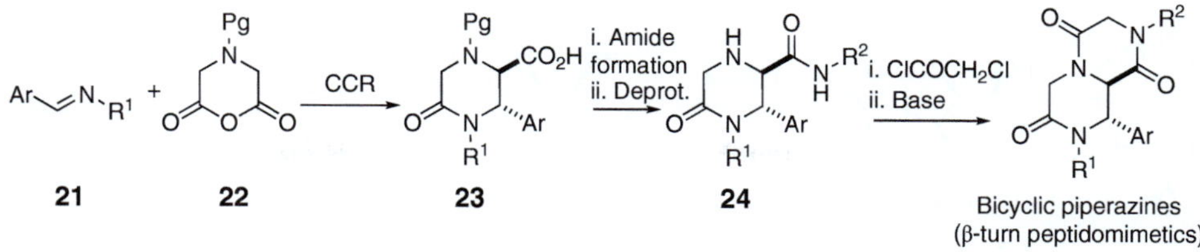

Figure 2.16 Synthesis of bicyclic piperazine peptide β-turn mimetics through CCR of morpholine-2,6-diones and post-condensation of resulting piperazines.

the CCR of morpholine-2,6-diones **22** and imines **21**, involving amide bond formation at the free carboxylic group resulting from CCR and amine deprotection to give **24**, followed by coupling of 2-chloroacetyl chloride and lactam ring formation under basic conditions (Figure 2.16). Similarly, 1,2,5-benzothiadiazepin-4-one-1,1-dioxide scaffolds were achieved by combining CCR with a post-condensation process [67, 68]. Moreover, the CCR of protected iminodiacetic anhydride yielded 5-oxopiperazine-2-carboxamides and their carboxylic acid counterparts, which were subsequently converted into *cis*- and *trans*-configured bicyclic piperazines with two stereogenic centers. These structures, valuable as β-turn mimetics and high-Fsp^3 drug design cores, were further expanded to include previously undescribed ring-expanded variants.

Overall, although CCR is not considered as a pure MCR, it is of significant interest for the development of peptidomimetic scaffolds since it produces structurally diverse lactams in 1–2 steps.

2.3.3 Petasis (Borono–Mannich) Reaction

The first example accounting for the application of Petasis–Borono–Mannich reaction to develop peptidomimetic scaffolds was reported by Petasis and Patel [69]. The Petasis reaction is a MCR of an amine, a carbonyl, and a vinyl- or aryl-boronic acid to form substituted amines. Amino acids and peptides used as the amine components in this reaction enable the development of peptidomimetic systems with a high degree of diastereocontrol. When diamine derivatives are employed, the Petasis three-component reaction yields peptidomimetic heterocycles including 2-oxopiperazines (piperazinones), benzopiperazinones, and benzodiazepines. Also, 2-oxopiperazines **25** are synthesized in a single step from a diamine, glyoxylic acid, and a boronic acid (Figure 2.17). It is proposed

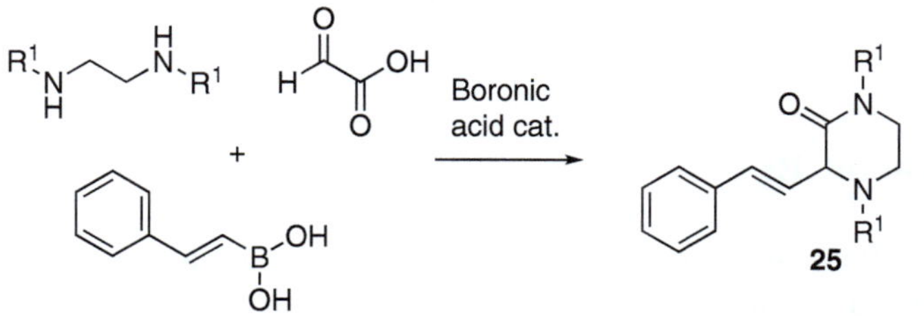

Figure 2.17 Petasis 3CR of diamines followed by boronic acid-catalyzed lactamization.

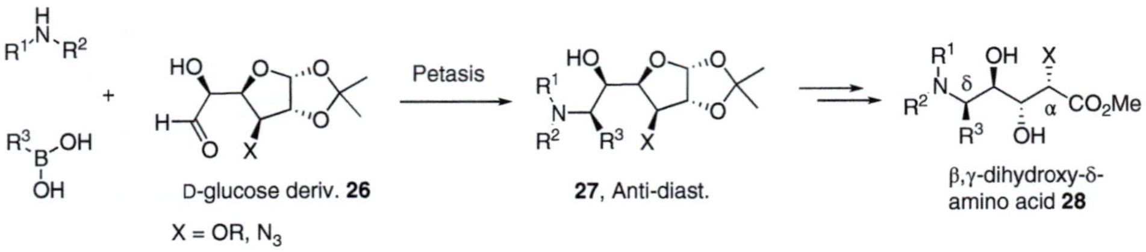

Figure 2.18 *Petasis 3CR on D-glucodialdofuranose to achieve polyhydroxylated δ-amino acid derivatives.*

that intermediate amino acids resulting from the multicomponent Petasis reaction undergo boronic acid-catalyzed lactamization [70] to form the corresponding piperazinone. Also, a two-step synthetic strategy exploiting similar chemical reactivity was utilized for the synthesis of benzopiperazinones [69].

Similarly, the synthesis of 1,4-benzodiazepin-3-ones [71] was achieved using 2-(aminomethyl)aniline in the Petasis reaction and subsequent deprotection and cyclization. The synthesis of isomeric 1,4-benzodiazepin-2-ones was achieved taking advantage of an intramolecular modification of the three-component process, using a bifunctional component possessing the amino and carbonyl functional groups in the same building block and the boronic acid component in a two-component reaction process.

A recent example was reported by Trabocchi et al., where the application of the Petasis reaction to sugar derivatives as the carbonyl components was exploited to expand the peptidomimetic toolbox with polyhydroxylated amino acids (Figure 2.18) [72]. In particular, attention was focused on δ-amino acids, as they represent relevant structural properties as dipeptide isosteres that are widely used for the synthesis of protease inhibitors [73]. The synthesis of a novel α-hydroxy aldehyde deriving from 1,2-*O*-isopropylidene D-glucoaldofuranose was achieved and conceived as a building block for the stereoselective synthesis of glucose-derived amines using the multicomponent Petasis reaction. Specifically, D-glucodialdofuranose was suitably achieved as a new α-hydroxy aldehyde **26** using TCICA/TEMPO (trichloroisocyanuric acid/2,2,6,6-tetramethylpiperidin-1-oxyl) for the chemoselective oxidation of the terminal deprotected diol moiety of 1,2-*O*-isopropylidene D-glucoaldofuranose. This optimized protocol enabled the convenient synthesis of this carbonyl precursor in good amount, thus enabling its application as the carbonyl component of the Petasis reaction with a range of amines and boronic acids, thus resulting in a stereoselective reaction thanks to the chiral environment of the sugar moiety. The synthetic utility of such Petasis product was demonstrated by synthesizing a protected polyhydroxylated δ-amino acid **28** starting from the coupling adduct **27** in three synthetic steps.

2.4 Tandem MCRs

Isocyanide MCRs have been exploited in MCR processes to prepare small-molecule compounds as candidate drugs and to expand the molecular diversity of biologically active compounds also by combining two MCR reactions together in a rapid and varied synthetic process. Accordingly, several possible combinations of isocyanide-based MCR have been described so far, including Ugi and Ugi, azido-Ugi and Ugi, Ugi and Passerini,

Groebke–Blackburn–Bienaymé (GBB-3CR) and Ugi, GBB-3CR and GBB-3CR, GBB-3CR and Passerini, Orru and Ugi, Orru and Passerini.

Peptidomimetics with a substituted imidazo[1,2-*a*]pyridine fragment were synthesized by the combination of two MCRs consisting of tandem Groebke–Blackburn–Bienaymé and Ugi reactions [74]. The target products contained substituted imidazo[1,2-*a*]pyridine and peptidomimetic moieties as pharmacophores with four diversity points introduced from readily available starting materials. A focused compound library of 20 Ugi products was prepared and screened for antibacterial activity, as a representative entry to medicinal chemistry applications.

2.5 Combination of MCRs with Other Processes

2.5.1 Enol–Ugi Condensation

A highly effective enol–Ugi reaction, employing a novel non-heterocyclic amide-stabilized enol derived from β,β-diketoamides, has been established [75]. This methodology offers a wide substrate scope, generating β-enaminoamide peptidomimetics **30** with defined conformations stabilized by CH-π interactions and hydrogen bonds. The use of a five-membered cyclic enol **29** was found critical for achieving high yields and stable products in this process (Figure 2.19). This research demonstrates the potential of the enol–Ugi reaction in the synthesis of diverse and densely functionalized peptidomimetic architectures, although this combined approach is not fully represented in the literature yet.

2.5.2 Interrupted Ugi–Passerini

Interrupting established reactions can lead to divergent pathways and the generation of unforeseen structures. Utilizing a building block with both acidic and oxo-functionalities on a single carbon allows for achieving interrupted Ugi and Passerini reaction variants

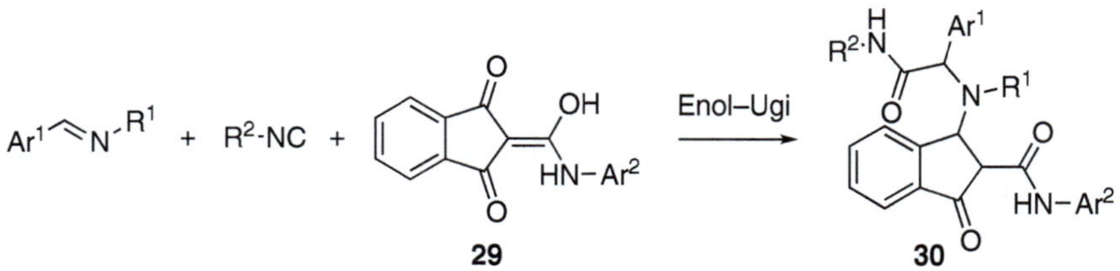

29 30

Figure 2.19 Enol–Ugi condensation of indanodione-derived enol.

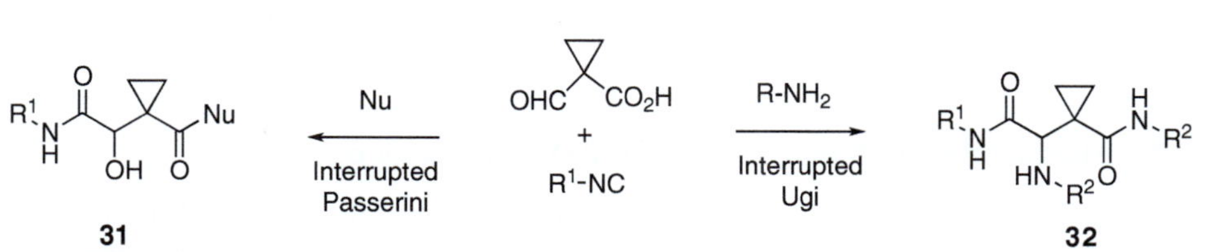

31
Nu = ROH, RSH
32

Figure 2.20 Interrupted Ugi and Passerini for the synthesis of peptidomimetic building blocks.

[76]. The synthesis of >25 peptide-like derivatives highlighted the broad applicability and flexibility of these interrupted processes (Figure 2.20). Furthermore, investigations into diverse nucleophiles and post-modification strategies have significantly expanded the resulting chemical diversity.

2.6 Biological Applications of MCRs in Peptidomimetic Design

As discussed, MCRs are a powerful tool in medicinal chemistry, enabling the rapid and convergent synthesis of structurally complex peptidomimetics. The ability to assemble diverse scaffolds from simple and readily available components enhances both the efficiency and the structural diversity of drug discovery programs. MCR strategies for peptide macrocyclization are highlighted in the literature, including stapling and ring-closing methods as key tools of structural diversity and metabolic stability [77]. This has been implemented in various fields of intervention, including cancer research, neurodegenerative and infectious diseases, to develop both small-molecule libraries and macrocyclic peptidomimetics, the latter achieved using MCRs as a stapling chemical tool. As representative studies in this context, applications in the field of infectious diseases and cancer research are given.

2.6.1 SARS-CoV-2 Mpro Inhibitors

Yang and collab. reported on the development of an orally available SARS-CoV-2 Mpro inhibitor with potent in vivo antiviral activity against wild-type and emerging variants of SARS-CoV-2 [78]. In analogy to similar approaches for developing covalent inhibitors [79], Ugi 4-CR was applied for rapidly achieving lead compounds, particularly using α-ketoamide and acrylamide as warhead groups at the R position for binding to the key Cys145 residue (Figure 2.21). These warheads were selected for their lower reactivity as compared to other chemotypes and for their reported clinical compliance in other compounds approved for use in the clinic [80].

The selected compound Y180, possessing the α-ketoamide as a warhead, showed potent SARS-CoV-2 Mpro inhibitory activity with an IC_{50} of 8.1 nM, as one of the most potent reported non-peptide peptidomimetic Mpro inhibitors, with demonstrated potency against wild-type SARS-CoV-2 and its variants, and also showing in vivo efficacy as an orally administered Mpro inhibitor in a lethal COVID-19 mouse model. A crystallographic study of the enzyme–inhibitor complex revealed the terminal acyl of the α-ketoamide warhead covalently interacting with Cys145 in the S1′ pocket (Figure 2.22). Also, the dibenzo(b,d)furyl group was found interacting in the S2 subsite, and the pyridine and p-F-phenyl ring at S1 and S4, respectively.

2.6.2 Macrocyclic Inhibitors of the p53-MDM2 Interaction

The application of MCRs for the development of peptidomimetics able to disrupt pathological PPIs is increasingly explored, especially for the generation of macrocyclic peptidomimetics. This is particularly relevant in cancer research, as over the years a large

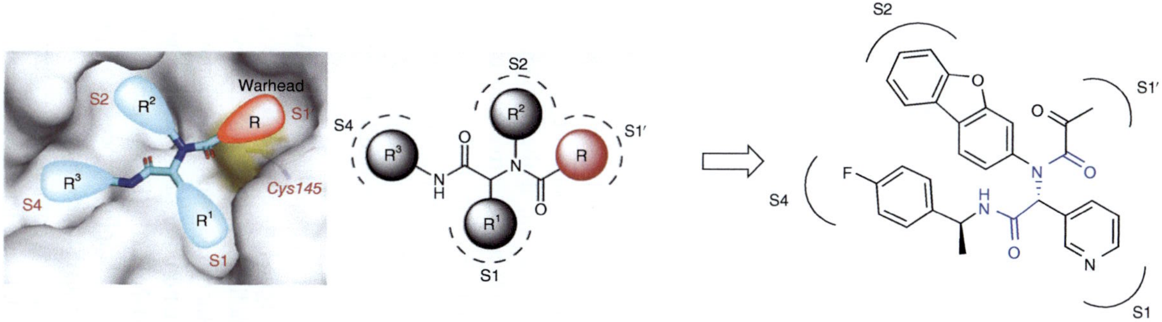

Figure 2.21 *Left: design of inhibitors for the Ugi-4CR conceiving fragments R, R1, R2, and R3 for the S1′, S1, S2, and S4 subsites of Mpro, respectively. Source: [78] / with permission of Springer Nature; published by Springer Nature, 2022; right: structure of lead compound Y180.*

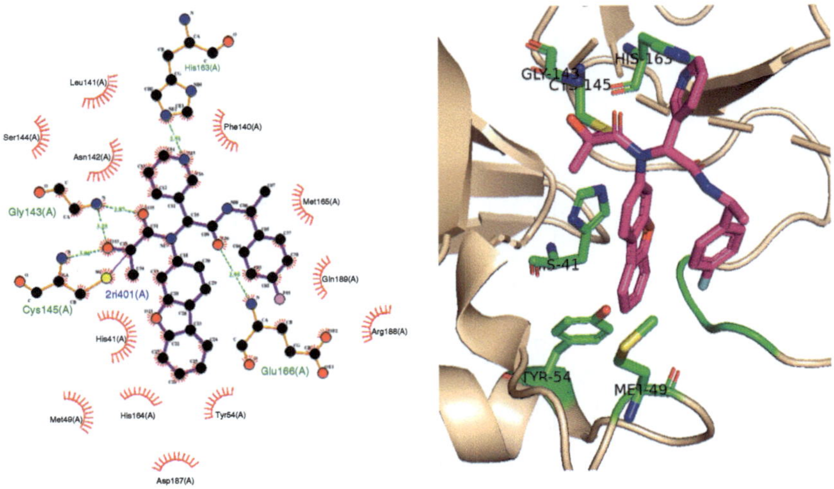

Figure 2.22 *Left: 2D binding mode of Y180 and Mpro with LigPlot+; right: 3D close-up of Y180 in the active site (PDB: 7FAZ).*

number of protein–protein interactions have been discovered with functional and therapeutic significance [81].

One of the most studied protein in cancer research is the p53 transcription factor. This protein is often referred to as "the guardian of the genome," playing a crucial role in cell cycle regulation and tumor suppression. It acts as a sensor of DNA damage, halting cell division to allow for repair or, if the damage is irreparable, inducing programmed cell death [82]. P53 exerts a protective role toward cancer progression, and mutations of this protein or inhibition of its pathway are recurrent in human cancers. The inactivation of p53 is mainly due to MDM2 and MDMX proteins acting as negative regulators. Thus, the binding inhibition of the interaction between the transcription factor p53 and MDM2 and MDMX is a successful strategy for p53 activation in tumor cells [83].

In this context, Dömling reported the application of MCR for the development of macrocyclic peptidomimetics designed to antagonize the p53-MDM2 protein–protein [84]. Specifically, a library of α-helical peptides was achieved by applying a double Ugi reaction on selected peptide sequences to obtain compounds possessing high activities for both MDM2 and MDMX in the nanomolar range, thus capable of displaying potent dual antagonistic action toward the MDM2/MDMX interaction with p53. The synthesis was accomplished by a double MCR for a stapling process on a presynthesized helical peptide sequence exploiting two Asp or Glu residues, and using *m*- or *p*-xylylene diisocyanides, methylamine, and paraformaldehyde as components (Figure 2.23, left). The interaction of the most active macrocyclic peptidomimetic with the target MDM2 protein was demonstrated by X-ray crystallography, highlighting the interaction of both the hydrophobic linker and the amide bonds of the Ugi-staple moiety in the binding process with MDM2 (Figure 2.23, right).

This work is an example of the potential of Ugi MCR for the rapid discovery of macrocyclic peptide inhibitors of protein–protein interactions.

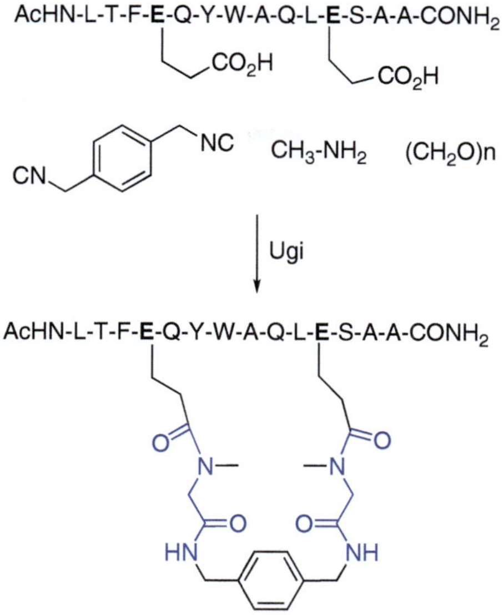

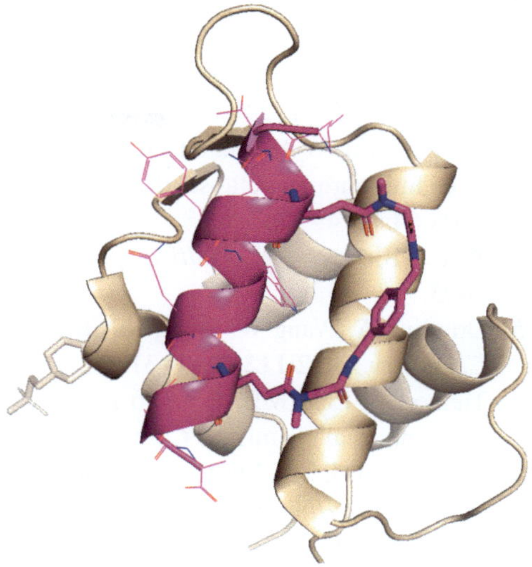

Figure 2.23 *Left: Ugi-4CR employing components; right: X-ray crystallographic structure of lead stapled peptidomimetic showing helix structure in complex with MDM2 (PDB: 6T2D).*

2.7 Conclusion

In recent decades, MCRs have become valuable tools for synthesizing peptidomimetic scaffolds. The effectiveness of these reactions is due to their ability to exploit the naturally occurring amino and carboxylic groups, as well as aldehyde derivatives, present in amino acid precursors. The structural diversity of peptidomimetic products generated through MCRs has increased significantly, particularly through their integration with cyclization reactions. This combination enables the construction of both small-molecule libraries and complex molecular frameworks and macrocyclic peptidomimetics, thus enabling rapid lead discovery in various fields of drug discovery. Recent developments on tandem MCR and the growing interest in imine-based MCR have allowed the expansion of the synthetic possibilities for the development of peptidomimetics. Furthermore, isocyanide-based MCRs, and Ugi in particular, have been consolidated in the use for the synthesis of macrocycles, especially from a couple-pair approach. It is expected that MCR approaches are being applied and studied in peptidomimetic chemistry by leveraging the complexity and diversity of the components and combining these reactions with transition-metal catalysis and other modern technologies in catalysis.

References

1. Dietrich SA, Banfi L, Basso A, Damonte G, Guanti G, Riva R. Application of tandem Ugi multi-component reaction/ring closing metathesis to the synthesis of a conformationally restricted cyclic pentapeptide. *Org Biomol Chem.* 2005;**3**:97–106.

2. Scheffelaar R, Paravidino M, Znabet A, Schmitz RF, de Kanter FJ, Lutz M, Spek AL, Fonseca Guerra C, Bickelhaupt FM, Groen MB, Ruijter E, Orru RV. Scope and limitations of an efficient four-component reaction for dihydropyridin-2-ones. *J Org Chem*. 2010;**75**:1723–1732.
3. Banfi L, Basso A, Guanti G, Merlo S, Repetto C, Riva R. A convergent synthesis of enantiopure bicyclic scaffolds through multicomponent Ugi reaction. *Tetrahedron*. 2008;**64**:1114–1134.
4. Zhu J, Bienayme H. *Multicomponent Reactions*. Weinheim, Germany: Wiley-VCH, 2005.
5. Dömling A, Wang W, Wang K. Chemistry and biology of multicomponent reactions. *Chem Rev*. 2012;**112**:3083–3135.
6. Tietze LF, Lieb ME. Domino reactions for library synthesis of small molecules in combinatorial chemistry. *Curr Opin Chem Biol*. 1998;**2**:363–371.
7. Tuch A, Wallé S. Multicomponent reactions. In: Nicolaou KC, Hanko R, Hartwig W, editors. *Handbook of Combinatorial Chemistry: Drugs, Catalysts, Materials*. Weinheim: Wiley-VCH Verlag GmbH, 2002. pp. 685–705.
8. Ros A, Matas I. Metal-catalyzed multicomponent reactions. In: Herrera RP, Lopez EM, editors. *Multicomponent Reactions: Concepts and Applications for Design and Synthesis*. Hoboken (NJ): John Wiley & Sons, 2015. pp. 72–126.
9. Cherian RM, Harry NA, Saranya S, Rohit KR, Anilkumar G. Copper-catalysed multicomponent syntheses of heterocycles. *Asian J Org Chem*. 2019;**8**:197–233.
10. Müller DS, Marek I. Copper mediated carbometalation reactions. *Chem Soc Rev*. 2016;**45**:4552–4566.
11. Zhu X, Chiba S. Copper-catalyzed oxidative carbon-heteroatom bond formation: a recent update. *Chem Soc Rev*. 2016;**45**:4504–4523.
12. Maaliki C, Thiery E, Thibonnet J. Emergence of copper-mediated formation of C—C bonds. *Eur J Org Chem*. 2017;209–228.
13. Strecker A. Ueber die künstliche Bildung der Milchsäure und einen neuen, dem Glycocoll homologen Körper. *Ann Chem*. 1850;**75**:27–45.
14. Hantzsch A. Condensationsprodukte aus Aldehydammoniak und ketonartigen Verbindungen. *Chem Ber*. 1881;**14**:1637–1638.
15. Ugi I. Recent progress in the chemistry of multicomponent reactions. *Pure Appl Chem*. 2001;**73**:187–191.
16. Choudhury LH, Parvin T. Recent advances in the chemistry of imine-based multicomponent reactions (MCRs). *Tetrahedron*. 2011;**67**:8213–8228.
17. Mannich C, Krösche W. Über ein Kondensationsprodukt aus Formaldehyd, Ammoniak und Antipyrin. *Arch Pharm*. 1912;**250**:647–667.
18. Petasis NA, Akritopoulou I. A new reaction of boronic acids with α-amino acids and carbonyl compounds. *Tetrahedron Lett*. 1993;**34**:583–586.
19. Dyatkin AB, Rivero RA. A novel three-component synthesis of β-amino ketones. *Tetrahedron Lett*. 1998;**39**:3647–3650.
20. Li CJ, Wei C. Highly efficient Grignard-type imine additions via C—H activation in water and under solvent-free conditions. *Chem Commun*. 2002;268–269.
21. Liu C, Voskressensky LG, Van der Eycken EV. Recent advances in the synthesis of peptidomimetics via Ugi reactions. *Chemistry*. 2024;**30**:e202303597.

22. Koopmanschap G, Ruijter E, Orru RV. Isocyanide-based multicomponent reactions towards cyclic constrained peptidomimetics. *Beilstein J Org Chem*. 2014;**10**:544–598.
23. Gulevich AV, Zhdanko AG, Orru RV, Nenajdenko VG. Isocyanoacetate derivatives: synthesis, reactivity, and application. *Chem Rev*. 2010;**110**:5235–5331.
24. Wessjohann LA, Rhoden CRB, Rivera DG, Vercillo OE. Multicomponent reactions in the synthesis of heterocycles. In: Orru RVA, Ruijter E, editors. *Synthesis of Heterocycles via Multicomponent Reactions I*. Berlin: Springer-Verlag, 2010. pp. 199–226.
25. Cankařová N, Krchňák V. Isocyanide multicomponent reactions on solid phase: state of the art and future application. *Int J Mol Sci*. 2020;**21**:9160.
26. Cheng JF, Chen M, Arrhenius T, Nadzan AA. A convenient solution and solid-phase synthesis of D5-2-oxopiperazines via *N*-acyliminium ion cyclization. *Tetrahedron Lett*. 2002;**43**:6293–6295.
27. Vishwanatha TM, Bergamaschi E, Dömling A. Sulfur-switch Ugi reaction for macrocyclic disulfide-bridged peptidomimetics. *Org Lett*. 2017;**19**:3195–3198.
28. Tahoori F, Balalaie S, Sheikhnejad R, Sadjadi M, Boloori P. Design and synthesis of anti-cancer cyclopeptides containing triazole skeleton. *Amino Acids*. 2014;**46**:1033–1046.
29. Zakharova EA, Shmatova OI, Kutovaya IV, Khrustalev VN, Nenajdenko VG. Synthesis of macrocyclic peptidomimetics via the Ugi-click-strategy. *Org Biomol Chem*. 2019;**17**:3433–3445.
30. For further information on β-turn mimetics, see: Trabocchi A, Guarna A. β-turn peptidomimetics. In: *Peptidomimetics in Organic and Medicinal Chemistry*. Chichester: Wiley, 2014. pp. 191–217.
31. Belvisi L, Gennari C, Mielgo A, Potenza D, Scolastico C. Conformational preferences of peptides containing reverse-turn mimetic bicyclic lactams: inverse γ-turns versus type-II′ β-turns – insights into β-hairpin stability. *Eur J Org Chem*. 1999;389–400.
32. Ball JB, Alewood PF. Conformational constraints: nonpeptide beta-turn mimics. *J Mol Recognit*. 1990;**3**:55–64.
33. Borthwick AD. 2,5-Diketopiperazines: synthesis, reactions, medicinal chemistry, and bioactive natural products. *Chem Rev*. 2012;**112**:3641–3716.
34. Hanessian S, McNaughton-Smith G, Lombart HG, Lubell WD. Design and synthesis of conformationally constrained amino acids as versatile scaffolds and peptide mimetics. *Tetrahedron*. 1997;**53**:12789–12794.
35. Ressurreição ASM, Delatouche R, Gennari C, Piarulli U. Bifunctional 2,5-diketopiperazines as rigid three-dimensional scaffolds in receptors and peptidomimetics. *Eur J Org Chem*. 2011;**2**:217–228.
36. Hulme C, Morrissette MM, Volz FA, Burns CJ. The solution phase synthesis of diketopiperazine libraries via the Ugi reaction: novel application of Armstrong's convertible isonitrile. *Tetrahedron Lett*. 1998;**39**:1113–1116.
37. Hulme C, Peng J, Morton G, Salvino JM, Herpin T, Labaudiniere R. Novel safety-catch linker and its application with a Ugi/De-BOC/Cyclization (UDC) strategy to access carboxylic acids, 1,4-benzodiazepines, diketopiperazines, ketopiperazines and dihydroquinoxalinones. *Tetrahedron Lett*. 1998;**39**:7227–7230.
38. Golebiowski A, Jozwik J, Klopfenstein SR, Colson AO, Grieb AL, Russell AF, Rastogi VL, Diven CF, Portlock DE, Chen JJ. Solid-supported synthesis of putative peptide

beta-turn mimetics via Ugi reaction for diketopiperazine formation. *J Comb Chem.* 2002;**4**:584–590.
39. Golebiowski A, Klopfenstein SR, Shao X, Chen JJ, Colson AO, Grieb AL, Russell AF. Solid-supported synthesis of a peptide beta-turn mimetic. *Org Lett.* 2000;**2**: 2615–2617.
40. Genin MJ, Gleason WB, Johnson RL. Design, synthesis, and x-ray crystallographic analysis of two novel spirolactam systems as beta-turn mimics. *J Org Chem.* 1993;**58**:860–863.
41. Maison W, Lützen A, Kosten M, Schlemminger I, Westerhoff O, Saak W, Martens J. Multicomponent synthesis of tripeptides containing pipecolic acid derivatives: selective induction of *cis*- and *trans*-imide bonds into peptide backbones. *J Chem Soc Perkin Trans.* 2000;**1**:1867–1871.
42. Rossetti A, Sacchetti A, Gatti M, Pugliese A, Roda G. Rapid access to reverse-turn peptidomimetics by a three-component Ugi reaction of 3,4-dihydroisoquinoline. *Chem Heterocycl Compd.* 2017;**53**:1214–1219.
43. Jiang G, Chen J, Huang JS, Che CM. Highly efficient oxidation of amines to imines by singlet oxygen and its application in Ugi-type reactions. *Org Lett.* 2009;**11**:4568–4571.
44. Schuster I, Sztojkov-Ivanov A, Lazar L, Fulop F. Synthesis of 1,2,3,4-tetrahydroisoquinoline-1-carboxylic acid derivatives via Ugi reactions. *Lett Org Chem.* 2007;**4**:102–106.
45. Banfi L, Basso A, Lambruschini C, Moni L, Riva R. The 100 facets of the Passerini reaction. *Chem Sci.* 2021;**12**:15445–15472.
46. Banfi L, Guanti G, Riva R. Passerini multicomponent reaction of protected α-aminoaldehydes as a tool for combinatorial synthesis of enzyme inhibitors. *Chem Commun.* 2000;**2000**:985–986.
47. Trabocchi A. *Diversity-Oriented Synthesis: Basics and Applications in Organic Synthesis, Drug Discovery, and Chemical Biology*. Hoboken, NJ: John Wiley & Sons, 2013.
48. Trost BM, Li C-J (eds.). *Modern Alkyne Chemistry: Catalytic and Atom-Economic Transformations*. Weinheim: Wiley-VCH Verlag GmbH & Co. KGaA, 2015.
49. Peshkov VA, Pereshivko OP, Van der Eycken EV. A walk around the A3-coupling. *Chem Soc Rev.* 2012;**41**:3790–3807.
50. Innocenti R, Lenci E, Baldini L, Faggi C, Menchi G, Trabocchi A. Copper-catalyzed A^3-coupling for the diversity-oriented synthesis of proline-derived alkynyl-substituted peptidomimetic scaffolds. Eur. *J Org Chem.* 2019;6203–6210.
51. Dobson CM. Chemical space and biology. *Nature.* 2004;**432**:824–828.
52. Sauer WH, Schwarz MK. Molecular shape diversity of combinatorial libraries: a prerequisite for broad bioactivity. *J Chem Inf Comput Sci.* 2003;**43**:987–1003.
53. Bi HP, Teng Q, Guan M, Chen WW, Liang YM, Yao X, Li CJ. Aldehyde- and ketone-induced tandem decarboxylation-coupling (Csp(3)-Csp) of natural alpha-amino acids and alkynes. *J Org Chem.* 2010;**75**:783–788.
54. Zhang J, Mulumba M, Ong H, Lubell WD. Diversity-oriented synthesis of cyclic azapeptides by A3 -macrocyclization provides high-affinity CD36-modulating peptidomimetics. *Angew Chem Int Ed Engl.* 2017;**56**:6284–6288.
55. Proulx C, Sabatino D, Hopewell R, Spiegel J, García Ramos Y, Lubell WD. Azapeptides and their therapeutic potential. *Future Med Chem.* 2011;**3**:1139–1164.

56. Tal-Gan Y, Freeman NS, Klein S, Levitzki A, Gilon C. Metabolic stability of peptidomimetics: *N*-methyl and aza heptapeptide analogs of a PKB/Akt inhibitor. *Chem Biol Drug Des*. 2011;**78**:887–892.
57. Ohm RG, Mulumba M, Chingle RM, Ahsanullah ZJ, Chemtob S, Ong H, Lubell WD. Diversity-oriented A3-macrocyclization for studying influences of ring-size and shape of cyclic peptides: CD36 receptor modulators. *J Med Chem*. 2021;**64**:9365–9380.
58. Ramiro JL, Martínez-Caballero S, Neo AG, Díaz J, Marcos CF. The Castagnoli–Cushman reaction. *Molecules*. 2023;**28**:2654.
59. Castagnoli N Jr. The condensation of succinic anhydride with benzylidinemethylamine. A stereoselective synthesis of trans- and cis-1-methyl-4-carbosy-5-phenyl-2-pyrrolidinone. *J Org Chem*. 1969;**34**:3187–3189.
60. Cushman M, Castagnoli N Jr. A novel approach to the synthesis of nitrogen analogs of the tetrahydrocannabinols. *J Org Chem*. 1973;**38**:440–448.
61. Haimova M, Mollov N, Ivanova S, Dimitrova A, Ognyanov V. A highly stereoselective synthesis of 3,4-dihydro-1(2*H*)-isoquinolinones and 8-oxoberbines from homophthalic anhydrides and azomethines. *Tetrahedron*. 1977;**33**:331–336.
62. Cushman M, Castagnoli N Jr. The condensation of succinic anhydrides with Schiff bases. Scope and mechanism. *J Org Chem*. 1971;**36**:3404–3411.
63. Calugi L, Lenci E, Innocenti R, Trabocchi A. Synthesis of morpholine derivatives using the Castagnoli–Cushman reaction as BACE1 inhibitors: unexpected binding activity of cyclic thioamides. *Bioorg Med Chem Lett*. 2020;**30**:127211.
64. MacDonald M, Aube J. Approaches to cyclic peptide β-turn mimics. *Curr Org Chem*. 2001;**5**:417–438.
65. Belov VN, Funke C, Labahn T, Es-Sayed M, de Meijere A. An easy access to bicyclic peptides with an octahydro[2*H*]pyrazino[1,2-*a*]pyrazine skeleton. *Eur J Org Chem*. 1999;1345–1356.
66. Kim HO, Nakanishi H, Lee MS, Kahn M. Design and synthesis of novel conformationally restricted peptide secondary structure mimetics. *Org Lett*. 2000;**2**:301–302.
67. Usmanova L, Dar'in D, Novikov MS, Gureev M, Krasavin M. Bicyclic piperazine mimetics of the peptide β-turn assembled via the Castagnoli–Cushman reaction. *J Org Chem*. 2018;**83**:5859–5868.
68. Usmanova L, Dar'in D, Krasavin M. A novel variant of the medicinally relevant 1,2,5-benzothiadiazepin-4-one-1,1-dioxide scaffold accessed via the downstream modification of Castagnoli-Cushman lactams. *Tetrahedron Lett*. 2019;**60**:151003.
69. Petasis NA, Patel ZD. Synthesis of piperazinones and benzopiperazinones from 1,2-diamines and organoboronic acids. *Tetrahedron Lett*. 2000;**41**:9607–9611.
70. Ishihara K, Yamamoto H. Arylboron compounds as acid catalysts in organic synthetic transformations. *Eur J Org Chem*. 1999;527–538.
71. Petasis NA. Multicomponent reactions with organoboron compounds. In: Zhu J, Bienaymé H, editors. *Multicomponent Reactions*. Weinheim: Wiley-VCH, 2005. pp. 199–223.
72. Lenci E, Menchi G, Guarna A, Trabocchi A. Skeletal diversity from carbohydrates: use of mannose for the diversity-oriented synthesis of polyhydroxylated compounds. *J Org Chem*. 2015;**80**:2182–2191.
73. Baldauf C, Günther R, Hofmann HJ. Delta-peptides and delta-amino acids as tools for peptide structure design--a theoretical study. *J Org Chem*. 2004;**69**:6214–6220.

74. Kolomiiets OV, Tsygankov AV, Kornet MN, Brazhko AA, Musatov VI, Chebanov VA. Synthesis of imidazo[1,2-a]pyridine-containing peptidomimetics by tandem of Groebke–Blackburn–Bienaymé and Ugi reactions. *Beilstein J Org Chem.* 2023;**19**:727–735.
75. Ramiro JL, Neo AG, Marcos CF. Amide-derived enols in enol-Ugi reactions: expanding horizons for peptidomimetic scaffold synthesis. *Org Biomol Chem.* 2024;**22**: 8300–8309.
76. Anastasiou PK, Fragkiadakis M, Thomaidi M, Froudas KG, Neochoritis CG. Navigating unexplored territories of the interrupted Ugi and Passerini reactions toward peptidomimetics. *Org Lett.* 2025;**27**:1829–1834.
77. Reguera L, Rivera DG. Multicomponent reaction toolbox for peptide macrocyclization and stapling. *Chem Rev.* 2019;**119**:9836–9860.
78. Quan BX, Shuai H, Xia AJ, Hou Y, Zeng R, Liu XL, Lin GF, Qiao JX, Li WP, Wang FL, Wang K, Zhou RJ, Yuen TT, Chen MX, Yoon C, Wu M, Zhang SY, Huang C, Wang YF, Yang W, Tian C, Li WM, Wei YQ, Yuen KY, Chan JF, Lei J, Chu H, Yang S. An orally available Mpro inhibitor is effective against wild-type SARS-CoV-2 and variants including Omicron. *Nat Microbiol.* 2022;**7**:716–725.
79. Kitamura N, Sacco MD, Ma C, Hu Y, Townsend JA, Meng X, Zhang F, Zhang X, Ba M, Szeto T, Kukuljac A, Marty MT, Schultz D, Cherry S, Xiang Y, Chen Y, Wang J. Expedited approach toward the rational design of noncovalent SARS-CoV-2 main protease inhibitors. *J Med Chem.* 2022;**65**:2848–2865.
80. Bauer RA. Covalent inhibitors in drug discovery: from accidental discoveries to avoided liabilities and designed therapies. *Drug Discov Today.* 2015;**20**:1061–1073.
81. Li Z, Ivanov AA, Su R, Gonzalez-Pecchi V, Qi Q, Liu S, Webber P, McMillan E, Rusnak L, Pham C, Chen X, Mo X, Revennaugh B, Zhou W, Marcus A, Harati S, Chen X, Johns MA, White MA, Moreno C, Cooper LA, Du Y, Khuri FR, Fu H. The OncoPPi network of cancer-focused protein–protein interactions to inform biological insights and therapeutic strategies. *Nat Commun.* 2017;**8**:14356.
82. Wang H, Guo M, Wei H, Chen Y. Targeting p53 pathways: mechanisms, structures, and advances in therapy. *Signal Transduct Target Ther.* 2023;**8**:92.
83. Hu B, Gilkes DM, Chen J. Efficient p53 activation and apoptosis by simultaneous disruption of binding to MDM2 and MDMX. *Cancer Res.* 2007;**67**:8810–8817.
84. Ricardo MG, Ali AM, Plewka J, Surmiak E, Labuzek B, Neochoritis CG, Atmaj J, Skalniak L, Zhang R, Holak TA, Groves M, Rivera DG, Dömling A. Multicomponent peptide stapling as a diversity-driven tool for the development of inhibitors of protein–protein interactions. *Angew Chem Int Ed Engl.* 2020;**59**:5235–5241.

3

Non-Coded Amino Acids: Cyclic Scaffolds and C—H Functionalization

3.1 Introduction

A broad range of techniques has been studied for developing novel non-coded amino acids, as well as for modifying natural amino acids by adding functional groups or cross-linking motifs. Given the relevance of non-coded amino acids within the peptidomimetic toolbox, over the last century, the stereoselective synthesis of novel amino acids, both linear and cyclic, has been largely explored to expand the chemical space around such class of compounds. Synthetic approaches include the application of multicomponent reactions (see Chapter 2), starting from the Strecker synthesis, and more generally the whole panorama of stereoselective processes to enable the achievement of optically pure non-coded amino acids, for example, through the enantioselective substitution of simple amino acid substrates such as glycine derivatives, or the hydrogenation of unsaturated amino acid analogues. In the past two decades, metal-mediated coupling chemistry has become a versatile option for peptide modification, thus paving the way for diverse bond types to be formed directly on amino acid side chains. Nevertheless, such methods generally depend on preactivated building blocks, like aryl halides or boron-containing partners, and often require multistep synthesis. These drawbacks have led researchers to study synthetic alternatives following the concepts of atom- and step-economy. One particularly effective solution has been the direct modification of C—H bonds. By applying directing group strategies, unreactive positions within amino acid or peptide frameworks can be selectively activated and functionalized, thus making this approach a powerful tool for generating structurally and functionally diverse molecules [1].

3.2 Cyclic Amino Acids

As discussed in Chapter 1, saturated azaheterocycles have emerged as prominent scaffolds in the design of peptidomimetics, particularly among nonaromatic cyclic frameworks. Over the decades, a variety of unnatural cyclic amino acids incorporating aziridine, azetidine, pyrrolidine, piperidine, morpholine, and piperazine rings have been synthesized and investigated [2]. These molecules are characterized by a covalent linkage between the α-carbon and the backbone nitrogen atom, thus generating a conformationally constrained motif resembling the typical structural bias of the naturally occurring proline. Such structural restriction arises from steric clashes with adjacent residues, thus influencing the local geometry around the modified amino acid. The variation in ring size provides a tunable mechanism for restraining the ψ dihedral angle, which is critical for modulating the peptide conformation in rational design. Therefore, a wide range of azaheterocyclic amino acids have been applied as enzymatic inhibitors and, more recently, as valuable elements to modulate protein–protein interactions within bioactive peptide sequences. In this view, several chemical methodologies have been studied to achieve cyclic amino acids, including lactamization as an entry of intramolecular polar cyclization, ring-closing metathesis (RCM), aldol-type cyclization reactions, intramolecular cross-coupling, and direct cyclization through C—H activation (Table 3.1).

In the field of peptidomimetics, several amino acid scaffolds have been extensively studied for their ability to mimic the structural and functional properties of natural peptides, offering tools for modulating the conformational preferences of peptides and influencing the bioactivity and stability of peptide-based drugs. As anticipated, saturated azaheterocycles, such as aziridine, azetidine, pyrrolidine, piperidine, morpholine, and piperazine, are particularly relevant for their conformational constraints, which influence peptide backbone rigidity and side-chain orientation. Aromatic and heteroaromatic scaffolds, including phenylalanine, tryptophan, histidine, and fused benzene or pyridine derivatives, are

Table 3.1 *Chemical methods exploited for the synthesis of cyclic amino acids.*

Method	Key transformation	Advantages	Limitations
Intramolecular cyclization (i.e. lactamization and condensation)	Cyclization of linear amino acid derivatives to ring structures	Simple; scalable	Ring-size control; side reactions
Ring-closing metathesis (RCM)	Olefin metathesis to form cyclic amino acid scaffolds	High efficiency; broad scope	Catalyst cost; *E*/*Z* control
Michael addition/aldol-type cyclization	Enolate-mediated intramolecular Michael addition–aldol domino to cyclic structures	Efficient stereocenter creation	Substrate-limited; side reactions
Metal-catalyzed cyclization (i.e. cross-coupling)	Intramolecular Pd/Ni/Cu-mediated C—C, C—N, and C—O bond formation	Versatile; amenable to late-stage diversification	Needs pre-functionalized precursors; catalyst cost
C—H activation/direct cyclization	Direct functionalization of inert C—H bonds using directing groups to cyclize	Atom- and step-economical; avoids pre-functionalization	Regioselectivity challenges; directing group steps

suitable for promoting π–π stacking and hydrogen-bond interactions. Among carbocyclic structures, cycloalkane scaffolds, ranging from cyclopropane to cyclohexane, have been studied and employed to control side-chain orientation and backbone rigidity. Additionally, β- and γ-amino acids extend the peptide backbone to stabilize secondary structures, whereas α,α-disubstituted amino acids, such as Aib, are instrumental to introduce steric hindrance, restrict conformational flexibility, and facilitate folding peptide sequences into helix structures.

3.2.1 Proline Analogues

Among saturated nitrogen-containing heterocycles, substituted pyrrolidines occupy a key role as versatile frameworks in clinically relevant molecules and, in the peptide field, are often elaborated as proline analogues. Proline occupies a unique position among the amino acids found in proteins due to its distinctive effect on the backbone architecture. In particular, the cyclic nature of proline leads to unique structural constraints, such as β- and γ-turns, which play a significant role in receptor recognition and affinity. Furthermore, proline exhibits *cis–trans* isomerism, where both isomers occur in solution. This property is characterized by a high rotational barrier and has important implications for protein folding and stability, as well as for the reactivity and recognition of proline-containing peptides and proteins. The utility of proline analogues spans from playing as fundamental building blocks in medicinal chemistry to acting as privileged fragments in fragment-based drug discovery [3]. Indeed, over last two decades, several drugs based on cyclic and bicyclic proline analogues have been approved, including saxagliptin (a dipeptidyl peptidase-4 (DPP-4) inhibitor for treating type 2 diabetes), icatibant (a short peptide bradykinin antagonist for treating hereditary angioedema), boceprevir, telaprevir, grazoprevir, and voxilaprevir (four drugs as inhibitors of NS3/4A hepatitis C protease, see Chapter 12), and nirmatrelvir (an Mpro inhibitor for treating COVID-19, see Chapter 8) (Figure 3.1) [3]. In particular, 4-hydroxy-proline was exploited to develop macrocyclic grazoprevir and voxilaprevir drugs (Figure 3.2).

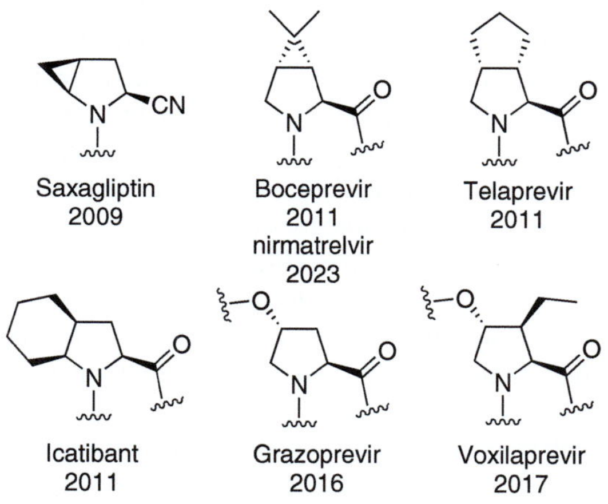

Figure 3.1 Representative FDA-approved drugs containing proline analogues.

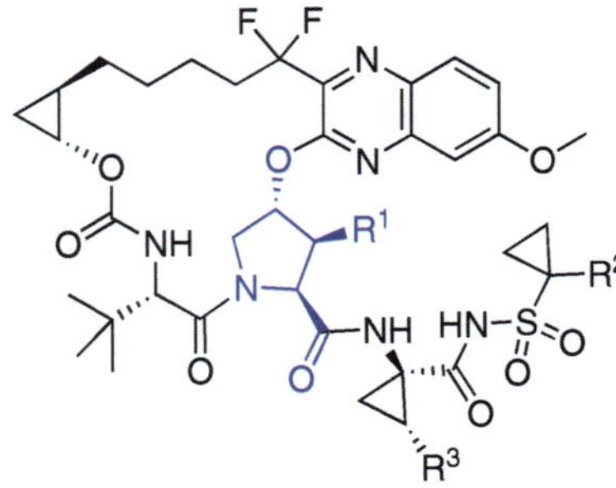

Figure 3.2 Structure of macrocyclic voxilaprevir and grazoprevir drugs containing 4-hydroxy-proline scaffold.

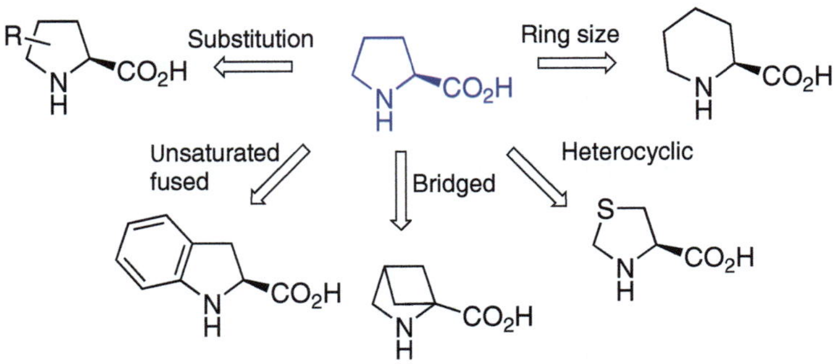

Figure 3.3 Typologies and examples of proline analogues.

The pyrrolidine scaffold of proline possesses seven nonequivalent hydrogen atoms on the four α- to δ-carbon atoms that can be substituted, thus enabling broad structural diversity and the development of numerous analogues taking advantage of methodologies in chemical synthesis, as reported previously [4] and in a comprehensive review by Kubyshkin and Rubini [5]. Various proline analogues proved to modulate the conformational and chemical behavior of this amino acid within polypeptide chains to accomplish diverse functional effects. Proline analogues can be arranged into several structural categories, including substituted, unsaturated, and fused ring homologues and heterocyclic and bridged analogues, guided by substitution on the pyrrolidine ring (Figure 3.3).

The synthetic approaches to develop molecular scaffolds with varying degrees of functional group diversity have led to the development of a wide range of 3-, 4-, and 5-substituted proline analogues, as well as other polyfunctional heterocyclic and homologated derivatives. These structural modifications are essential in generating peptidomimetics with enhanced conformational bias and chemical diversity. For example, one of the key strategies for developing proline analogues involves 3-substituted prolines, which are considered conformationally restricted analogues of natural α-amino acids. These analogues contain an ethylene bridge linking the α-nitrogen and β-carbon to

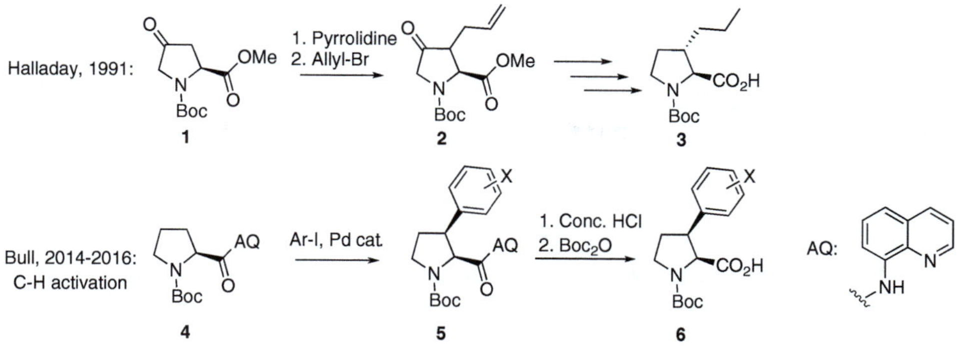

Figure 3.4 Synthetic approaches to 3-substituted proline analogues.

promote a unique structural constraint. In an early work by Holladay et al. [6], the synthesis of 3-substituted prolines was described with diverse substituents at position 3 of the pyrrolidine ring through a multistep process taking advantage of enamine formation from 4-oxo-proline **1** and subsequent allylation and reduction (Figure 3.4, top). Some decades later, in a different approach, Bull and co-workers reported a two-step synthesis of *cis*-3-substituted proline analogues **6** through Pd-catalyzed C—H functionalization to stereospecifically add (hetero)aryl and vinyl substituents (Figure 3.4, bottom), allowing for the preparation of enantiopure *cis*-2,3-disubstituted pyrrolidines (and expanded to pipecolic acids, too) useful for increasing the fragment- and drug-like chemical space around cyclic amino acids [7, 8].

Recent entries about the generation of substituted prolines include the synthesis of compounds taking advantage of a chiral pool approach, asymmetric synthesis based on organometallic chemistry, and C—H activation.

Chiral pool approach starting from natural proline derivatives was undertaken using the commercially available *cis*-4-hydroxy-L-proline to achieve a proline analogue embedding an alkyl group at position 4 through standard functional group manipulation and introducing specific substituents with stereocontrolled formation of C—C bonds to expand the library of substituted prolines [9]. The relevance of substitution at position 4 is exemplified in a work by Liu et al. on the improvement of the development of 4,4-disubstituted proline analogues as selective inhibitors of Factor XIa (Figure 3.5), as oral anticoagulants for potential prevention and treatment of thrombotic diseases [10].

Another entry to functionalized prolines consisted of the application of organometallic or catalytic asymmetric methods. In this context, the Rh-catalyzed conjugate addition to dehydroproline electrophiles **7** was reported to achieve 3-substituted prolines **8** with high stereoselectivity (Figure 3.6) [11]. Specifically, Rh-catalyzed conjugate addition was applied for the synthesis of 3-substituted proline analogues from dehydroproline electrophiles and boronic acid nucleophiles with high enantio- and diastereoselectivity, following a high-throughput experimentation approach.

Regarding the synthesis of fused bicyclic α-amino acids, these molecules have attracted significant attention due to their pharmacological potential and their role as building blocks for the creation of conformationally constrained peptides. These amino acids, often derived

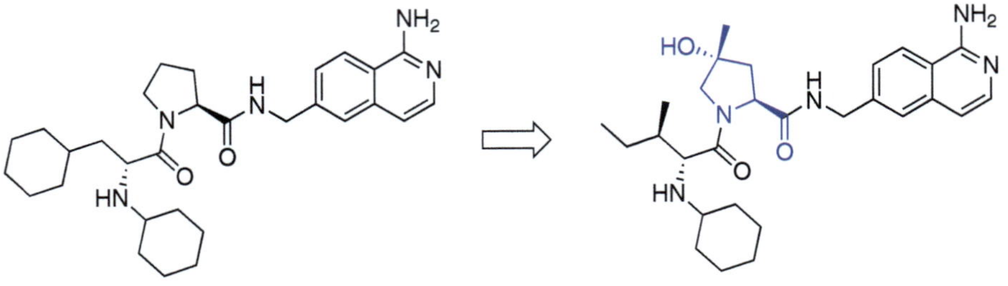

Figure 3.5 Optimization of proline-based compounds for achieving selective inhibitors of FXIa.

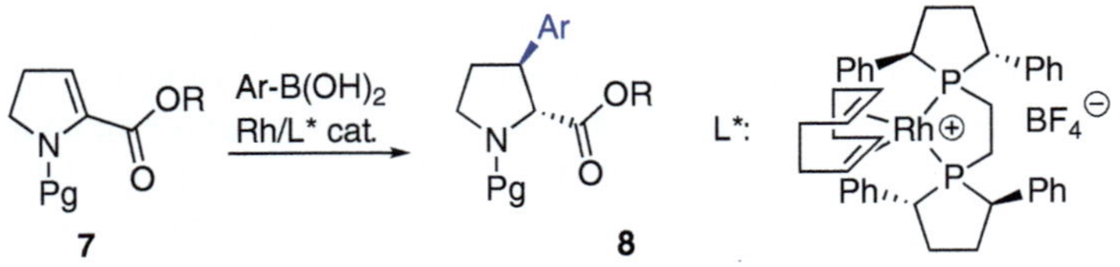

Figure 3.6 Synthesis of 3-substituted proline analogues based on Rh-catalyzed conjugate addition.

from proline, are particularly valued for their ability to introduce chemical diversity in peptide structures. Synthetic efforts have focused on generating heterocyclic species with varying ring sizes, thus enabling the exploration of more complex structural variants within the α-amino acid class. Bicyclic α-amino acids can be classified into three main groups based on their structural features:

- bicyclic proline analogues,
- α,α-disubstituted bicyclic α-amino acids,
- bicyclic derivatives of homoproline, pipecolic acid, and morpholine-based α-amino acids.

These groups have been exploited for the development of suitable synthetic methods, each designed to enhance the functionality and flexibility of peptides. The family of pyrrolidine-containing bicyclic proline analogues is particularly diverse in terms of chemical composition. Numerous studies have focused on synthesizing bicyclic compounds with distinct structural modifications around the five-membered proline ring. In particular, synthetic strategies have been developed for creating bicyclic systems with ring junctions at various positions within the proline ring, which contribute to a range of conformational constraints. This chemical diversity is crucial for optimizing the bioactivity and stability of peptides, making these bicyclic proline analogues valuable tools for advancing drug discovery and the design of biologically active compounds.

A recent review reported the synthetic approaches to apply rigidification through bicyclic or bridged structures. Specifically, a comprehensive summary of synthetic methodologies developed from 2010 for the construction of functionalized fused azabicyclolactam frameworks was reported, with particular reference to the 6,5-, 7,5-, and 8,5-fused systems

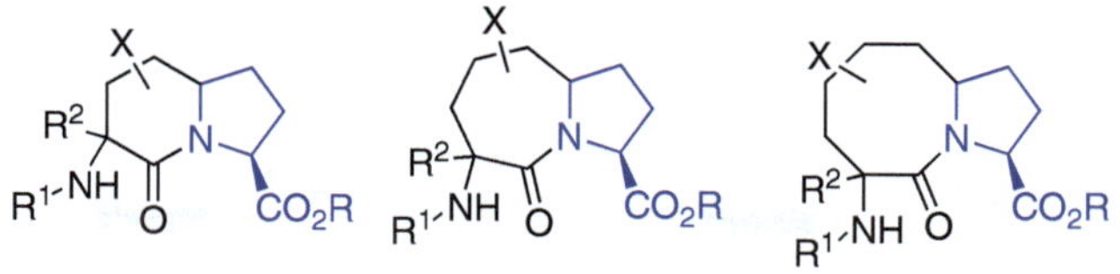

Figure 3.7 *Azabicyclo[X.Y.0]alkanone-derived amino acids.*

and applications of azabicyclo[X.Y.0]alkanone-derived amino acids in drug discovery (Figure 3.7) [12].

These bicyclic amino acids, recognized as a pivotal class of conformationally restricted dipeptide mimetics, continue to attract considerable interest within the fields of medicinal chemistry and pharmaceutical research. This is attributed to their high chemical stability and their ability to mimic diverse β-turn motifs, which are modulated by size, stereochemistry, and three-dimensional conformation of their rigid bicyclic cores. Since the initial synthesis of 6,5-fused bicyclic lactams in 1994, numerous research efforts have been directed toward the development of efficient and scalable synthetic strategies for accessing 6,5-, 7,5-, and 8,5-fused derivatives. These approaches have enabled the production of target compounds on gram scales with improved synthetic economy, particularly by applying methods for introducing substituents and optimizing the stereoselectivity.

3.2.2 Pipecolic Acid Derivatives

L-Pipecolic acid is a naturally occurring, non-proteinogenic cyclic α-amino acid that has been identified in a wide range of biological sources, including plants, fungi, and microorganisms [13]. Structurally analogous to proline, the pipecolic acid scaffold has attracted considerable attention in both organic and peptide chemistry due to its significant role in modulating the three-dimensional conformation of peptides and proteins. Such conformational properties make pipecolic acid a valuable synthetic target, particularly in the design of structurally constrained peptides with enhanced biological activity [14]. A variety of pharmacologically relevant natural products contain the pipecolic acid motif within their molecular frameworks, such as the immunosuppressive agent rapamycin and FK506. In addition, hydroxy-substituted derivatives of pipecolic acid, both naturally occurring and synthetically accessible, have emerged as important molecular entities in medicinal chemistry, due to their potential therapeutic applications. Given its structural rigidity and biological relevance, pipecolic acid is used as an effective α-amino acid surrogate in the synthesis of bioactive compounds. It is not only a fundamental component of numerous natural products but also a versatile building block for generating conformationally restricted analogues. Consequently, research efforts have focused on the development of functionalized pipecolic acid derivatives, many of which maintain the characteristic six-membered ring structure that is central to their chemical and biological properties [15]. One significant contribution to the field was reported by De Kimpe, who designed tricyclic α-amino acids incorporating both cyclopropane and benzo-fused pipecolic acid structures [16]. These compounds were developed as constrained phenylalanine analogues,

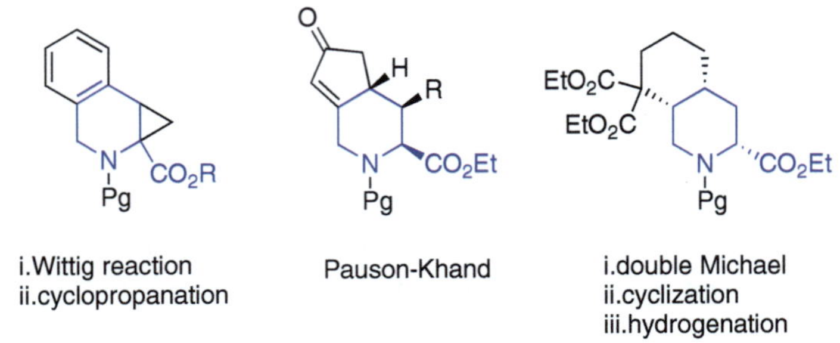

Figure 3.8 *Structure of fused pipecolic acids.*

which led to the synthesis of tricyclic structures using a Wittig reaction followed by cyclopropanation, resulting in constrained peptides. Additionally, pipecolic-based bicyclic α-amino acids with a hexahydro-cyclopenta[*c*]pyridine structure were synthesized using a Pauson–Khand cycloaddition reaction, thus highlighting the versatility and complexity of pipecolic derivatives in medicinal chemistry [17]. Grossman et al. explored the synthesis of saturated *cis*- and *trans*-fused bicyclic α-amino acids as conformationally restricted analogues of pipecolic acid [18]. They employed a double Michael reaction followed by cyclization and hydrogenation, obtaining perhydroisoquinoline systems, showing how the modification of the substituent patterns of pipecolic acid could lead to conformationally constrained analogues with potential therapeutic applications (Figure 3.8).

Hydroxylated derivatives of pipecolic acid, particularly those containing two or more hydroxyl groups, remain relatively rare in nature, with only some naturally occurring examples reported to date. Nevertheless, these compounds are increasingly recognized for their relevance in medicinal chemistry, where they serve as structural motifs in pharmacologically active agents, contributing to plant defense mechanisms or as bioactive intermediates. Thus, numerous efforts have been made for the asymmetric synthesis of hydroxy-pipecolic acids, especially those substituted at the 3-, 4-, and 5-positions. Occhiato and Prandi reported synthetic approaches based on the strategic transformation of lactam-derived vinyl triflates and phosphates, enabling the enantioselective construction of these molecules through the use of Suzuki–Miyaura cross-coupling and palladium-catalyzed methoxycarbonylation reactions, achieving high levels of regio- and stereocontrol and giving access to a broad range of hydroxylated pipecolic acid derivatives [13]. As an example, the preparation of both enantiomers of *trans*-3-hydroxypipecolic acid **13** was achieved through a multistep sequence (Figure 3.9). The route consists of the use of a vinyl phosphate intermediate obtained from commercially available *N*-Boc-δ-valerolactam. This species was subjected to a Suzuki–Miyaura coupling with furanyl boronic acid, forming the key C–C bond. Subsequent hydroboration–oxidation of the coupling product yielded racemic alcohol **10**, which was subjected to enzymatic kinetic resolution using CAL-B lipase to separate the enantiomers. The enantioenriched intermediate **11** was then manipulated *via* oxidative cleavage of the furanyl moiety using $RuCl_3$ catalysis to introduce the carboxylic acid functionality required for the pipecolic acid derivative. Final hydrolysis of the ester intermediate **12** yielded the desired (+)-**13** in 17% overall yield, demonstrating the viability of this route for the enantioselective synthesis of hydroxylated pipecolic acids.

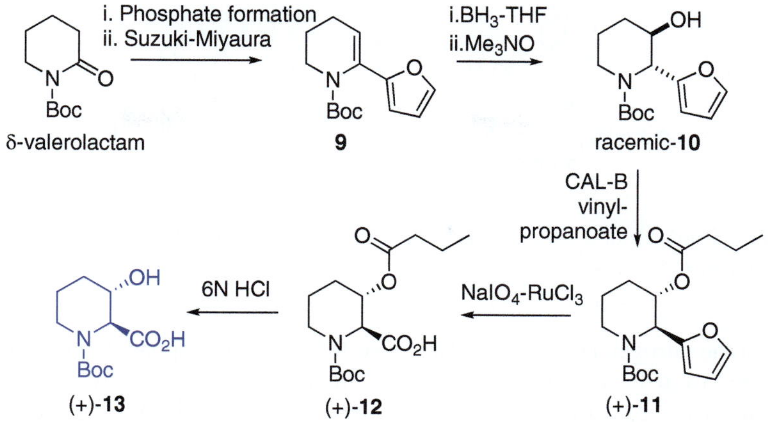

Figure 3.9 *Synthesis of enantiopure* trans-*3-hydroxypipecolic acid.*

Also, the synthesis of various stereoisomers of 4-hydroxypipecolic acid, specifically *cis*-(2*R*,4*S*), *cis*-(2*S*,4*R*), and *trans*-(2*R*,4*R*) derivatives, was accomplished through two complementary strategies. Similarly to the previous synthetic approach, the first route employed a chemoenzymatic approach starting from δ-valerolactam [19]. Following Pd-catalyzed methoxycarbonylation and allylic oxidation, a racemic alcohol intermediate was obtained and subsequently resolved by lipase-catalyzed kinetic resolution. This enzymatic step yielded enantiomerically enriched alcohols and esters, which were converted into the desired *cis* and *trans* isomers by stereoselective hydrogenation and hydrolysis. An alternative synthetic pathway involved ethyl (3*R*)-4-cyano-3-hydroxybutanoate **14** (Figure 3.10) [20]. This molecule was protected at the hydroxy group as a *tert*-butyl ester and hydrogenated to yield an enantiopure lactam **15**. Then, the compound was

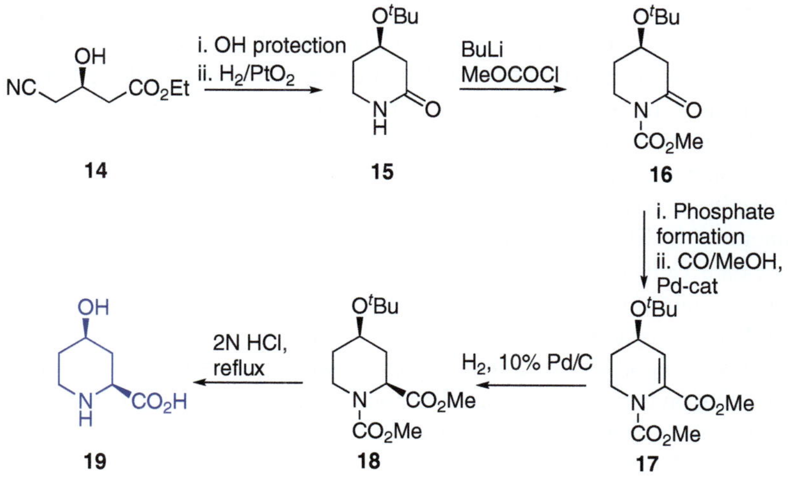

Figure 3.10 *Synthesis of enantiopure* cis-*4-hydroxypipecolic acid.*

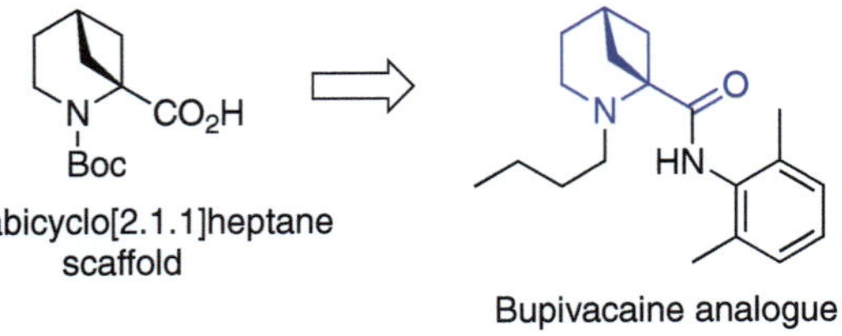

Figure 3.11 *Structure of azabicyclic analogue of pipecolic acid and application for the synthesis of a Bupivacaine analogue.*

protected at the nitrogen atom to give **16**, converted to the corresponding vinyl phosphate, and subjected to methoxycarbonylation under Pd catalysis, followed by stereocontrolled reduction to give access to *cis*-(2*S*,4*R*)-4-hydroxypipecolic acid **19** in 66% overall yield over seven steps. Additionally, the *trans*-(2*R*,4*R*) isomer was synthesized *via* conjugate hydride addition to a shared α,β-unsaturated ester intermediate, affording the product in 42% overall yield across eight steps.

A recent work on piperidine isosteres and their application to the generation of constrained pipecolic acid analogues was reported by Mykhailiuk and co-workers [21]. In this work, azabicyclic compounds have been investigated as structural mimics of piperidine, and two nonchiral amide analogues of the anesthetic drug Bupivacaine were synthesized to explore their potential as bioisosteres (Figure 3.11). The study demonstrated a scalable and versatile method for synthesizing azabicycles through KOtBu/I_2/CO_2-mediated cyclization of alkenyl amines. This approach was extended to other azabicyclic systems, including 1,2-disubstituted bicyclo[1.1.1]pentanes that are otherwise difficult to obtain.

3.2.3 Morpholine-Based Amino Acids

Among the heterocyclic structures that have most attracted the interest of researchers over the years, morpholine derivatives [22] play a primary role. Morpholine is an organic heterocycle with the formula $O(CH_2CH_2)_2NH$ featuring an ether and an amine functionality, which makes the molecule a weak base (p*K*a about 8.7, lower than that of piperidine). It is an excellent solvent and is widely used in organic synthesis and at the industrial level [23]. Morpholines are useful starting materials in the production of enantiopure α-amino acids [24, 25], β-amino alcohols [26], and peptides [27], and in the synthesis of biologically active compounds [28]. Numerous papers have addressed the synthesis and biological functionality of variously substituted morpholines [29] and their derivatives, including morpholinones [30]. The most common starting materials for the synthesis of morpholine heterocycles are vicinal amino alcohols, generally with a nitrogen-protecting group, and their *N*- and *O*-substituted derivatives containing allylic, propargylic, or oxirane groups. Also, β-azidoalkylpropargyl ethers, bis(2-hydroxyethyl)-amines, 2,2′-dihalodiethyl ethers, and their mesylated and silylated homologues, as well as oxiranes and aziridines, can be used. Depending on the starting material, the achievement of the morpholine ring requires the participation of a cyclizing agent, a metal catalyst, a stoichiometric base, or a dehydrating agent (Figure 3.12).

Recently, several efficient strategies addressing the challenges associated with carbon–carbon bond formation in morpholine frameworks have been developed. In

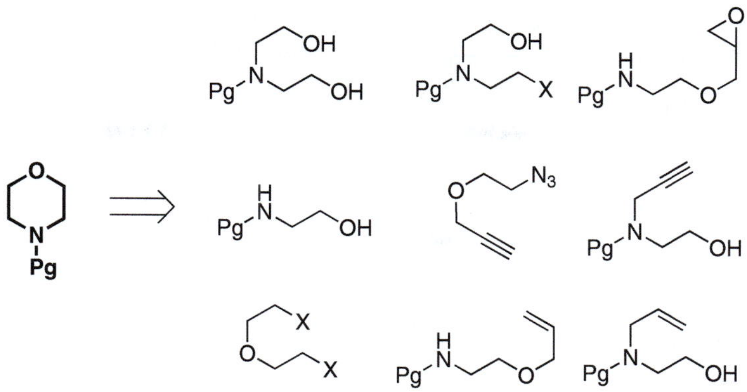

Figure 3.12 Common starting material for the synthesis of morpholine compounds.

particular, the use of SnAP reagents by Bode and co-workers proved to exhibit favorable reactivity toward aldehyde and ketone substrates and to promote the formation of morpholine rings and related saturated heterocyclic structures [31]. Furthermore, photoredox methodologies have been increasingly employed for the synthesis of morpholines [32], as well as for the direct modification of morpholine precursors. These approaches include decarboxylative functionalization and hydrogen atom transfer (HAT) mechanisms [33]. Natural α-amino acid derivatives represent another important starting point for the production of morpholines, as shown in Figures 3.13. Specifically, starting from L-threonine methyl ester hydrochloride (pathway A), it is possible to easily obtain a tri-substituted morpholine heterocycle in three synthetic steps [34]. In Pathway B, through the application of a diversity-oriented synthesis (DOS) approach to a complex L-threonine-derived skeleton, it was possible to obtain a remarkable range of different morpholine-based heterocyclic products [35].

3.2.4 Piperazine-Based Amino Acids

Piperazine-2-carboxylic acid plays an important role in the synthesis of peptides and peptidomimetics. Its rigid molecular structure and chemical properties are central to mimicking proline in the context of local modification of peptides and to serving as a constrained scaffold to modulate the conformational preferences of the peptide backbone. Specifically, the use of piperazine–carboxylic acid scaffolds in peptide sequences proved to be useful for generating conformationally constrained peptidomimetics. This substitution is particularly effective in inducing specific secondary turn structures within peptide chains [36]. In the context of drug discovery and development, piperazine-2-carboxylic acid and its derivatives have been applied in the synthesis of several therapeutic agents (Figure 3.14) [37], including those aimed at treating anxiety, depression, and viral infections, such as the HIV protease inhibitor indinavir [38].

The synthesis of this compound, particularly the asymmetric synthesis, is still of interest due to its utility in drug design and as a building block for complex heterocyclic frameworks. Early syntheses of piperazine-2-carboxylic acid were based on multistep routes starting from diamines, resulting in moderate yields and low stereoselectivity. For

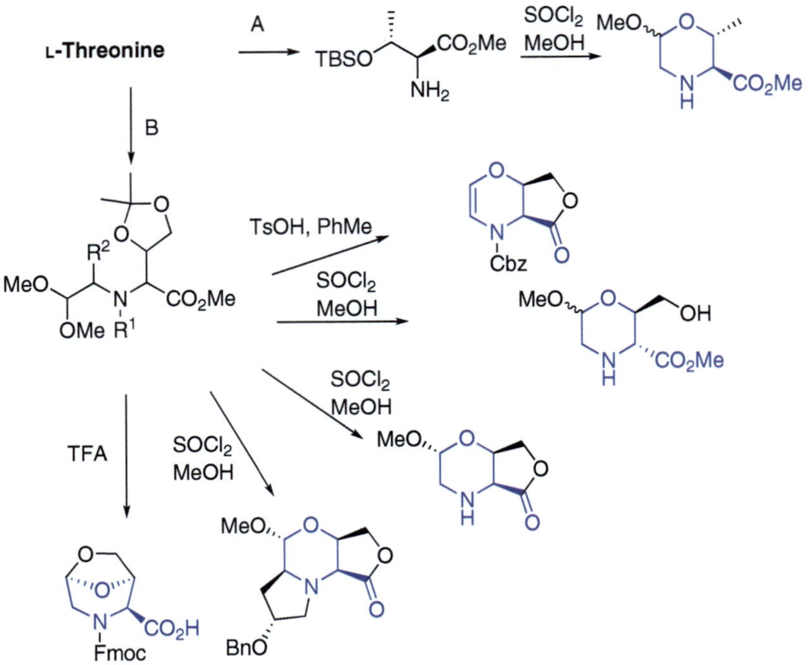

Figure 3.13 Diversity-oriented synthesis (DOS) of morpholine-2-carboxylic acids from threonine.

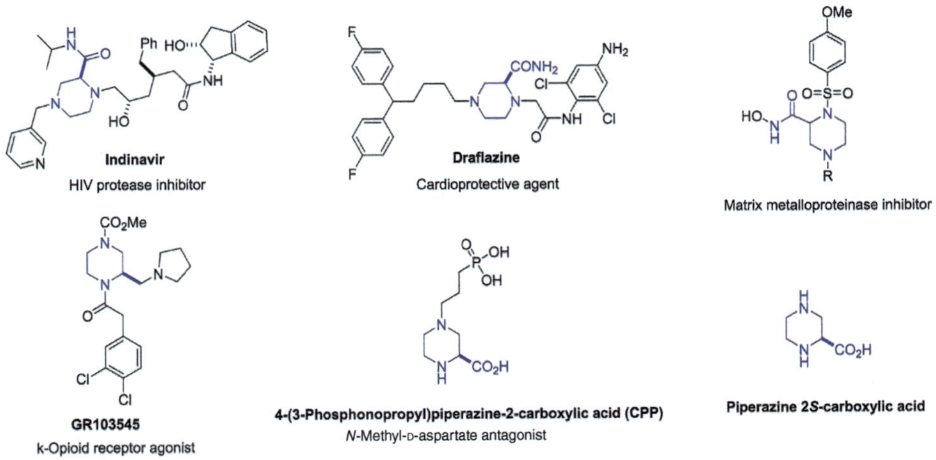

Figure 3.14 Pharmacologically active molecules containing piperazine 2-carboxylic acid scaffold. Source: Reprinted with permission from ref. 37. Copyright 2024 American Chemical Society.

example, a work on the synthesis of unsubstituted piperazine-2-carboxylic acid derivatives consisted of the cyclization of *N,N′*-bis(trifluoroacetyl)ethylenediamine with glycine derivatives, followed by hydrolysis and esterification [39]. Despite moderate yields, this route provides direct access to the parent scaffold and has served as a method for further synthetic applications. In a different approach, the synthesis of *N*-orthogonally protected (*S*)-piperazine-2-carboxylic acid was reported, enabling regioselective modifications for peptide synthesis [40]. The orthogonal protection strategy at the nitrogen atoms facilitates their sequential functionalization, thus expanding the synthetic utility of the molecule. Recent developments described the stereoselective synthesis of azabicyclic isosteres of (*S*)-piperazine-2-carboxylic acid, as reported by Vaillancourt et al. [37]. Specifically, two complementary approaches were developed, the first one using L-pyroglutamic acid and involving a sequential Michael–aza-Michael reaction cascade to install the diazabicyclic framework, and the second one based on the enolate chemistry of a 4-keto-L-proline derivative, leading directly to the diazabicyclo[3.3.0]octane-2-carboxylic acid scaffold.

Another contribution was reported by Chamakuri et al. in 2022, describing a concise synthetic method for the synthesis of 3-substituted piperazine-2-acetic acid esters structurally related to piperazine-2-carboxylic acid [41], based on the annulation of enantiopure 1,2-diamines derived from amino acids, followed by cyclization and substitution steps to access functionalized piperazines.

3.3 Principles of C—H Functionalization

Over the past decades, there has been an exponential growth in the development of methods for C—H functionalization [42, 43]. Indeed, the possibility of directly transforming C—H bonds is attractive for avoiding the need for tedious functional group manipulation and/or interconversion. In this way, it is possible to dramatically simplify multistep chemical processes, offering novel disconnections of complex compounds from simple starting materials. C—H functionalization strategies are broadly categorized into two mechanistic approaches, namely directed and innate pathways [44]. Directed C—H functionalization is based on the use of coordinating groups or transiently associated auxiliaries, either covalently or non-covalently attached, that guide the installation of new functional moieties by selectively activating specific C—H bonds. In a different way, innate C—H functionalization works independently of such guiding elements, proceeding through the intrinsic reactivity dictated by the electronic and steric environment of the substrate. A milestone contribution to the field was made by Daugulis in 2005, who reported the first catalytic system capable of achieving highly regioselective arylation of inactive aliphatic $C(sp^3)$—H bonds [45]. In this approach, 8-aminoquinoline (8-AQ) and 2-picolinic acid (2-PA) were employed as bidentate directing groups for the arylation of carboxylic acids and primary amines, respectively (Figure 3.15). The transformation proceeded through the formation of a five-membered palladacycle intermediate, enabling β-arylation of carboxylic acid derivatives and γ-arylation of amine substrates [46]. Following this study, a wide array of methodologies have been reported exploiting structurally diverse directing groups to achieve site-selective C—H activation at the β- and γ-positions of amino acid derivatives, thus significantly expanding the scope of C—H functionalization in complex molecule synthesis.

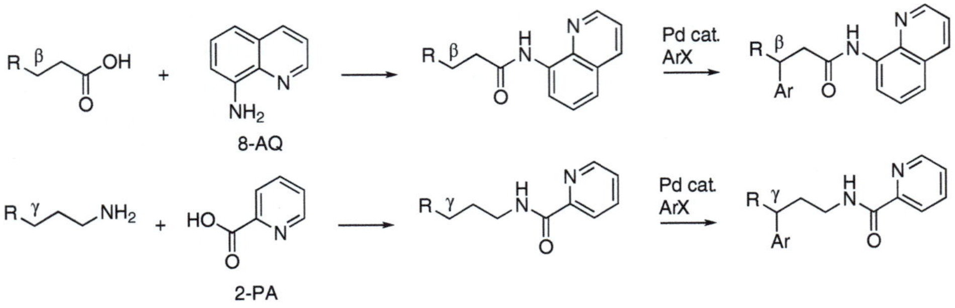

Figure 3.15 *Regioselective arylation of inactive $C(sp^3)$—H bonds of carboxylic acids and primary amines.*

$C(sp^3)$—H bond functionalization is helpful for the synthesis of structurally complex molecules, which contain a high ratio of sp^3-hybridized carbon atoms. Typically, such kind of bonds can be activated by exploiting their acidity, taking advantage of the presence of an adjacent electron-withdrawing group. They can be converted under ionic conditions, such as deprotonation with a base to generate a reactive carbanion and subsequent reaction with an electrophile. This strategy is somewhat limited by the availability of such bonds, and it is often associated with racemization of any existing stereocenter. The most widespread component is nonacidic C—H bonds in general, although such C—H bonds ($C(sp^3)$—H bonds more than $C(sp^2)$—H bonds) are inert because of their low reactivity under ionic conditions. Accordingly, the functionalization at nonacidic $C(sp^3)$—H bonds remains challenging, due to the lack of strategies for activating such unreactive C—H bonds [47]. In this view, transition metal-catalyzed direct functionalization of nonacidic unreactive $C(sp^3)$—H bonds has been broadly studied. The energy difference between their HOMO and LUMO orbitals can be circumvented through the use of palladium catalysts, giving access to relatively low-energy C—Pd bonds and allowing transformations to be conducted, taking advantage of the redox properties of palladium [48]. To promote the formation of a C—Pd bond, a strongly coordinating bidentate directing group (DG) can be employed, to chelate the transition metal ion, to bring it close to the C—H bond of the substrate, and to use proximity as a key factor for controlling site selectivity. Additionally, directing groups shall meet some other criteria to be suitable for the catalytic process:

- stability during the functionalization reaction
- reversible coordination to the catalytic metal center
- failure to participate in the reaction in the absence of the catalyst
- be readily removable

The last criterion is a relevant issue for the first generation of designed DGs, also named "covalent-type" DGs, since they are introduced and removed by additional synthetic steps before and after the C—H activation reaction.

Concerning the mechanism, the coordination of the transition metal with the DG is a crucial step in the catalytic C—H bond activation cycle, which first leads to the formation of the corresponding palladacycle and then to the desired product through an oxidative addition step. For example, the arylation reaction through an aryl iodide occurs through a

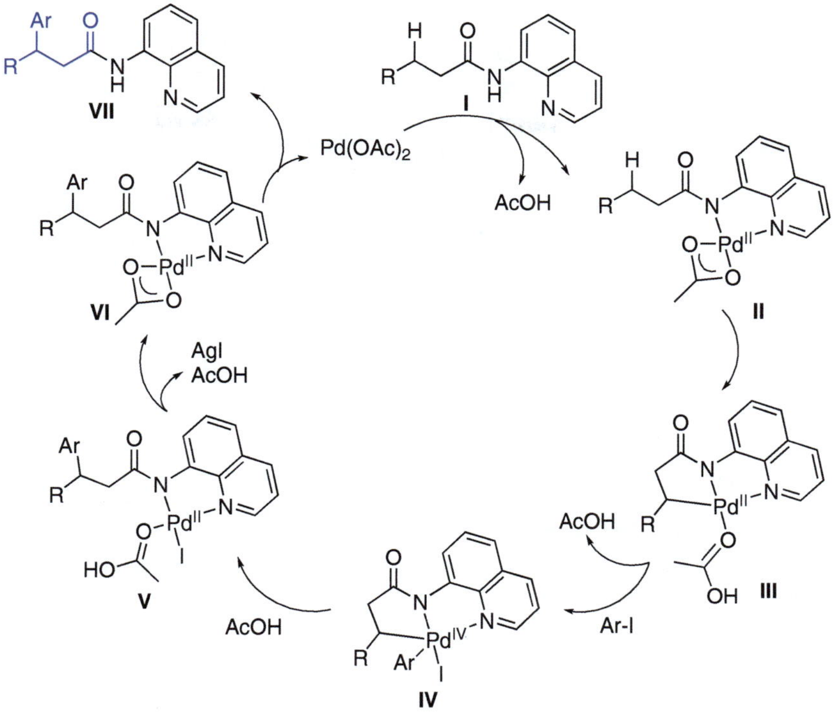

Figure 3.16 *Catalytic cycle of the 8-AQ-directed β-C(sp^3)—H arylation of compound.*

five-membered Pd(IV) complex involving the iodine atom and the aryl functional group coming from the aryl iodide employed, followed by a reductive elimination step that leads to the formation of the new C—C bond (Figure 3.16).

Nowadays, 8-AQ is a well-established DG thanks to its reliability and efficiency. Accordingly, a wide number of examples are reported in the literature on the C—H functionalization carried out taking advantage of this DG, such as the synthesis of pipercyclobutanamide A (Figure 3.17), a natural compound isolated from the fruits of the black pepper plant, reported by Baran and co-workers [49], that demonstrated to possess an inhibitory activity toward cytochrome P450 2D6 [50]. After achieving the cyclobutane ring **20** through a photochemical 4π-electrocyclization with subsequent hydrogenation, 8-AQ was introduced through amide bond formation and exploited to carry out sequential β-arylation and β′-olefination on **21**. Then, the desired pipercyclobutanamide A was obtained following four additional synthetic steps for functional group manipulation with an overall yield of 5%.

Several other covalent-type DGs beyond 8-AQ, as reported by Daugulis et al. [45], have been employed for C—H activation reactions, exploiting also the amino functional groups (Figure 3.18) [51–58]. They are all based on chemical moieties that can be inserted through amide coupling reactions and can establish a bidentate chelation of the catalytic metal ion, thus favoring the C—H activation step.

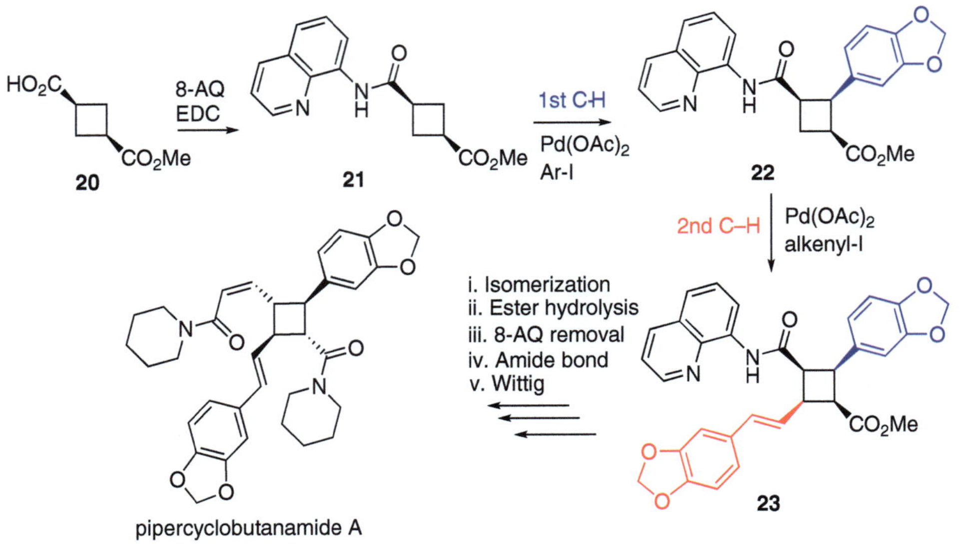

Figure 3.17 *Sequential C—H activation on carbomethoxycyclobutane-carboxylic acid using 8-AQ as DG.*

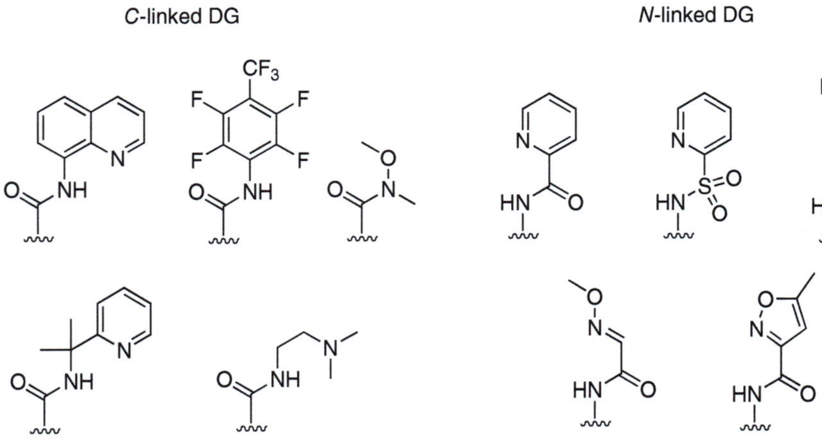

Figure 3.18 *Representative* C- *and* N-*linked DGs.*

Concerning the DGs that exploit the amino functional group, it is possible to identify a different group of DGs applicable through imine formation employing an aldehyde [59], which can be installed and cleaved *in situ*. In this view, the application of imine to install a DG can be conceived with the aldehyde coming from the substrate to be functionalized and exploiting the amino group as a chemical motif of a DG to generate the corresponding imine-based DG. The reversible character of the imine bond allows for considering this approach as a transient directing group (TDG) strategy, since the DG is a chemical species introduced immediately before the catalytic C—H functionalization and is cleaved

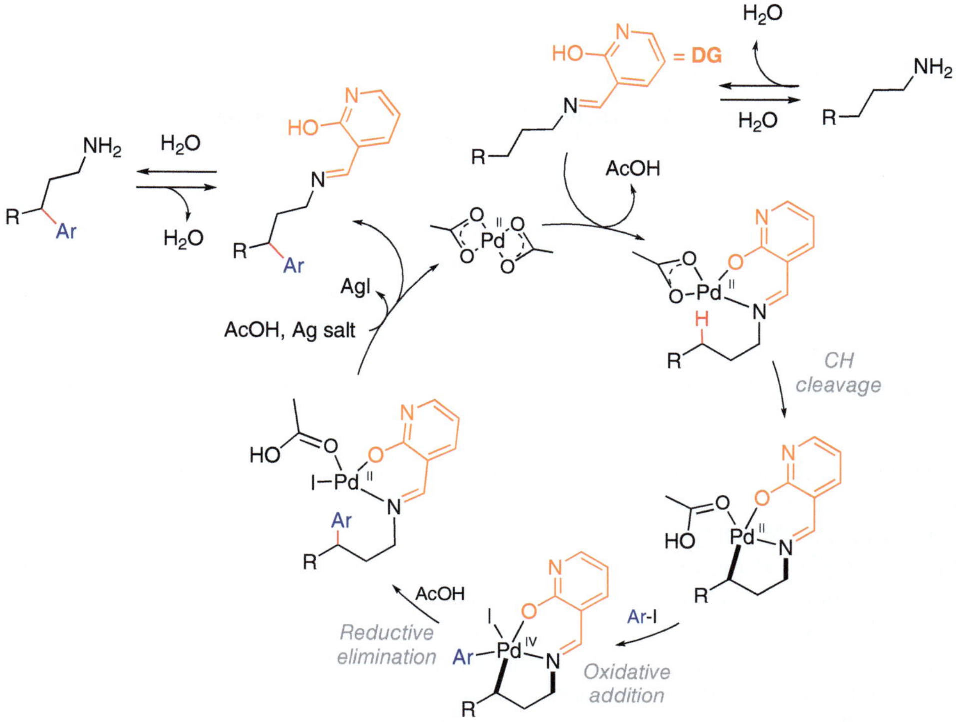

Figure 3.19 Catalytic cycle for the 2-hydroxynicotinaldehyde-based imine-directed γ-C(sp^3)—H arylation.

immediately after, in one pot. Indeed, sub-stoichiometric amounts of the exogenous chemical species is employed, usually 30 mol%. In such TDGs, the bidentate chelation mode is generally carried out by the lone pair of the nitrogen atom forming the imine and an additional functional group of the aromatic aldehyde (i.e. a hydroxyl functional group). As in the catalytic cycle reported in Figure 3.19 for the 8-AQ-directed C—H functionalization, the mechanistic pathway of the reaction consists of the following steps:

- formation of the desired palladacycle
- oxidative addition step
- reductive elimination step
- release of the catalytic palladium ion, to afford the regeneration of the catalyst and the formation of the functionalized product

A third category of DG is represented by the so-called "innate"-DG. Procedures that exploit the innate capability of unaltered functional groups such as carboxylate, amine, and thioamide to act as DGs have been developed. The presence of atoms bearing lone pairs enables the chelation of the catalytic transition metal. Nevertheless, the chelation mode is often monodentate, thus less efficient than the bidentate chelation that most of the "covalent-type" DGs establish with the metal catalyst [60]. To overcome this lack of efficiency, it is possible to combine the use of exogenous ligands or electrophiles that

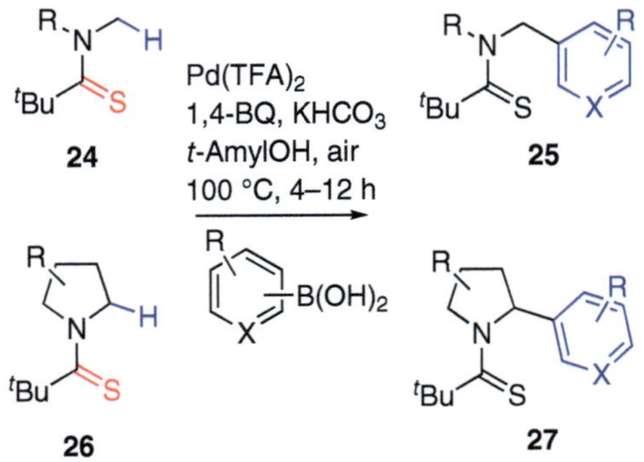

Figure 3.20 Thioamide-directed α-(hetero)arylation through C(sp^3)—H activation.

contribute to the formation of a more stable metal complex, particularly concerning the C—H activation of free carboxylates. Indeed, several research groups have published studies on the Pd-catalyzed activation of proximal (α and β) as well as distal (γ and higher) C(sp^3)—H positions, concerning reactions such as β- and γ- arylation, β- and γ-lactonization, β-olefination, and β-acetoxylation [61–67]. Thioamide is an interesting native functional group with innate capability of chelating transition metals that has been exploited to promote the C(sp^3)—H activation. Accordingly, Yu's research group developed α-mono- and di-(hetero)arylation of linear and cyclic aliphatic thioamides, as shown in Figure 3.20 for compounds **24** and **26** to achieve the corresponding arylated products **25** and **27** [68].

Pd-catalyzed directed C(sp^3)—H activation reactions have been studied over the last decades since they have been demonstrated to be viable synthetic routes for the stereoselective introduction of functional groups. Indeed, many examples are reported concerning the functionalization of saturated carbocycles [69] and α- and β-amino acids [70, 71]. The specific application of C—H activation strategies to amino acids and peptides has been explored widely over the last decade and has been subjected to thorough review papers accordingly [1].

The utility of C—H activation using directing group strategies is feasible only if the DG is readily removable after the C—H functionalization process. In this view, efforts have been devoted to studying the chemistry of directing groups and a wide range of

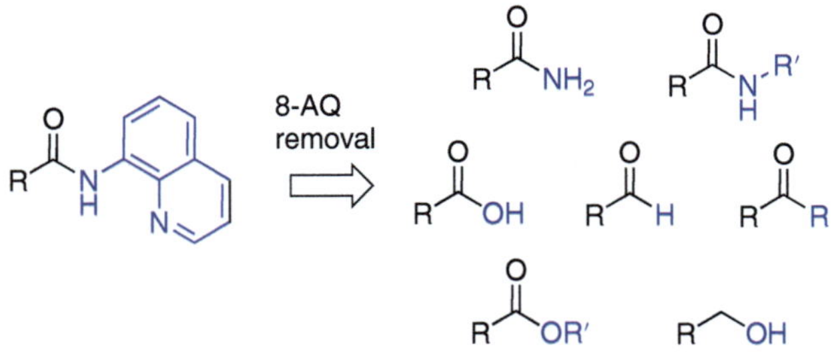

Figure 3.21 Strategies for the removal of 8-aminoquinoline DG.

strategies have been proposed to selectively remove the DG. For example, a review by Fitzgerald and O'Duill reported all the methods for removing the widely *N,N*-bidentate 8-aminoquinoline directing group useful for activating adjacent positions to a carboxylic group to which is bonded as an amide derivative [72]. These approaches can be classified in three groups, namely hydrolytic-, oxidative-, reductive-, and metal Lewis acid-promoted removal, resulting in the achievement of the title compounds as free carboxylic acid, ester, amide, aldehyde, ketone, or alcohol (Figure 3.21).

3.4 C—H Functionalization of Amino Acids

The first example of C—H functionalization on amino acid substrates is represented by the C(sp^3)—H arylation on C—H bonds of a protected amino acid [46]. In this application, the α-amino group was protected with a phthaloyl group (Phth) and carboxylic acid functionalized with 8-aminoquinolinamide as a directing group. Both the acetoxylation and arylation reactions on secondary β-C(sp^3)—H of phenylalanine and leucine derivatives proved successful in good yields and high diastereoselectivity, as represented in Figure 3.22 for the achievement of compounds **29** and **30** from leucine derivative **28**.

Since then, numerous applications of C—H functionalization have been studied in the field of amino acids and peptides, resulting in a significant improvement of the toolbox for expanding the chemical diversity in peptidomimetic chemistry with atom- and step-economy approaches, and several reviews report a comprehensive description of this chemistry [1, 73]. In this view, various C—H positions within α-amino acids and peptides that have been successfully subjected to functionalization to date (Figure 3.23) and can be broadly categorized in two groups:

- α-C—H functionalization at the backbone
- side-chain modification, that can be further divided into (hetero)aryl functionalization [C(sp^2)—H] and transformation at β-, γ-, and δ-positions [C(sp^3)—H]

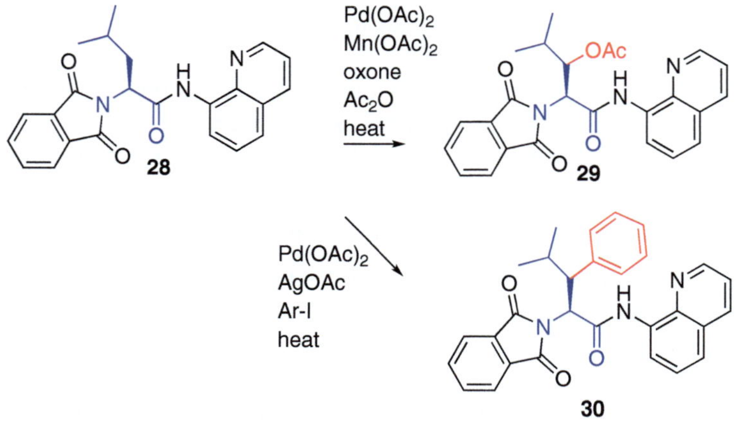

Figure 3.22 8-AQ-directed C—H functionalization of Phth-protected leucine.

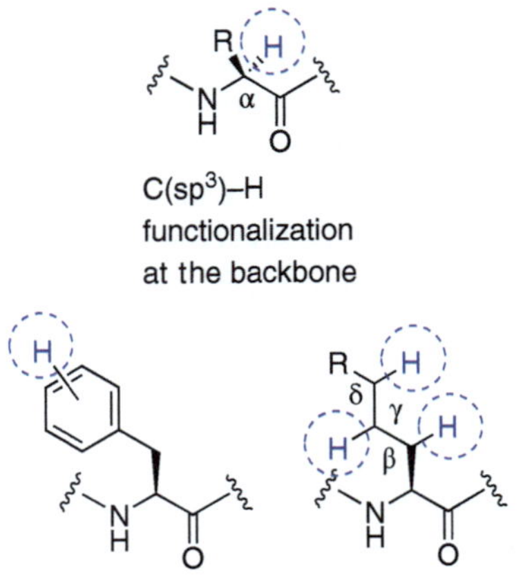

Figure 3.23 Directed C—H bond positions in amino acids and peptides.

In general, C—H functionalization strategies face two principal challenges:

- the intrinsic difficulty of activating inert C—H bonds, which is associated with their high bond dissociation energies
- the need to achieve selective functionalization of a single C—H bond in the presence of numerous chemically distinct C—H bonds within the same molecule

In contrast to traditional synthetic methodologies, direct C—H bond functionalization for the preparation and modification of α-amino acids and peptides presents significant advantages:

- reduction of reaction steps, given the decreased requirement for protection and deprotection;
- utilization of chiral α-amino acids as an inexpensive and abundant pool of starting materials;
- facile generation of a structurally diverse array of derivatives (particularly valuable for compound library construction and pharmaceutical screening);
- potential for late-stage functionalization of structurally complex peptides and proteins.

3.4.1 α-C—H Functionalization at the Backbone

The α-C—H functionalization approach enables the direct transformation of simple amino acid precursors into more complex α-substituted amino acids, often with high stereocontrol and reduced need for protecting group strategies. Several strategies have been studied and reported over the years to promote the functionalization of α-C—H bond of amino acid residues, including transition metal and photoredox catalysis, enabling the direct modification of the central carbon atom adjacent to the amide nitrogen of amino acid residues that are traditionally considered chemically inert due to their resonance-stabilized nature

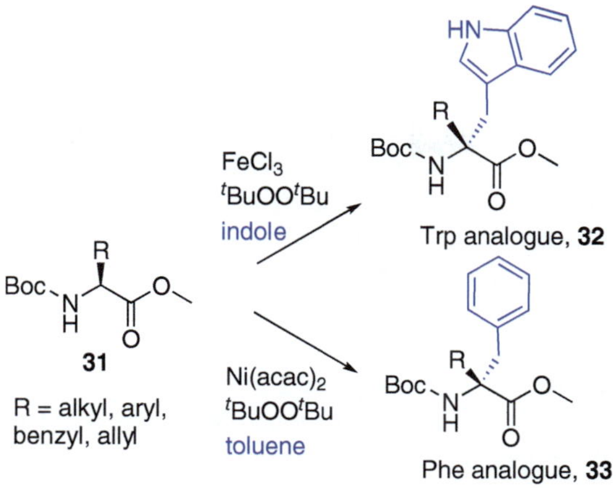

Figure 3.24 C—H functionalization at the α-position of α-amino acids to achieve quaternary amino acids.

and low intrinsic reactivity, paving the way for late-stage peptide diversification, macrocyclization, and the installation of functional probes or pharmacophores. Specifically, the α-position of α-amino acids can be functionalized in the unsubstituted α-position of glycine and in the α-position of substituted amino acids, which can be converted into pharmaceutically relevant quaternary α-amino acids. A prominent strategy in C—H functionalization involves the selective oxidation of $C(sp^3)$—H bonds adjacent to heteroatoms like nitrogen, which stabilizes the reactive intermediates and enables site-selective transformations via radical or cationic species, as represented by the cross-dehydrogenative coupling of glycine residues [74]. An additional example is represented by the Co-catalyzed $C(sp^3)$—H oxidative coupling of *N*-aryl glycine with indoles or cyclic ethers [75].

Concerning the functionalization of the α-position of substituted amino acids, picolinamide-protected α-amino acids were transformed to achieve quaternary α-amino acid derivatives as reported by You et al. using an iron-catalyzed Friedel–Crafts alkylation as a first entry to oxidative $C(sp^3)$—H functionalization on α-substituted amino acids (Figure 3.24), demonstrating the suitability of the process on phenylalanine, tryptophan, and other unnatural amino acids in the reaction with indoles and other electron-rich arenes [76]. Similarly, a related nickel-catalyzed method was reported by the same group for coupling picolinamide-protected α-amino acids with benzylic compounds such as represented with toluene to give **33** (Figure 3.24), following an oxidative process generating a benzylic radical reacting with the Ni-bound substrate [77].

3.4.2 C—H Functionalization on the Side Chain

$C(sp^2)$—H functionalization. Aromatic α-amino acids, such as phenylalanine, tyrosine, tryptophan, and histidine, play a fundamental role in modulating the biological activity of proteins. Beyond their natural functions, both these amino acids and their non-natural analogues have emerged as key scaffolds in medicinal chemistry. The presence of an aromatic side chain renders them particularly attractive for site-selective modifications via $C(sp^2)$—H bond activation. This powerful strategy enables diverse functionalization

reactions, including acetoxylation, arylation, alkenylation, alkynylation, halogenation, and borylation, thereby expanding their chemical and therapeutic utility [73]. Accordingly, the most extensively studied substrates for C(sp^2)—H arylation of α-amino acids and peptide-related molecules are the indole moiety in tryptophan. Several methodologies have been developed for the regioselective functionalization at the C2-position of the indole ring, such as for **34** (Figure 3.25), employing diverse arylating agents and oxidative conditions and using palladium acetate as the catalyst of choice in these transformations [78, 79]. Other aromatic α-amino acids, such as phenylalanine, tyrosine, and histidine, have been employed in C(sp^2)—H arylation protocols. For example, Mitchell et al. reported a rhodium-catalyzed ortho-arylation of tyrosine derivatives with bromoarenes [80], where the hydroxyl group on the phenol ring proved to control the site selectivity through an *ortho*-directing effect to the catalyst (see compound **36**, Figure 3.25). Concerning histidine, Jain and co-workers developed a Pd(0)-catalyzed arylation of *N*-benzyl-protected histidine derivatives under microwave-assisted conditions (see compound **38**, Figure 3.25) [81], enabling the selective arylation at the C5-position of the imidazole ring with a broad range of electronically and sterically diverse iodoarenes.

As an alternative to synthesis of borylated arenes via metal-catalyzed borylation of halides, direct C—H borylation has attracted interest, in view of accessing organoboron compounds in the amino acid field as an important class of intermediates for Suzuki–Miyaura cross-couplings. Accordingly, Smith extended the application of iridium-catalyzed borylation methodologies to tryptophan (Figure 3.26), achieving the

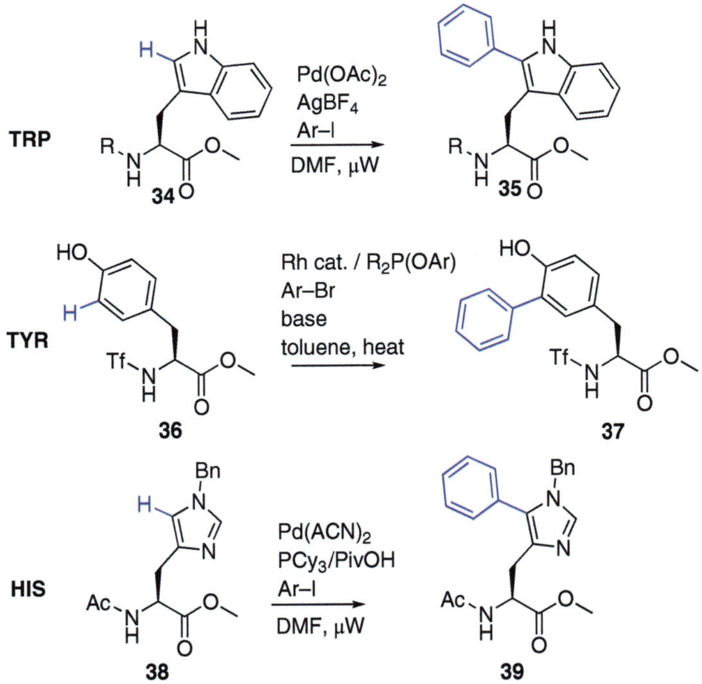

Figure 3.25 Pd-catalyzed C(sp^2)—H arylation of amino acids at the aromatic side chain.

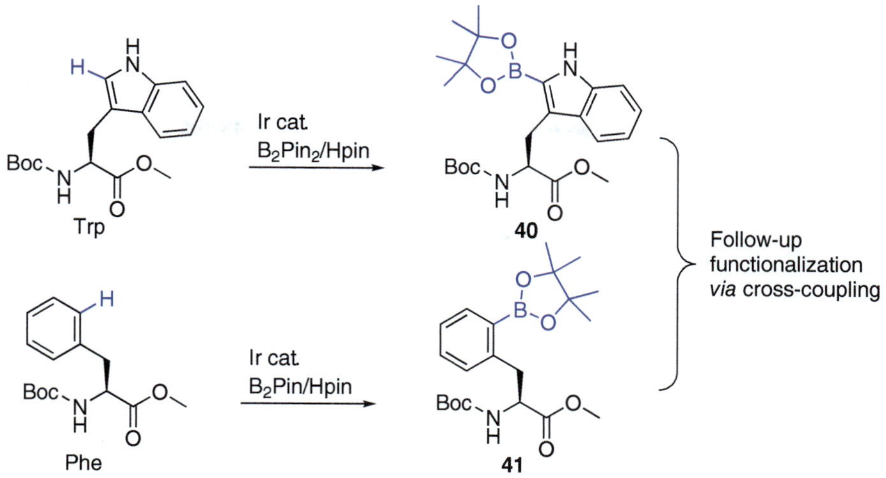

Figure 3.26 Iridium-catalyzed C—H borylation of phenylalanine and tryptophane.

mono-borylated product **40** in moderate yields, thus highlighting selectivity issues with respect to competing formation of diborylated products [82]. Similarly, a borylation reaction of phenylalanine derivatives was reported by applying the same iridium catalyst and ligand (Figure 3.26), resulting in higher selectivity for the meta-substituted borylated product **41** [83], and proving functional group tolerance with both electron-donating and electron-withdrawing properties.

C(sp^3)—H functionalization. Significant advances have been made in the challenging area of alkyl C(sp^3)—H bond activation. The direct functionalization of these otherwise inert bonds opens new synthetic strategies for constructing valuable molecular scaffolds and enables late-stage diversification of complex molecules. Despite its potential, this approach presents a major challenge for achieving high site selectivity in the activation of a single aliphatic C—H bond among numerous similar bonds within the same molecule. Recently, numerous strategies have been developed to achieve site-selective C—H functionalization of α-amino acids and peptides. These include both covalently installed directing groups and ligand-enabled approaches, which enable precise activation at β-, γ-, or δ-C(sp^3)—H bonds, as well as at aromatic C(sp^2)—H sites within side chains. Together, these methodologies provide a powerful platform for the predictable and stereocontrolled installation of diverse functional groups at specific positions within peptide frameworks.

Recent advances in the β-C—H activation of amino acids exploited bidentate directing groups to enable selective functionalization. Shi pioneered the use of 2-(pyridin-2-yl) isopropyl (PIP) as a directing group for the Pd-catalyzed β-arylation of alanine derivatives [52]. Following this approach, Shi developed a β-selective fluorination protocol for alkyl- and aryl-substituted alanines (compound **42–43**, Figure 3.27), using Selectfluor as a fluorinating reagent [84]. Similarly, Ackermann introduced the triazolyldimethylmethyl (TAM) directing group for β-C—H arylation of alanine and phenylalanine derivatives (compound **44–45**, Figure 3.27), thus enabling the coupling with a series of iodoarenes [85]. When incorporated into peptide backbones, TAM directed regioselective β-functionalization at

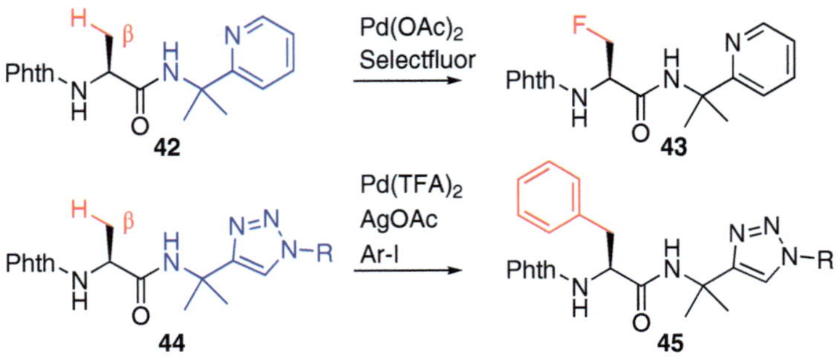

Figure 3.27 Representative C(sp^3)—H functionalization reactions at the β-position exploiting TAM and PIP as directing groups at the carboxylic acid end.

neighboring residues, demonstrating the potential of this approach for the late-stage diversification of peptide-like molecules.

The γ-C(sp^3)—H arylation of α-amino acids with high selectivity has been accomplished through the use of nitrogen-based directing groups on the amino group to achieve five-membered metallacycle intermediates. This approach typically involves installing the directing group, as attempts to use C-terminal auxiliaries remain limited due to the lower stability of six-membered palladacycles required for γ-functionalization. In this context, a 2-pyridylsulfonyl directing group was reported by Carretero et al. to direct the γ-arylation of aliphatic amino acids [55]. For valine substrates **46** (Figure 3.28), the monoarylation to **47** could be achieved in moderate to good yields using sterically hindered aryl iodides, whereas isoleucine analogues showed exclusive monoarylation at the γ-methyl position. Achieving γ-C(sp^3)—H borylation of amino acid derivatives presents significant synthetic challenges, primarily due to the incompatibility between palladium catalysts and the

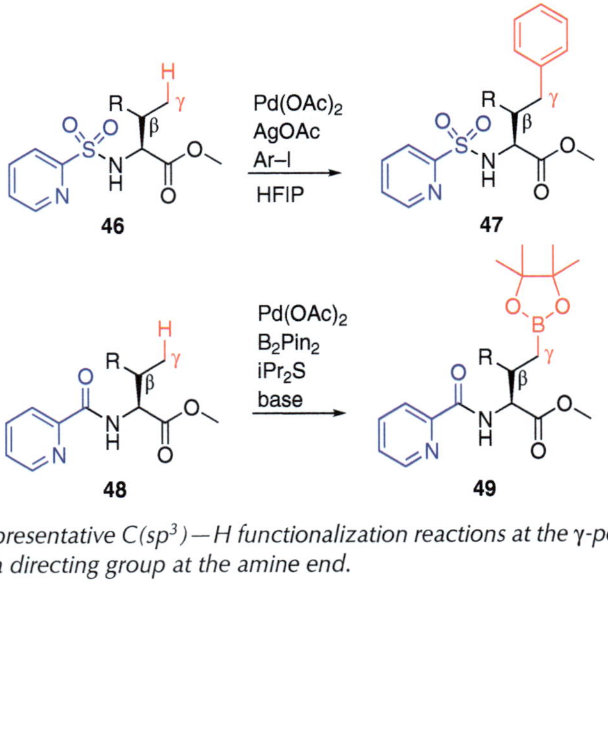

Figure 3.28 Representative C(sp^3)—H functionalization reactions at the γ-position exploiting picolinamide as a directing group at the amine end.

highly reactive C—B bond. In this context, Shi reported a Pd-catalyzed γ-borylation strategy for valine and isoleucine derivatives **48** using picolinamide as a directing group (Figure 3.28) [86]. Reaction optimization identified $Pd(OAc)_2$ in combination with an excess of diisopropyl sulfide as an effective catalytic system and a mixture of lithium carbonate and sodium bicarbonate as the base system to promote the mono-borylation selectively at the γ-position of both valine and isoleucine with good efficiency.

While the employment of directing groups has proven effective in achieving high regioselectivity and efficiency in C—H activation, their use presents a fundamental limitation, as the necessity to install and subsequently remove these external auxiliaries adds synthetic complexity, particularly in the preparation of modified α-amino acids and peptides. As a result, strategies that eliminate the need for such auxiliaries are of great synthetic interest. Accordingly, Yu introduced an innovative approach, consisting of the use of ligands to modulate the reactivity and selectivity of palladium-catalyzed β-C—H activation at α-amino acid side chain [51]. Specifically, pyridine- and quinoline-based ligands have been employed to control catalyst-mediated $C(sp^3)$—H arylation (Figure 3.29), with pyridine selectively facilitating monoarylation, and quinoline enhancing catalyst activity to enable diarylation, also in one pot. By applying these ligands sequentially, a methyl group in an alanine derivative **50** can be arylated to **51** and then **52** with two distinct aryl iodides with >20:1 diastereoselectivity. Notably, this enabled the β-arylation of alanine derivatives in the absence of any appended directing group, highlighting the potential of auxiliary-free methodologies. In a similar approach, the same authors applied this concept for the double arylation of alanine as an *N*-methoxyamide auxiliary to direct β-C—H activation assisted by pyridine-type ligands to significantly improve the β-arylation of carboxylic acid derivatives [53].

Also, the same research group proposed the method of ligand-assisted C—H activation, where an external ligand is introduced to overcome the weak directing ability of the carboxyl group in α-amino acids [87]. Specifically, Yu et al. reported the acid-directed β-$C(sp^3)$—H arylation of α-amino acids enabled by pyridine-type ligands (see Figure 3.30 for alanine **53**). Interestingly, this transformation proceeds without the need for an exogenous directing group and proved to be scalable, thus allowing the synthesis of Fmoc-protected unnatural amino acids as **55** in three steps. The use of pyridine-type ligands proved essential to facilitate this $C(sp^3)$—H arylation strategy.

A ligand-enabled approach for the γ-arylation of valine derivatives, analogous to β-arylation protocol, was developed using a quinoline-based ligand (Figure 3.31) [88].

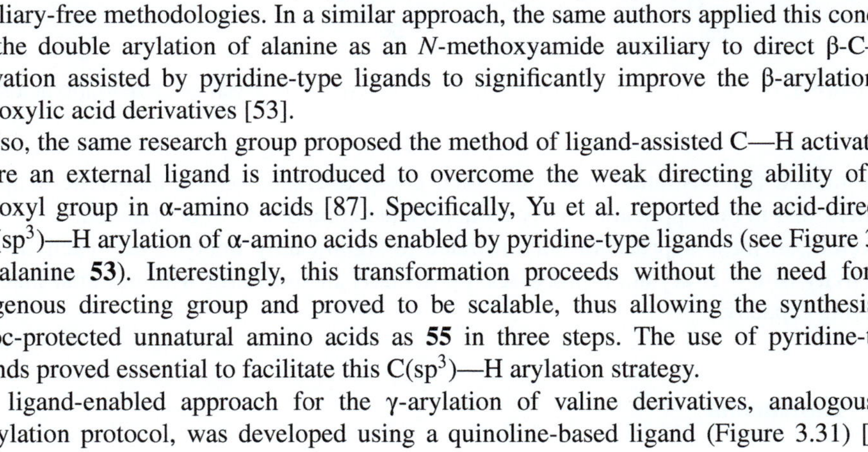

Figure 3.29 Sequential ligand-directed double β-C—H arylation of alanine with high diastereoselectivity as a method for synthesizing chiral β-Ar-β-Ar′-α–amino acids.

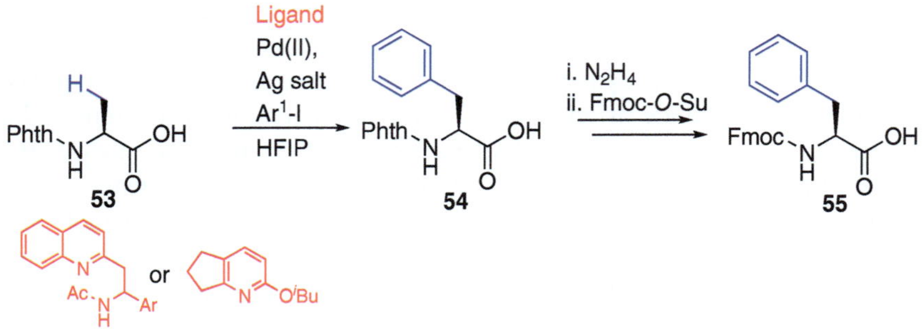

Figure 3.30 Pd(II)-catalyzed β-C(sp^3)—H arylation of α-amino acids as free carboxylic acids enabled by pyridine- or quinoline-based ligands.

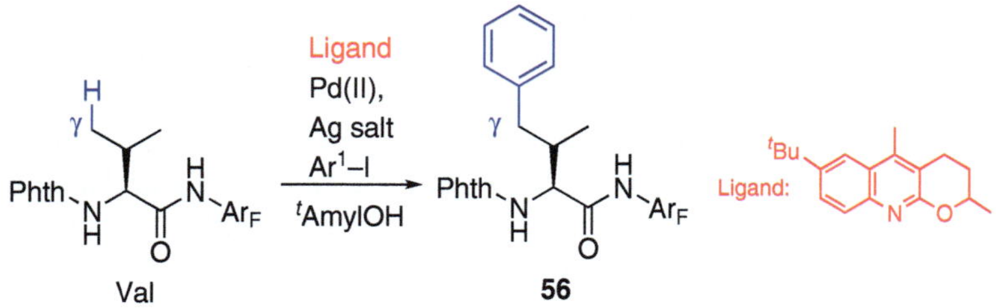

Figure 3.31 Ligand-enabled Pd(II)-catalyzed γ-C(sp^3)—H arylation of α-amino acids.

This method demonstrates broad functional group tolerance on the iodoarene coupling partners, and in this application, only monoarylated compounds were obtained.

Overall, all these approaches demonstrated to significantly improve the reactivity and selectivity of C—H arylation processes, particularly in substrates with minimal inherent directing capacity.

3.5 Conclusion

Over the past century, significant efforts have been devoted toward the stereoselective synthesis of linear and cyclic non-coded amino acids for peptidomimetic design. Particularly, in the last twenty years synthetic strategies experienced a significant innovation, especially in the context of heterocyclic scaffolds. Such molecules, including azaheterocycles, aromatic moieties, and proline derivatives, have played a key role in modulating the conformational rigidity, metabolic stability, and biological activity of peptides. Specifically, metal-catalyzed cross-coupling reactions have emerged as powerful tools for the functionalization of the side chains. However, considering the limited efficiency of these approaches due to the need of pre-functionalized substrates and multistep synthetic sequences, direct C—H bond functionalization emerged as an efficient and atom-economical alternative. This methodology enables the site-selective functionalization of otherwise inert C—H bonds, thus minimizing the need for protecting groups strategies. As a result, C—H activation methods now offer a straightforward tool for the late-stage modification of amino acids and peptides, facilitating access to structurally complex analogues using chiral and commercially available precursors. Such advances are improving the synthetic efficiency and expanding the landscape of bioactive molecules available for drug discovery applications. In particular, direct C—H functionalization is a promising strategy for the efficient synthesis of structurally different α-amino acid and peptide derivatives. Although examples of asymmetric functionalization are still limited, future advances are likely to deliver protocols of general use at different synthetic scales, capable of operating on complex molecular systems and specifically achieve non-coded amino acids and peptide-based therapeutics for academic research and industrial drug development.

References

1. Shabani S, Wu Y, Ryan HG, Hutton CA. Progress and perspectives on directing group-assisted palladium-catalysed C—H functionalisation of amino acids and peptides. *Chem Soc Rev*. 2021;**50**:9278–9343.
2. Singh MK, Lakshman MK. Recent developments in the utility of saturated azaheterocycles in peptidomimetics. *Org Biomol Chem*. 2022;**20**:963–979.
3. Kubyshkin V, Mykhailiuk PK. Proline analogues in drug design: current trends and future prospects. *J Med Chem*. 2024;**67**:20022–20055.
4. Trabocchi A, Guarna A. Cyclic α-amino acids as proline mimetics peptidomimetics. In: *Organic and Medicinal Chemistry: The Art of Transforming Peptides in Drugs*. Chichester, West Sussex, United Kingdom: John Wiley & Sons Inc., 2014. pp. 165–190.
5. Kubyshkin V, Rubini M. Proline analogues. *Chem Rev*. 2024;**124**:8130–8232.
6. Holladay MW, Lin CW, May CS, Garvey DS, Witte DG, Miller TR, Wolfram CA, Nadzan AM. trans-3-*n*-propyl-L-proline is a highly favorable, conformationally restricted replacement for methionine in the C-terminal tetrapeptide of cholecystokinin. Stereoselective synthesis of 3-allyl- and 3-*n*-propyl-L-proline derivatives from 4-hydroxy-L-proline. *J Med Chem*. 1991;**34**:455–457.
7. Affron DP, Davis OA, Bull JA. Regio- and stereospecific synthesis of C-3 functionalized proline derivatives by palladium catalyzed directed C(sp^3)—H arylation. *Org Lett*. 2014;**16**:4956–4959.
8. Affron DP, Bull JA. Palladium-catalyzed directed C(sp^3)—H arylation of saturated heterocycles at C-3 using a concise optimization approach. *Eur J Org Chem*. 2016;**2016**:139–149.
9. Mandal GH, Choudhary S, Kelley SP, Sathyamoorthi S. Stereoselective preparation of (4*S*)-1-methyl-4-propyl-L-proline commencing from (*cis*)-4-hydroxy-L-proline. *Molbank*. 2025;**2025**:M2003.
10. Hussain Z, Cooke AJ, Neelamkavil S, Brown L, Carswell E, Geissler WM, Guo Z, Hawes B, Kelly TM, Kiyoi Y, Lai K, Lesburg C, Pow E, Zang Y, Wood HB, Edmondson SD, Liu W. Design and synthesis of novel proline based factor XIa selective inhibitors as leads for potential new anticoagulants. *Bioorg Med Chem Lett*. 2020;**30**:127072.

11. Edelstein EK, Rankic DA, Dudley CC, McMinn SE, Adpressa DA. Synthesis of proline analogues via Rh-catalyzed asymmetric conjugate addition. *ACS Catal.* 2021;**11**:743–749.
12. Serra M, Terreni M, Bernardi E, Colombo L. Synthesis of functionalized proline-derived azabicycloalkane amino acids and their applications in drug discovery: recent advances. *Eur J Org Chem.* 2023;**26**:e202201394.
13. Prandi C, Occhiato EG. From synthetic control to natural products: a focus on N-heterocycles. *Pest Manag Sci.* 2019;**75**:2385–2402.
14. Kadouri-Puchot C, Comesse S. Recent advances in asymmetric synthesis of pipecolic acid and derivatives. *Amino Acids.* 2005;**29**:101–130.
15. Cant AA, Sutherland A. Asymmetric synthesis of pipecolic acid and derivatives. *Synthesis.* 2012;**44**:1935–1950.
16. Czombos J, Aelterman W, Tkachev A, Martins JC, Tourwé D, Péter A, Tóth G, Fülöp F, De Kimpe N. Convenient synthesis of 3,4-methano-1,2,3, 4-tetrahydroisoquinoline-3-carboxylic acid and its derivatives as doubly constrained nonproteinogenic amino acid derivatives. *J Org Chem.* 2000;**65**:5469–5475.
17. Günter M, Gais HJ. Asymmetric synthesis of fused bicyclic alpha-amino acids having a hexahydro-cyclopenta[c]pyridine skeleton via intramolecular Pauson-Khand reaction of 1-sulfonimidoyl-substituted 5-azaoct-1-en-7-ynes. *J Org Chem.* 2003;**68**:8037–8041.
18. Hattori K, Grossman RB. Functionalized cis- and trans-fused bicyclic alpha-amino acids via stereoselective double annulation and dequaternization reactions. *J Org Chem.* 2003;**68**:1409–1417.
19. Bartali L, Casini A, Guarna A, Occhiato EG, Scarpi D. Enantiodivergent chemoenzymatic synthesis of 4-hydroxypiperidine alkaloids. *Eur J Org Chem.* 2010;**2010**: 5831–5840.
20. Occhiato EG, Scarpi D, Guarna A, Tabasso S, Deagostino A, Prandi C. A short and convenient synthesis of enantiopure *cis*- and *trans*-4-hydroxypipecolic acid. *Synthesis.* 2009;**21**:3611–3616.
21. Stashkevych O, Kokhalskyi V, Mynak Y, Levterov V, Shablykin O, Pishel I, Mykhailiuk PK. Bicyclic isosteres of pyridine/piperidine: from synthesis to applications. *Angew Chem Int Ed Engl.* 2025;e202517814.
22. Pal'chikov VA. Morpholines. Synthesis and biological activity. Russ. *J Org Chem.* 2013;**49**:787–814.
23. Guo M, Zhu Z, Huang H, Zhang Q. Phosphine-free palladium catalyzed Suzuki coupling of aryl chlorides in morpholine under aerobic conditions. *Catal Commun.* 2009;**10**:865.
24. Abellan T, Chinchilla R, Galindo N, Najera C, Sansano JM. New oxazinone and pyrazinone derivatives as chiral reagents for the asymmetric synthesis of α-amino acids. *J Heterocyclic Chem.* 2000;**37**:467.
25. Bera S, Panda G. I2-mediated diversity oriented diastereoselective synthesis of amino acid derived trans-2,5-disubstituted morpholines, piperazines, and thiomorpholines. *ACS Comb Sci.* 2012;**14**:1–4.
26. Segat-Dioury F, Lingibé O, Graffe B, Sacquet M-C, Lhommet G. A general synthesis of enantiopure 1,2-aminoalcohols via chiral morpholinones. *Tetrahedron.* 2000;**56**:233.

27. Trabocchi A, Krachmalnicoff A, Menchi G, Guarna A. Synthesis and conformational studies of a hybrid β-alanine–morpholine tetramer. *Tetrahedron*. 2012;**68**:9701.
28. Walker DP, Eklov BM, Bedore MW. Practical synthesis of 3-Oxa-6-azabicyclo[3.1.1] heptane hydrotosylate; a novel morpholine-based building block. *Synthesis*. 2012;**44**: 2859–2862.
29. Wijtmans R, Vink MKS, Schoemaker HE, van Delft FL, Blaauw RH, Rutjes FPJT. Biological relevance and synthesis of C-substituted morpholine derivatives. *Synthesis*. 2004;**5**:641–662.
30. Trstenjak U, Ilaš J, Kikelj D. Advances in the synthesis of morpholin-3-ones and morpholin-2-ones. *Synthesis*. 2012;**44**:3551–3578.
31. Luescher MU, Bode JW. Catalytic synthesis of N-unprotected piperazines, morpholines, and thiomorpholines from aldehydes and SnAP reagents. *Angew Chem Int Ed Engl*. 2015;**54**:10884–10888.
32. Murray PRD, Leibler IN, Hell SM, Villalona E, Doyle AG, Knowles RR. Radical redox annulations: a general light-driven method for the synthesis of saturated heterocycles. *ACS Catal*. 2022;**12**:13732–13740.
33. Grainger R, Heightman TD, Ley SV, Lima F, Johnson CN. Enabling synthesis in fragment-based drug discovery by reactivity mapping: photoredox-mediated cross-dehydrogenative heteroarylation of cyclic amines. *Chem Sci*. 2018;**10**: 2264–2271.
34. Sladojevich F, Trabocchi A, Guarna A. Stereoselective cyclopropanation of serine- and threonine-derived oxazines to access new morpholine-based scaffolds. *Org Biomol Chem*. 2008;**6**:3328–3336.
35. Lalli C, Trabocchi A, Sladojevich F, Menchi G, Guarna A. Diversity-oriented synthesis of morpholine-containing molecular scaffolds. *Chemistry*. 2009;**15**:7871–7875.
36. Golebiowski A, Klopfenstein SR, Chen JJ, Shao X. Solid supported high-throughput organic synthesis of peptide β-turn mimetics via tandem Petasis reaction/ diketopiperazine formation. *Tetrahedron Lett*. 2000;**41**:4841–4844.
37. Vaillancourt B, Mishra P, Duchamp J, Hanessian S. Synthesis of azabicyclic isosteres of piperazine 2*S*-carboxylic acid: access and physicochemical properties. *J Org Chem*. 2024;**89**:1944–1955.
38. Rossen K, Weissman SA, Sager J, Reamer RA, Askin D, Volante RP, Reider PJ. Asymmetric hydrogenation of tetrahydropyrazines: synthesis of (*S*)-piperazine-2-*tert*-butylcarboxamide, an intermediate in the preparation of the HIV protease inhibitor indinavir. *Tetrahedron Lett*. 1995;**36**:6419–6422.
39. Petrova VM, Ioffe SL. Synthesis and conversions of piperazine-2-carboxylic acid. *Chem Heterocycl Compd*. 1985;**21**:803–807.
40. Koch SS, Ducray P. Synthesis and N-orthogonally protected (*S*)-piperazine-2-carboxylic acid. *J Org Chem*. 1997;**62**:5308–5309.
41. Chamakuri S, Tang Y, Tran T, Guduru R, Bolin TD, MacKenzie KR, Young DW. Concise synthesis of 3-substituted piperazine-2-acetic acid esters from 1,2-diamines. *Molecules*. 2022;**27**:3419.
42. Rogge T, Kaplaneris N, Chatani N, Kim J, Chang S, Punji B et al. C—H activation. *Nature Reviews Methods Primers*. 2021;**1**:43.
43. Bergman RG. Organometallic chemistry: C—H activation. *Nature*. 2007;**446**(7134): 391–393.

44. Brückl T, Baxter RD, Ishihara Y, Baran PS. Innate and guided C—H functionalization logic. *Acc Chem Res*. 2012;**45**:826–839.
45. Zaitsev VG, Shabashov D, Daugulis O. Highly regioselective arylation of sp^3 C—H bonds catalyzed by palladium acetate. *J Am Chem Soc*. 2005;**127**:13154–13155.
46. Reddy BV, Reddy LR, Corey EJ. Novel acetoxylation and C—C coupling reactions at unactivated positions in alpha-amino acid derivatives. *Org Lett*. 2006;**8**:3391–3394.
47. Kamijo S. C—H activation via radical processes using photo-excited ketones. In: Landais Y, editor. *Free-Radical Synthesis and Functionalization of Heterocycles. Topics in Heterocyclic Chemistry*, vol. **54**. Cham: Springer, 2018.
48. Suseelan AS, Dutta A, Lahiri GK, Maiti D. Organopalladium intermediates in coordination-directed $C(sp^3)$—H functionalizations. *Trends Chem*. 2020;**3**:188–203.
49. Gutekunst WR, Gianatassio R, Baran PS. Sequential $C(sp^3)$—H arylation and olefination: total synthesis of the proposed structure of pipercyclobutanamide A. *Angew Chem Int Ed Engl*. 2012;**51**:7507–7510.
50. Tsukamoto S, Tomise K, Miyakawa K, Cha BC, Abe T, Hamada T, Hirota H, Ohta T. CYP3A4 inhibitory activity of new bisalkaloids, dipiperamides D and E, and cognates from white pepper. *Bioorg Med Chem*. 2002;**10**:2981–2985.
51. He J, Li S, Deng Y, Fu H, Laforteza BN, Spangler JE, Homs A, Yu JQ. Ligand-controlled $C(sp^3)$—H arylation and olefination in synthesis of unnatural chiral α-amino acids. *Science*. 2014;**343**:1216–1220.
52. Zhang Q, Chen K, Rao W, Zhang Y, Chen FJ, Shi BF. Stereoselective synthesis of chiral α-amino-β-lactams through palladium(II)-catalyzed sequential monoarylation/amidation of C(sp(3))—H bonds. *Angew Chem Int Ed Engl*. 2013;**52**:13588–13592.
53. Chen G, Shigenari T, Jain P, Zhang Z, Jin Z, He J, Li S, Mapelli C, Miller MM, Poss MA, Scola PM, Yeung KS, Yu JQ. Ligand-enabled β-C—H arylation of α-amino acids using a simple and practical auxiliary. *J Am Chem Soc*. 2015;**137**:3338–33351.
54. Lou J, Wang Q, He Y, Yu Z. A simple aliphatic diamine auxiliary for palladium-catalyzed arylation of unactivated β-$C(sp^3)$—H bonds. *Adv Synth Catal*. 2018;**360**: 4571–4584.
55. Rodrıguez N, Romero-Revilla JA, Fernandez-Ibanez MA, Carretero JC. Palladium-catalyzed *N*-(2-pyridyl)sulfonyl-directed $C(sp^3)$—H γ-arylation of amino acid derivatives. *Chem Sci*. 2013;**4**:175–179.
56. Zhang G, Xie X, Zhu J, Li S, Ding C, Ding P. Pd(II)-catalyzed C(sp(3))—H arylation of amino acid derivatives with click-triazoles as removable directing groups. *Org Biomol Chem*. 2015;**13**:5444–5449.
57. Fan M, Ma D. Palladium-catalyzed direct functionalization of 2-aminobutanoic acid derivatives: application of a convenient and versatile auxiliary. *Angew Chem Int Ed Engl*. 2013;**52**:12152–12155.
58. Pasunooti KK, Banerjee B, Yap T, Jiang Y, Liu CF. Auxiliary-directed Pd-catalyzed γ-C(sp(3))—H bond activation of α-aminobutanoic acid derivatives. *Org Lett*. 2015;**17**:6094–6097.
59. Higham JI, Bull JA. Transient imine directing groups for the C—H functionalisation of aldehydes, ketones and amines: an update 2018–2020. *Org Biomol Chem*. 2020;**18**:7291–7315.

60. Engle KM, Mei TS, Wasa M, Yu JQ. Weak coordination as a powerful means for developing broadly useful C—H functionalization reactions. *Acc Chem Res*. 2012;**45**: 788–802.
61. Giri R, Maugel N, Li JJ, Wang DH, Breazzano SP, Saunders LB, Yu JQ. Palladium-catalyzed methylation and arylation of sp^2 and sp^3 C—H bonds in simple carboxylic acids. *J Am Chem Soc*. 2007;**129**:3510–3511.
62. Zhu Y, Chen X, Yuan C, Li G, Zhang J, Zhao Y. Pd-catalysed ligand-enabled carboxylate-directed highly regioselective arylation of aliphatic acids. *Nat Commun*. 2017;**8**:14904.
63. Liu L, Liu YH, Shi BF. Synthesis of amino acids and peptides with bulky side chains via ligand-enabled carboxylate-directed γ-C(sp^3)—H arylation. *Chem Sci*. 2020;**11**:290–294.
64. Lee JM, Chang S. Pt-Catalyzed sp^3 C—H bond activation of *o*-alkyl substituted aromatic carboxylic acid derivatives for the formation of aryl lactones. *Tetrahedron Lett*. 2006;**47**:1375–1379.
65. Fraunhoffer KJ, Narayanasamy P, Sirois LE, White MC. Macrolactonization via hydrocarbon oxidation. *J Am Chem Soc*. 2006;**128**:9032–9033.
66. Zhuang Z, Yu CB, Chen G, Wu QF, Hsiao Y, Joe CL, Qiao JX, Poss MA, Yu JQ. Ligand-enabled β-C(sp^3)—H olefination of free carboxylic acids. *J Am Chem Soc*. 2018;**140**:10363–10367.
67. Zhuang Z, Herron AN, Fan Z, Yu JQ. Ligand-enabled monoselective β-C(sp^3)—H acyloxylation of free carboxylic acids using a practical oxidant. *J Am Chem Soc*. 2020;**142**:6769–6776.
68. Spangler JE, Kobayashi Y, Verma P, Wang DH, Yu JQ. α-Arylation of saturated azacycles and *N*-methylamines via palladium(II)-catalyzed C(sp(3))—H coupling. *J Am Chem Soc*. 2015;**137**:11876–11879.
69. Wu Y, Chen YQ, Liu T, Eastgate MD, Yu JQ. Pd-catalyzed γ-C(sp^3)—H arylation of free amines using a transient directing group. *J Am Chem Soc*. 2016;**138**:14554–14557.
70. Wang Z, Fu Y, Zhang Q, Liu H, Wang J. Pd(II)-catalyzed direct γ-C(sp^3)—H arylation between free β2-amino esters and β3-amino esters and aryl iodides using a catalytic transient directing group. *J Org Chem*. 2020;**85**:7683–7693.
71. Zhan BB, Li Y, Xu JW, Nie XL, Fan J, Jin L, Shi BF. Site-selective δ-C(sp^3)—H alkylation of amino acids and peptides with maleimides via a six-membered palladacycle. *Angew Chem Int Ed Engl*. 2018;**57**:5858–5862.
72. Fitzgerald LS, O'Duill ML. A guide to directing group removal: 8-aminoquinoline. *Chemistry*. 2021;**27**:8411–8436.
73. Brandhofer T, Mancheño OG. Site-selective C—H bond activation/functionalization of alpha-amino acids and peptide-like derivatives. *Eur J Org Chem*. 2018;**2018**: 6050–6067.
74. Zhao L, Li CJ. Functionalizing glycine derivatives by direct C—C bond formation. *Angew Chem Int Ed Engl*. 2008;**47**:7075–7078.
75. San Segundo M, Guerrero I, Correa A. Co-catalyzed C(sp^3)—H oxidative coupling of glycine and peptide derivatives. *Org Lett*. 2017;**19**:5288–5291.
76. Li K, Tan G, Huang J, Song F, You J. Iron-catalyzed oxidative C—H/C—H cross-coupling: an efficient route to α-quaternary α-amino acid derivatives. *Angew Chem Int Ed Engl*. 2013;**52**:12942–12945.

77. Li K, Wu Q, Lan J, You J. Coordinating activation strategy for C(sp(3))—H/C(sp(3))—H cross-coupling to access β-aromatic α-amino acids. *Nat Commun*. 2015;**6**:8404.
78. Reay AJ, Hammarback LA, Bray JTW, Sheridan T, Turnbull D, Whitwood AC, Fairlamb IJS. Mild and regioselective Pd(OAc)2-catalyzed C—H arylation of tryptophans by [ArN2]X, promoted by tosic acid. *ACS Catal*. 2017;**7**:5174–5179.
79. Reay AJ, Williams TJ, Fairlamb IJ. Unified mild reaction conditions for C2-selective Pd-catalysed tryptophan arylation, including tryptophan-containing peptides. *Org Biomol Chem*. 2015;**13**:8298–8309.
80. Bedford RB, Haddow MF, Webster RL, Mitchell CJ. The catalytic ortho-arylation of tyrosine. *Org Biomol Chem*. 2009;**7**:3119–3127.
81. Mahindra A, Bagra N, Jain R. Palladium-catalyzed regioselective C-5 arylation of protected L-histidine: microwave-assisted C—H activation adjacent to donor arm. *J Org Chem*. 2013;**78**:10954–10959.
82. Kallepalli VA, Shi F, Paul S, Onyeozili EN, Maleczka RE Jr, Smith MR 3rd. Boc groups as protectors and directors for Ir-catalyzed C—H borylation of heterocycles. *J Org Chem*. 2009;**74**:9199–9201.
83. Meyer FM, Liras S, Guzman-Perez A, Perreault C, Bian J, James K. Functionalization of aromatic amino acids via direct C—H activation: generation of versatile building blocks for accessing novel peptide space. *Org Lett*. 2010;**12**:3870–3873.
84. Zhang Q, Yin XS, Chen K, Zhang SQ, Shi BF. Stereoselective synthesis of chiral β-fluoro α-amino acids via Pd(II)-catalyzed fluorination of unactivated methylene C(sp(3))—H bonds: scope and mechanistic studies. *J Am Chem Soc*. 2015;**137**: 8219–8226.
85. Bauer M, Wang W, Lorion MM, Dong C, Ackermann L. Internal peptide late-stage diversification: peptide-isosteric triazoles for primary and secondary $C(sp^3)$—H activation. *Angew Chem Int Ed Engl*. 2018;**57**:203–207.
86. Zhang LS, Chen G, Wang X, Guo QY, Zhang XS, Pan F, Chen K, Shi ZJ. Direct borylation of primary C—H bonds in functionalized molecules by palladium catalysis. *Angew Chem Int Ed Engl*. 2014;**53**:3899–3903.
87. Chen G, Zhuang Z, Li GC, Saint-Denis TG, Hsiao Y, Joe CL, Yu JQ. Ligand-enabled β-C—H arylation of α-amino acids without installing exogenous directing groups. *Angew Chem Int Ed Engl*. 2017;**56**:1506–1509.
88. Li S, Zhu RY, Xiao KJ, Yu JQ. Ligand-enabled arylation of γ-C—H bonds. *Angew Chem Int Ed Engl*. 2016;**55**:4317–4321.

4

Late-Stage Functionalization of Peptides

4.1 Introduction

The chemical modification of peptides at late stages of synthesis has emerged to expand the molecular diversity and modulate the physicochemical, conformational, and biological properties of peptide-based molecules, as reported in several comprehensive reviews [1–5]. The modification of peptides usually relies on the direct incorporation of non-canonical amino acids or on lengthy multistep procedures consisting of functional group manipulation, whereas late-stage functionalization enables the direct modification of assembled peptides, thus minimizing synthetic complexity while maintaining the structural and functional versatility. This strategy is inspired by the biosynthetic logic of many natural systems, in which complex peptide frameworks are generated through post-translational modifications rather than through *de novo* peptide assembly. Hydroxylation, halogenation, acylation, alkylation, arylation, oxidative coupling, and other covalent transformations are frequently employed in nature to decorate and constrain peptide scaffolds. As a result, a large number of natural products exhibit complex molecular shapes and enhanced biological activity that arise directly from site-specific chemical alterations. This is particularly evident in cyclic and bicyclic natural peptides, where covalent linkages between side chains lead to conformationally constrained and densely functionalized molecular frameworks, as represented by compounds such as asperipin-2a, streptide, darobactin, and celogentin [6–9] (Figure 4.1). Their structures are characterized by C—C and C–heteroatom linkages formed through enzymatic or spontaneous processes, which contribute to their remarkable stability and potency. These structural motifs have inspired chemists to devise methods that mimic or reproduce these transformations *in vitro*, thus enabling access to complex architectures through selective late-stage modification.

One of the central challenges in the chemical modification of peptides and proteins is the achievement of site selectivity. Peptides often display multiple identical functional groups distributed across their sequence. For example, lysine residues have primary amines at various positions, while cysteine residues contain thiols that are chemically equivalent but spatially distinct. Traditional strategies targeting such side chains frequently result in heterogeneous mixtures of products due to uncontrolled modification at multiple sites. Also, the scope of chemical methods compatible with aqueous buffers and at physiological pH for the manipulation of peptides/proteins is narrower than for classical

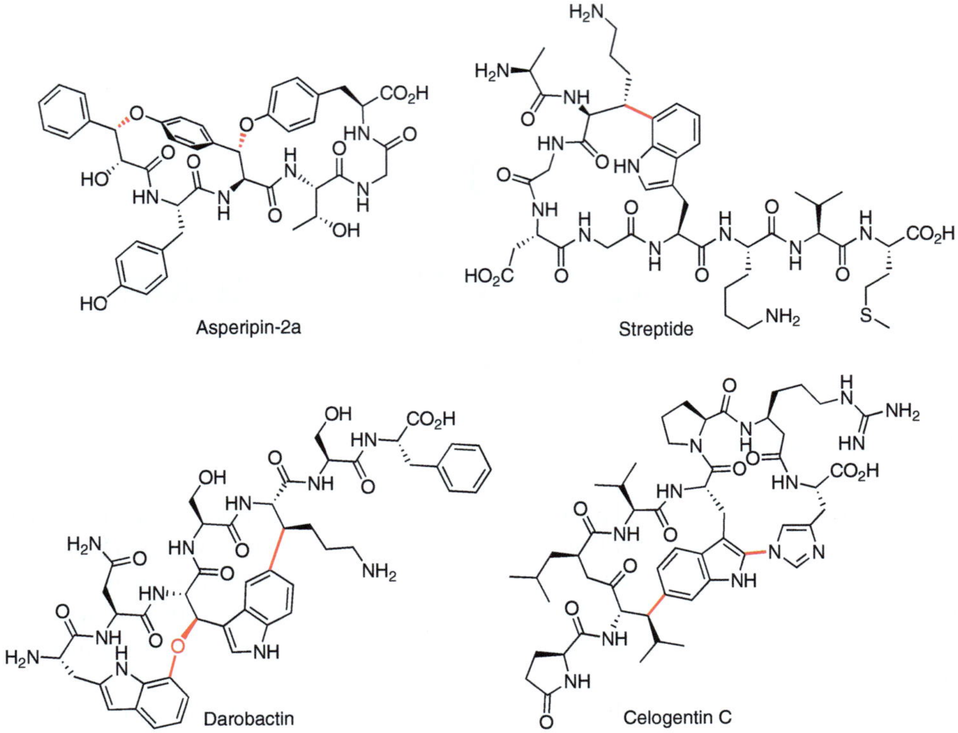

Figure 4.1 Representative cyclic and bicyclic peptides containing covalent C—C or C–heteroatom linkages between side chains (highlighted in red).

organic synthesis. These limitations have driven the development of chemoselective, site-specific functionalization methods that run under mild, biocompatible conditions. A straightforward strategy focuses on *N*-terminal modification [10], which exploits the different acidity of the α-amino group as compared to the ε-amine of the lysine side chain, taking advantage of different p*K*a values, with the former being more basic. Furthermore, the *N*-terminus is often solvent-exposed and can be exploited by introducing additional residues without interfering with the overall protein structure or activity, such as proline through oxidative coupling with amino phenols [11], tryptophan *via* Pictet–Spengler reaction [12, 13], cysteine through native chemical ligation [14], serine/threonine through oxidation to corresponding aldehydes and subsequent conjugation with α-nucleophiles [15]. In addition to residue-specific transformations, several general labeling strategies have been developed for the *N*-terminal modification irrespective of the peptide sequence, including diazotransfer using imidazole-1-sulfonyl azide [16], alkyne installation *via* phenyl ketene derivatives [17], transamination using 2-pyridinecarboxaldehydes [18], and reductive alkylation [19]. Finally, beyond side-chain reactivity, backbone functionalization represents a frontier in peptide chemistry [20]. Modifying the peptide backbone has profound implications for the molecular structure, as it directly influences its folding, conformation, solubility, and aggregation. While traditional approaches have primarily relied on functional group transformations, a powerful advance has emerged from direct

C—H bond functionalization. This approach enables the direct transformation of otherwise inert bonds, thus expanding significantly the chemical space accessible through late-stage modification. Nature itself has demonstrated the value of such transformations through enzyme-promoted C—H activation mechanisms, including C2-mannosylation of tryptophan, C5-alkylation of histidine, and C3-hydroxylation of proline in collagen [21]. Transition-metal catalysis is particularly useful for enabling the C—H activation of various molecules, including peptides [22]. In this context, methods originally developed for small molecules have been adapted to work under aqueous and biocompatible conditions. For example, Pd-catalyzed C(sp^3)—H arylation directed by amidoxime ether has allowed the selective modification of internal residues in peptides with high stereocontrol, followed by traceless removal of the directing group (Figure 4.2) [23], thus demonstrating the feasibility of performing highly selective transformations within complex peptide backbones.

Nevertheless, the development of methods for the C—H functionalization of peptides remains at an early stage due to several challenges, such as the maintenance of the stereochemistry at the α-carbon, and the lack of strongly differentiating features between different C—H bonds of both backbone and side chains. The complexity of peptide substrates amplifies the complexity of C—H activation chemistry, as peptides display multiple C—H bonds with similar bond dissociation energies, numerous stereogenic centers susceptible to racemization, and functional groups sensitive to harsh conditions, such as thiols. Thus, all these issues require synthetic methods for peptide C—H functionalization possessing high site- and stereoselectivity, and optimal catalytic performance. In this context, emerging methods that use transient directing groups or exploit the coordination to amide nitrogen or carbonyl oxygen atoms are demonstrating their potential, particularly

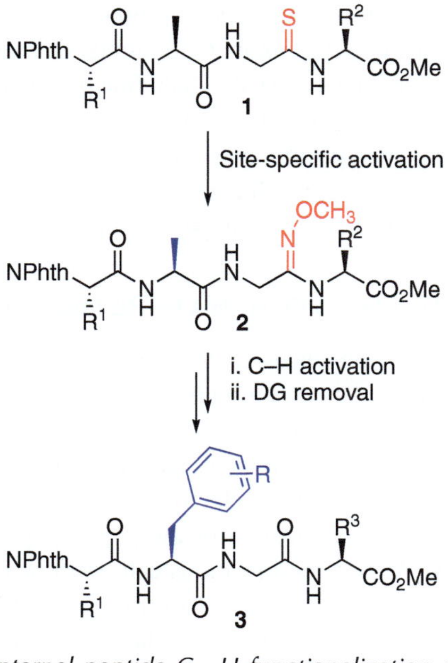

Figure 4.2 *Site-specific internal peptide C—H functionalization using a backbone-inserted directing group.*

for the generation of peptidomimetics without pre-functionalization, thus expanding the toolkit available for specific molecular editing of peptides. Next generation of late-stage peptide modification strategies should combine several requirements to achieve structurally diverse peptidomimetic libraries with reduced synthetic steps, such as:

- direct C—H activation at side chains and backbone
- orthogonal reactivity to enable multiple sequential modifications
- compatibility with physiological media
- site selectivity through directing elements or engineered motifs

4.2 Classification of Late-Stage Functionalization of Peptides

Traditional methods for the modification of peptides mainly consist of coupling reactions involving cysteine and lysine as polar nucleophilic residues, whereas convenient methods exploiting hydrophobic residues, both aliphatic and aromatic, have been less developed [20]. Accordingly, the use of transition-metal-catalyzed C—H activation in peptide chemistry has brought remarkable advances [24]. Initially developed for small-molecule synthesis, these methods have been adapted to biomolecules, enabling the late-stage diversification without the need of preactivated substrates. In particular, the metal-catalyzed directed C—H functionalization of the side chain of alanine through directed approaches has been developed extensively over the last few years. Among transition metals, palladium is currently the preferred catalyst for this reaction for the formation and functionalization of the palladacycle intermediate under mild conditions. The typologies of peptide functionalization can be broadly grouped into backbone and side-chain modification as two major categories, encompassing several approaches that have been proposed for a range of post-assembly functionalization methods at the *N*-terminus, *C*-terminus, or internal site of the peptide, taking advantage of both amide bonds and side-chain functional groups to achieve site- and stereoselectivity [4]. Concerning C—H activation strategies, a number of directing groups and strategies for employing site-specific activating groups have been proposed over the years to address the challenging issue of achieving precise control of various types of selectivity over multiple C—H bonds present in the peptide molecules.

4.2.1 Classification of Peptide Modifications by Residue Type

As highlighted by Baran [20], advances in bioconjugation chemistry and the precise modification of proteins through targeted amino acid functionalization have been instrumental in enabling site-selective protein engineering for designing therapeutically relevant biomolecules. The growing interest in the pharmaceutical potential of peptides and proteins has strengthened the need for chemoselective late-stage functionalization methods, which allow for the structural diversification of the parent peptide without affecting the conformation or bioactivity. Peptide modifications can occur at various sites of the target peptide structure, namely at the *C*-terminus, *N*-terminus, backbone, or specific amino acid residues, thus requiring specific chemical methods to achieve the desired chemical modification and biological compatibility. In general, successful bioconjugation or peptide modification protocols should proceed under mild and physiologically relevant conditions,

Table 4.1 Classification of functionalization strategies based on residue types.

Residue type	Representative residues	Key functionalization strategies	Challenges
Aliphatic	Ala, Val, Leu, Ile	C—H bond activation; Pd-catalyzed arylation of *N*-terminal Ala	Limited substrate scope
Aromatic	Tyr, His, Trp, Phe	pH-dependent reactions (ene-type, diazonium, Mannich); metal-catalyzed C—H arylation or alkynylation	Control of regioselectivity
Polar/ionizable	Cys, Lys, Asp, Glu, Arg	Thiol alkylation/acylation (Cys); selective acylation/arylation (Lys, Arg); selection of reactivity at α- vs. side-chain groups (Lys, Asp, Glu)	Chemoselectivity and competing reactivity
Backbone/terminal sites	*N*-terminus, *C*-terminus, and peptide bond	Direct coupling, acylation, amidation, and C—H activation	Maintenance of structural integrity

such as neutral pH, aqueous medium, and ambient temperature, in order to preserve the peptide/protein structure and function. Concerning the chemical reactivity of amino acid residues, the modification at the side chain generally depends on the chemical structure (Table 4.1).

Traditionally, aliphatic residues have not been exploited as sites of functionalization in the past being generally inert. Nevertheless, advances in C—H bond activation have recently provided access to direct functionalization on such a type of side chain. For example, palladium-catalyzed C–H arylation of *N*-terminal alanine residues exemplifies a novel approach for hydrocarbon modification, although still restricted to simple peptide substrates.

Aromatic residues such as tyrosine, histidine, tryptophan, and phenylalanine offer versatile modification sites. Their reactivity can be modulated by the control of pH or through metal-catalyzed C—H functionalization. For example, tyrosine can be exploited for a range of transformations, including ene-type, diazonium, and Mannich-type reactions, whereas tryptophan generally participates in C-2 indole functionalization through both metal-catalyzed and metal-free radical methods.

Polar and ionizable residues, such as cysteine, lysine, arginine, aspartic acid, and glutamic acid, exhibit high reactivity and have been exploited over the decades as sites for peptide and protein bioconjugation, with cysteine being one of the most exploited residues taking advantage of the thiol functionality.

Concerning acidic and basic amino acids, an issue in this approach is the achievement of site selectivity among competing reactive groups, for example, in distinguishing α- vs.

side-chain functionalities in carboxylic or amine-containing residues when such amino acids are at the *C*- or *N*-terminus, respectively.

4.2.2 Classification of C—H Functionalization by Site of Activation in Peptides

Concerning the application of C—H activation strategies in peptide modification, these approaches can be classified into backbone and side-chain groups, the latter with additional breakdown, as follows:

- Backbone C—H activation
- Side-chain C—H activation
  - *N*-terminal C—H functionalization
  - internal C—H functionalization
  - *C*-terminal C—H functionalization

Backbone C—H activation. Modification of the peptide backbone is more challenging but allows for achieving a higher level of structural modification. Unlike side-chain functionalization, backbone modification has a direct effect on the conformation, solubility, and aggregation of the peptide, thus influencing the biological activity more deeply. Nevertheless, only a limited number of methodologies currently exist for backbone C—H activation, due to the need to maintain the stereochemical integrity at α-stereocenters and to the lack of specific sites of activation. Emerging strategies have been proposed to exploit transient directing groups or metal coordination to amide oxygen or nitrogen atoms to selectively activate backbone positions, thus facilitating cross-coupling or oxidative transformations.

Side-chain C—H activation. These methods enable the late-stage incorporation of aryl or alkyl substituents without extensive pre-functionalization. In this context, transition-metal catalysis has been widely explored to allow for the $C(sp^3)$—H or $C(sp^2)$—H arylation, alkylation, and oxidative transformations on residues such as tryptophan, histidine, tyrosine, phenylalanine, and leucine. For example, the palladium-catalyzed $C(sp^3)$—H arylation directed by amidoxime ethers allows site and stereoselective modification of internal residues in native peptides, with traceless removal of the directing group after functionalization (Figure 4.2) [23]. The terminal positions offer unique electronic and steric environments, making them attractive for selective C—H functionalization. Beyond classical *N*-terminal labeling *via* amine reactivity, transition-metal catalysis and photoredox methods are increasingly applied to terminal positions, particularly at the α-carbon. *C*-terminal modifications remain less explored but represent an emerging field with potential for introducing handles for bioconjugation or macrocyclization. Unlike directing groups (DGs) at *C*-terminus, the exploitation of the *N*-terminal position for attaching DGs has seen modest development for the C—H functionalization. This is mainly due to the challenge of generating an appropriate five-membered palladacycle, as the amide nitrogen can significantly inhibit Pd-catalyzed C—H functionalization processes through the coordination to palladium to form a stable five-membered *N,N*-chelated complex. With respect to internal side-chain modification, successful strategies have been based on exploiting specific residues or backbone functional motifs capable to orient the catalytic system toward target C—H bonds.

4.2.3 Solid-Phase Approaches for the Late-Stage Functionalization of Peptides

Since early phases of its development, peptide chemistry has been largely advanced on the solid phase, and this technology has been taken into account for late-stage functionalization of peptides, too, allowing for the achievement of chemical modification of the synthesized peptides while remaining immobilized on the solid-phase resin. This approach is instrumental for the rapid diversification of peptide libraries, providing a convenient platform for post-synthetic modifications without the need for intermediate purification steps or repeated synthesis. Thus, advantages of solid-phase late-stage functionalization allow for the development of efficient strategies of late-stage diversification and can be summarized as follows:

- improved stability toward aggregation or degradation of the immobilized peptides
- use of excess quantities of reagents readily removed by simple washing
- minimization of undesired side reactions thanks to the resin matrix

The successful application of late-stage functionalization techniques on solid phase requires careful consideration of resin selection (e.g. Rink amide, TentaGel), solvent compatibility (e.g. DMF, DMSO), and the steric environment around the site of modification, as resin swelling behavior and peptide backbone accessibility can significantly influence the reaction efficiency. Several chemoselective methodologies have been developed for the late-stage modification of peptides on solid support, including side-chain modification, bioconjugation, and peptide cyclization and stapling. For example, functional groups such as the ε-amino group at the side-chain of lysine, the thiol of cysteine, and the hydroxyl groups of serine and threonine can be selectively modified to address side-chain functionalization on the solid support, including acylation, alkylation, PEGylation, phosphorylation, and glycosylation. Also, click chemistry, particularly referred to as the copper(I)-catalyzed azide–alkyne cycloaddition (CuAAC), is one of the most robust and widely used reactions for on-resin late-stage functionalization. The incorporation of azido- or alkyne-functionalized amino acids during peptide elongation enables the post-synthesis selective conjugation with small molecules or fluorophores. Finally, on-resin cyclization strategies, including lactam ring formation, disulfide bond formation, and ring-closing metathesis (RCM), can be employed to enhance the structural rigidity and bioactivity, taking advantage of orthogonally protected side chains to guide intramolecular reactions.

Other late-stage functionalization strategies developed on solid phase include the use of immobilized peptides for site-specific enzymatic modifications, such as phosphorylation, acetylation, and glycosylation, all carried out under near-physiological conditions.

4.3 Backbone Modification

The most common strategies for backbone modifications are based on glycine residues, which are converted into corresponding α-alkyl, aryl-, or α-allyl derivatives [25]. The advances made over the last years are remarkable [26], as the use of chiral catalysts enabled the insertion of side chains to the carbon backbone. Other strategies have also been reported, such as fragmentation of serine, threonine, hydroxyproline, glutamic acid, or *C*-terminal residues, followed by the *in situ* addition of new chains, thus providing α,β-, α,γ-, or α,δ-peptide hybrids to peptides with a variety of *N*-substituents [27].

4.3.1 Backbone Functionalization at α-Carbon Atom

Concerning the modification of peptides at the α-carbon atoms of the backbone, glycine is the most frequently employed residue [4, 28], although other amino acids, such as serine, threonine, glutamic acid, proline, hydroxyproline, and dehydroamino acid units have also been exploited. The first significant advance in this area was reported by Seebach, who exploited the chemistry of enolates to install functional groups at the α-carbon. In a first report, *N*-protected glycine enolates generated at low temperatures were reacted with a range of electrophilic substrates in the presence of other amino acids, demonstrating the feasibility of this process in a peptide sequence [29]. Successively, improvements in the context of this synthetic process were reported by the same group, consisting of the use of an aminomalonate unit as a substrate to perform the alkylation under basic conditions followed by ester deprotection and *in situ* decarboxylation, ultimately yielding an aromatic residue as a hydrogen-to-side-chain replacement (Figure 4.3) [30].

In a similar strategy exploiting the chemistry of enolates, the insertion of appendages to the α-carbon of glycine in small peptides was achieved with high regio- and diastereoselectivity, taking advantage of palladium-catalyzed alkylation or allylation (Figure 4.4) [31]. The stereoselectivity was guided by the adjacent amino, leading to a *C*-terminal amino acid with inverted configuration (i.e. (*S*)-amino acid inducing the generation of (*R*)-stereocenter at the adjacent amino acid on the *C*-terminal side).

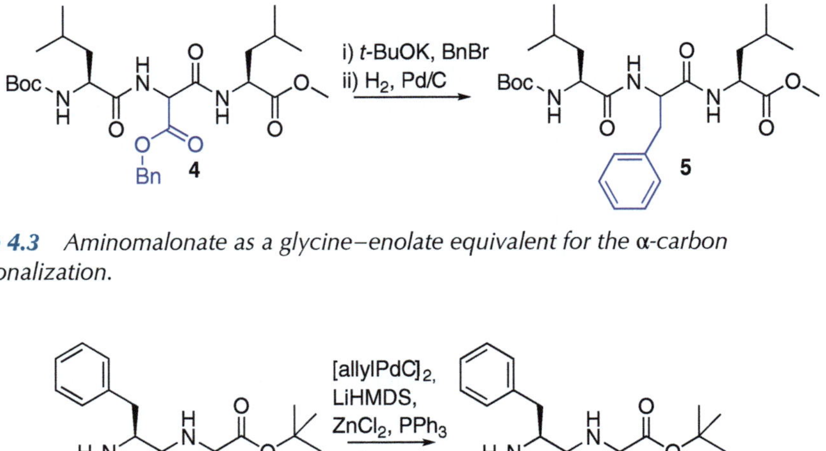

Figure 4.3 Aminomalonate as a glycine–enolate equivalent for the α-carbon functionalization.

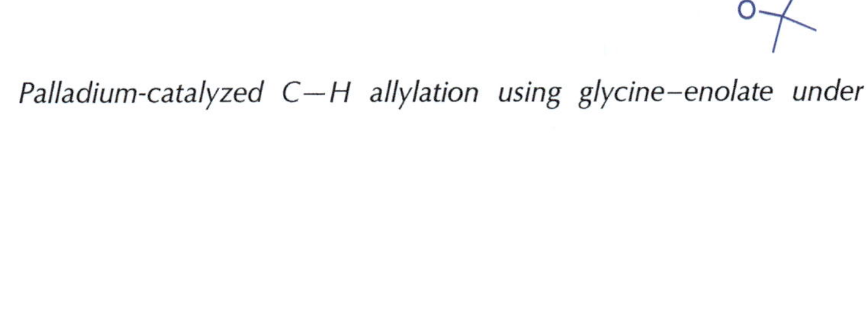

Figure 4.4 Palladium-catalyzed C—H allylation using glycine–enolate under palladium catalysis.

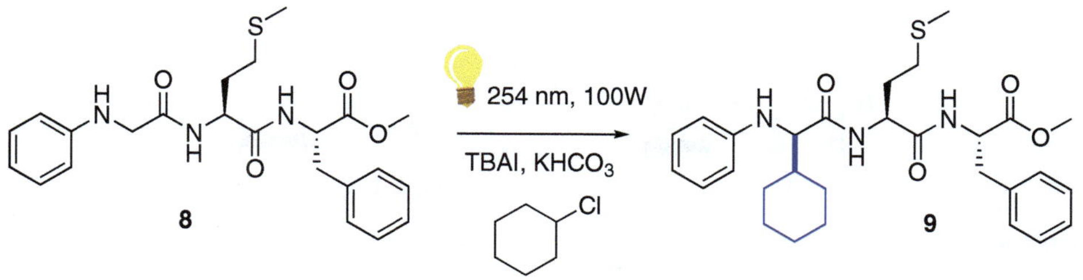

Figure 4.5 *Transition-metal-free alkylation of* N*-arylated glycine in [Leu5]-enkephalin analogues.*

More recently, a transition-metal-free alkylation of *N*-arylated glycine in peptides was demonstrated through the synthesis of enkephalin analogues (Figure 4.5) [32]. In this photo-induced approach, UV irradiation in the presence of tetrabutylammonium iodide (TBAI) enabled the functionalization of the α-carbon of glycine using alkyl chlorides or iodides. The scope of this approach is characterized by the wide occurrence of alkyl chlorides and iodides in the development of the corresponding α-alkyl glycines within peptide frameworks.

Another approach exploiting widely accessible substrates for the functionalization of glycine at the α-carbon is represented by the report by Li et al. [33], describing the use of aryl, alkenyl, or alkynyl boronic acids in a radical process promoted by *tert*-butyl hydroperoxide and copper catalysts, such as CuBr, allowing for maintaining the configuration of other stereocenters in the peptide.

A general limitation of these approaches is related to peptides possessing several similar chemical moieties, as differentiating among them is often difficult. In particular, for peptides containing multiple glycine units, the selective enolization of only one residue is problematic, thus requiring other approaches for mimicking the functionalization of the α-carbon of a glycine residue, for example using side chain of specific amino acids. An example is given by the cleavage of serine residues using lead tetraacetate to generate an active acetoxy glycine derivative to furnish unnatural α-alkylglycines in peptides [34].

4.3.2 Peptide Editing

A powerful strategy to modulate the chemical and biological properties of peptides is the introduction of heterocycles in the backbone, as described in [4]. For example, oxazoles and thiazoles are recurrent in natural peptides as amide surrogates that significantly affect the molecular structure, metabolic stability, and biological activity, especially in the context of antimicrobial or cytotoxic peptides [35–37]. Oxazole and oxazolidine rings typically arise from serine residues via intramolecular dehydrocyclization, while thiazole and thiazoline motifs generally originate from cysteine. Several methods have been successfully developed to replicate these transformations, enabling access to diverse heterocycles such as triazoles, oxazoles, thiazoles, and oxazolidines. Indeed, these heterocycles have been employed as chemical handles in pharmaceutical drugs [38], such as the inclusion of thiazole in ritonavir [39] and largazole [40].

Although modifications involving C—N bond formation are still poorly described, following initial studies on radical-transfer reactions [41], a transition-metal-free approach

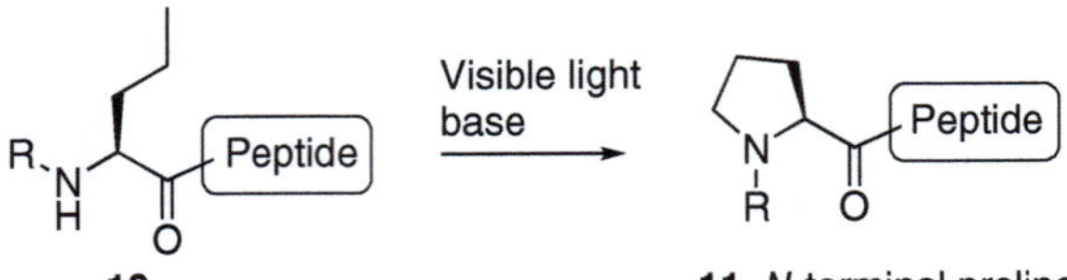

Figure 4.6 *Site-selective δ-C(sp^3)—H functionalization under visible light for the intramolecular cyclization to* N*-terminal proline.*

for late-stage peptide modification was reported [42], consisting of an *N*-terminal linear aliphatic side chain undergoing a site-selective δ-C(sp^3)—H functionalization under visible light, thus enabling the intramolecular cyclization to generate a proline skeleton (Figure 4.6). This protocol particularly highlighted the following advantages:

- metal-free conditions to promote skeletal editing through C—N bond formation;
- broad compatibility with diverse amino acid residues;
- absence of any requirement for directing groups to achieve selectivity or reactivity.

Although thiazole derivatives are capable of coordinating transition metals, their application as directing groups has been limited so far. A highly versatile strategy has been recently developed for late-stage palladium-catalyzed C(sp^3)—H arylation of peptides, exploiting thiazole units as internal directing groups [43]. This approach enables the regio- and site-selective arylation of β- and γ-C(sp^3)—H bonds in peptide side chains and displays high biocompatibility toward a range of aryl donors, biomolecules, and fluorophores, thus facilitating the synthesis of thiazole-containing peptide backbones. In the context of synthesizing peptides with heterocyclic motifs, a method was reported for the synthesis of a cyclic peptide Trunkamide A precursor exploiting the serine residue [44]. Specifically, the selective conversion of one amide bond to the corresponding thioamide in precursor **12** was achieved by treating the cyclopeptide with diethylaminosulfur trifluoride (DAST) followed by hydrogen sulfide, in a process consisting of cyclodehydration to the corresponding oxazoline and subsequent thiolysis (Figure 4.7). Then, the thioamide was subjected to intramolecular cyclization with an adjacent serine residue upon further treatment with DAST to the corresponding thiazoline derivative **13**. The same group employed the same approach using DAST for the total synthesis of thiazoline-containing cyclopeptides of biological interest [45]. With aim to expand such cyclization–dehydration method for the generation of thiazolines, the same group reported the treatment of linear peptide thioamide **14** with Burgess' reagent to form a substituted thiazoline moiety in the corresponding compound **15**, thus allowing to achieve the Trunkamide precursor **13** following cyclization steps (Figure 4.7).

Cyclization strategies to install heterocycles in the backbone are not limited to serine or cysteine derivatives. Accordingly, a method for introducing an oxazole ring at the *C*-terminal position of a dipeptide without relying on hydroxylated residues has been reported (Figure 4.8a) [46]. In this process, decarboxylation of dipeptide **16** with PhI(OAc)$_2$ and iodine yields an iminium ion, which isomerizes to an enamide before undergoing intramolecular cyclization with the adjacent amide carbonyl group and final aromatization to provide the corresponding oxazole in **17**. In another example (Figure 4.8b), dipeptide **18** consisting of *N*-terminal glutamic acid linked to leucine *via* the side-chain carboxylic

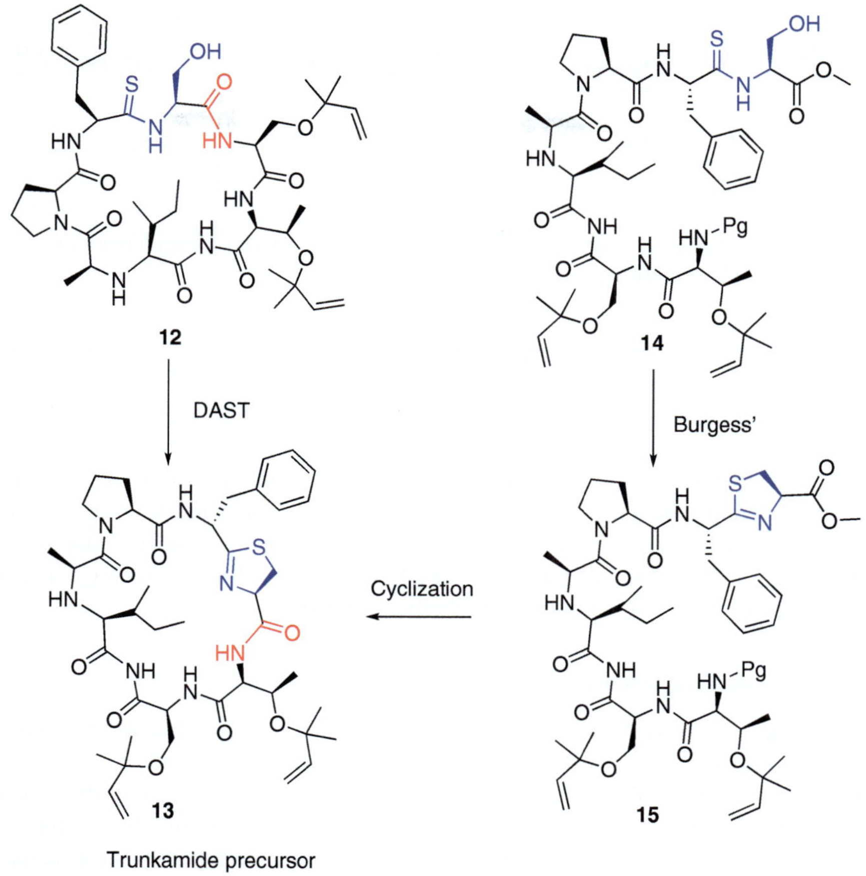

Figure 4.7 *Cyclization–dehydration method for the generation of thiazoline within the synthesis of a Trunkamide precursor.*

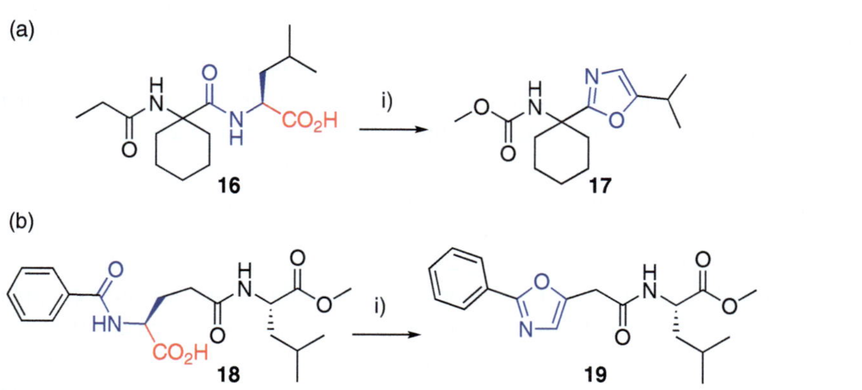

Figure 4.8 *Introduction of oxazole rings through decarboxylation; (a) functionalization at terminal C-position of a dipeptide; (b) functionalization at N-terminal amino acid of a dipeptide. Conditions: (i) PhI(OAc)$_2$, PhCO$_2$H, DCE, 2 hours, then I$_2$, 80 °C, 18 hours.*

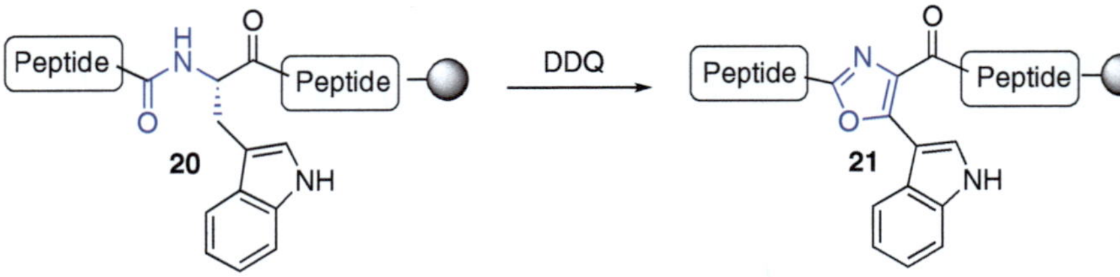

Figure 4.9 *Chemoselective modification of Trp-containing peptides through oxidative formation of oxazole in the backbone.*

group furnished the corresponding oxazole-derived peptidomimetic **19**, following a similar process starting from decarboxylation to cyclization.

As another entry to the selective insertion of oxazole in peptide backbones, tryptophan was reported as a versatile residue suitable for the modification of large peptides [47]. Specifically, a chemoselective method was reported for the insertion of oxazole in tryptophan-containing peptides using DDQ (2,3-dichloro-5,6-dicyano-1,4-benzoquinone) to oxidize the indole moiety and to successively promote cyclodehydration for the formation of an indolyl-oxazole moiety as in **21** (Figure 4.9). Such a moiety was found particularly interesting for the fluorescent labeling of peptides.

4.3.3 Backbone Functionalization at the Nitrogen Atom

The *N*-functionalization of backbone nitrogen atoms in peptides is significant in the context of structural properties [5]. In fact, the addition of substituents at amide nitrogen atoms contributes to increasing the steric hindrance at peptide bonds, perturbing the *cis/trans* amide equilibrium typically present in tertiary amide bonds and altering the spatial interactions between backbone *N*-substituents and side chains, consequently influencing the peptide conformation. Also, the modulation of amide nitrogen substitution is reflected in the capability to establish hydrogen-bonding and modulate the conformational preferences of biological relevance, stabilizing hydrogen-bonded structures by restricting backbone flexibility. The modulation of *N*-substitution may confer advantageous properties such as enhanced resistance to enzymatic hydrolysis, improved membrane permeability, and increased oral bioavailability. Indeed, the introduction and variation of *N*-substituents at backbone nitrogen atoms represent an important strategy to expand the diversity of peptide libraries, although selective modification of these substituents remains challenging [4]. Hydroxyproline has been proposed as a source of diversity through its insertion in the peptide, enabling subsequent scission and functionalization to afford *N*-alkylated peptides [48]. Specifically, the addition of different nucleophiles resulted in precursor **23** to achieve *N*-methyl peptidomimetic **24** *via* reduction, or other *N*-alkylated derivatives **25**–**27** through reactions with *C*-nucleophile. This methodology was further applied to the preparation of highly functionalized diketopiperazine scaffolds (Figure 4.10).

An overview of chemical methods for residue-directed *N*-functionalization has been reported by Baran [20], giving useful highlights in this context. Direct modification of peptide and protein backbones directed through side chain has been explored only sparingly, although an interesting report described a histidine-directed *N*-functionalization strategy [49], representing an elegant and chemoselective method for direct backbone

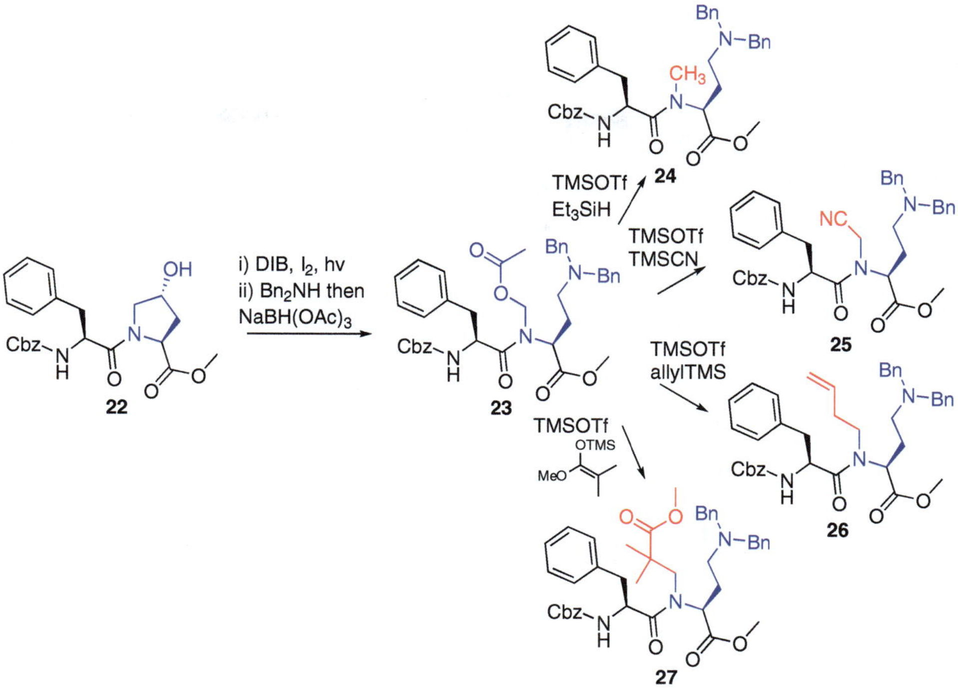

Figure 4.10 *Scission and functionalization of 4-hydroxyproline to afford* N*-alkylated peptides.*

editing. Using arylboronic or alkenylboronic acids, the nitrogen atom of the peptide bond undergoes selective C—N bond formation. It was demonstrated that the treatment of peptides with aryl- and alkenylboronic acids under copper catalysis enables the selective installation of *N*-substituents at the residue immediately preceding histidine (Figure 4.11a), resulting from the formation of a copper complex between the histidine N—H, its imidazole side chain, and the backbone N—H of the adjacent amino acid. Such an *N*-functionalization process displays broad substrate scope, as the boronic acid reagents generally perform efficiently with minimal sensitivity to electronic effects, as well as being widely available as chemical commodities. The utility of this method is highlighted in Figure 4.11b, where it is reported that neuromedin B undergoes selective alkenylation exclusively at the glycine preceding histidine to give peptide **31** [50].

4.4 Side-Chain Functionalization

In recent years, amino acid side chains have emerged as useful directing groups for promoting C(sp^3)—H functionalization within peptide frameworks. Early studies demonstrated that asparagine [51] and aspartic acid [52] residues can direct β-C(sp^3)—H arylation and alkynylation at *N*-terminal alanine. Subsequent advances expanded the chemical toolbox of directing residues, as methionine was reported as a suitable residue for directing the γ- and β-C(sp^3)—H arylation of *N*-terminal positions in both linear and cyclic peptides [53]. More

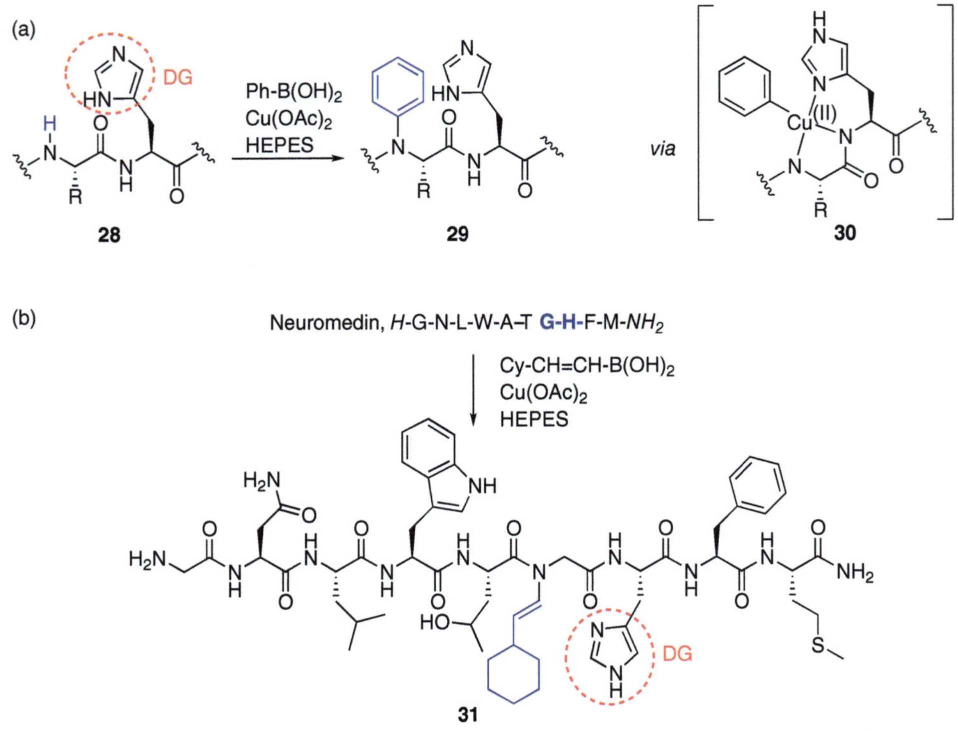

Figure 4.11 (a) Cu-catalyzed histidine-directed N-arylation using boronic acids; (b) application of the strategy to the His-directed N-alkenylation of neuromedin B.

recently, the application of histidine as an effective directing group was described, enabling the highly site-selective β-C(sp^3)—H arylation of alanine embedded within linear peptide sequences.

4.4.1 *N*-Terminal C(sp^3)—H Functionalization

The *N*-terminus of peptides represents a unique and accessible site for chemical modification, offering the advantage of single-site conjugation without significantly perturbing the three-dimensional structure or function of the peptide or protein. Concerning bioconjugation in proteins, unlike residue-selective approaches, *N*-terminal modification provides predictable conjugation and typically occurs at a region of the protein that is distant from the bioactive domain, minimizing potential effects on stability or activity. For these reasons, strategies targeting the *N*-terminal α-amino group have attracted considerable attention in recent years.

Several methods have been developed to selectively functionalize the *N*-terminus, and aldehydes, ketenes [17], or side chain of polar amino acids can be exploited to achieve selective conjugation [13, 15]. As a first example, the use of pyridine-2-carboxaldehyde-based reagents that form transient imines with the *N*-terminal amine was described, which subsequently cyclize to imidazolidinones (Figure 4.12a) [18]. Similarly, carboxylic

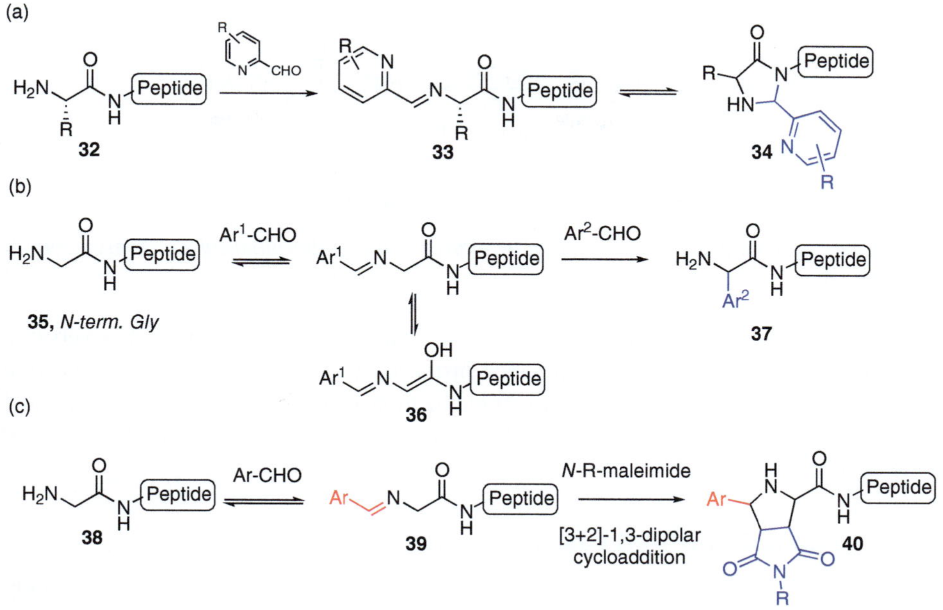

Figure 4.12 *Synthetic strategies for* N*-terminal side-chain modification.*

acid- or amide-linked aldehydes have been used to convert *N*-terminal glycine into amino alcohols via imine formation followed by aldol-type reactions, thus providing robust C—C bond formation and *N*-terminal glycine selectivity (Figure 4.12b) [54]. In a recent report, a three-component [3+2] cycloaddition strategy with aldehydes and maleimides under mild copper catalysis was demonstrated as an entry to a modular dual functionalization (Figure 4.12c) [55]. This method allows for the selective reaction at the amino group at *N*-terminal glycine to give iminopeptides in the presence of free lysine ε-amines and a broad range of other functional groups, thus representing a significant advancement toward flexible, site-selective *N*-terminal modifications suitable for incorporating multiple functional moieties.

Transition-metal-catalyzed approaches have also been successfully applied to *N*-terminal functionalization. Specifically, Pd-catalyzed C(sp^3)—H activation enables the selective arylation or alkynylation at the *N*-terminal residue, with the peptide backbone itself often serving as a bidentate directing group. In this context, Yu et al. reported the *N*-terminal-selective arylation of oligopeptides [56]. These methods demonstrate excellent diastereoselectivity, minimal epimerization, and tolerance of a range of *N*-terminal residues, although certain coordinating or reducible residues may be incompatible (Figure 4.13).

Additional strategies include the Pd-catalyzed intramolecular C—H amination for modifying *N*-terminal residues such as phenylalanine, and the use of *N*-protected amino acids within the peptide as bidentate directing groups for synthesizing functionalized dipeptides. The mechanism typically involves coordination of the Pd catalyst, C—H activation, oxidation by hypervalent iodine reagents, and selective C—N reductive elimination, thus regenerating the active catalyst in this catalytic cycle.

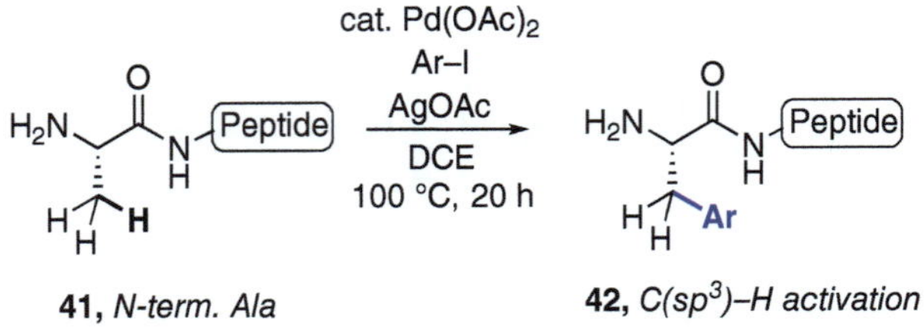

Figure 4.13 Pd-catalyzed C(sp^3)—H activation for the N-terminal-selective arylation of oligopeptides.

Beyond transition-metal catalysis, photochemical approaches have emerged as versatile tools for *N*-terminal functionalization. For instance, a visible light-promoted reaction was reported to achieve δ-C(sp^3)—H functionalization with high diastereoselectivity and broad substrate scope, further expanding the toolkit for late-stage peptide modification for biomedical applications [42].

4.4.2 *C*-Terminal C(sp^3)—H Functionalization

The direct activation of aliphatic C(sp^3)—H bonds remains one of the most conceptually and practically demanding challenges in modern synthesis. These bonds are ubiquitous and typically undifferentiated, making selective activation of a single site within a complex molecular framework inherently difficult. Although the installation of exogenous directing groups has provided a reliable means to impose regioselectivity, their required introduction and removal introduce undesired synthetic steps, especially in the context of modified α-amino acids and peptide substrates. Consequently, strategies that circumvent the use of external auxiliaries are of substantial synthetic value. Starting from their earlier studies on peptide arylation, Yu et al. extended the scope of backbone-directed C—H activation to achieve a site-selective alkynylation of dipeptides with free *C*-termini (Figure 4.14a) [57].

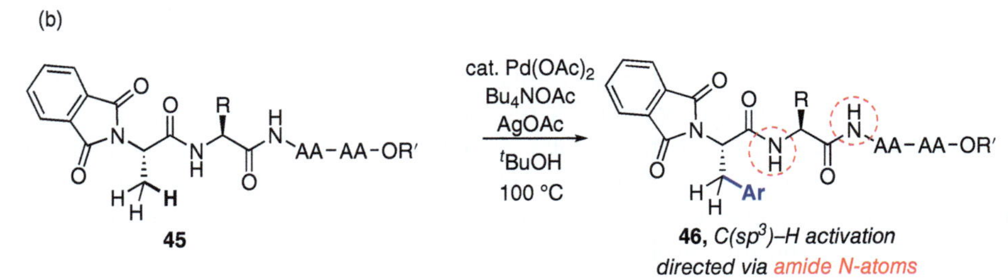

Figure 4.14 C-terminus- and amide-directed β-C(sp^3)—H functionalization: (a) alkynylation; (b) arylation.

This transformation proved well-suited for bioconjugation chemistry, thus illustrating the applicability of C(sp^3)—H functionalization strategies to the late-stage modification of biologically relevant peptide–drug constructs. Also, the same group reported that structural elements of α-amino acids embedded within peptides can function as internal coordination sites for palladium to promote β-C—H activation without any preinstalled directing group [56]. In this context, an auxiliary-free approach enabled the β-arylation of alanine derivatives (Figure 4.14b), thus providing a clear example of how backbone functionality can be leveraged to achieve selective C—H activation without the need for synthetic handles. Complementary to this strategy, Yu and collab. introduced a ligand-assisted approach conceived to overcome the intrinsically weak directing ability of the α-carboxyl group in amino acids. In these systems, an external ligand enhances metal coordination, thus improving both reactivity and positional selectivity in C—H arylation processes. This approach is particularly valuable for substrates with minimal innate directing capacity to extend the scope of C(sp^3)—H functionalization to a broader range of peptide and amino-acid-based scaffolds.

4.4.3 Internal Functionalization of Peptides

Internal functionalization is particularly valuable, as it enables the modification of peptides at positions inaccessible through *N*- or *C*-terminal derivatization. Two main classes of transformations have emerged, namely site-selective *C(sp^2)—H functionalization* of aromatic residues and *C(sp^3)—H functionalization* of unactivated aliphatic residues.

4.4.3.1 Site-Selective Late-Stage Peptide Modification via *C(sp^2)—H Functionalization*

Among the proteinogenic amino acids, tryptophan, tyrosine, phenylalanine, and histidine provide innate aryl units with distinct reactivity profiles [3]. Tryptophan and tyrosine possess electron-rich aromatic systems, whereas histidine contains an electron-poor imidazole ring, and phenylalanine features an electronically neutral phenyl ring. Such intrinsic differences facilitate the feasibility and selectivity of C(sp^2)—H activation on peptides.

Tryptophan functionalization. The C2—H bond of tryptophan is the most acidic C(sp^2)—H position and constitutes a privileged site for activation. Accordingly, the late-stage C(sp^2)—H functionalization of tryptophan residues has been extensively studied, exploiting several transition-metal catalysts. Early methods exploited the electrophilic aromatic substitution, although poor selectivity toward other nucleophilic residues limited their utility. Radical-mediated pathways enabled C2—H alkylation with electrophilic radicals. The first metal-catalyzed C—H functionalization in proteins employed Rh-catalyzed carbenoid insertion [58]. Nevertheless, palladium catalysis has played a dominant role in developing general arylation protocols, and the first reported Pd-catalyzed undirected C2—H arylation of tryptophan in short peptides demonstrated the compatibility of the reactions with numerous unprotected amino acids, although requiring high temperatures [59]. Successively, milder Cu/Pd co-catalyzed arylations with boronic acids were described in the literature [60], followed by studies on diaryliodonium-based methodologies, which proceed under mild conditions and in aqueous media [61, 62], and arylation methods with aryldiazonium salts, that has been extended to protein conjugation in aqueous solution [63]. Further refinement of Pd catalysis was achieved by Albericio et al. [59], who reported a racemization-free C2—H arylation of tryptophan in both free amino acids and short

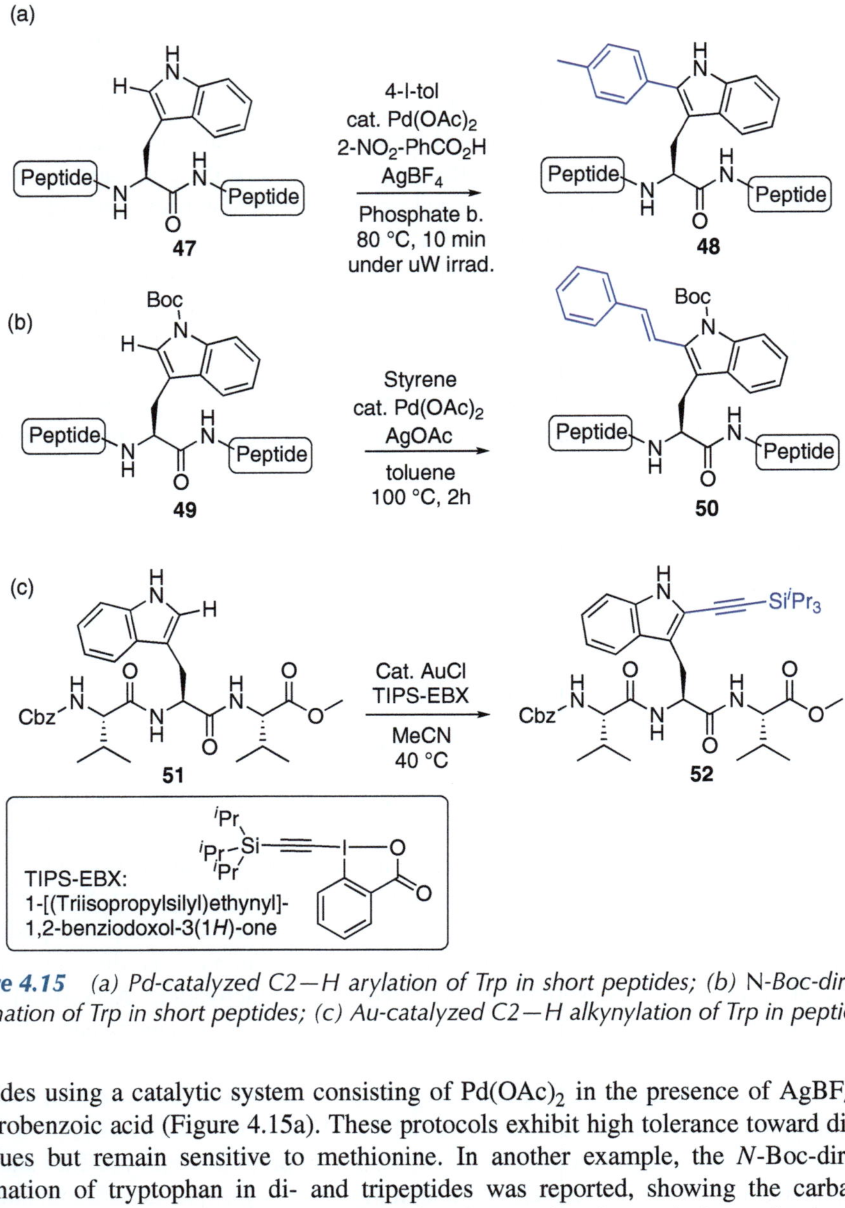

Figure 4.15 *(a) Pd-catalyzed C2—H arylation of Trp in short peptides; (b) N-Boc-directed olefination of Trp in short peptides; (c) Au-catalyzed C2—H alkynylation of Trp in peptides.*

peptides using a catalytic system consisting of $Pd(OAc)_2$ in the presence of $AgBF_4$ and 2-nitrobenzoic acid (Figure 4.15a). These protocols exhibit high tolerance toward diverse residues but remain sensitive to methionine. In another example, the *N*-Boc-directed olefination of tryptophan in di- and tripeptides was reported, showing the carbamate acting as a directing group (Figure 4.15b) [64]. Concerning the catalytic mechanism, two pathways are associated with Pd-catalyzed $C(sp^2)$—H arylation of peptides [65]:

- Pd(II)/Pd(IV) pathway, where the C—H palladation is followed by the oxidative addition of aryl iodides or diaryliodonium salts to form Pd(IV) intermediates that reductively eliminate arylated peptides.
- Transmetalation pathway, that is effective with arylboronic acids, in which a Pd(II) intermediate undergoes transmetalation to form an aryl–Pd(II) species followed by reductive elimination.

Besides palladium, a broad collection of other metals, namely Cu, Ru, Co, Au, Zn, and Mn, is now known to mediate direct C2 functionalization of tryptophan residues, often without the need for directing groups. For example, gold catalysis also enables the efficient C2—H alkynylation in short peptides, as reported by Hansen [66] and Waser [67] (Figure 4.15c).

Tyrosine functionalization. The phenolic side chain of tyrosine exists in equilibrium between phenol and phenolate, enabling classical electrophilic aromatic substitution, such as halogenation, Mannich reactions, and diazonium coupling [68], as well as allowing for radical-based C3 functionalization *via* tyrosyl radicals. Despite its inherent reactivity, metal-catalyzed directed or undirected C—H activation of tyrosine remains limited, highlighting the challenge of achieving selective arylation at this residue.

Phenylalanine functionalization. The lack of innate coordinating ability or suitable reactivity handles makes phenylalanine difficult to functionalize through C—H activation. Few examples have been reported, including Pd-catalyzed diolefination of Phe-containing peptides, and oxazole-directed δ-C(sp^2)—H olefination, where an internal oxazole installed on the peptide backbone acts as a directing group.

Histidine functionalization. Late-stage modification of histidine is rare due to the electron-deficient nature of the imidazole ring. As a significant entry in this context, the C2 alkylation and fluoroalkylation of histidine using sulfinate salts under radical conditions was achieved with high selectivity over tyrosine and tryptophan (Figure 4.16) [69].

4.4.3.2 *Site-Selective Late-Stage Peptide Modification* via *C(sp^3)—H Functionalization*

Unactivated C(sp^3)—H bonds, especially the methyl groups of alanine, valine, isoleucine, and leucine, exhibit minimal innate reactivity [3], although high site selectivity is difficult to achieve because peptides contain many electronically and sterically similar aliphatic positions. Directed C—H activation using bidentate directing groups represents a powerful strategy, whereas undirected hydrogen-atom-transfer approaches enable the selective targeting of tertiary C—H sites, particularly in valine, isoleucine, and leucine side chain. Proline, bearing relatively activated δ-C—H bonds adjacent to nitrogen, constitutes a special case of increased intrinsic reactivity.

Triazole as a directing group. Triazoles act as amide isosteres with strong *N,N*-chelating capability. In this context, being available by copper-catalyzed dipolar cycloaddition, triazoles have been applied as amide isosteres in bioactive peptidomimetics due to similarities

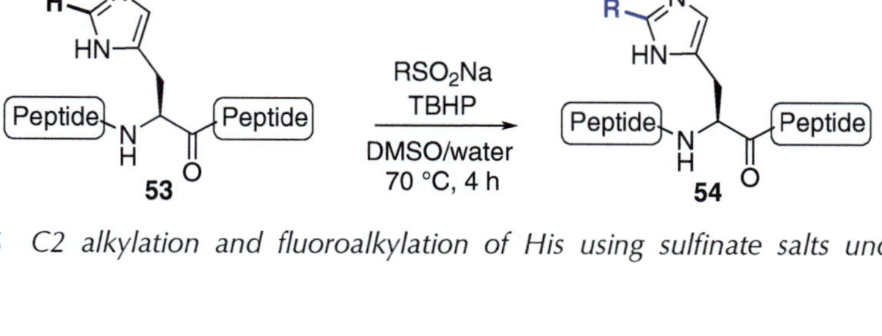

Figure 4.16 *C2 alkylation and fluoroalkylation of His using sulfinate salts under radical conditions.*

in the electronic and topological properties [70]. Ackermann and collab. pioneered the use of triazoles for the C(sp^3)—H arylation of amino acids and peptides [71], displaying the following features:

- *C*-terminal, *N*-terminal, and internal β-C(sp^3)—H functionalization;
- high diastereoselectivity;
- chemoselective monoarylation of peptides containing multiple alanine residues;
- chemical ligation to construct higher-order peptides.

In particular, Ackermann first demonstrated the Tzl-directed β-C(sp^3)—H arylation of alanine, with the directing group installed on the carbonyl group (Figure 4.17) and expanding this strategy to C—H alkenylation with 2-iodoglycals at alanine residues [72].

Endogenous directing groups. More recently, the exploitation of innate amino acid side chains has provided auxiliary-free strategies. Asparagine was shown to function as an internal bidentate directing group [51], enabling the β-C(sp^3)—H arylation through the formation of a 5,6-fused palladacycle. Comparable reactivity has been demonstrated for aspartic acid [52] and methionine [53, 73], which can promote C(sp^3)—H activation through endogenous coordination motifs (Figure 4.18).

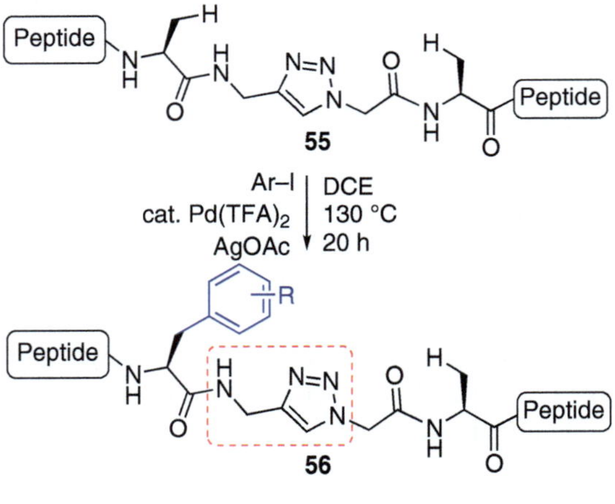

Figure 4.17 Triazole-directed β-C(sp^3)—H arylation of alanine.

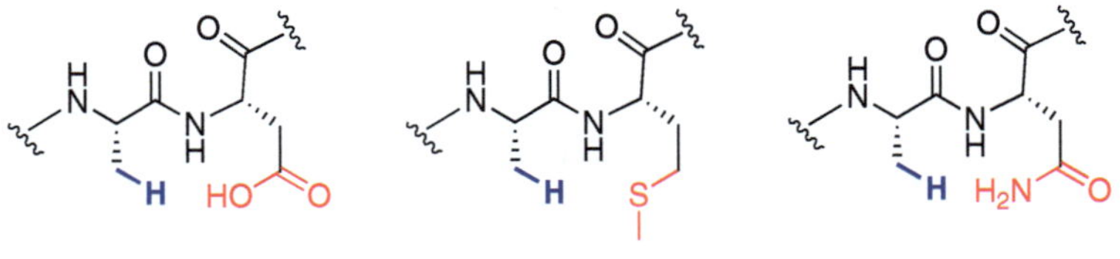

Figure 4.18 β-C(sp^3)—H functionalization of alanine directed by adjacent aspartic acid (Asp), methionine (Met), and asparagine (Asn) side chains [20].

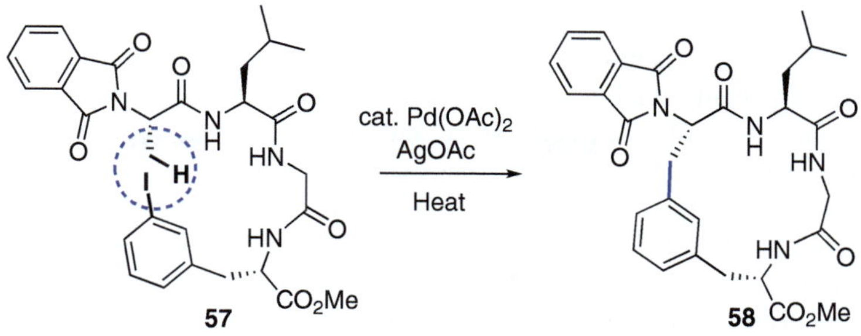

Figure 4.19 Pd-catalyzed C(sp^3)—H intramolecular arylation.

Sequential C—H *arylation.* The synthetic potential of these methods is exemplified by two-step arylation sequences [42]. Using thiazole-enabled activation, alanine residues can be arylated first, followed by peptide elongation with valine and subsequent selective γ-C—H arylation of such amino acid, yielding doubly arylated products with high regioselectivity.

Macrocyclization. Yu demonstrated that the native peptide backbone can direct the β-C(sp^3)—H arylation of *N*-terminal residues. Subsequent reports showed that intramolecular C(sp^3)—H arylation can induce peptide macrocyclization (Figure 4.19) [74, 75].

4.5 Conclusion

Significant advances have been made in the field of site-selective peptide modification over the last decade. Although most progress consists of side-chain manipulation, a growing number of methodologies now are centered toward peptide backbone to modulate conformation, folding, physicochemical behavior, and biological activity. The increasing synthetic toolbox for such transformations is represented by the synthetic approaches enabling the selective installation of substituents at α-C or *N* positions, ranging from the use of "customizable" amino acids to residue-directed transformations. Recent progress with triazole directing groups, Asn/Asp/Met endogenous coordination, and sulfur-directed glycosylation is beginning to expand the scope toward internal residues and more diverse transformations. The cleavage and functionalization of C—C and C—H bonds allow for the efficient diversification at the α-carbon, with glycine serving as a versatile platform for introducing alkyl, allyl, or aryl substituents, in some cases with excellent stereocontrol. The modification of the peptide bond, including thioamide substitution and heterocycle formation, allows for tuning the backbone rigidity, the hydrogen-bonding capacity, and enzymatic recognition. These developments apply not only to internal residues but also to *N*- and *C*-terminal positions, thus broadening the scope of backbone engineering.

Despite major progress in Pd-catalyzed C(sp^3)—H functionalization of amino acids, the application of this chemistry to site-selective modification of peptides remains still narrow, and it is limited to *N*-terminal alanine residues, short peptide substrates, and peptidic systems requiring preinstalled directing groups. A major obstacle to expand the functionalization to more amino acids is the dependence on preinstalled directing groups, thus reducing the overall atom- and step-economy of peptide modification due to the need for their installation and subsequent removal. Although directing-group-free strategies have emerged, especially for free amine- and carboxylate-directed $C(sp^3)$—H activation, these approaches are still at an early stage and still experience regioselectivity issues, epimerization, or modest yields. Approaches exploiting native functionalities, such as the asparagine side-chain amide, enabled the functionalization of peptides through internal residue-directed C—H activation but remain limited by the intrinsic coordinating ability of proteinogenic side chains. Consequently, while C—H functionalization at *N*-terminal residues is relatively well advanced, the selective activation at *C*-terminal residues is underdeveloped, as well as methods enabling site-selective $C(sp^3)$—H activation at internal peptide positions. Backbone-directed activation is actually restricted to short peptides containing simple residues and must overcome the competition from multiple backbone amide ligands.

Taken together, these advances emphasize that selective modification of the peptide backbone is a rapidly evolving area with significant implications for controlling peptide structure and function.

References

1. Wang W, Lorion MM, Shah J, Kapdi AR, Ackermann L. Late-stage peptide diversification by position-selective C—H activation. *Angew Chem Int Ed Engl*. 2018;**57**: 14700–14717.
2. Shabani S, Wu Y, Ryan HG, Hutton CA. Progress and perspectives on directing group-assisted palladium-catalysed C—H functionalisation of amino acids and peptides. *Chem Soc Rev*. 2021;**50**:9278–9343.
3. Tong HR, Li B, Li G, He G, Chen G. Postassembly modifications of peptides via metal-catalyzed C—H functionalization. *CCS Chem*. 2021;**3**:1797–1820.
4. Boto A, González CC, Hernández D, Romero-Estudillo I, Saavedra CJ. Site-selective modification of peptide backbones. *Org Chem Front*. 2021;**8**:6720–6759.
5. Shi J, Suna T, Yang M. Site-selective editing of peptides via backbone modification. *Org Chem Front*. 2024;**11**:1623–1640.
6. Nagano N, Umemura M, Izumikawa M, Kawano J, Ishii T, Kikuchi M, Tomii K, Kumagai T, Yoshimi A, Machida M, Abe K, Shin-Ya K, Asai K. Class of cyclic ribosomal peptide synthetic genes in filamentous fungi. *Fungal Genet Biol*. 2016;**86**:58–70.
7. Schramma KR, Bushin LB, Seyedsayamdost MR. Structure and biosynthesis of a macrocyclic peptide containing an unprecedented lysine-to-tryptophan crosslink. *Nat Chem*. 2015;**7**:431–437.
8. Imai Y, Meyer KJ, Iinishi A, Favre-Godal Q, Green R, Manuse S, Caboni M, Mori M, Niles S, Ghiglieri M, Honrao C, Ma X, Guo JJ, Makriyannis A, Linares-Otoya L, Böhringer N, Wuisan ZG, Kaur H, Wu R, Mateus A, Typas A, Savitski MM, Espinoza JL, O'Rourke A, Nelson KE, Hiller S, Noinaj N, Schäberle TF, D'Onofrio A, Lewis K. A new antibiotic selectively kills Gram-negative pathogens. *Nature*. 2019;**576**:459–464.
9. Kobayashi J, Suzuki H, Shimbo K, Takeya K, Morita H. Celogentins A–C, new antimitotic bicyclic peptides from the seeds of Celosia argentea. *J Org Chem*. 2001;**66**(20): 6626–6633.

10. Rosen CB, Francis MB. Targeting the N terminus for site-selective protein modification. *Nat Chem Biol.* 2017;**13**:697–705.
11. Obermeyer AC, Jarman JB, Francis MB. N-terminal modification of proteins with o-aminophenols. *J Am Chem Soc.* 2014;**136**:9572–9579.
12. Agarwal P, van der Weijden J, Sletten EM, Rabuka D, Bertozzi CR. A Pictet–Spengler ligation for protein chemical modification. *Proc Natl Acad Sci USA.* 2013;**110**:46–51.
13. Li X, Zhang L, Hall SE, Tam JP. A new ligation method for N-terminal tryptophan-containing peptides using the Pictet–Spengler reaction. *Tetrahedron Lett.* 2000;**41**:4069–4073.
14. Dawson PE, Muir TW, Clark-Lewis I, Kent SB. Synthesis of proteins by native chemical ligation. *Science.* 1994;**266**:776–779.
15. Geoghegan KF, Stroh JG. Site-directed conjugation of nonpeptide groups to peptides and proteins via periodate oxidation of a 2-amino alcohol. Application to modification at N-terminal serine. *Bioconjug Chem.* 1992;**3**:138–146.
16. Schoffelen S, van Eldijk MB, Rooijakkers B, Raijmakers R, Heckc AJR, van Hest JCM. Metal-free and pH-controlled introduction of azides in proteins. *Chem Sci.* 2011;**2**:701–705.
17. Chan AO, Ho CM, Chong HC, Leung YC, Huang JS, Wong MK, Che CM. Modification of N-terminal α-amino groups of peptides and proteins using ketenes. *J Am Chem Soc.* 2012;**134**:2589–2598.
18. MacDonald JI, Munch HK, Moore T, Francis MB. One-step site-specific modification of native proteins with 2-pyridinecarboxyaldehydes. *Nat Chem Biol.* 2015;**11**: 326–331.
19. Chen D, Disotuar MM, Xiong X, Wang Y, Chou DH. Selective N-terminal functionalization of native peptides and proteins. *Chem Sci.* 2017;**8**:2717–2722.
20. deGruyter JN, Malins LR, Baran PS. Residue-specific peptide modification: a chemist's guide. *Biochemistry.* 2017;**56**:3863–3873.
21. Lewis JC, Coelho PS, Arnold FH. Enzymatic functionalization of carbon-hydrogen bonds. *Chem Soc Rev.* 2011;**40**:2003–2021.
22. Lu X, He SJ, Cheng WM, Shi J. Transition-metal-catalyzed C—H functionalization for late-stage modification of peptides and proteins. *Chin Chem Lett.* 2018;**29**:1001–1008.
23. Wu Y, Zhu B, Fan H, Bernard H, Hutton CA. Late-stage Pd(II)-catalyzed C(sp^3)—H functionalization of peptides directed by a removable, backbone-inserted amidoxime ether. *Angew Chem Int Ed Engl.* 2025;**64**:e202423979.
24. Zhan BB, Jiang MX, Shi BF. Late-stage functionalization of peptides via a palladium-catalyzed C(sp^3)—H activation strategy. *Chem Commun.* 2020;**56**: 13950–13958.
25. Xie Z, Liu X, Liu L. Copper-catalyzed aerobic enantioselective cross-dehydrogenative coupling of N-aryl glycine esters with terminal alkynes. *Org Lett.* 2016;**18**:2982–2985.
26. King TA, Mandrup Kandemir J, Walsh SJ, Spring DR. Photocatalytic methods for amino acid modification. *Chem Soc Rev.* 2021;**50**:39–57.
27. Hernández D, Carro C, Boto A. "Doubly Customizable" unit for the generation of structural diversity: from pure enantiomeric amines to peptide derivatives. *J Org Chem.* 2021;**86**:2796–2809.

28. Osberger TJ, Rogness DC, Kohrt JT, Stepan AF, White MC. Oxidative diversification of amino acids and peptides by small-molecule iron catalysis. *Nature*. 2016;**537**: 214–219.
29. Seebach D, Bech AK, Studer A. In: Ernst B, Leumann C, editors. *Modern Synthetic Methods*, vol. **7**. Weinheim: VCH, 1995.
30. Bossler HG, Waldmeier P, Seebach D. Preparation and C-alkylation of peptides with aminomalonate building blocks. *Angew Chem Int Ed Engl*. 1994;**33**:439–440.
31. Deska J, Kazmaier U. Highly stereoselective peptide modifications through Pd-catalyzed allylic alkylations of chelated peptide enolates. *Chemistry*. 2007;**13**: 6204–6211.
32. Xue H, Guo M, Wang C, Shen Y, Qi R, Wu Y, Xu Z, Chang M. Photo-induced preparation of unnatural α-amino acids: synthesis and characterization of novel Leu5-enkephalin analogues. *Org Chem Front*. 2020;**7**:2426–2431.
33. Zhao L, Baslé O, Li CJ. Site-specific C-functionalization of free-(NH) peptides and glycine derivatives via direct C—H bond functionalization. *Proc Natl Acad Sci USA*. 2009;**106**:4106–4111.
34. Ricci M, Blakskjær P, Skrydstrup T. Selective side chain introduction onto small peptides mediated by samarium diiodide: a potential route to peptide libraries. *J Am Chem Soc*. 2000;**122**:12413–12421.
35. Mhlongo JT, Brasil E, de la Torre BG, Albericio F. Naturally occurring oxazole-ontaining peptides. *Mar Drugs*. 2020;**18**:203.
36. Dahiya R, Dahiya S, Fuloria NK, Kumar S, Mourya R, Chennupati SV, Jankie S, Gautam H, Singh S, Karan SK, Maharaj S, Fuloria S, Shrivastava J, Agarwal A, Singh S, Kishor A, Jadon G, Sharma A. Natural bioactive thiazole-based peptides from marine resources: structural and pharmacological aspects. *Mar Drugs*. 2020;**18**:329.
37. Roy RS, Gehring AM, Milne JC, Belshaw PJ, Walsh CT. Thiazole and oxazole peptides: biosynthesis and molecular machinery. *Nat Prod Rep*. 1999;**16**:249–263.
38. Smolyar IV, Yudin AK, Nenajdenko VG. Heteroaryl rings in peptide macrocycles. *Chem Rev*. 2019;**119**:10032–10240.
39. Hsu A, Granneman GR, Bertz RJ. Ritonavir. Clinical pharmacokinetics and interactions with other anti-HIV agents. *Clin Pharmacokinet*. 1998;**35**:275–291.
40. Hong J, Luesch H. Largazole: from discovery to broad-spectrum therapy. *Nat Prod Rep*. 2012;**29**:449–456.
41. Meng D, Tang Y, Wei J, Shi X, Yang M. Copper-catalyzed remote (δ) C(sp^3)—H bond amination: a practical strategy to construct pyrrolidine derivatives. *Chem Commun*. 2017;**53**:5744–5747.
42. Han Y, Shi J, Li S, Dan T, Yang W, Yang M. Selective editing of a peptide skeleton via C—N bond formation at N-terminal aliphatic side chains. *Chem Sci*. 2022;**13**: 14382–14386.
43. Bai Z, Chen Q, Gu J, Cai C, Zheng J, Sheng W, Yi S, Liu F, Wang H. Late-stage functionalization and diversification of peptides by internal thiazole-enabled palladium-catalyzed C(sp^3)—H arylation. *ACS Catal*. 2021;**11**:15125–15134.
44. McKeever B, Pattenden G. Total synthesis of the prenylated cyclopeptide trunkamide A, a cytotoxic metabolite from Lissoclinum sp. *Tetrahedron Lett*. 2001;**42**:2573–2577.

45. Deeley J, Bertram A, Pattenden G. Novel polyoxazole-based cyclopeptides from *Streptomyces* sp. Total synthesis of the cyclopeptide YM-216391 and synthetic studies towards telomestatin. *Org Biomol Chem.* 2008;**6**:1994–2010.
46. Romero-Estudillo I, Batchu VR, Boto A. One-Pot conversion of amino acids into 2,5-disubstituted oxazoles: no metals needed. *Adv Synth Catal.* 2014;**356**:3742–3748.
47. Petersen J, Christensen KE, Nielsen MT, Mortensen KT, Komnatnyy VV, Nielsen TE, Qvortrup K. Oxidative modification of tryptophan-containing peptides. *ACS Comb Sci.* 2018;**20**:344–349.
48. Saavedra CJ, Carro C, Hernández D, Boto A. Conversion of "Customizable Units" into *N*-alkyl amino acids and generation of *N*-alkyl peptides. *J Org Chem.* 2019;**84**:8392–8410.
49. Mangubat-Medina AE, Martin SC, Hanaya K, Ball ZT. A vinylogous photocleavage strategy allows direct photocaging of backbone amide structure. *J Am Chem Soc.* 2018;**140**:8401–8404.
50. Ball ZT. Protein substrates for reaction discovery: site-selective modification with boronic acid reagents. *Acc Chem Res.* 2019;**52**:566–575.
51. Weng Y, Ding X, Oliveira JCA, Xu X, Kaplaneris N, Zhu M, Chen H, Chen Z, Ackermann L. Peptide late-stage $C(sp^3)$—H arylation by native asparagine assistance without exogenous directing groups. *Chem Sci.* 2020;**11**:9290–9295.
52. Chen Z, Zhu M, Cai M, Xu L, Weng Y. Palladium-catalyzed $C(sp^3)$—H arylation and alkynylation of peptides directed by aspartic acid (Asp). *ACS Catal.* 2021;**11**:7401–7410.
53. Chen X, Li B, Tong H, Qi L, He G, Chen G. Palladium-catalyzed methionine-facilitated β and γ $C(sp^3)$—H arylation of N-Terminal aliphatic amino acids of peptides. *Chin J Chem.* 2022;**40**:2502–2506.
54. Purushottam L, Adusumalli SR, Singh U, Unnikrishnan VB, Rawale DG, Gujrati M, Mishra RK, Rai V. Single-site glycine-specific labeling of proteins. *Nat Commun.* 2019;**10**:2539.
55. Machida H, Kanemoto K. N-terminal-specific dual modification of peptides through copper-catalyzed [3+2] cycloaddition. *Angew Chem Int Ed Engl.* 2024;**63**: e202320012.
56. Gong W, Zhang G, Liu T, Giri R, Yu JQ. Site-selective $C(sp^3)$—H functionalization of di-, tri-, and tetrapeptides at the N-terminus. *J Am Chem Soc.* 2014;**136**:16940–16946.
57. Liu T, Qiao JX, Poss MA, Yu JQ. Palladium(II)-catalyzed site-selective $C(sp^3)$—H alkynylation of oligopeptides: a linchpin approach for oligopeptide-drug conjugation. *Angew Chem Int Ed Engl.* 2017;**56**:10924–10927.
58. Antos JM, Francis MB. Selective tryptophan modification with rhodium carbenoids in aqueous solution. *J Am Chem Soc.* 2004;**126**:10256–10257.
59. Ruiz-Rodríguez J, Albericio F, Lavilla R. Postsynthetic modification of peptides: chemoselective C-arylation of tryptophan residues. *Chemistry.* 2010;**16**:1124–1127.
60. Williams TJ, Reay AJ, Whitwood AC, Fairlamb IJ. A mild and selective Pd-mediated methodology for the synthesis of highly fluorescent 2-arylated tryptophans and tryptophan-containing peptides: a catalytic role for Pd(0) nanoparticles? *Chem Commun.* 2014;**50**:3052–3054.

61. Reay AJ, Williams TJ, Fairlamb IJ. Unified mild reaction conditions for C2-selective Pd-catalysed tryptophan arylation, including tryptophan-containing peptides. *Org Biomol Chem*. 2015;**13**:8298–8309.
62. Zhu Y, Bauer M, Ackermann L. Late-stage peptide diversification by bioorthogonal catalytic C—H arylation at 23 °C in H_2O. *Chemistry*. 2015;**21**:9980–9983.
63. Reay AJ, Hammarback LA, Bray JTW, Sheridan T, Turnbull D, Whitwood AC, Fairlamb IJS. Mild and regioselective Pd(OAc)2-catalyzed C—H arylation of tryptophans by [ArN2]X, promoted by tosic acid. *ACS Catal*. 2017;**7**:5174–5179.
64. Terrey MJ, Holmes A, Perry CC, Cross WB. C—H olefination of tryptophan residues in peptides: control of residue selectivity and peptide-amino acid cross-linking. *Org Lett*. 2019;**21**:7902–7907.
65. Sandtorv AH. Transition metal-catalyzed C—H activation of indoles. *Adv Synth Catal*. 2015;**357**:2403–2435.
66. Hansen MB, Hubálek F, Skrydstrup T, Hoeg-Jensen T. Chemo- and regioselective ethynylation of tryptophan-containing peptides and proteins. *Chemistry*. 2016;**22**: 1572–1576.
67. Tolnai GL, Brand JP, Waser J. Gold-catalyzed direct alkynylation of tryptophan in peptides using TIPS-EBX. *Beilstein J Org Chem*. 2016;**12**:745–749.
68. Jones MW, Mantovani G, Blindauer CA, Ryan SM, Wang X, Brayden DJ, Haddleton DM. Direct peptide bioconjugation/PEGylation at tyrosine with linear and branched polymeric diazonium salts. *J Am Chem Soc*. 2012;**134**:7406–7413.
69. Noisier AFM, Johansson MJ, Knerr L, Hayes MA, Drury WJ 3rd, Valeur E, Malins LR, Gopalakrishnan R. Late-stage functionalization of histidine in unprotected peptides. *Angew Chem Int Ed Engl*. 2019;**58**:19096–19102.
70. Pedersen DS, Abell A. 1,2,3-Triazoles in peptidomimetic chemistry. *Eur J Org Chem*. 2011;2399–2411.
71. Bauer M, Wang W, Lorion MM, Dong C, Ackermann L. Internal peptide late-stage diversification: peptide-isosteric triazoles for primary and secondary $C(sp^3)$—H activation. *Angew Chem Int Ed Engl*. 2018;**57**:203–207.
72. Wu J, Kaplaneris N, Ni S, Kaltenhäuser F, Ackermann L. Late-stage $C(sp^2)$—H and $C(sp^3)$—H glycosylation of C-aryl/alkyl glycopeptides: mechanistic insights and fluorescence labeling. *Chem Sci*. 2020;**11**:6521–6526.
73. Li G, Yuan F, Yao B. Post-assembly modification of head-to-tail cyclic peptides by methionine-directed β-$C(sp^3)$—H arylation. *Org Lett*. 2022;**24**:5767–5771.
74. Noisier AF, García J, Ionuţ IA, Albericio F. Stapled peptides by late-stage $C(sp^3)$—H activation. *Angew Chem Int Ed Engl*. 2017;**56**:314–318.
75. Tang J, He Y, Chen H, Sheng W, Wang H. Synthesis of bioactive and stabilized cyclic peptides by macrocyclization using $C(sp^3)$—H activation. *Chem Sci*. 2017;**8**: 4565–4570.

Part II

Peptidomimetic Compounds

5

Macrocyclic Peptides and Peptidomimetics

5.1 Introduction

Macrocycles play a key role in modulating protein–protein interactions (PPIs) and are considered optimal molecular frameworks due to their balanced flexibility, as represented by numerous bioactive natural products possessing cyclic or polycyclic molecular scaffolds. Their constrained architectures, stabilized by intramolecular hydrogen-bonding, enable the adoption of defined conformations such as helices and turns; thus, macrocycles can selectively interact with large protein surfaces with high affinity and specificity [1]. In this context, cyclic peptides have emerged as a relevant class of molecules in modern drug discovery due to their structural and pharmacological advantages over linear peptides. Depending on their size, peptides possess large and chemically diverse molecular surfaces that enable them to mimic specific fragments of protein structures, thus being capable of modulating protein function and disrupting PPIs, which are frequently inaccessible to conventional small-molecule therapeutics [2]. The introduction of a covalent ring confers conformational rigidity, which enhances metabolic stability and reduces liability to enzymatic degradation, thus improving the *in vivo* half-life. Also, the cyclic structure reduces the entropic cost associated with target binding, often resulting in improved affinity and selectivity toward biological receptors. Cyclic peptides are widespread among naturally occurring compounds and display a wide array of biological activities, including antimicrobial, antiviral, anti-inflammatory, antitumor, and immunosuppressive effects [3]. Clinically important examples such as insulin, oxytocin, cyclosporin A, daptomycin, and gramicidin S highlight the therapeutic value of this class of molecules (Figure 5.1).

Inspired by naturally occurring macrocyclic peptides, synthetic strategies have been developed to generate novel compounds with specific properties, and a large number of macrocycles have been developed accordingly, using different types of amino acids and applying cyclization methods, such as head-to-tail linkage or side-chain cross-linking, to generate macrocyclic peptidomimetics able to improve proteolytic resistance, membrane permeability, and target engagement. A particular category of macrocyclic peptidomimetic is represented by molecules designed to stabilize peptide secondary structures, such as

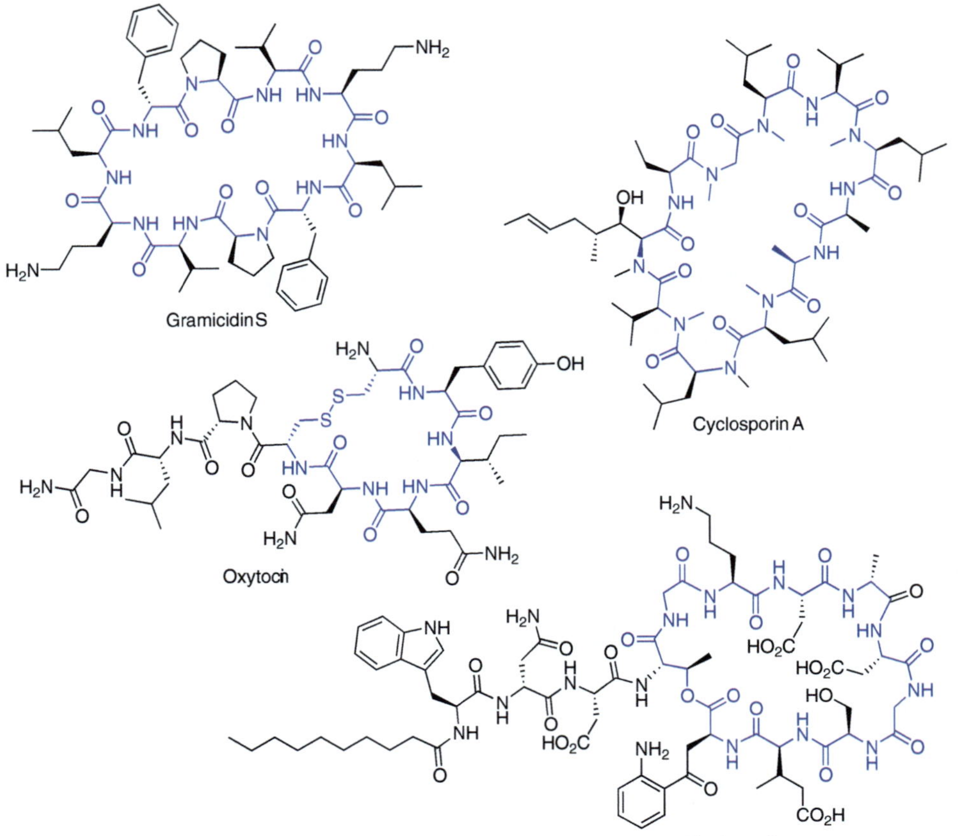

Figure 5.1 *Representative bioactive natural peptide macrocycles (the macrocyclic backbone is highlighted in blue).*

α-helices, β-sheets, and β-turns, that are of special interest for modulating PPIs. When peptides adopt defined secondary structures, the use of peptidomimetic approaches that stabilize such folded conformations can preserve or enhance the biological activity. Consequently, numerous design strategies and peptidomimetic frameworks have been developed over the years with the goal of selectively stabilizing specific peptide secondary structural elements. One approach is represented by the development of stapled peptides, as discussed in Chapter 6. Overall, the success of macrocyclic peptides is reflected in their growing clinical impact, as represented by a significant proportion of newly approved peptide therapeutics, with over 40 FDA-approved drugs derived from or inspired by macrocyclic scaffolds in the past two decades [4].

Several recent reviews have been published over the last decade to give a comprehensive overview of the relevance of macrocyclic peptides and peptidomimetics and modern synthetic approaches for their development, and representative titles are suggested for further information [5–9].

5.2 Classification of Macrocyclic Peptides

In drug discovery, a molecule possessing a 12-membered ring system or larger is classified as a macrocyclic compound, and those in particular comprising 8–11 atoms are classified as medium-sized macrocycles [10]. Macrocycles can be generally classified based on the chemical nature and their occurrence, such as peptide/non-peptide macrocycles, natural products, and synthetic macrocycles. As a further classification of macrocyclic peptides, these compounds can be categorized based on ring size, topology, cyclization strategy, and chemical composition. Variations in ring size influence the conformational rigidity and synthetic complexity, whereas different topologies and cyclization sites enable precise control over molecular geometry. Additionally, the extent of peptide vs. non-peptide content determines how closely these structures resemble natural peptides or incorporate small-molecule features, thus affecting their structural and functional properties:

- Ring size
  - Small macrocycles (<12)
  - Medium macrocycles (12–20)
  - Large macrocycles (>20)
- Topology
  - Monocyclic
  - Bicyclic/polycyclic
  - Branched macrocycles
- Cyclization site
  - Head-to-tail
  - Side chain-to-side chain
  - Backbone-to-side chain
- Peptide character
  - Peptide-like
  - Peptidomimetic
  - Hybrid scaffold

5.3 Peptide Macrocyclization Chemistry

As discussed in the introductory chapter, cyclizing linear peptides into macrocyclic molecular architectures represents a powerful approach for the development of conformationally constrained peptidomimetics. In contrast to strategies that rely on local cyclic elements, macrocyclization affects the overall conformational profile of the peptide, thus limiting flexibility and favoring the stabilization of defined secondary structural motifs, resulting in particular utility for therapeutic strategies targeting PPIs [11]. As well as enabling improved resistance toward metabolic degradation and access to stable and well-organized three-dimensional structures as compared with their linear counterparts, macrocyclization demonstrates enhanced membrane permeability and cytoplasmic delivery ability [12]. These properties arise from favorable structural features such as reduced backbone flexibility, elimination of free *N*- and *C*-termini, and an increased occurrence of *cis* amide bonds.

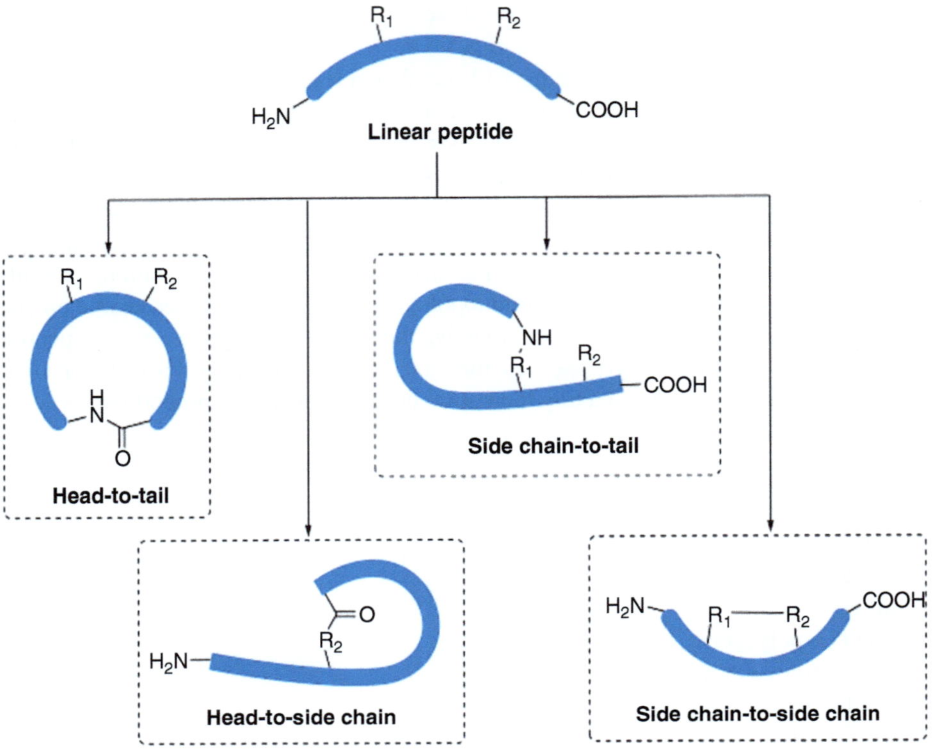

Figure 5.2 Peptide cyclization strategies. Source: Reprinted with permission from Ref. [6]. Copyright 2019 American Chemical Society.

As anticipated, peptide cyclization strategies can be generally arranged into four categories (Figure 5.2) based on the connectivity used to impose conformational restriction [13]:

- *head-to-tail cyclization* involving bond formation between the *N*- and *C*-termini of the peptide;
- *side chain-to-side chain cyclization* exploiting functional groups present on the side chains of either natural or modified amino acids;
- *head-to-side chain cyclization* approaches including the *N*-terminus reacting with a complementary side-chain functional group;
- *Side chain-to-tail cyclization* linking a side-chain moiety to the *C*-terminus.

Macrocyclization reactions are generally performed under high-dilution conditions to suppress competing intermolecular processes such as dimerization or polymer formation, and this issue is often addressed by anchoring the peptide to a solid support to achieve a pseudo-dilution effect. Ring size plays a crucial role in cyclization efficiency, as peptides containing more than seven amino acids typically cyclize readily, whereas shorter sequences frequently encounter kinetic and entropic barriers. Thus, the presence of turn-promoting motifs as conformational inducers is useful to pre-organize the peptide backbone and to reduce the strain associated with the ring closure. Accordingly, a wide range of synthetic techniques has been established to transform linear sequences into

Table 5.1 *Comparative analysis of common chemical methods for peptide macrocyclization.*

Method	Bond	Stability	Rigidity	Typical use
Lactam	Amide	High	Moderate–high	Helix stapling and backbone cyclization
Disulfide	S—S	Low (redox sensitive)	Moderate	Natural scaffolds and reversible constraints
Thioether	S—C	High	Moderate	Stable disulfide surrogates
RCM	C=C	High	High	Hydrocarbon staples and tertiary structure mimicry
CuAAC/click chemistry	Triazole	Very high	Moderate–high	Side chain-to-side chain cyclization and bioorthogonal stapling (Cu-free)

macrocyclic systems. Commonly applied methods include lactam and lactone formation, disulfide bond formation between cysteine residues, copper-catalyzed azide–alkyne cycloaddition (CuAAC) to form triazoles, and ring-closing metathesis (Table 5.1). Challenges such as the requirement of pre-functionalized starting materials enabled the study of more atom- and step-economic alternatives, with C—H bond functionalization emerging as a powerful tool for the functionalization of C—H bonds in amino acids and peptides [14]. Such cyclization strategies have been developed over the years, inspired by natural product diversity observed in naturally occurring macrocyclic peptides, showing unusual linkages between amino acids such as those involving the indole moiety of tryptophan as a linking scaffold [15].

5.3.1 Lactam Cyclization

Lactam ring formation is the most commonly used cyclization motif among structurally characterized peptide loops and is a covalent crosslink in numerous bioactive natural products, including cyclosporin A and gramicidin S [6, 11]. This strategy encompasses multiple linkage approaches, such as head-to-tail, side chain-to-tail, and side chain-to-side chain connections, often involving complementary basic and acidic residues such as lysine and glutamate or aspartate, respectively. Many naturally occurring cyclic peptides adopt a head-to-tail topology, which confers enhanced resistance to enzymatic degradation by eliminating free *N*- and *C*-termini as substrates for the action of exopeptidases. Despite its widespread use, the structural diversity in ring size and cyclization geometry precludes direct correlations between a specific lactamization strategy and the corresponding conformational outcome. Lactam macrocyclization, proceeding through intramolecular amide bond formation between an amino and a carboxylate, is typically performed using established peptide coupling reagents, such as carbodiimides, phosphonium salts, or uranium-based systems, under conditions that favor the intramolecular ring closure. High-dilution techniques are commonly employed to suppress competing intermolecular reactions, and solid-phase cyclization offers an effective alternative by providing a pseudo-dilution effect and simplifying the purification.

From a design perspective, lactam macrocyclization offers several advantages, including robust and biocompatible chemistry, as well as flexibility in the arrangement of the

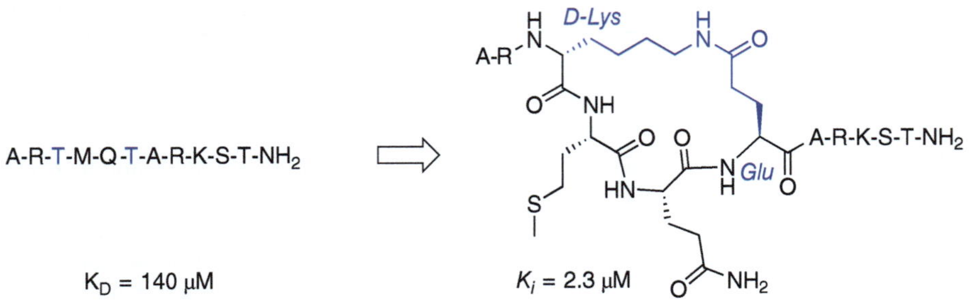

Figure 5.3 Macrocyclic peptide designed from the linear peptide $H3_{1-11}$ K4M.

peptide link, thus enabling its widespread use in stabilizing secondary structural elements, such as α-helices and β-hairpins. Constrained peptides generated through a lactam bond have proven effective as proteomimetics and PPI inhibitors, with optimized molecular architectures displaying enhanced potency and reduced molecular size as compared to less constrained analogues. An example of the successful application of lactam bond formation as a macrocyclization link is represented by an H3-derived peptide, showing lactam cyclization between D-Lys and Glu residues that was reported to achieve >40-fold improvement in the binding affinity to lysine-specific demethylase 1 (LSD1), an epigenetic enzyme overexpressed in certain cancers that regulates the methylation of Lys4 of histone 3 (Figure 5.3) [16].

5.3.2 Disulfide Cyclization

Beyond head-to-tail closure, disulfide bond formation is a side chain-to-side chain typology that represents the most prevalent cyclization motif among naturally occurring cyclic peptides. Many disulfide-bridged peptides preferentially adopt β-hairpin conformations, in which cysteine residues are positioned at non-hydrogen-bonding sites, whereas shorter sequences that lack the capacity to form extended β-sheet structures typically fold into irregular loop conformations. As representative entries to significant macrocyclic peptides of this series, somatostatin and its analogue drug octreotide are two well-known and characterized macrocyclic peptides, possessing a disulfide bond as a tethering moiety (Figure 5.4).

Disulfide bonds are formed through the oxidation of cysteine thiol groups under mild conditions, commonly using air, dimethyl sulfoxide, iodine, or other oxidizing agents. The high nucleophilicity of the cysteine thiolate group makes this residue particularly useful for chemical modification and has contributed to the widespread occurrence of disulfides as stabilizing elements in proteins and peptide natural products. Despite their synthetic simplicity and efficiency, disulfide linkages confer only moderate conformational rigidity and are labile under reducing conditions or in the presence of competing thiols, thus promoting potential exchange reactions. As a result, disulfides are frequently employed as temporary constraints in peptide design to be modified in a second stage or as native motifs in extracellular peptides. Thus, to overcome stability limitations, disulfide bonds have often been replaced with more robust isosteres, including lactams, thioethers, triazoles, selenium-containing linkages, or dicarba analogues, although such substitutions typically require substantial modifications to peptide building blocks.

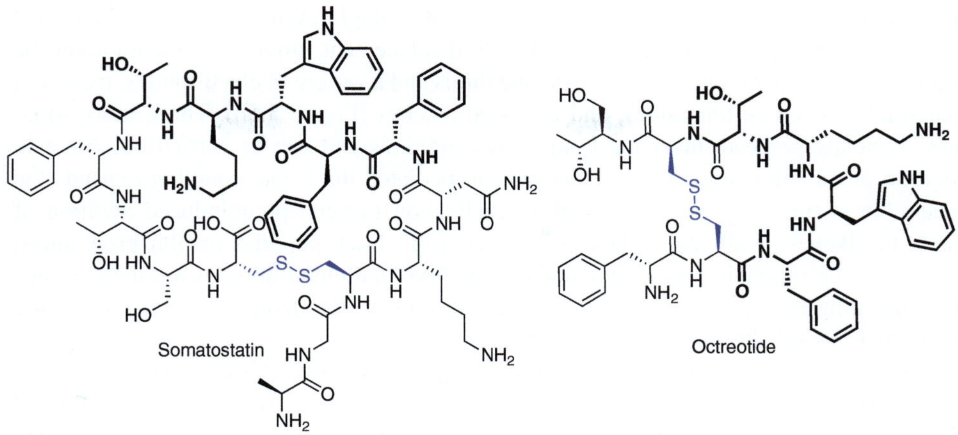

Figure 5.4 Structure of disulfide-linked natural macrocyclic peptide somatostatin and its analogue drug octreotide.

5.3.3 Thioether Cyclization

Thioether linkages constitute another widely applied strategy for the cyclization of peptides with non-regular conformations and are a significant structural moiety of several natural product classes. Thioether cyclization is characterized by the chemical stability of the linkage, as it is non-reducible and resistant to thiol exchange. Consequently, thioethers are frequently employed as permanent disulfide surrogates in peptide design, as well as in peptide stapling and for the development of hybrid macrocyclic or bicyclic molecules. Thioether macrocyclization most commonly proceeds through nucleophilic substitution reactions in which cysteine thiolates displace activated halides (Figure 5.5a).

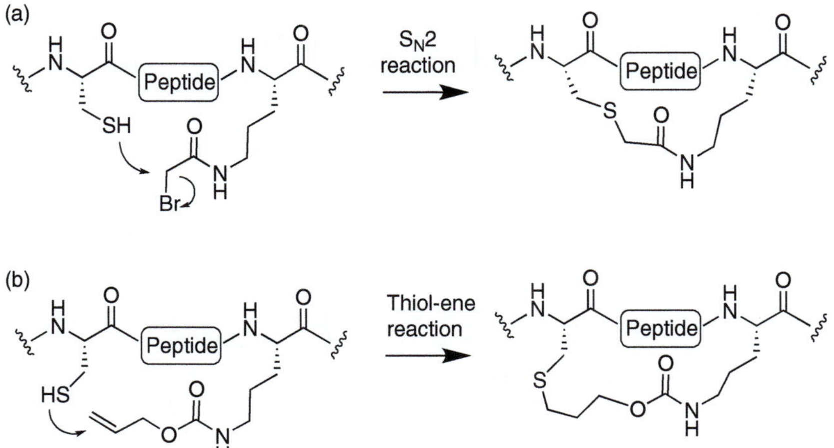

Figure 5.5 Macrocyclization via thioether formation: (a) S_N2 with an α-bromoacetate moiety; (b) thiol–ene reaction with an O-allyl moiety.

Representative approaches include the reaction of a single cysteine with an *N*-terminal chloroacetyl group [17]. Beyond alkyl halide displacement, thioether formation can be achieved through reactions between cysteine thiols and a variety of electrophiles, including maleimides, vinyl sulfonamides, and activated alkenes (Figure 5.5b). These transformations typically proceed under mild, aqueous conditions and can be carried out either in solution or on solid support. Also, the radical-mediated thiol–ene coupling expands the accessible chemical space, thus enabling cyclization through photoinduced addition of thiols to alkene- or alkyne-containing residues [18]. Such peptides exhibit pronounced resistance to proteolytic degradation and display high binding affinity toward their targets, demonstrating the value of thioether-based constraints in stabilizing noncanonical peptide conformations suitable for protein recognition.

An extension of this strategy involves the incorporation of multiple cysteine pairs within a single sequence to generate distinct thioether crosslinks [19]. Specifically, bifunctional alkylation using α,α′-dibromoxylene derivatives connects two cysteine residues [20], and trifunctional alkylation with α,α′,α″-tribromomesitylene allows to generate bridged bicyclic structures [21]. The primary limitation of this approach is the need to introduce suitable electrophilic functionalities. In this view, a variety of rigid electrophilic scaffolds have been developed for cysteine-mediated cyclization (Figure 5.6). Nevertheless, in this approach, the rapid and chemoselective nucleophilic substitution of electrophiles by cysteine thiolates proceeds efficiently in mildly basic aqueous conditions and achieves

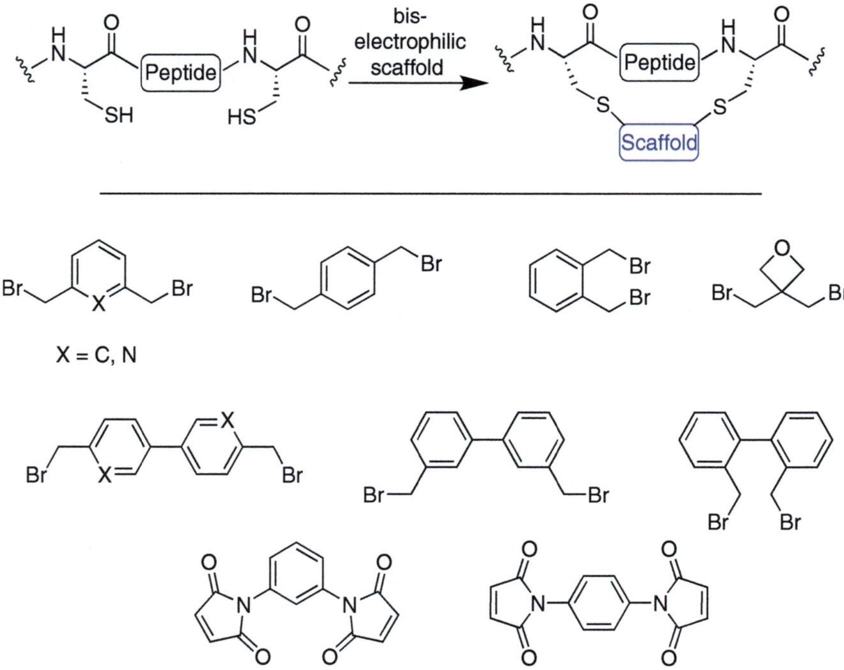

Figure 5.6 Top: application of bis-electrophilic tethering scaffolds for peptide macrocyclization through double thioether bond with two cysteine residues; bottom: representative bis-electrophilic scaffolds for thioether formation based on bromomethylene or maleimide units.

high conversion at ambient temperature to achieve the cyclization of fully unprotected peptides with a simple synthetic procedure and preserving the structural integrity of the peptide [22].

The application of bis-electrophilic scaffolds for thiol-to-thiol stapling in thioether bond formation as a crosslink strategy was applied for the generation of a library of thioether-cyclized peptides, too. A two-step combinatorial synthesis was conceived for synthesizing the libraries, in which "m" random peptides were cyclized by "n" bis-electrophilic linkers to obtain thioether-cyclized peptides and successively acylated at *N*-terminal with "o" carboxylic acids to yield m × n × o macrocyclic peptides [23].

5.3.4 Ring-Closing Metathesis

Olefin-based cross-linking is a widely used strategy for stabilizing α-helical peptides, with important applications in PPI inhibition and peptide-based drug design (see Chapters 6 and 10). This approach is based on alkene ring-closing metathesis (RCM) to form rigid hydrocarbon crosslinks or "staples" that strengthen the secondary structure while preserving the peptide backbone arrangement [24]. In addition to enhancing α-helicity, olefin stapling often improves the proteolytic stability and cellular uptake. The application of RCM to peptides has been extensively reviewed, and some comprehensive reports are suggested for insight into this methodology [25–27].

Early examples of olefin stapling were reported by Blackwell and Grubbs, who connected by RCM terminal olefins incorporated as *O*-allyl serine or homoserine at the i and $i+4$ positions of α-helices [28]. Significant improvements were achieved by Verdine through the use of α,α-disubstituted olefin-bearing amino acids, taking advantage of the intrinsic helix-promoting properties of α-aminoisobutyric acid (Aib) (Figure 5.7). Optimization of side-chain length and stereochemistry enabled effective $i-i+4$ and $i-i+7$ staples, thus achieving peptides with increased α-helical content and enhanced resistance to proteolytic degradation.

Subsequent studies demonstrated that monosubstituted olefinic amino acids can provide comparable stabilization of α-helices, although it has also been reported that its use can strengthen other secondary structural motifs, such as β-sheets and β-turns. Moreover, it was shown that hydrocarbon linkage often increases cellular uptake, particularly when the crosslink increases a hydrophobic surface on the peptide, further enhancing their utility for intracellular targets.

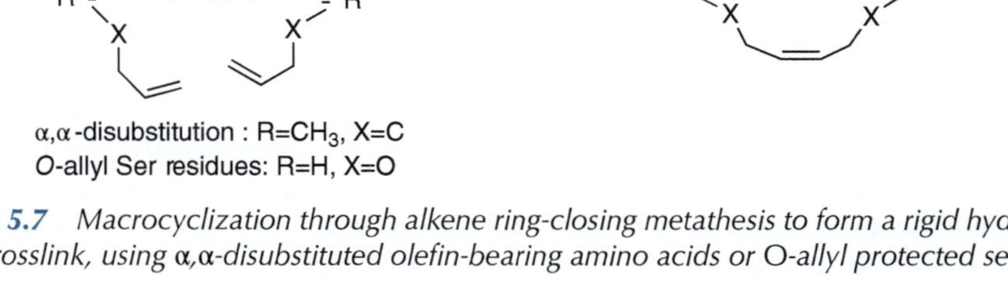

Figure 5.7 Macrocyclization through alkene ring-closing metathesis to form a rigid hydrocarbon crosslink, using α,α-disubstituted olefin-bearing amino acids or O-allyl protected serine.

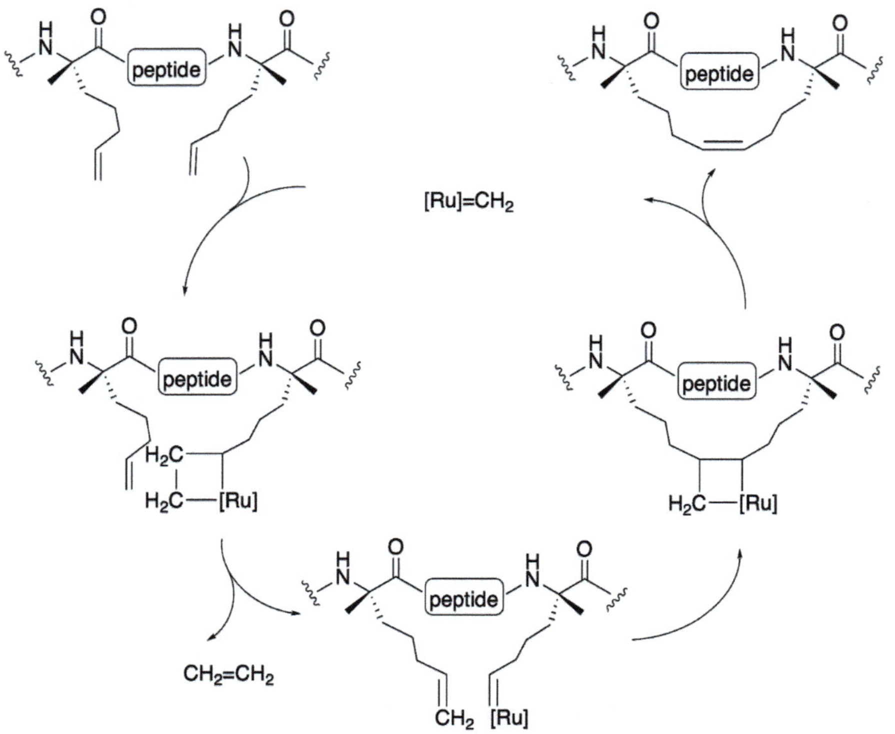

Figure 5.8 Mechanism of the Ru-catalyzed RCM.

Concerning the chemistry of olefin cyclization, this is based on alkene metathesis, a reaction popularized in the nineties following the development of stable and chemoselective molybdenum- and ruthenium-based catalysts. The reaction proceeds through a catalyst-bound carbene or alkylidene that forms a metallacyclobutane intermediate through a reversible [2+2] cycloaddition (Figure 5.8). Because the process is under thermodynamic control, unstrained systems generally favor *E*-alkenes, although macrocyclic peptides frequently yield mixtures of *E/Z* isomers. Ruthenium-based Grubbs and Hoveyda–Grubbs catalysts are commonly employed for peptide cyclization due to their functional-group tolerance and stability [29]. Successively, second- and third-generation Ru catalysts proved to improve both reactivity and substrate scope, including access to catalysts that favor the *Z*-alkene formation. RCM can be performed either in solution or on solid support [30], with low concentrations typically used in solution to minimize dimerization.

Olefinic functionalities for RCM have been introduced through a variety of positions and functional groups, including carbon-bound side chains, α-nitrogen substituents, alcohols, phenols, acids, carbamates, and cysteine thiols. Frequently, α-methyl-α-alkenyl side chains are employed to further enhance helix stability. As a representative application of RCM for the generation of bioactive macrocyclic peptidomimetics, it was reported the structure-based macrocyclization of the *C*-terminal hexapeptide fragment of neurotensin NT (8–13) *via* RCM was capable of preserving the binding and agonist activity of the

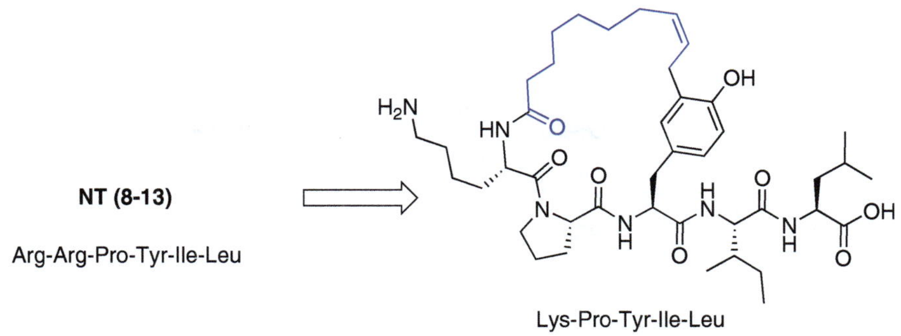

Figure 5.9 *RCM macrocyclization of C-terminal hexapeptide fragment of neurotensin NT (8–13).*

parent peptide, yielding a novel macrocyclic scaffold with sub-micromolar binding activity and markedly enhanced plasma stability (Figure 5.9) [31]. The cyclization strategy was conceived from the analysis of the structure of NT (8–13) in the binding pocket of NTS1 that showed proximity between residues Tyr11 and Arg8, thus applying the RCM on solid support between an *ortho*-allylated tyrosine and a nonenoic acid in place of Arg8 for a head-to-side chain macrocyclization. The resulting lead compound was considered a promising starting point for developing stable, potent neurotensin analogues.

5.3.5 Click Chemistry

The copper(I)-catalyzed azide–alkyne cycloaddition (CuAAC) has emerged as a versatile method for peptide cross-linking and also for secondary structure stabilization, particularly for α-helices, as reported in several reviews and accounts (e.g. see [32, 33]). The CuAAC reaction is a catalytic variant of the Huisgen 1,3-dipolar cycloaddition and is the classical "click reaction" as defined by Sharpless and co-workers [34], where Cu(I) catalysis allows for achieving 1,4-disubstituted 1,2,3-triazoles from terminal alkynes and azides under mild, regioselective, and aqueous-compatible conditions. The resulting triazole is metabolically stable and is considered a peptide bond isostere on the basis of hydrogen-bonding topology and dipole moment. Similarly to RCM, these properties have enabled triazoles to function as surrogates for amide bonds and disulfides and to stabilize diverse peptide motifs, including helices, β-hairpins, and disulfide bridges [35, 36].

The catalytic cycle consists of the coordination of Cu(I) species to the alkyne, followed by binding of the azide to the resulting complex (Figure 5.10). Subsequent addition of the internal alkyne carbon to the terminal azide nitrogen leads to the formation of a six-membered metallacycle. Finally, reductive ring contraction and copper elimination release the triazole.

CuAAC can be performed either in solution or on solid support and does not require protecting groups. Solution-phase reactions are often facilitated due to the greater conformational flexibility of the peptide substrate. Common catalytic systems include $CuSO_4$ with sodium ascorbate or Cu(I) salts such as CuI or $Cu(MeCN)_4PF_6$. Oxidative side reactions involving histidine or cysteine residues may occur [38] but can be diminished through the use of tris(triazolylmethyl)amine-based ligands, which stabilize Cu(I) and

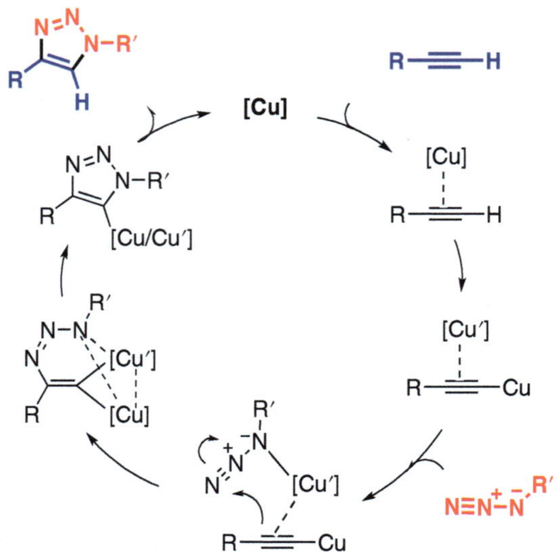

Figure 5.10 Mechanism of the CuAAC reaction proposed by Fokin et al. [37].

accelerate the reaction rates. Also, Wang et al. designed and synthesized two cyclopeptidic Smac mimetics by replacing amide bonds in the lead cyclopeptide with triazoles by means of click chemistry. Specifically, the alkyne component was derived from phenylalanine or phenylglycine, whereas the azide counterpart consisted of azido-Lys bearing the N_3 group on the side chain in place of the ε-amino group (Figure 5.11). In particular, one compound showed enhanced binding to inhibitor of apoptosis proteins (IAPs) and greater anticancer activity, making it a promising lead for developing new IAP-targeted anticancer agents [39].

Recent advances have further expanded the synthetic scope of CuAAC in peptide chemistry, including on-resin click chemistry combined with cross-coupling strategies, enabling late-stage diversification of stapled peptides (Chapter 6) [40].

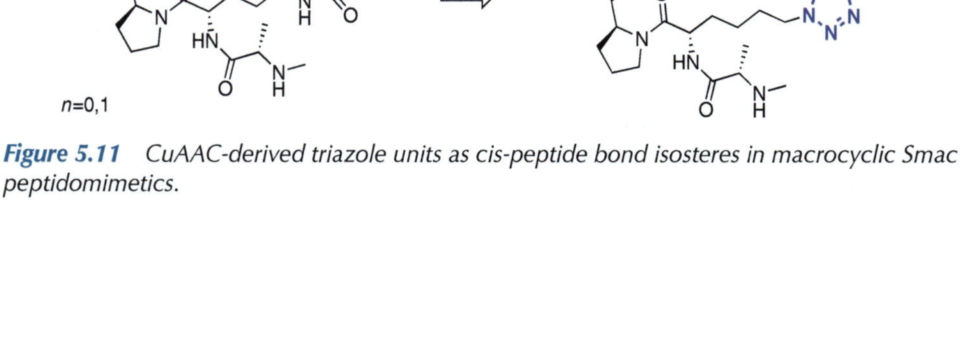

Figure 5.11 CuAAC-derived triazole units as cis-peptide bond isosteres in macrocyclic Smac peptidomimetics.

Overall, reports on the use of CuAAC for peptide macrocyclization provided evidence of this technique as a robust and adaptable platform for peptide stabilization, functionalization, and chemical biology applications.

5.3.6 C—H Activation

As anticipated in Chapter 4, late-stage macrocyclization of peptides can be achieved through C—H activation strategies. In recent years, metal-catalyzed cross-coupling reactions have been widely applied to peptide functionalization to form new bonds on amino acid side chains. These methods have enabled access to a wide range of modified peptides. Nevertheless, they typically rely on pre-functionalized substrates such as aryl halides or boronic acids and often require multistep synthetic sequences; thus, the need for greater atom economy and step efficiency has pushed toward the development of alternative approaches. In this context, C—H bond functionalization strategies employing directing groups have emerged as powerful tools for the selective activation of inert C—H bonds in amino acids and peptides. Such methodologies enable late-stage diversification and macrocyclization without prior installation of reactive chemical moieties, thus expanding the synthetic toolbox for peptide modification significantly. Among these approaches, indole C—H cross-linking has attracted particular attention as a subcategory of peptide macrocyclization strategies, as documented extensively in the literature. In this context, several authoritative and comprehensive reviews have been reported to account for such approaches, which can be broadly classified in peptide macrocyclization *via* C(sp^2)—H and C(sp^3)—H activation [14, 41–43].

5.3.6.1 *Peptide Macrocyclization* via *C(sp^2)—H Activation*

Late-stage peptide macrocyclization via C(sp^2)—H activation has emerged as a powerful strategy for developing conformationally constrained and structurally complex peptides. Among heteroaromatic residues, tryptophan has proven particularly useful to such transformations due to the reactivity of its indole moiety, allowing for site-selective macrocyclization through C—H activation at C-2, C-4, and C-6 atoms of the heterocyclic ring. One of the earliest examples of peptide macrocyclization *via* indole C—H activation was reported by James in 2012 [44]. In this work, a palladium-catalyzed C2—H arylation of the tryptophan side chain was developed using *m*- and *p*-iodophenylalanine as aryl iodides, allowing for the formation of macrocyclic peptidomimetics with variable ring size (Figure 5.12). The screening of catalysts and additives during method development allowed the identification of $Pd(OAc)_2$ in combination with silver salts as the optimal catalytic system.

Successively, Albericio et al. developed Pd-catalyzed indole C2—H arylation for peptide macrocyclization both in solution and on solid support [45]. Intramolecular C—C bond formation between the C2—H bond of the indole side chain of tryptophan and iodoaryl-containing amino acids was achieved under microwave irradiation, affording macrocycles under relatively mild conditions. This approach was extended to generate bicyclic architectures through double arylation of peptides containing a central di-iodinated tyrosine and two tryptophan side residues, delivering double-stapled bicyclic peptides (Figure 5.13). While effective, these strategies were limited by the need

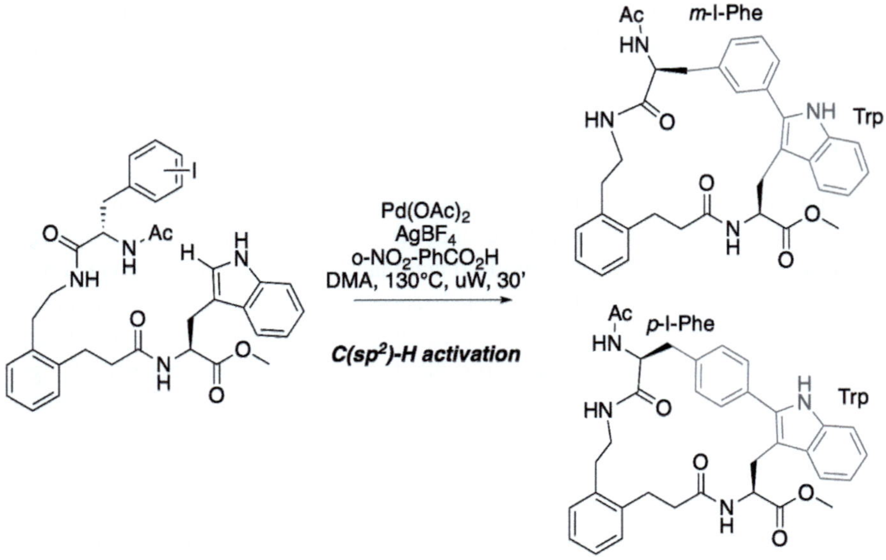

Figure 5.12 Peptide macrocyclization through Pd-catalyzed C2−H arylation of tryptophan side chain using m- or p-iodo-phenylalanine.

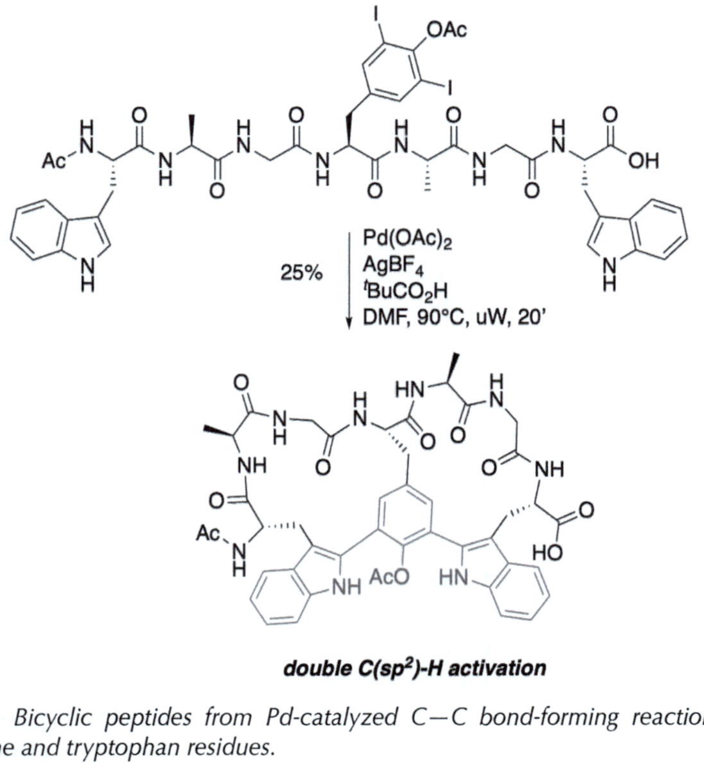

Figure 5.13 Bicyclic peptides from Pd-catalyzed C−C bond-forming reactions between phenylalanine and tryptophan residues.

for pre-functionalized aryl halides, high dilution, and sensitivity to certain amino acid residues, such as methionine, cysteine, and histidine.

Beyond arylation, olefination reactions have become established tools for peptide macrocyclization. In this context, Wang reported a series of Pd(II)-catalyzed intramolecular olefination of tryptophan residues, achieving site-selective macrocyclization at either the C2 or C4 position of the indole ring [46]. For C2-selective olefination, the peptide backbone amides acted as endogenous directing groups. Peptides bearing acrylate-modified serine or threonine residues were cyclized under oxidative conditions, and the general applicability of the cyclization process to afford 15–26-membered ring macrocycles was demonstrated (Figure 5.14, top). The C-4 atom of the indole ring of tryptophan was also exploited for the selective peptide macrocyclization, though requiring a different directing strategy. Specifically, a trifluoromethanesulfonyl group at the *N*-terminus directed Pd-catalyzed olefination to C4—H of tryptophan, while a bulky triisopropylsilyl group suppressed competing C2 activation. Using this approach, the first examples of tryptophan–alkene crosslinks at the indole C4 atom were obtained, affording 16–23-membered ring macrocycles (Figure 5.14, bottom). Thus, such strategy enabled access to indole positions that are difficult to functionalize using conventional peptide coupling chemistry.

The Wang group also extended Pd-catalyzed C—H olefination to phenylalanine residues [47], developing a series of macrocycles through backbone-assisted intramolecular olefination of phenylalanine side chains under relatively mild conditions. Also, peptide-guided macrocyclization of bioactive peptidosulfonamides was achieved through Pd-catalyzed C—H olefination at the *ortho* position of the aryl ring, thus giving 14–28-membered ring macrocycles [48].

Alternative transition metals have been explored to address limitations associated with palladium, including cost and toxicity. Rhodium(III)-catalyzed C(sp^2)—H olefination

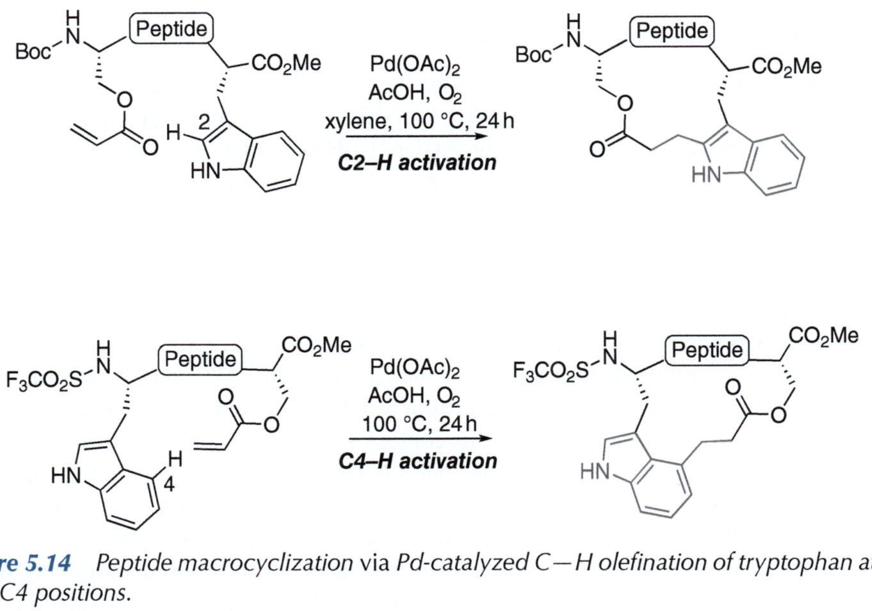

Figure 5.14 Peptide macrocyclization via Pd-catalyzed C—H olefination of tryptophan at C2 and C4 positions.

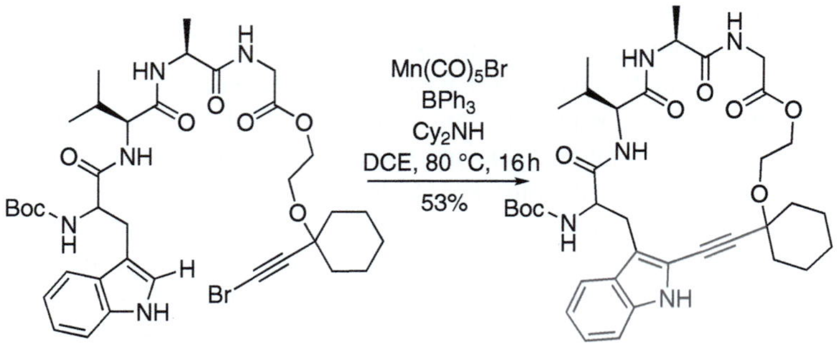

Figure 5.15 *Manganese-catalyzed intramolecular C(sp^2)—H alkynylation.*

of tryptophan residues bearing pyridyl or pyrimidinyl directing groups has enabled the selective C2 functionalization and macrocycle formation under mild conditions [49]. In this context, earth-abundant metals have attracted attention as sustainable entries to metal-catalyzed C—H functionalization reactions. Accordingly, Ackermann and co-workers reported a manganese(I)-catalyzed intramolecular C(sp^2)—H alkynylation of tryptophan-containing peptides using pyrimidine directing groups to activate selectively the C2 position of the indole ring [50]. This racemization-free process enabled late-stage macrocyclization to afford alkynyl-bridged macrocycles in good yields (Figure 5.15), highlighting the potential of base-metal catalysis in peptide macrocyclization.

Overall, C(sp^2)—H activation has become a versatile and powerful platform for peptide macrocyclization exploiting the side chain of aromatic amino acids, such as tryptophan, phenylalanine, and tyrosine, thus giving access to structurally diverse and conformationally constrained peptide macrocycles.

5.3.6.2 *Macrocyclization* via *C(sp^3)—H activation*

Metal-catalyzed direct C(sp^3)—H activation enables C—C bond construction at inert C—H bonds to achieve a versatile route to peptide macrocycles. In this context, peptide macrocyclization *via* intramolecular arylation of the β-C(sp^3)—H bond of *N*-terminal amino acids in oligopeptides without directing groups was reported by Wang and co-workers [51]. Specifically, tetrapeptides bearing various *N*-terminal residues and meta-iodophenylalanine at the *C*-terminal underwent C—C coupling between the β-carbon of the *N*-terminal amino acid and the *meta*-position of phenylalanine (Figure 5.16, top). Also, the use of tryptophan for Pd-catalyzed C(sp^3)—H activation was applied to couple the β-carbon of terminal amino acids with the C6 position of the indole side chain of tryptophan, reproducing a crosslink found in several natural product macrocycles such as celogentin C (Figure 5.16, bottom).

A significant entry to peptide macrocyclization by C(sp^3)—H activation was achieved by taking advantage of directing groups. In this context, Chen et al. reported the Pd-catalyzed 8-aminoquinoline (AQ)-directed intramolecular β-C(sp^3)—H arylation by linking modified *C*-terminal amino acids to an *N*-terminal alkyl chain to synthesize cyclophane-based peptide macrocycles [52]. This approach overcame size- and composition-dependence, tolerated various iodinated aromatic amino acids, and produced macrocycles of varied ring sizes (Figure 5.18, top). As another directed approach, Pd-catalyzed picolinic acid (PA)-directed

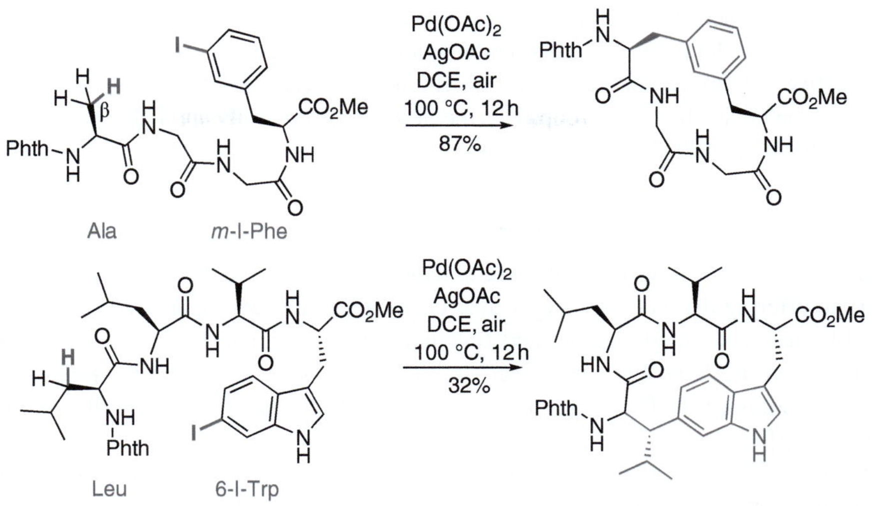

Figure 5.16 Peptide macrocyclization using Pd-catalyzed C(sp^3)—H functionalization.

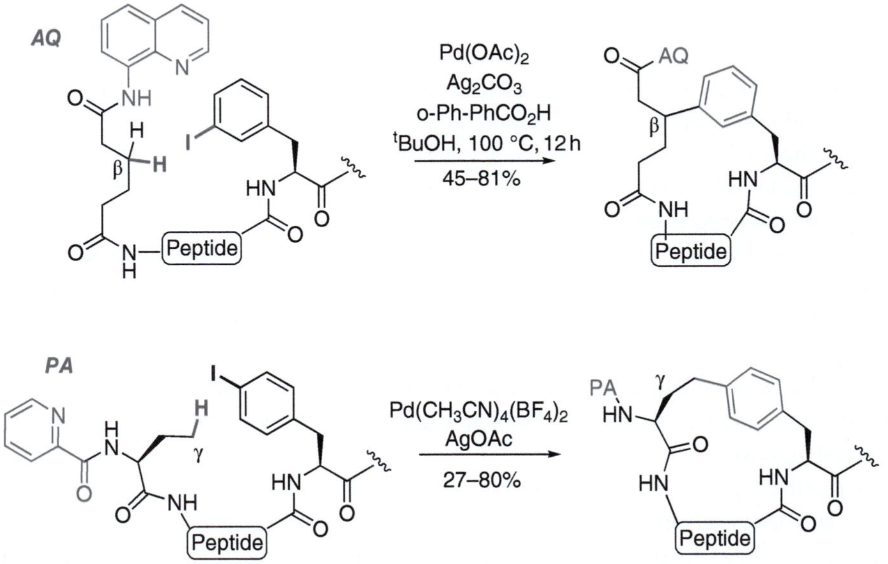

Figure 5.17 Directed Pd-catalyzed intramolecular C—H arylation. Top: AQ-directed intramolecular β-C(sp^3)—H arylation; bottom: PA-directed intramolecular γ-C(sp^3)—H arylation.

intramolecular γ-C(sp^3)—H arylation allowed for the macrocyclization from *C*-terminal aryl iodides, such as *p*-iodo-phenylalanine, to the γ-position of *N*-terminal aliphatic amino acids, affording cyclized peptides with good diastereoselectivity (Figure 5.17, bottom). Unprotected peptides cyclized effectively in aqueous media, tolerating free amines, alcohols, carboxylic acids, and amide bonds [53].

Beyond C—H activation as an established approach for tryptophan C2 functionalization at the indole side chain, Yu et al. also demonstrated the applicability to C(sp^3)—H bonds for the selective *N*-terminal arylation using stoichiometric Ag(I) salts, with the peptide moiety acting as a bidentate directing group. This method was successfully applied to tetrapeptides, demonstrating its utility in terminal peptide diversification [54].

5.3.7 Multicomponent Reactions

Multicomponent reactions (MCRs) have emerged as powerful tools for peptide macrocyclization, offering distinct advantages over classical ligation strategies by enabling rapid assembly, backbone modification, and functional diversification. Several comprehensive accounts have been recently published in this area, and the review by Reguera and Ribera is recommended for insight into macrocyclization strategies using multicomponent reactions [55].

Among MCRs applied for the cyclization of peptides, Ugi- and Petasis-type reactions have been particularly significant [56, 57]. An interesting contribution in this context was the development of a Ugi-type four-component macrocyclization based on aziridine aldehydes [58]. In this strategy, an aziridine aldehyde, an isocyanide, the *N*-terminal amine, and the *C*-terminal carboxylic acid of a peptide fragment were exploited for participating in the cyclization process. The mechanism of this reaction consists of the condensation of the aldehyde with the *N*-terminal amine, generating an iminium species, which undergoes synergistic attack by the isocyanide to form an imido-anhydride intermediate [59]. This concerted addition pathway was found beneficial for achieving an unusually high stereocontrol in the MCR. Subsequent intramolecular attack by the *C*-terminal carboxylate closes the macrocycle, followed by an *O*-to-*N* acyl transfer that inserts an acyl aziridine moiety directly into the peptide backbone.

Beyond head-to-tail cyclization, Ugi-based MCRs have been adapted for side chain-to-side chain and side chain-to-terminus macrocyclizations, which are often faster and more efficient (Figure 5.18). These variants significantly expand structural diversity by exploiting side-chain functional groups. For example, the use of two carboxylate side chains or unnatural amino acids enables the coupling with bifunctional linkers to facilitate the conjugation of diverse moieties and fine-tune the peptide conformation and biological activity [7].

Importantly, Ugi-type macrocyclizations proceed without epimerization of the terminal carboxylate, a key advantage over many traditional coupling-based approaches. Although asymmetric carbonyl components can lead to epimeric mixtures, this issue can be avoided through the use of simple aldehydes such as paraformaldehyde or acetone, although reducing the chemical diversity.

This methodology has proven particularly effective for peptides bearing an *N*-terminal proline residue and has been successfully extended to oligoproline macrocycles. Moreover, Ugi reactions have been taken into account to generate turn-inducing *N*-alkylated dipeptide motifs, as represented in the synthesis of gramicidin S analogues (Figure 5.19) [60]. In this solid-phase strategy, an acyclic decapeptide synthesized with orthogonal protecting groups undergoes Ugi macrocyclization to afford novel cyclic peptides bearing exocyclic *N*-functionalities.

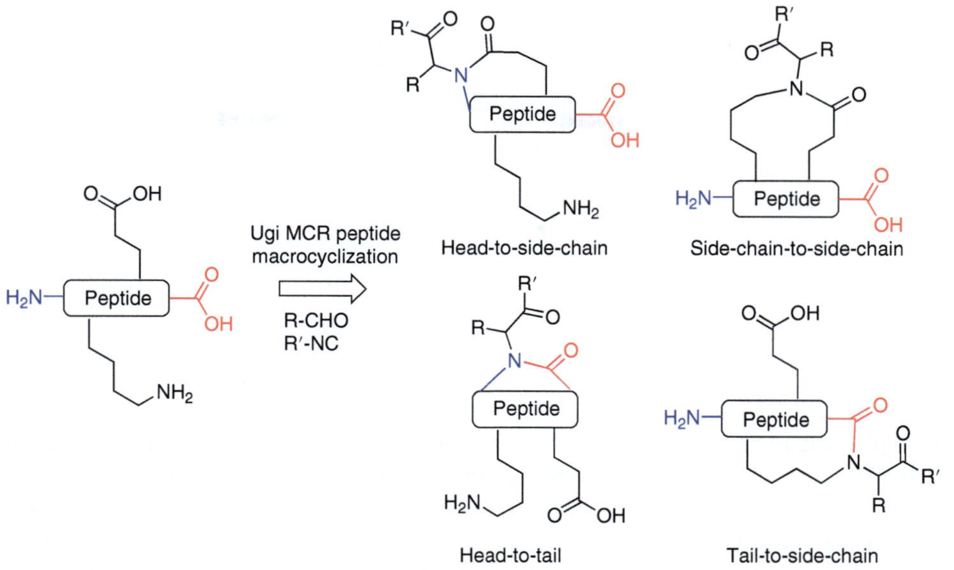

Figure 5.18 *Different approaches to Ugi-type peptide macrocyclization.*

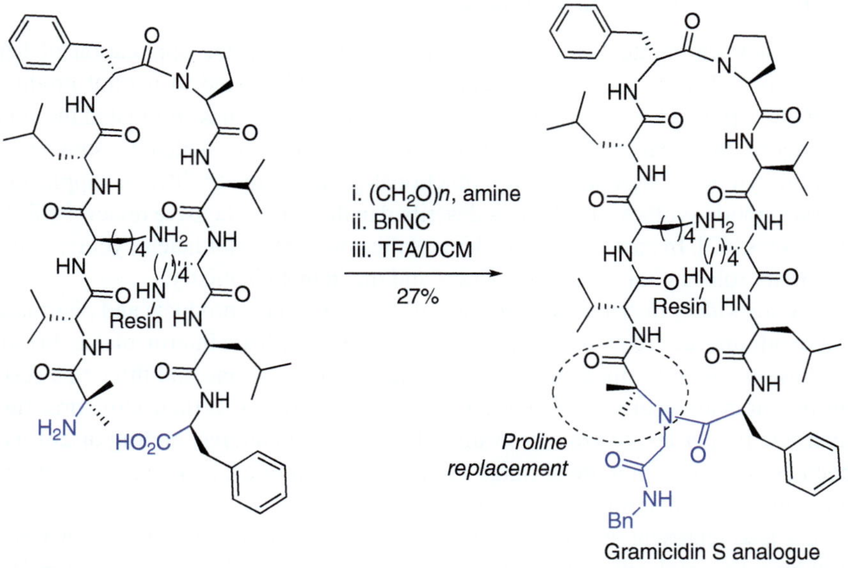

Figure 5.19 *Solid-phase synthesis of cyclic peptide gramicidin S analogue using on-resin multicomponent macrocyclization in a head-to-tail approach.*

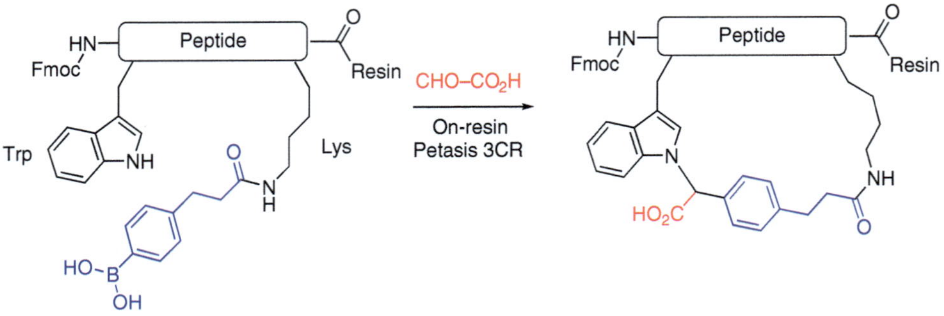

Figure 5.20 Tryptophan-mediated Petasis reaction-based peptide macrocyclization strategy.

Complementary to Ugi chemistry, the Petasis reaction (also called the boron–Mannich reaction) has been introduced as a three-component alternative for peptide macrocyclization. The reaction requires an aldehyde or ketone, an amine, and a boronic acid and has better mechanistic control and reduced pathway complexity as compared to four-component systems. In a paper by Spring describing the application of tryptophan-mediated Petasis reaction for the efficient on-resin peptide stapling and labeling (Figure 5.20), the late-stage incorporation of biologically relevant tags such as fluorophores and cell-penetrating motifs was achieved with broad substrate scope and high functional-group tolerance [61].

5.3.8 Other Heterocycles in Macrocyclization

A relevant chemical strategy for peptide macrocyclization is the application of heterocyclic skeletons. This is a cyclization approach commonly found in natural products as a biosynthetic strategy for constraining peptides. Such evidence has stimulated the study of a wide array of chemical methods applying several heterocyclic skeletons, showing challenges in the synthesis, stereochemistry, functional-group compatibility, and application to solid-phase peptide synthesis (SPPS), as described in the comprehensive review by Balalaie and co-workers [62]. In particular, oxazoles and thiazoles are major heterocyclic motifs that play important roles in both natural products and medicinal chemistry.

Oxazoles are widely distributed in biologically active natural products and pharmaceutical compounds and are frequently employed as versatile synthetic intermediates. In biosynthetic pathways, oxazole rings are commonly derived from serine and threonine residues through post-translational or enzymatic modifications. Owing to their structural rigidity and electronic properties, oxazole derivatives have attracted interest in drug discovery and have been associated with a broad range of biological activities, including hypoglycemic, anti-inflammatory, and antibacterial effects.

Thiazoles also represent essential heterocyclic scaffolds found in numerous natural and synthetic bioactive molecules. They occur in natural compounds such as vitamin B_1 and are important scaffolds found in the structure of many drugs, including antibiotics such as penicillin, antimicrobial agents like sulfazole, antiretrovirals such as ritonavir, and antifungal agents including abafungin. The discovery of oxazole and thiazole motifs inserted in cyclopeptide backbones has highlighted their importance in peptide chemistry. Their incorporation into cyclic peptides introduces conformational constraints that can enhance structural rigidity, metabolic stability, and target selectivity. As a result, substantial efforts have been devoted to understanding the formation, reactivity, and functional consequences of these heterocycles in peptide frameworks.

As a representative example, starting from naturally occurring cyclic peptides biosynthesized through peptide aldehyde intermediates, Baran and co-workers reported an innovative macrocyclization strategy that exploits the intrinsic reactivity between the *N*-terminal amino group and the *C*-terminal aldehyde of an unprotected linear peptide [63]. Specifically, this approach was inspired by non-ribosomal peptide synthetase (NRPS) pathways, in which peptide aldehydes are key intermediates for generating noncanonical cyclic peptide architectures. In analogy with many head-to-tail cyclization methods, the reaction is carried out under highly dilute conditions to suppress intermolecular oligomerization. The cyclization reaction mechanism is supposed to proceed through spontaneous intramolecular imine formation between the *N*-terminal amine and the *C*-terminal aldehyde in aqueous solution. Then, the resulting imine intermediate is subjected to nucleophilic addition, thus giving a stabilized cyclic product. The structural outcome at the ring-closure site depends on the nucleophile type. Specifically, nucleophiles such as potassium cyanide or sodium cyanoborohydride afford α-aminonitrile or secondary amine linkages, respectively. In contrast, internal nucleophiles positioned proximal to the *N*-terminus, including indole, imidazole, thiol, and selenol functionalities, participate intramolecularly to generate a range of heterocyclic motifs at the cyclization junction. Such chemical diversity enables the direct incorporation of non-amide backbone elements into cyclic peptides, closely mimicking the structural motifs found in natural products. The synthesis of linear peptide precursors bearing *C*-terminal aldehydes is readily compatible with standard Fmoc SPPS. In this protocol, a Fmoc-protected amino aldehyde is reacted with the 1,2-hydroxyamine moiety of a threonine residue anchored on Rink amide resin as a Thr–Gly dipeptide, forming an oxazolidine intermediate (Figure 5.21a). Following Boc protection of the oxazolidine nitrogen atom, peptide chain elongation proceeds under conventional Fmoc-SPPS conditions. Global deprotection with trifluoroacetic acid simultaneously removes protecting groups and hydrolyzes the oxazolidine ring, thus giving the unprotected linear peptide with a free *C*-terminal aldehyde in a single step. Using this approach, Baran and co-workers demonstrated a broad scope comprising several examples with ring sizes ranging from 5 to 10 amino acids. The method was further validated through the synthesis of analogues of naturally occurring cyclic peptides containing thiazolidine or thiazole linkages, including sanguinamide A (Figure 5.21b).

A major strength of this cyclization strategy lies in its modularity and versatility. As the imine intermediate can be reacted with a variety of nucleophiles, multiple cyclic peptide analogues can be generated from a common linear precursor, thus facilitating rapid structure–activity relationship studies and library construction. Also, the absence of an activated *C*-terminal ester eliminates any side reactions such as hydrolysis that frequently complicate traditional macrolactamization reactions.

5.3.9 Macrocyclization with SuFEx Chemistry

Since its introduction by Sharpless as a next-generation click reaction [64], sulfur fluoride exchange (SuFEx) chemistry has emerged as a powerful chemical strategy for tyrosine-selective peptide cyclization fulfilling all the click chemistry requirements,

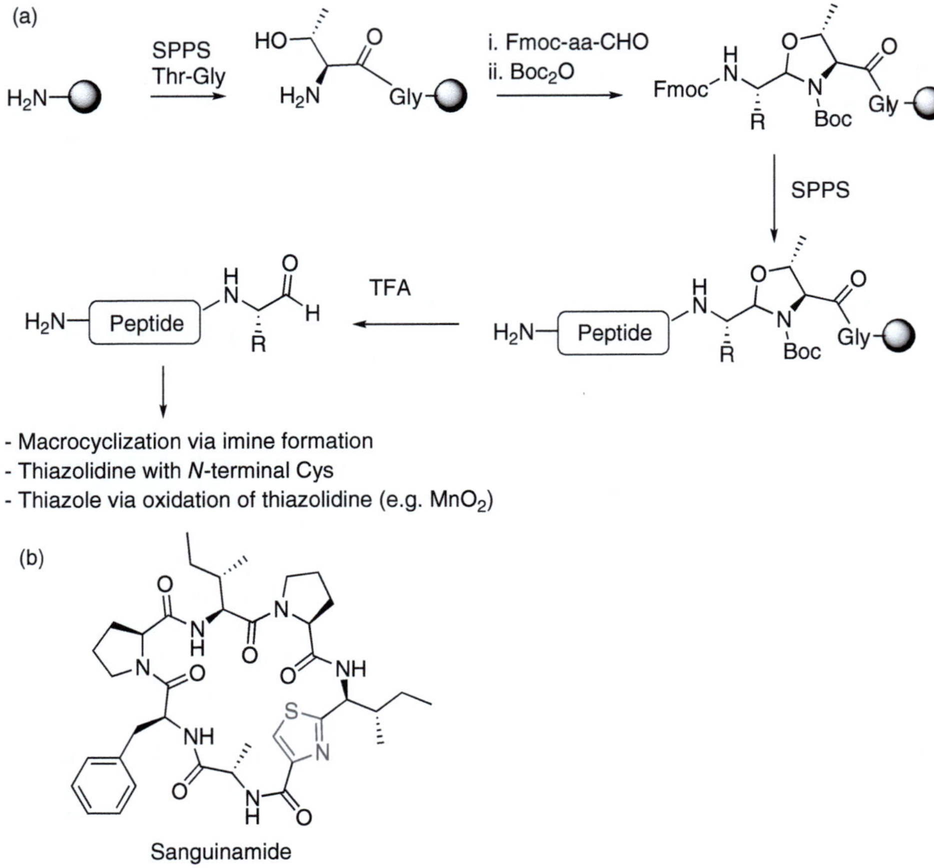

Figure 5.21 (a) Macrocyclization strategy based on the reactivity between the N-terminal amino group and the C-terminal aldehyde of an unprotected linear peptide; (b) naturally occurring macrocyclic peptide sanguinamide containing the thiazole heterocycle.

proceeding efficiently under biologically compatible conditions. Central to this chemistry are sulfonyl fluorides, which display exceptional stability toward hydrolysis and reduction while retaining reactivity toward nucleophiles under mild conditions. These features have driven the broad adoption of SuFEx chemistry across chemical biology and materials science [65, 66].

Accordingly, a SuFEx-based peptide macrocyclization strategy encompassing the side chain of tyrosine was reported (Figure 5.22a), thus giving access to sulfonate-linked macrocyclic peptides (STEMtides) over a wide range of peptides with structural diversity [67]. High chemoselectivity and excellent functional-group tolerance make this approach a robust and orthogonal platform for generating a new class of conformationally constrained peptide therapeutics. The generality and utility of the approach were demonstrated through the application of this chemistry for the synthesis of several bioactive peptides, such as leuprorelin, β-melanocyte-stimulating hormone

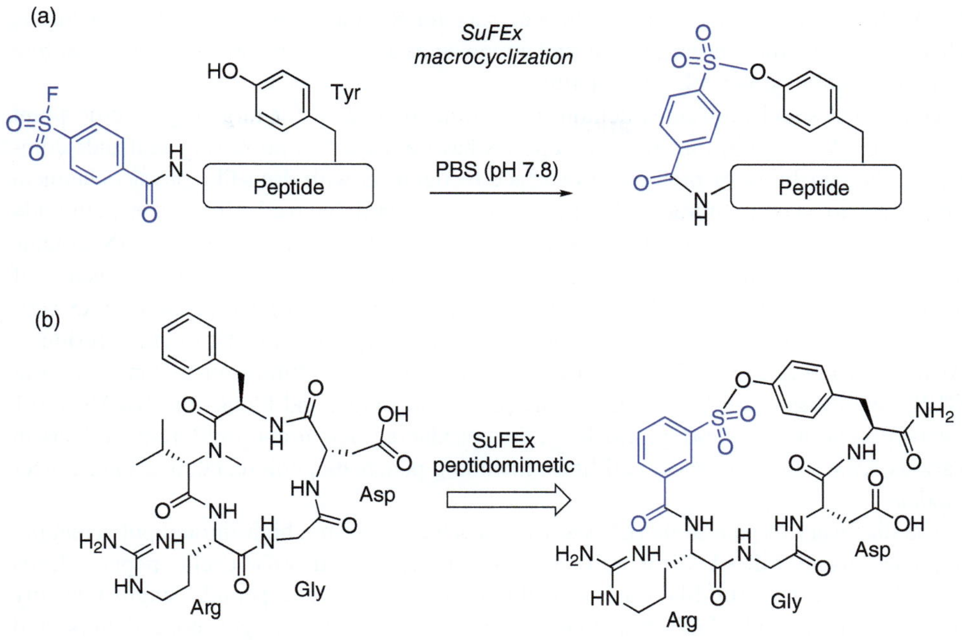

Cilengitide

Figure 5.22 SuFEx reaction for peptide macrocyclization: (a) chemical methods exploiting tyrosine side chain; (b) synthesis of a SuFEx RGD peptidomimetic.

(β-MSH), liraglutide, and cyclic Arg-Gly-Asp (RGD) tripeptide derived from cilengitide (Figure 5.22b).

5.4 Case Studies of Macrocyclic Peptides in Drug Discovery Applications

Macrocyclic peptides and peptidomimetics have been largely applied in drug discovery programs, resulting in the identification of a significant number of compounds that have found approval as drugs [68], as also reported in the introductory paragraph of this chapter. Also, in Chapter 1, the successful development of macrocyclic peptidomimetics as third-generation NS3/4A protease inhibitors for the treatment of *Herpes simplex* virus (HCV) infections was reported, with particular reference to voxilaprevir.

5.4.1 Paluratide (LUNA18): An Orally Bioavailable Cyclic Peptide KRAS Inhibitor

Paluratide (development code: LUNA18) is an oral 11-mer macrocyclic peptide with extensive *N*-alkylations capable of disrupting the KRAS–SOS1 interaction and inhibiting oncogenic KRAS signaling, that is indicated for the treatment of KRAS-driven solid tumors such as pancreatic and colorectal cancers. It represents a milestone in this field as an orally bioavailable cyclic peptide KRAS inhibitor designed to directly interfere with

KRAS function by blocking key PPIs necessary for RAS activation [69]. The structure is characterized by the presence of unusual amino acids and *N*-methylated residues, enabling oral bioavailability and cell permeability.

Traditional small molecules demonstrated limited success for targeting specific point mutations. As a result, macrocyclic peptides have emerged as promising scaffolds combining the capability of biologics in surface recognition with drug-like physicochemical properties to acquire intracellular penetration and oral bioavailability. The paluratide (LUNA18) mechanism consists of disrupting the interaction between mutant RAS proteins and guanine nucleotide exchange factors such as SOS1, which catalyze the exchange of guanosine diphosphate (GDP) for guanosine triphosphate (GTP) and promote RAS activation. This is distinct from covalent RAS inhibitors targeting specific cysteine residues. At the molecular level, paluratide (LUNA18) exhibits high affinity for mutant KRAS at G12D, with dissociation constants in the picomolar range, and blocks the KRAS-SOS1 interaction with a single-digit nM IC_{50}, thus producing potent antiproliferative effects in various KRAS-mutant cancer cell lines, including pancreatic, colorectal, and lung cancer models.

The discovery of paluratide (LUNA18) was achieved from a high-throughput screening approach using mRNA display libraries engineered to identify cyclic peptide leads according to early drug-like criteria, including moderate cyclic peptide length typically of 9–11 amino acids, high N-alkylation content to reduce hydrogen bond donors, and physicochemical profiles beneficial to membrane permeability [70]. Following this approach, the hit molecule AP8784 was identified, exhibiting significant inhibition of the PPI between KRAS with the G12D mutation and the guanine nucleotide exchange factor SOS1 (Figure 5.23). Subsequent structure-based optimization and improvement of membrane permeability allowed the transformation of the hit into the clinical candidate paluratide (LUNA18) (Figure 5.23). Specifically, drug-likeness criteria were established

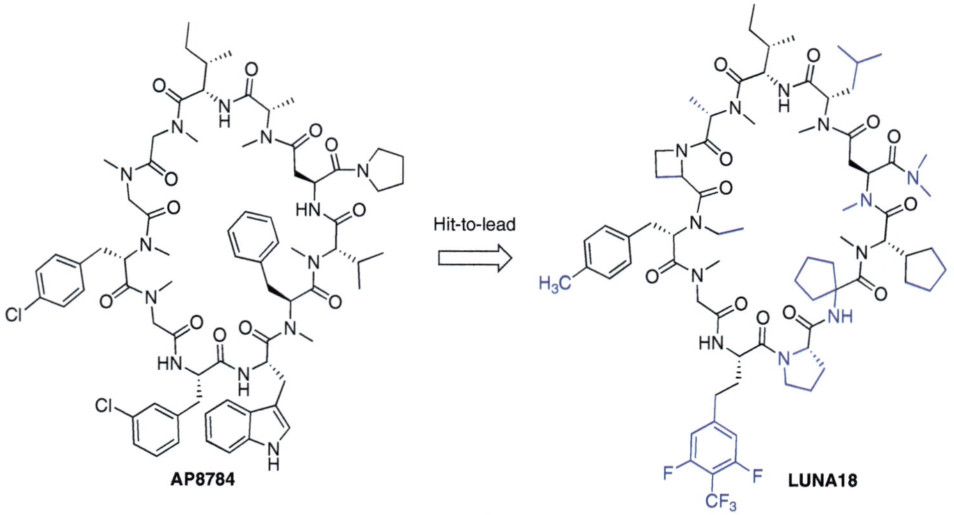

Figure 5.23 Structure of hit ligand AP8784 (left) and drug candidate LUNA18 (right). The modifications at side chains and at nitrogen atoms are highlighted in blue.

for cyclic peptides by having the following four characteristics, particularly devoted to improving cell membrane permeability:

- a total of 9–11 of amino acids;
- 6–7 *N*-alkylated amino acids;
- ClogP of 12.7–12.9;
- no charged functional groups, such as carboxylic acid or amine.

Paluratide (LUNA18) is synthesized at industrial level using a novel liquid-phase peptide synthesis (LPPS) method, in place of traditional solid-phase methods, to overcome challenges with *N*-alkylated cyclic peptides. This process involves a convergent route of 24 optimized synthetic steps and a final crystallization step [71]. Although resulting from innovative design and showing a potent preclinical profile, the development of paluratide (LUNA18) was discontinued in 2025 due to a narrow therapeutic window with respect to other KRAS inhibitors. Nevertheless, paluratide (LUNA18) represents a milestone in cyclic peptide drug discovery by demonstrating that rational chemical optimization can produce an orally bioavailable PPI inhibitor capable of targeting an intracellular oncogenic protein. Its development paved the way for navigating the "beyond rule of five" chemical space, showing practical criteria for permeability and bioavailability that may benefit future peptide drug programs. Specifically, the design principles behind the development of paluratide (LUNA18), including targeted *N*-alkylation, conformational control, and permeability optimization, remain informative for the design of next-generation macrocyclic peptides targeting intracellular PPIs in oncology and other therapeutic areas (Figure 5.24).

5.4.2 Zosurabalpin (ZAB): A Cyclic Peptide Antibiotic Targeting the LPS Transport

Zosurabalpin (development codes RG6006, RO7223280, or Abx-MCP) is a benzoic acid-derived tethered macrocyclic peptide, representing a new class of peptide antibiotics with zwitterionic character, designed to enhance pharmacokinetic properties such as plasma compatibility while retaining potent antibacterial activity. It specifically targets the

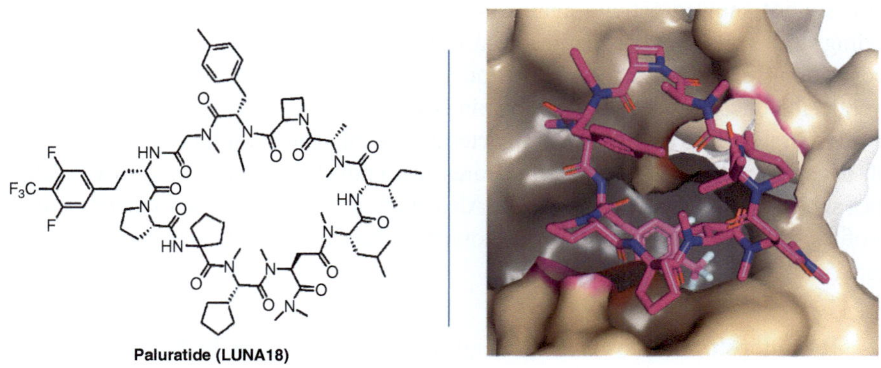

Paluratide (LUNA18)

Figure 5.24 Close-up of the X-ray crystallographic structure of human K-Ras G12D (GDP-bound) in complex with cyclic peptide inhibitor paluratide (LUNA18, magenta, PDB: 7YV1).

lipopolysaccharide (LPS) transport system in *Acinetobacter baumannii* by disrupting the outer membrane assembly and overcoming resistance mechanisms typical of infections due to carbapenem-resistant *A. baumannii* (CRAB), which is classified as a high-priority pathogen by the World Health Organization, thus marking a significant breakthrough in the field of antibiotics and offering new possibilities for treating this difficult-to-target bacterium. Its distinct mechanism, combined with potent *in vitro* and *in vivo* efficacy and improved pharmacokinetic characteristics, makes zosurabalpin a promising candidate in the fight against critical Gram-negative infections.

Carbapenem-resistant *A. baumannii* is considered a critical pathogen due to limited treatment options and high mortality rates. Conventional antibiotics, including β-lactams, colistin, and tigecycline, often show reduced or unreliable efficacy against CRAB, thus the discovery of new antibacterials is a clear medical need. Traditional small-molecules have failed to target the outer membrane and resistance mechanisms of Gram-negative bacteria, thus encouraging the exploration of alternative molecular scaffolds and targets. In this context, zosurabalpin was identified through whole-cell phenotypic screening of 44,985 macrocyclic peptides against a collection of strains of both Gram-negative and Gram-positive human pathogens. Initial hits were subsequently optimized through an iterative hit-to-lead approach to improve all key endpoints required for a new drug application, namely potency, selectivity, and tolerability, thus resulting in the identification of zosurabalpin as a lead candidate [72]. The optimization process included the modification of physicochemical properties to reduce undesirable interactions with plasma proteins, taking into account the development of zwitterionic derivatives, such as zosurabalpin, to substantially diminish plasma precipitation and improve *in vivo* tolerability profile. Unlike conventional antibiotics that target cell wall biosynthesis, protein translation, or DNA replication, zosurabalpin exhibits a unique mechanism of action by targeting the bacterial LPS transport machinery, a process essential for the assembly and maintenance of the Gram-negative outer membrane. Specifically, zosurabalpin inhibits the $LptB_2FGC$ complex, an ATP-binding cassette (ABC) transporter in the inner membrane, that is responsible for extracting LPS from the inner membrane and delivering it to the outer membrane. Such inhibition disrupts the structural asset of the outer membrane, leading to a lethal accumulation of LPS intermediates within the cell envelope and ultimately making the pathogen susceptible to lethal stress. *in vitro* studies showed activity against several panels of clinical *A. baumannii* isolates, including strains resistant to colistin and carbapenems, with promising minimum inhibitory concentration (MIC) values and efficacy on a range of resistant phenotypes.

Overall, the development of zosurabalpin is a successful example of the application of innovative strategies to expand the antibacterial arsenal against multidrug-resistant (MDR) Gram-negative pathogens, particularly addressing a novel target with innovative molecular entities such as zwitterionic macrocycles. Also, with improved pharmacokinetics and toxicity profiles, zosurabalpin represents a prototype for future classes of selective antibacterials (Figure 5.25).

5.4.3 Romidepsin: A Bicyclic Depsipeptide HDAC Inhibitor in T-Cell Lymphoma

Romidepsin (brand name: Istodax) is an approved chemotherapy drug acting as a potent and selective histone deacetylase (HDAC) inhibitor for the treatment of cutaneous T-cell lymphomas and peripheral blood T-cell lymphomas that have failed other therapies [73].

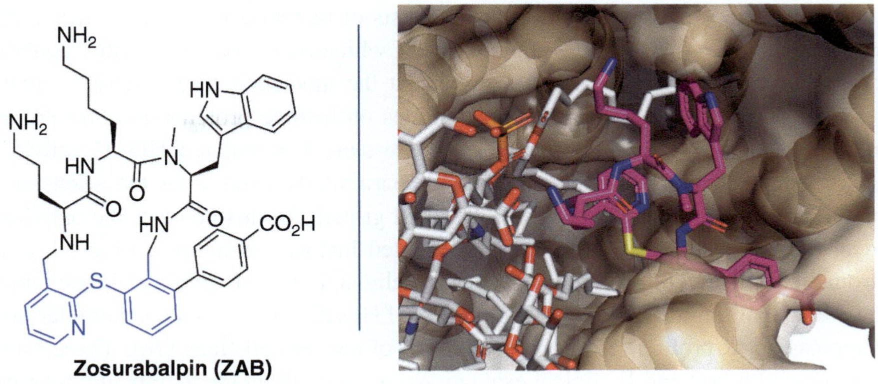

Figure 5.25 *Left: structure of zosurabalpin (the linking moiety is highlighted in blue); right: close-up of the X-ray crystallographic structure of* Acinetobacter baylyi *LptB2FG bound to lipopolysaccharide (gray) and zosurabalpin (ZAB, magenta, PDB: 8FRN).*

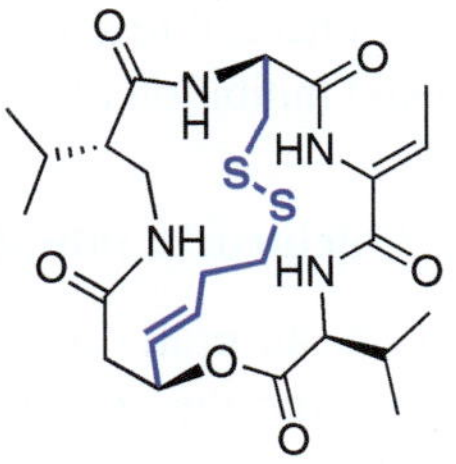

Figure 5.26 *Structure of romidepsin.*

It works by blocking key enzymes in tumor cells, leading to their death through the induction of apoptosis, altering gene expression, and arresting the cell cycle [74]. This compound is a complex naturally derived bicyclic depsipeptide prodrug characterized by a unique macrocyclic structure with an internal disulfide bond, containing L-valine, D-valine, D-cysteine, and unusual amino acid residues (Figure 5.26), which is reduced inside cells to expose a reactive thiol that binds to zinc in HDAC enzymes and inhibits their function, thus restoring histone acetylation, altering gene expression, and causing cell cycle arrest and tumor cell death [75].

Romidepsin was originally isolated from cultures of the soil bacterium *Chromobacterium violaceum* and subsequently developed as an antineoplastic agent with significant clinical activity in hematologic malignancies. The compound received U.S. FDA approval in 2009 for the treatment of cutaneous T-cell lymphoma and later in 2011 for peripheral T-cell lymphoma in patients who have received at least one prior systemic therapy. Romidepsin remains among the few epigenetic therapies with regulatory approval in hematologic oncology.

It is a hybrid peptidemacrocyclic scaffold formed by a bicyclic ring system containing a disulfide bridge.

The inhibition mechanism of romidepsin consists of targeting HDAC enzymes, particularly class I HDACs (e.g. HDAC1 and HDAC2), which play a key role in the regulation of chromatin structure and gene transcription. At the molecular level, HDACs catalyze the removal of acetyl groups from lysine residues on histone proteins, thus resulting to chromatin condensation and transcriptional repression. Unregulated HDAC activity is frequently implicated in oncogenic alteration, particularly promoting the silencing of tumor suppressor genes and the uncontrolled cell growth. In this context, the activity of romidepsin leads to the accumulation of acetylated histones, resulting to the relaxation of chromatin and growth arrest and death of malignant cells. Preclinical investigations demonstrated that romidepsin exhibits inhibition of HDAC activity in the nanomolar range and suppression of proliferation in a wide range of cancer cell lines [76]. These effects pushed toward Phase I and II clinical evaluations, particularly in the therapeutic treatment of hematologic malignancies where epigenetic dysregulation is of key importance. Following primary indications of the efficacy of this drug for targeting T-cell lymphomas, additional research was extended to other applications, for example launching preclinical and clinical studies to explore the effects of romidepsing toward acute myeloid leukemia, myelodysplastic syndromes, and a range of solid tumors. All in all, Romidepsin represents a successful entry to the translation of a natural macrocyclic depsipeptide into a clinically useful anticancer agent through targeted inhibition of histone deacetylases.

5.4.4 Voclosporin: A Novel Oral Calcineurin Inhibitor for Lupus Nephritis

Voclosporin (brand name: Lupkynis) is an orally available calcineurin inhibitor developed as an immunosuppressive agent for the treatment of lupus nephritis, a severe kidney-damaging manifestation of systemic lupus erythematosus [77]. Within this chronic autoimmune condition, 40–60% of patients develop lupus nephritis during the course of their disease, which significantly contributes to mortality rates if inadequately treated. The need for more effective and targeted therapies has encouraged research that resulted to the development of voclosporin, that received US FDA approval in January 2021 for adults with active class III, IV, or V lupus nephritis in combination with standard immunosuppressive therapy. The approval of voclosporin was the first new oral treatment specifically indicated for lupus nephritis after decades, and the European Medicines Agency (EMA) also approved voclosporin in 2022 for the same indication.

Concerning the chemical structure, voclosporin is a cyclic undecapeptide structurally related to cyclosporine A, an established calcineurin inhibitor with decades of clinical use in transplantation and for the treatment of autoimmune diseases. However, voclosporin contains a modification at the amino acid-1 position, which improves the binding affinity to its intracellular target, cyclophilin A, and results in an increased inhibition of calcineurin with an improved pharmacokinetic profile as compared to older calcineurin inhibitors (Figure 5.27).

The binding of voclosporin to cyclophilin A forms a complex that inhibits calcineurin, thus effectively preventing the dephosphorylation of nuclear factor of activated T cells, leading to reduced T-cell activation and cytokine production. In addition to its immunomodulatory effects, evidence suggests a non-immunological action in the kidney. This dual mechanism is hypothesized to contribute to both systemic immunosuppression and direct

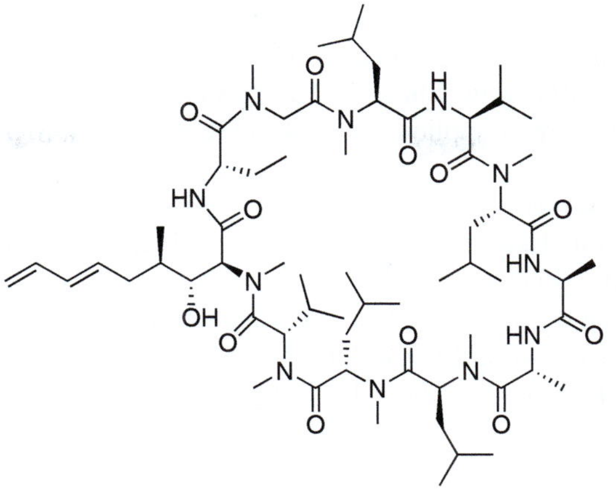

Figure 5.27 Structure of voclosporin.

renal protection, which is therapeutically advantageous in lupus nephritis. As with other calcineurin inhibitors, voclosporin's safety profile is characterized by its immunosuppressive activity and effects on renal hemodynamics.

5.5 Conclusion

Macrocyclic peptides represent a powerful and versatile class of molecules in contemporary drug discovery, bridging the gap between small molecules and biologics, due to their enhanced structural and pharmacological properties compared with linear peptides. Their conformational rigidity, structural diversity, and enhanced pharmacokinetic properties enable effective targeting of challenging biological interfaces, particularly PPIs, also taking advantage of the possibility of stabilizing defined secondary structures such as α-helices, β-sheets, and β-turns. A wide range of cyclization methods has been developed, including lactam or lactone formation, disulfide bonding, CuAAC, and RCM. Recent advances such as C—H functionalization, multicomponent reactions, heterocycle incorporation, and SuFEx chemistry have further expanded the structural diversity. Continued innovation in cyclization strategies and synthetic methodologies has greatly expanded their chemical space and accessibility. As a result, macrocyclic peptides have achieved significant clinical success, and ongoing advances in design and synthesis are expected to further strengthen their impact on therapeutic development across a broad range of disease areas.

References

1. Yudin AK. Macrocycles: lessons from the distant past, recent developments, and future directions. *Chem Sci*. 2015;**6**:30–49.
2. Cardote TA, Ciulli A. Cyclic and macrocyclic peptides as chemical tools to recognise protein surfaces and probe protein–protein interactions. *ChemMedChem*. 2016;**11**:787–794.

3. Bockus AT, McEwen CM, Lokey RS. Form and function in cyclic peptide natural products: a pharmacokinetic perspective. *Curr Top Med Chem.* 2013;**13**:821–836.
4. Wang L, Wang N, Zhang W, Cheng X, Yan Z, Shao G, Wang X, Wang R, Fu C. Therapeutic peptides: current applications and future directions. *Signal Transduct Target Ther.* 2022;**7**:48.
5. Swenson CS, Mandava G, Thomas DM, Moellering RE. Tackling undruggable targets with designer peptidomimetics and synthetic biologics. *Chem Rev.* 2024;**124**:13020–13093.
6. Chow HY, Zhang Y, Matheson E, Li X. Ligation technologies for the synthesis of cyclic peptides. *Chem Rev.* 2019;**119**:9971–10001.
7. Fang P, Pang WK, Xuan S, Chan WL, Leung KC. Recent advances in peptide macrocyclization strategies. *Chem Soc Rev.* 2024;**53**:11725–11771.
8. Bechtler C, Lamers C. Macrocyclization strategies for cyclic peptides and peptidomimetics. *RSC Med Chem.* 2021;**12**:1325–1351.
9. Ji X, Nielsen AL, Heinis C. Cyclic peptides for drug development. *Angew Chem Int Ed Engl.* 2024;**63**:e202308251.
10. Driggers EM, Hale SP, Lee J, Terrett NK. The exploration of macrocycles for drug discovery-an underexploited structural class. *Nat Rev Drug Discov.* 2008;**7**:608–624.
11. Merritt HI, Sawyer N, Arora PS. Bent into shape: folded peptides to mimic protein structure and modulate protein function. *Pept Sci.* 2020;**112**:e24145.
12. Park SE, Sajid MI, Parang K, Tiwari RK. Cyclic cell-penetrating peptides as efficient intracellular drug delivery tools. *Mol Pharm.* 2019;**16**:3727–3743.
13. Hayes HC, Luk LYP, Tsai YH. Approaches for peptide and protein cyclisation. *Org Biomol Chem.* 2021;**19**:3983–4001.
14. Das S, Pradhan TK, Samanta R. Recent progress on transition metal catalyzed macrocyclizations based on C—H bond activation at heterocyclic scaffolds. *Chem Asian J.* 2024;**19**:e202400397.
15. Matinkhoo K, Pryyma A, Todorovic M, Patrick BO, Perrin DM. Synthesis of the Death-cap mushroom toxin α-amanitin. *J Am Chem Soc.* 2018;**140**:6513–6517.
16. Yang J, Talibov VO, Peintner S, Rhee C, Poongavanam V, Geitmann M, Rossi Sebastiano M, Simon B, Hennig J, Dobritzsch D, Danielson UH, Kihlberg J. Macrocyclic peptides uncover a novel binding mode for reversible inhibitors of LSD1. *ACS Omega.* 2020;**5**:3979–3995.
17. Goto Y, Ohta A, Sako Y, Yamagishi Y, Murakami H, Suga H. Reprogramming the translation initiation for the synthesis of physiologically stable cyclic peptides. *ACS Chem Biol.* 2008;**3**:120–129.
18. Tian Y, Yang D, Ye X, Li Z. Thioether-derived macrocycle for peptide secondary structure fixation. *Chem Rec.* 2017;**17**:874–885.
19. Kale SS, Villequey C, Kong XD, Zorzi A, Deyle K, Heinis C. Cyclization of peptides with two chemical bridges affords large scaffold diversities. *Nat Chem.* 2018;**10**:715–723.
20. Peraro L, Siegert TR, Kritzer JA. Conformational restriction of peptides using dithiol Bis-alkylation. *Methods Enzymol.* 2016;**580**:303–332.
21. Chen S, Morales-Sanfrutos J, Angelini A, Cutting B, Heinis C. Structurally diverse cyclisation linkers impose different backbone conformations in bicyclic peptides. *ChemBioChem.* 2012;**13**:1032–1038.

22. Timmerman P, Beld J, Puijk WC, Meloen RH. Rapid and quantitative cyclization of multiple peptide loops onto synthetic scaffolds for structural mimicry of protein surfaces. *ChemBioChem*. 2005;**6**:821–824.
23. Merz ML, Habeshian S, Li B, David JGL, Nielsen AL, Ji X, Il Khwildy K, Duany Benitez MM, Phothirath P, Heinis C. De novo development of small cyclic peptides that are orally bioavailable. *Nat Chem Biol*. 2024;**20**:624–633.
24. Schafmeister CE, Po J, Verdine GL. An all-hydrocarbon cross-linking system for enhancing the helicity and metabolic stability of peptides. *J Am Chem Soc*. 2000;**122**:5891–5892.
25. Walensky LD, Bird GH. Hydrocarbon-stapled peptides: principles, practice, and progress. *J Med Chem*. 2014;**57**:6275–6288.
26. Lau YH, de Andrade P, Wu Y, Spring DR. Peptide stapling techniques based on different macrocyclisation chemistries. *Chem Soc Rev*. 2015;**44**:91–102.
27. Ali AM, Atmaj J, Van Oosterwijk N, Groves MR, Dömling A. Stapled peptides inhibitors: a new window for target drug discovery. *Comput Struct Biotechnol J*. 2019;**17**:263–281.
28. Blackwell HE, Grubbs RH. Highly efficient synthesis of covalently cross-linked peptide helices by ring-closing metathesis. *Angew Chem Int Ed Engl*. 1998;**37**:3281–3284.
29. Trnka TM, Grubbs RH. The development of L2X2Ru=CHR olefin metathesis catalysts: an organometallic success story. *Acc Chem Res*. 2001;**34**:18–29.
30. Kim YW, Grossmann TN, Verdine GL. Synthesis of all-hydrocarbon stapled α-helical peptides by ring-closing olefin metathesis. *Nat Protoc*. 2011;**6**:761–771.
31. Sousbie M, Besserer-Offroy É, Brouillette RL, Longpré JM, Leduc R, Sarret P, Marsault É. In search of the optimal macrocyclization site for neurotensin. *ACS Med Chem Lett*. 2018;**9**:227–232.
32. Li H, Aneja R, Chaiken I. Click chemistry in peptide-based drug design. *Molecules*. 2013;**18**:9797–9817.
33. Holub JM, Kirshenbaum K. Tricks with clicks: modification of peptidomimetic oligomers via copper-catalyzed azide-alkyne [3+2] cycloaddition. *Chem Soc Rev*. 2010;**39**:1325–1337.
34. Kolb HC, Finn MG, Sharpless KB. Click chemistry: diverse chemical function from a few good reactions. *Angew Chem Int Ed Engl*. 2001;**40**:2004–2021.
35. Jacobsen Ø, Maekawa H, Ge NH, Görbitz CH, Rongved P, Ottersen OP, Amiry-Moghaddam M, Klaveness J. Stapling of a 3(10)-helix with click chemistry. *J Org Chem*. 2011;**76**:1228–1238.
36. Gori A, Wang CI, Harvey PJ, Rosengren KJ, Bhola RF, Gelmi ML, Longhi R, Christie MJ, Lewis RJ, Alewood PF, Brust A. Stabilization of the cysteine-rich conotoxin MrIA by using a 1,2,3-triazole as a disulfide bond mimetic. *Angew Chem Int Ed Engl*. 2015;**54**:1361–1364.
37. Worrell BT, Malik JA, Fokin VV. Direct evidence of a dinuclear copper intermediate in Cu(I)-catalyzed azide-alkyne cycloadditions. *Science*. 2013;**340**:457–460.
38. Li S, Cai H, He J, Chen H, Lam S, Cai T, Zhu Z, Bark SJ, Cai C. Extent of the oxidative side reactions to peptides and proteins during the CuAAC reaction. *Bioconjug Chem*. 2016;**27**:2315–2322.
39. Sun H, Liu L, Lu J, Qiu S, Yang CY, Yi H, Wang S. Cyclopeptide Smac mimetics as antagonists of IAP proteins. *Bioorg Med Chem Lett*. 2010;**20**:3043–3046.

40. Shepperson OA, Malone MA, Arnott KIM, Brown RJ, Jamieson AG. On-resin synthesis and late-stage functionalization of macrocyclic atosiban mimetics via 5-iodo-1,4-triazoles. *Org Lett*. 2025;**27**:11249–11253.
41. Wang W, Lorion MM, Shah J, Kapdi AR, Ackermann L. Late-stage peptide diversification by position-selective C—H activation. *Angew Chem Int Ed Engl*. 2018;**57**: 14700–14717.
42. Rivera DG, Ojeda-Carralero GM, Reguera L, Van der Eycken EV. Peptide macrocyclization by transition metal catalysis. *Chem Soc Rev*. 2020;**49**:2039–2059.
43. Shabani S, Wu Y, Ryan HG, Hutton CA. Progress and perspectives on directing group-assisted palladium-catalysed C—H functionalisation of amino acids and peptides. *Chem Soc Rev*. 2021;**50**:9278–9343.
44. Dong H, Limberakis C, Liras S, Price D, James K. Peptidic macrocyclization via palladium-catalyzed chemoselective indole C-2 arylation. *Chem Commun*. 2012;**48**: 11644–11646.
45. Mendive-Tapia L, Preciado S, García J, Ramón R, Kielland N, Albericio F, Lavilla R. New peptide architectures through C—H activation stapling between tryptophan-phenylalanine/tyrosine residues. *Nat Commun*. 2015;**6**:7160.
46. Bai Z, Cai C, Sheng W, Ren Y, Wang H. Late-stage peptide macrocyclization by palladium-catalyzed site-selective C—H olefination of tryptophan. *Angew Chem Int Ed Engl*. 2020;**59**:14686–14692.
47. Bai Z, Cai C, Yu Z, Wang H. Backbone-enabled directional peptide macrocyclization through late-stage palladium-catalyzed δ-C(sp^2)—H olefination. *Angew Chem Int Ed Engl*. 2018;**57**:13912–13916.
48. Tang J, Chen H, He Y, Sheng W, Bai Q, Wang H. Peptide-guided functionalization and macrocyclization of bioactive peptidosulfonamides by Pd(II)-catalyzed late-stage C—H activation. *Nat Commun*. 2018;**9**:3383.
49. Liu J, Liu X, Zhang F, Qu J, Sun H, Zhu Q. A peptide stapling strategy with built-in fluorescence by direct late-stage C(sp^2)—H olefination of tryptophan. *Chemistry*. 2020;**26**:16122–16128.
50. Ruan Z, Sauermann N, Manoni E, Ackermann L. Manganese-catalyzed C—H alkynylation: expedient peptide synthesis and modification. *Angew Chem Int Ed Engl*. 2017;**56**:3172–3176.
51. Tang J, He Y, Chen H, Sheng W, Wang H. Synthesis of bioactive and stabilized cyclic peptides by macrocyclization using C(sp^3)—H activation. *Chem Sci*. 2017;**8**: 4565–4570.
52. Zhang X, Lu G, Sun M, Mahankali M, Ma Y, Zhang M, Hua W, Hu Y, Wang Q, Chen J, He G, Qi X, Shen W, Liu P, Chen G. A general strategy for synthesis of cyclophane-braced peptide macrocycles via palladium-catalysed intramolecular sp^3 C—H arylation. *Nat Chem*. 2018;**10**:540–548.
53. Li B, Li X, Han B, Chen Z, Zhang X, He G, Chen G. Construction of natural-product-like cyclophane-braced peptide macrocycles via sp^3 C—H arylation. *J Am Chem Soc*. 2019;**141**:9401–9407.
54. Gong W, Zhang G, Liu T, Giri R, Yu JQ. Site-selective C(sp3)-H functionalization of di-, tri-, and tetrapeptides at the *N*-terminus. *J Am Chem Soc*. 2014;**136**:16940–16946.
55. Reguera L, Rivera DG. Multicomponent reaction toolbox for peptide macrocyclization and stapling. *Chem Rev*. 2019;**119**:9836–9860.

56. Domling A, Ugi I. Multicomponent reactions with isocyanides. *Angew Chem Int Ed.* 2000;**39**:3168–3210.
57. Wu P, Givskov M, Nielsen TE. Reactivity and synthetic applications of multicomponent petasis reactions. *Chem Rev.* 2019;**119**:11245–11290.
58. Hili R, Rai V, Yudin AK. Macrocyclization of linear peptides enabled by amphoteric molecules. *J Am Chem Soc.* 2010;**132**:2889–2891.
59. Zaretsky S, Hickey JL, Tan J, Pichugin D, St Denis MA, Ler S, Chung BKW, Scully CCG, Yudin AK. Mechanistic investigation of aziridine aldehyde-driven peptide macrocyclization: the imidoanhydride pathway. *Chem Sci.* 2015;**6**:5446–5455.
60. Rivera DG, Vasco AV, Echemendía R, Concepción O, Pérez CS, Gavín JA, Wessjohann LA. A multicomponent conjugation strategy to unique *N*-steroidal peptides: first evidence of the steroidal nucleus as a β-turn inducer in acyclic peptides. *Chemistry.* 2014;**20**:13150–13161.
61. Krajcovicova S, Spring DR. Tryptophan in multicomponent petasis reactions for peptide stapling and late-stage functionalisation. *Angew Chem Int Ed Engl.* 2023;**62**:e202307782.
62. Hamdan F, Tahoori F, Balalaie S. Synthesis of novel cyclopeptides containing heterocyclic skeletons. *RSC Adv.* 2018;**8**:33893–33926.
63. Malins LR, deGruyter JN, Robbins KJ, Scola PM, Eastgate MD, Ghadiri MR, Baran PS. Peptide macrocyclization inspired by non-ribosomal imine natural products. *J Am Chem Soc.* 2017;**139**:5233–5241.
64. Dong J, Krasnova L, Finn MG, Sharpless KB. Sulfur(VI) fluoride exchange (SuFEx): another good reaction for click chemistry. *Angew Chem Int Ed Engl.* 2014;**53**:9430–9448.
65. Liu Z, Li J, Li S, Li G, Sharpless KB, Wu P. SuFEx click chemistry enabled late-stage drug functionalization. *J Am Chem Soc.* 2018;**140**:2919–2925.
66. Barrow AS, Smedley CJ, Zheng Q, Li S, Dong J, Moses JE. The growing applications of SuFEx click chemistry. *Chem Soc Rev.* 2019;**48**:4731–4758.
67. Seyrani H, Heidarzadeh Vazifehkhorani H, Outlaw VK. Chemoselective sulfonyl fluoride exchange (SuFEx)-induced macrocyclization of tyrosine-containing peptides in aqueous media. *Chem Sci.* 2025;**16**:21359–21367.
68. Zhang H, Chen S. Cyclic peptide drugs approved in the last two decades (2001–2021). *RSC Chem Biol.* 2021;**3**:18–31.
69. Tanada M, Tamiya M, Matsuo A, Chiyoda A, Takano K, Ito T, Irie M, Kotake T, Takeyama R, Kawada H, Hayashi R, Ishikawa S, Nomura K, Furuichi N, Morita Y, Kage M, Hashimoto S, Nii K, Sase H, Ohara K, Ohta A, Kuramoto S, Nishimura Y, Iikura H, Shiraishi T. Development of orally bioavailable peptides targeting an intracellular protein: from a hit to a clinical KRAS inhibitor. *J Am Chem Soc.* 2023;**145**:16610–16620.
70. Kage M, Hayashi R, Matsuo A, Tamiya M, Kuramoto S, Ohara K, Irie M, Chiyoda A, Takano K, Ito T, Kotake T, Takeyama R, Ishikawa S, Nomura K, Furuichi N, Morita Y, Hashimoto S, Kawada H, Nishimura Y, Nii K, Sase H, Ohta A, Kojima T, Iikura H, Tanada M, Shiraishi T. Structure-activity relationships of middle-size cyclic peptides, KRAS inhibitors derived from an mRNA display. *Bioorg Med Chem.* 2024;**110**:117830.

71. Hou Z, Komiya S, Iwasaki K, Kondo Y, Momose Y, Sasakura K, Tanaka K, Wadamoto M, Chia P, Dan K, Ejima H, Fujisawa H, Fukuzawa A, Hashimoto S, Ishizawa T, Ito H, Nakagawa Y, Sato H, Sato N, Sekita H, Serizawa H, Tamura S, Enomoto T, Honma A, Ikeda T, Murakata M, Shiina J, Yabuzaki T, Tsukazaki M, Iwamura H, Maeda K. Major progress in the liquid-phase synthesis of *N*-alkyl-rich cyclic peptides: kilogram-scale GMP production of the KRAS inhibitor LUNA18. *Org Process Res Dev*. 2025;**29**:2764–2782.
72. Pahil KS, Gilman MSA, Baidin V, Clairfeuille T, Mattei P, Bieniossek C, Dey F, Muri D, Baettig R, Lobritz M, Bradley K, Kruse AC, Kahne D. A new antibiotic traps lipopolysaccharide in its intermembrane transporter. *Nature*. 2024;**625**:572–577.
73. Alpdogan O, Kartan S, Johnson W, Sokol K, Porcu P. Systemic therapy of cutaneous T-cell lymphoma (CTCL). *Chin Clin Oncol*. 2019;**8**:10.
74. VanderMolen KM, McCulloch W, Pearce CJ, Oberlies NH. Romidepsin (Istodax, NSC 630176, FR901228, FK228, depsipeptide): a natural product recently approved for cutaneous T-cell lymphoma. *J Antibiot*. 2011;**64**:525–531.
75. El Omari N, Lee LH, Bakrim S, Makeen HA, Alhazmi HA, Mohan S, Khalid A, Ming LC, Bouyahya A. Molecular mechanistic pathways underlying the anticancer therapeutic efficiency of romidepsin. *Biomed Pharmacother*. 2023;**164**:114774.
76. Klimek VM, Fircanis S, Maslak P, Guernah I, Baum M, Wu N, Panageas K, Wright JJ, Pandolfi PP, Nimer SD. Tolerability, pharmacodynamics, and pharmacokinetics studies of depsipeptide (romidepsin) in patients with acute myelogenous leukemia or advanced myelodysplastic syndromes. *Clin Cancer Res*. 2008;**14**:826–832.
77. McArn AC, Nixon AR, Jarrell KL. Voclosporin: a novel calcineurin inhibitor for the treatment of lupus nephritis. *Ann Pharmacother*. 2022;10600280221075331.

6

Stapled Peptides

6.1 Introduction

Protein–protein interactions (PPIs) are fundamental to almost all biological processes, including signal transduction, transcriptional regulation, cell metabolism, and apoptosis. Many of these interactions are mediated by relatively large and flat protein surfaces rather than deep binding pockets; thus, their targeting with conventional therapeutic approaches is generally difficult. Traditional small-molecule drugs (<500 Da) typically exhibit good oral bioavailability and synthetic accessibility, but they are generally suited for binding to well-defined cavities and often lack selectivity when applied to complex PPIs [1]. In contrast, biologics such as antibodies and recombinant proteins (>5,000 Da) can effectively recognize extended protein surfaces but suffer from poor cell permeability and oral bioavailability, metabolic instability, and limitation to extracellular targets [2]. Consequently, a large fraction of the proteome, such as transcription factors or signaling proteins, remains essentially undruggable.

Peptides have emerged as promising candidates to bridge the gap between small molecules and biologics. Their intermediate molecular weight allows them to combine the high target specificity typical of proteins with the synthetic accessibility and modularity of smaller compounds. Importantly, many PPIs are mediated by short secondary-structure motifs, especially α-helices, which are the most abundant structural elements at protein interaction interfaces controlling cellular functions. However, when α-helical segments are removed from their native protein context, they typically lose their secondary structure in solution and are found in a random-coil arrangement, thus reducing the binding affinity and biological activity. While α-helices are widely used to mediate PPIs, β-hairpins are recurrent in biomolecular recognition systems such as antibodies and T-cell receptors and are highly attractive for peptidomimetic design. A β-hairpin consists of two antiparallel β-strands connected by a loop, and despite such a simple structure, they show a remarkable degree of conformational diversity. In fact, β-hairpins vary substantially in length, hydrogen-bonding arrangement, and loop size, spanning five or fewer residues, and distinct β-turn types can be found depending on loop length, all characterizing the versatility of β-hairpins as recognition motifs [3].

The main limitation of small peptides in establishing bioactive conformations is the high entropy loss due to folding. Unconstrained peptides experience a large conformational space; thus, the stabilization of a single folded conformational state is unfavorable, overall resulting in linear peptides being rapidly degraded by proteases and experiencing poor membrane permeability. To overcome these limitations, numerous chemical strategies have been developed to stabilize peptide secondary structure, to improve stability, and to enhance the pharmacological profile [4].

Among these strategies, peptide stapling has emerged as one of the most powerful and versatile approaches [5]. It refers to the covalent tethering of two amino acid side chains within a peptide sequence to form a macrocycle to constrain the peptide backbone into a defined conformation. By reducing the conformational flexibility, stapling reduces the entropic loss associated with folding and promotes the formation of stable secondary structures, generally α-helices [6]. Helical stapled peptides often display enhanced helicity, increased binding affinity to their protein targets, improved resistance to proteolytic degradation and often display increased cellular uptake as compared with their unmodified counterparts.

Early efforts to stabilize α-helical peptides were reported the eighties and included covalent cross-links such as lactam bridges between lysine and aspartate or glutamate residues, as well as disulfide bonds. Although these approaches demonstrated that side-chain cross-linking could stabilize helical structures, they were often limited by chemical lability or insufficient improvements in helicity.

A major breakthrough occurred with the development of all-hydrocarbon stapling. Specifically, Blackwell and Grubbs demonstrated the feasibility of forming hydrocarbon cross-links on peptide templates using olefin metathesis [7]. Based on such initial report, a few years later Verdine and co-workers introduced α,α-disubstituted non-natural amino acids bearing olefinic side chains and applied ruthenium-catalyzed ring-closing metathesis (RCM) to generate robust all-hydrocarbon staples [8]. This strategy produced peptides with significantly enhanced α-helical content and stability and established the modern concept of stapled peptides. Since this seminal work, peptide-stapling chemistry has expanded significantly, and peptide-stapling approach has been applied to investigate numerous PPIs, including targets ranging from cancer to infectious diseases. In addition to hydrocarbon staples formed *via* RCM, a wide range of macrocyclization chemistries has been developed, including lactam bridges, triazole linkages formed by copper(I)-catalyzed azide–alkyne cycloaddition (CuAAC), thioether and thioacetal bridges, cysteine arylation, alkylation strategies, and transition metal-catalyzed C—H activation [9, 10]. Such chemical diversity enables the fine-tuning of peptide conformation, physicochemical properties, and biological activity, thus allowing stapled peptides to be optimized for different targets and applications.

Another breakthrough in the field was the demonstration that stapled peptides could function inside cells. Specifically, Walensky reported the development of hydrocarbon-stapled BH3 α-helical peptides targeting BCL-2 family proteins, which regulate mitochondrial apoptosis [11]. These stapled peptides, as well as showing enhanced protease resistance and binding affinity, proved to possess remarkable cellular permeability, thus enabling the modulation of intracellular PPIs and demonstrating the application of stapled peptides as feasible intracellular therapeutics. Following advances of this class of compounds to improve the PPI targeting, the therapeutic potential of stapled peptides was further supported by their progression into clinical development. A representative example is ALRN-6924, a hydrocarbon-stapled peptide that mimics the α-helical transactivation domain of p53 and acts as a dual inhibitor of MDM2 and MDMX. This compound advanced into Phase 1 and Phase 2 clinical trials for hematological malignancies and solid tumors, providing an early indication of the translational potential of stapled α-helical peptides [12].

Despite these advances, some challenges are still under study, as not all stapled peptides exhibit sufficient cellular uptake, and strategies such as the bioconjugation of stapled peptides with cell-penetrating sequences can increase molecular weight and add toxicity issues. Nevertheless, macrocyclization itself has been shown to enable cell permeability beyond traditional "rule-of-five" limitations, as exemplified by natural products like cyclosporin and recent macrocyclic peptide inhibitors. Thus, continued study for the optimization of staple type, position, and overall physicochemical properties is of primary interest.

6.2 Strategic Approaches for Peptide-Stapling Design

As anticipated, when removed from their native protein environment, folded peptide epitopes frequently lose their secondary structure, leading to reduced binding affinity, low proteolytic stability, and diminished biological activity. Thus, a broad range of chemical strategies has been developed to constrain peptides into well-defined secondary conformations, particularly denoted as peptide-stapling approaches [13]. In this context, structure-based approaches are a fundamental tool in modern stapled peptide design. Specifically, high-resolution structures of protein–protein complexes can reveal secondary-structure epitopes, often α-helical in shape, that are directly involved in interacting surface areas on target proteins, thus enabling the identification of candidate peptide epitopes for stabilization. Once a core interacting folded peptide epitope is selected, computational modeling and molecular dynamics simulations can provide an additional contribution to guide the choice of stapling positions and linker length compatible with the native binding geometry. In the context of a hierarchical peptidomimetic approach, alanine scanning is often used experimentally to identify residues that tolerate substitution without loss of activity in the selected peptide epitope, thus identifying the optimal sites for staple installation. Concerning bioactivity read-out methods in such design workflow, biophysical characterization is essential to confirm successful conformational stabilization.

Concerning secondary-structure mimicking, circular dichroism spectroscopy is routinely employed to assess secondary structure, as α-helical and β-strand peptides exhibit characteristic spectral profiles distinct from unfolded peptides displaying random-coil conformations. This can be a quantitative measure, suggesting that an increased signal intensity at diagnostic wavelengths directly reflects enhanced helicity and provides a quantitative measure of stapling efficiency [14]. Unstructured peptides typically display a single pronounced minimum near 195 nm, reflecting a random-coil state. In contrast, α-helical peptides exhibit a distinctive spectral pattern characterized by a positive band around 190–195 nm, together with two negative minima at 208 and 222 nm. Concerning β-strands, CD profiles are characterized by a negative band in the 210–220 nm or 200 nm regions for β-I and β-II subtypes, respectively, and a positive band around 195 nm [15]. In the context of helical peptide stapling, CD measurements routinely reveal substantial gains in helicity relative to linear analogues, in particular, when chemical staples connect more distant residues (distal stapling) along the sequence or when multiple staples are introduced (i.e. double linking) with respect to proximal stapling (Figure 6.1), as these constraints more effectively lock the peptide into a stable α-helical conformation [17].

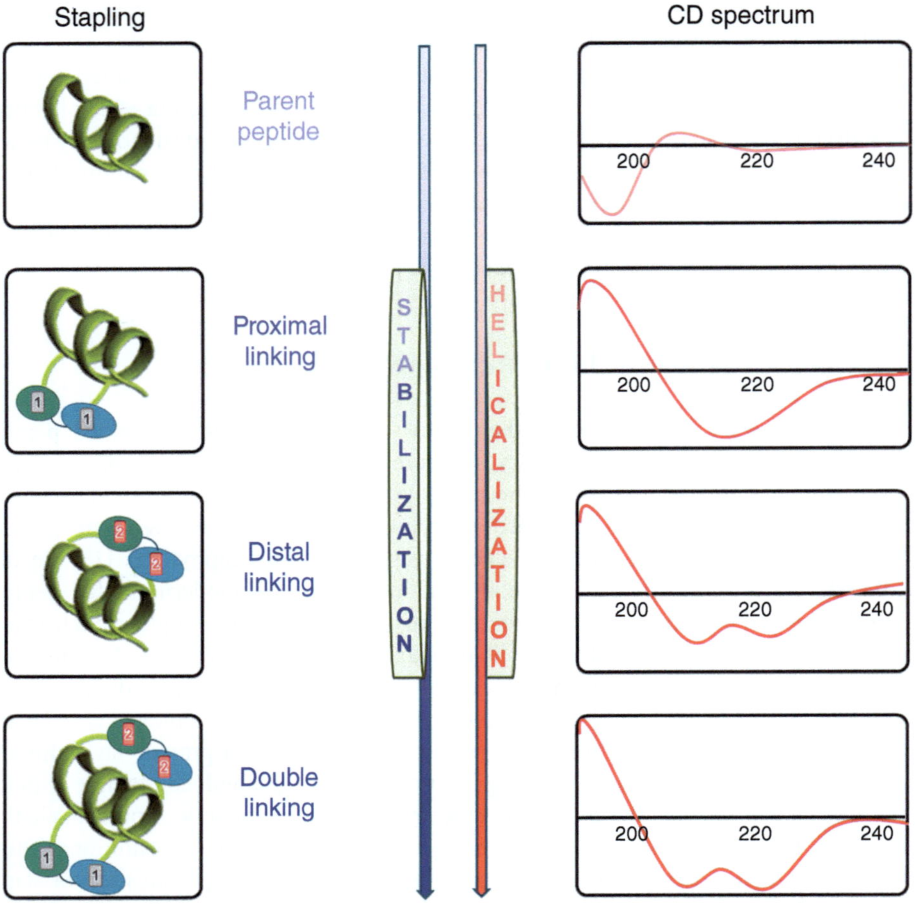

Figure 6.1 *CD signatures for helicity and types of peptide stapling. As stabilization progresses, the CD spectrum exhibits two distinct minima. Source: [16] / MDPI / CC BY 4.0.*

Based on these concepts, a general workflow for stapled peptide design emerged [18], typically beginning with the identification of a α-helical peptide epitope derived from a native PPI interface. Next, a systematic "staple scan" is performed to identify optimal stapling positions, often starting with a well-established chemistry. A classical and widely used strategy for helix peptides relies on covalently linking side chains positioned on the same face of an α-helix, and starting with i to $i + 4$ or i to $i + 7$ staples, the latter being preferred to achieve improved helical stabilization and proteolytic resistance [19]. Peptide length is then optimized through *N*- and *C*-terminal truncations or extensions, the latter usually by adding a sequence containing several alanine residues to increase hydrophobicity that is beneficial to improve the helical propensity. Subsequent refinement involves varying staple type, rigidity, and number to improve affinity, stability, and cellular uptake [20]. Finally, fine-tuning optimization at the level of single residues is carried out to improve target engagement and minimize off-target effects. All these combined approaches have transformed peptide

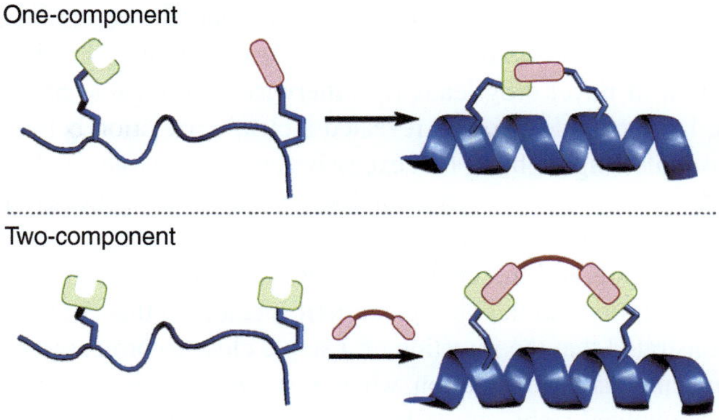

Figure 6.2 One- and two-component strategy for peptide stapling. Source: [9] / with permission of the Royal Society of Chemistry.

stapling into a mature and versatile platform for stabilizing peptide motifs, especially those displaying α-helical character, thus enabling the discovery of peptide-based modulators of challenging PPIs.

Concerning the chemistry of peptide stapling, synthetic methods have been classified into one-component and two-component strategies (Figure 6.2) [9]. In one-component stapling, two non-natural amino acids possessing complementary reactive groups on their side chains are incorporated into the peptide at specific sequence positions as required by the staple and directly linked through the paired functional groups by a covalent bond. This approach is particularly common for hydrocarbon stapling, where olefin-containing residues undergo intramolecular cyclization. In contrast, two-component stapling employs an external bifunctional linker to connect two modified side chains, thus offering greater modularity although requiring an increased synthetic complexity. Despite the availability of both approaches, one-component stapling remains the most widely adopted, especially for i to $i + 4$ and i to $i + 7$ tethers.

Early work focused on i to $i + 4$ linkages that mimic native salt bridges found in protein helices. One of the first demonstrations involved lactam bridges between lysine and aspartate or glutamate residues to stabilize helix mimics of human growth hormone-releasing factor [21]. Since then, this side-chain cross-linking concept has expanded to include other chemical tethers, such as disulfide bonds, thioethers, triazoles, and olefin-based linkers, enabling fine control over helix stability and physicochemical properties.

Staple position and topology strongly influence peptide structure and function, and the choice of linker is crucial, as it must be compatible with the desired α-helical structure and the target site [22]. Shorter staples typically stabilize a single helical turn, whereas longer staples encompassing i to $i + 7$ or i to $i + 11$ residues can reinforce extended helices and provide enhanced resistance to proteolytic degradation [23]. Longer and more rigid constraints often lead to improved cellular activity, because stronger conformational locking reduces backbone exposure to solvent and lowers the energetic cost of membrane translocation. Also, peptides containing multiple staples, such as double-stapled and stitched or cross-stitched molecular structures, have been reported to restrict conformational flexibility and enhance biological activity [24].

High-throughput screening strategies have been introduced to optimize stapled helical peptides, taking advantage of the development of various stapling chemistry techniques, for the identification of novel drug leads for otherwise undruggable protein targets [25]. In this approach, large peptide libraries revealed a clear correlation between lipophilicity and cellular uptake, also highlighting that excessive cationic charge can lead to membrane disruption and off-target toxicity, whereas anionic residues are generally better tolerated and can improve overall safety profiles.

Concerning the position and the chemical nature of a peptide staple, this may impact both the structural stability and the biological performance. In this context, Walensky and co-workers demonstrated that the insertion of a staple close to protease cleavage sites can effectively prevent enzymatic degradation, whereas staples positioned outside these regions may induce good α-helical stabilization but contribute to a lesser extent to reducing the proteolysis [26].

Concerning the role of chemical staples in physicochemical properties, the hydrophobic surface area introduced by the staple is a critical determinant of aqueous solubility and membrane permeability, thus contributing to overall bioactivity, especially regarding intracellular PPI [27]. This is a critical issue for successful stapled peptide design, as the efficient delivery of stapled peptides to intracellular compartments such as the cytosol or nucleus remains challenging, and studies necessitate careful integration of staple positioning, chemical composition, and sequence-level modifications to achieve optimal intracellular efficacy.

6.3 Chemical Methods for Peptide Stapling

Concerning synthetic methods for peptide stapling, similar chemistry as described in Chapter 5 for peptidomimetic macrocycles has been applied for the development of stapled peptides, taking into account the choice of suitable linker strategy to mimic the secondary structure of the precursor peptide. One-component peptide-stapling strategies [5, 9] have been developed to stabilize bioactive conformations using a single intramolecular linkage formed directly between side chains within the peptide sequence, to overcome limitations of early peptide stapling that relied on non-covalent salt bridges between oppositely charged residues, most commonly lysine and glutamic acid [28]. The structural outcome of one-component stapling is strongly influenced by the length and geometry of the employed side-chain tethers. These parameters determine whether the staple spans one or two helical turns, thus modulating α-helical stability and overall peptide conformation.

Early strategies focused on side-chain cyclization between naturally occurring amino acids, most notably lactam formation *via* amide bond formation between lysine and glutamic or aspartic acid residues, that although effective, offered limited structural diversity [21]. Successively, a broad range of stapling methodologies has been introduced, employing natural amino acids, unnatural amino acids, or combinations of both [1]. These strategies encompass diverse chemical reactions that expand the structural and functional repertoire of stapled peptides, enabling improved conformational control and facilitating the discovery of α-helix-based inhibitors of PPIs. Representative approaches (Figure 6.3) include disulfide bridge formation, thioether linkages, RCM, Cu(I)-catalyzed azide–alkyne cycloaddition, Pd-catalyzed C—H activation, and other residue-specific

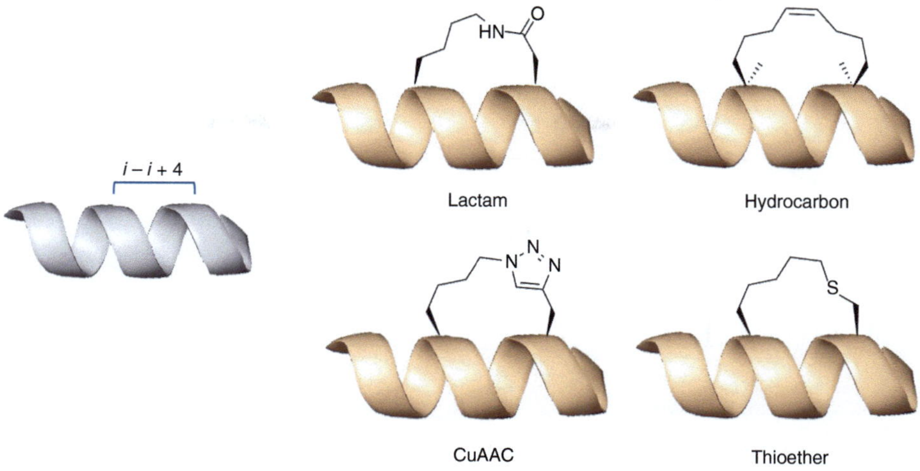

Figure 6.3 *Representative one-component stapling strategies of helical peptides with* i – i + 4 *spacing.*

pairing strategies employing linkers in a two-component approach (e.g., Lys–Lys, Cys–Cys, or Lys–Cys) [9, 29, 30]. Among these, all-hydrocarbon stapling has had the greatest impact, incorporating olefin-bearing unnatural amino acids followed by RCM to generate conformationally constrained helices with enhanced stability, cell permeability, and proteolytic resistance [8]. Other synthetic strategies encompassing the aromatic side chain of natural residues, such as the intramolecular Pd-catalyzed C—H activation process of a iodo-tyrosine and tryptophan to afford the corresponding Tyr–Trp-stapled peptides, and cross-coupling employing borylated and halogenated phenylalanine derivatives.

Concerning two-component peptide stapling, this is a modular macrocyclization strategy in which an external bifunctional linker connects two side chains within a peptide, thus expanding the chemical toolbox of peptide stapling by enabling independent variation of the staple structure and peptide sequence [5, 9], thus allowing for fine control over conformational constraint and functionality.

A wide range of two-component stapling reactions has been reported, including photo-controllable macrocyclization reactions [31] and chemical staples based on alkyl, aromatic, perfluoroaryl, and tetrazine linkers [32–35]. Cysteine is the most commonly employed anchoring residue because the thiol side chain is highly nucleophilic and can be applied for thiol alkylation with α-halocarbonyl electrophiles, Michael addition to activated alkenes, and bis-alkylation or bis-arylation reactions. Such chemoselective transformations proceed efficiently under mild conditions and have made cysteine cross-linking a central platform for two-component stapling. Alternative anchoring strategies using lysine (*via* amine alkylation or acylation), tryptophan (through C—H functionalization), and non-canonical amino acids bearing alkyne groups have also been developed.

As to the reaction pathway, many two-component stapling strategies follow a stepwise process in which a bifunctional linker first reacts with one residue to generate an acyclic intermediate, followed by a second intramolecular reaction to close the macrocycle. Representative examples include double CuAAC, double thiol–ene coupling,

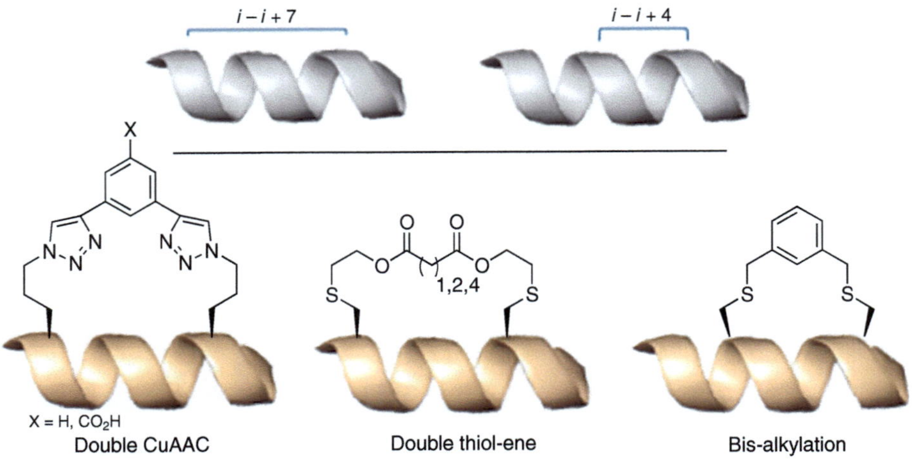

Figure 6.4 *Representative examples of chemical approaches in two-component stapling.*

and sequential bis-alkylation reactions (Figure 6.4). An important subclass in this series incorporates linkers bearing additional functional handles. For example, dialkyne linkers enable post-stapling diversification through further CuAAC reactions, whereas carbonyl-containing bridges allow subsequent modification *via* oxime ligation, providing a highly adaptable framework for stabilizing peptide secondary structure while enabling functionalization for biological and therapeutic applications.

6.3.1 Disulfide Bond: A Reversible Stapling Linker

Disulfide bonds are a fundamental covalent strategy in proteins, formed through the oxidation of proximal cysteine thiols. They stabilize secondary to quaternary protein structures and maintain conformational integrity by tethering distant regions in α-helices, β-sheets, and other secondary structures. Beyond structural stabilization, disulfide bridges influence protein function, including enzymatic activity, by preserving the spatial arrangement of active sites.

Accordingly, disulfide chemistry found in naturally occurring peptides has been used in short peptides to stabilize both β-hairpins and α-helices. For example, the bee venom peptide apamin contains a crossed disulfide pattern connecting a *C*-terminal helix to an *N*-terminal loop (Figure 6.5) that enables maintaining the helical structure under extreme conditions, including high temperature and low pH [36]. Replacement studies of cysteine with alanine showed significantly reduction of helicity, and that an i to $i + 7$ disulfide linkage more effectively promotes α-helical structure, demonstrating the importance of disulfide positioning for helix stabilization.

Pioneering work on peptide stapling consisted of introducing i to $i + 7$ disulfide bridges between D- and L-cysteine residues, yielding stapled peptides with increased helicity, as confirmed by circular dichroism [37]. Despite being among the first stapling strategies, disulfide-based stapling is rarely employed for intracellular targets due to instability under reducing conditions. To overcome these limitations, alternative covalent stapling strategies

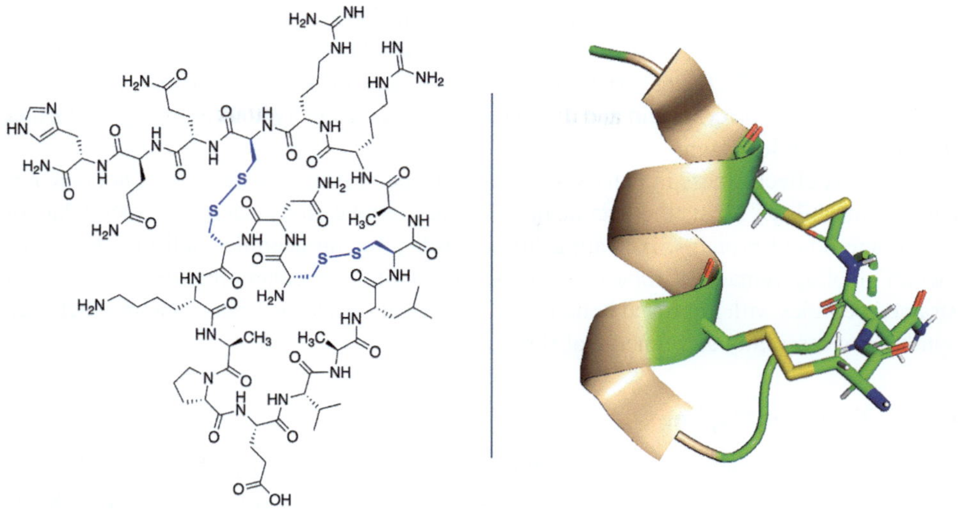

Figure 6.5 *Left: chemical structure of naturally occurring bee venom peptide apamin containing a crossed disulfide as a peptide staple; right: solution NMR structure of bee apamin, the staple is shown in green (PDB: 7OXF).*

have been developed, and while reversible disulfide stapling remains challenging, ongoing efforts aim to integrate dynamic covalent chemistry for intracellular applications.

6.3.2 Lactamization

Lactamization is a widely used peptide-stapling strategy that stabilizes α-helices through the formation of intramolecular amide bonds between side chains of amino acids spaced at i to $i+4$ positions. This approach typically employs proteinogenic residues such as lysine and aspartic acid or glutamic acid, allowing for the cyclization without the need for non-canonical amino acids. The technique was first introduced in the eighties by the already cited work by Felix and colleagues [21], who cyclized lysine and aspartic acid side chains in short growth hormone-releasing factor analogues. Structural analyses using NMR and circular dichroism confirmed that helicity and biological activity were preserved in the stapled peptide. Systematic studies by Fairlie demonstrated that even short peptides could be stabilized using a single lactam bridge [38], and in another work, it was reported that consecutive lactam staples further reinforced α-helicity in aqueous environments [39]. This work extended such an approach to biologically relevant systems, for example, doubly lactam-stapled peptides as helix-constrained nociceptin derivatives acting as potent agonists and antagonists of ERK phosphorylation and thermal analgesia in mice [40]. Also, Norton reported stapled peptides of μ-conotoxin KIIIA exhibiting varying levels of helicity and inhibitory activity, thus showing that the biological outcome is sensitive to staple placement and composition [41].

From a chemical perspective, lactamization is a straightforward approach due to the use of proteogenic amino acids, although some limitations are present. In fact, selective orthogonal protection is required for the amino acid side chains prior to cyclization, requiring additional steps in the synthetic process, and the optimal spacing for Lys–Asp/Glu staples is i to $i+4$ [1, 9], as longer linkages such as i to $i+7$ are not achievable without extended side chains or additional linkers [27]. Other variations include Lys–Lys and Glu–Glu or Asp–Asp staples formed using dicarboxylic acid and diamine linkers, respectively, thus allowing for further flexibility in staple design.

Despite its effectiveness in helix stabilization, lactamization generally does not enhance cell permeability. Most therapeutically relevant applications target extracellular or membrane-bound proteins, indicating limited utility for intracellular delivery. However, lactam stapling remains a robust and versatile method for generating structurally constrained peptides with improved helicity and biological activity, offering a balance between synthetic accessibility and functional stabilization.

6.3.3 Thioether Stapling

Thioether-based stapling strategies exploit the high nucleophilicity of cysteine thiols to generate conformationally constrained peptides with enhanced α-helicity. A remarkable one-component approach consisted of the introduction of peptide stapling by reacting cysteine thiols with α-bromoamide groups [42]. This linkage was designed to reproduce the ring size of classical lactam staples by replacing the amide bond with a thioether and was successfully applied to gp41-derived peptides in the context of HIV vaccine development, enabling the efficient macrocyclization to optimized stapled peptides [43].

This chemical strategy has been widely applied in two-component stapling design through bis-thioether formation. In these approaches, peptides containing two cysteine residues are first synthesized and deprotected to expose free thiols, which are maintained in a reduced state using reagents such as *tris*(2-carboxyethyl)phosphine (TCEP), dithiothreitol (DTT), or glutathione [44]. Cyclization is then achieved in solution by reaction with bifunctional electrophilic linkers, such as 1,3-dibromomethylbenzene, to afford bis-thioether staples (Figure 6.4). Although this method requires an additional solution-phase step, it avoids the need for non-canonical or orthogonally protected amino acids, and more importantly, such two-component cysteine stapling offers high modularity and functionalization, as variations in the cross-linking agent allow for the straightforward tuning of staple length and chemical structure, thus expanding the utility of thioether staples in peptide design.

6.3.4 Click Chemistry

Copper(I)-catalyzed azide–alkyne cycloaddition (CuAAC), independently discovered by Sharpless [45] and Meldal [46], and commonly referred to as "click chemistry," is a robust and widely applied method for peptide stapling due to its high efficiency, chemoselectivity, and compatibility with diverse functional groups. Soon, CuAAC gained importance in the context of peptide chemistry, as the biocompatibility and tolerance of aqueous conditions of this reaction enabled the application of this reaction in the context of bioconjugation strategies and stapled peptides as a mature and versatile platform, especially for stabilizing α-helical peptides.

One of the earliest examples of CuAAC-mediated peptide stapling was reported in 2010 using ω-azido- and ω-alkynyl-functionalized amino acids placed at i to $i+4$ positions

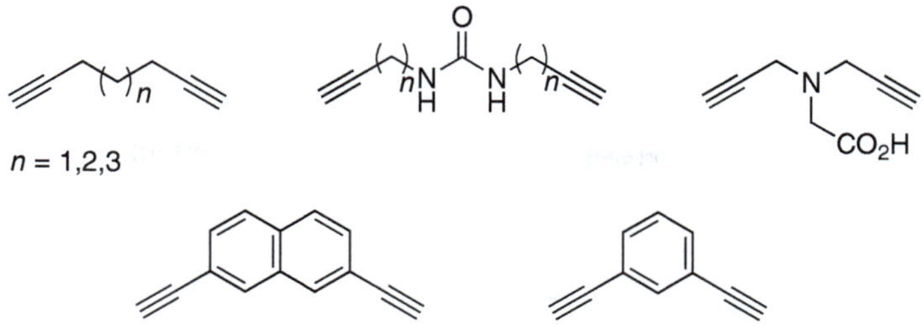

Figure 6.6 Bis-alkynyl linkers applied for the two-component CuAAC double-click reaction.

in peptides derived from parathyroid hormone-related protein [47]. In this entry, the macrocyclization reaction was performed in solution on fully deprotected peptides, yielding triazole-containing staples reproducing the conformational constraints and α-helicity imposed by lactam bridges, as confirmed by circular dichroism and NMR spectroscopy. Such an approach was soon applied for developing stapled bioactive peptides, as exemplified by a report on triazole-stapled peptides as inhibitors of the oncogenic interaction between BCL9 and β-catenin [48]. In this work, by systematically varying linker length, five methylene units were identified as optimal, and the incorporation of two triazole staples within a single peptide was proved feasible. These optimized constructs showed enhanced *in vitro* binding affinity and improved proteolytic stability relative to unstapled analogues, emphasizing the functional relevance of triazole cross-links.

More complex architectures have also been explored, particularly referred to as two-component stapling strategies, as initially reported by Spring and collab. with a "double-click" approach, in which bis-alkynyl linkers react with two azide-bearing residues to form dual triazole constraints (Figure 6.6) [49, 50]. Specifically, two-component triazole staples spanning two turns of the helix ($i - i + 7$) were synthesized using benzene cores bearing two alkyne functionalities. Meta-substitution on the aromatic linker enabled further functionalization, increasing the modularity of the system. For example, the addition of arginine residues enhanced the cellular uptake, thus demonstrating that triazole-based staples can be engineered to modulate both structural and pharmacokinetic properties.

Further innovation was introduced by the same researchers, who employed strained diynes to enable copper-free cycloaddition to address issues associated with copper toxicity, resulting in strain-promoted azide–alkyne cycloaddition (SPAAC). This biorthogonal alternative enables the *in situ* stapling and rapid library generation directly in assay formats, such as 96-well plates, and was applied in a first example of *in situ* peptide cross-linking in cell culture to achieve a stapled peptide inhibitor of the p53-MDM2 interaction (Figure 6.7) [51].

6.3.5 All-Hydrocarbon Stapling *via* RCM

All-hydrocarbon stapling is considered the gold standard among one-component peptide-stapling strategies, due to its robust ability to stabilize α-helical conformations and enhance the pharmacological properties of peptide-based ligands. This approach

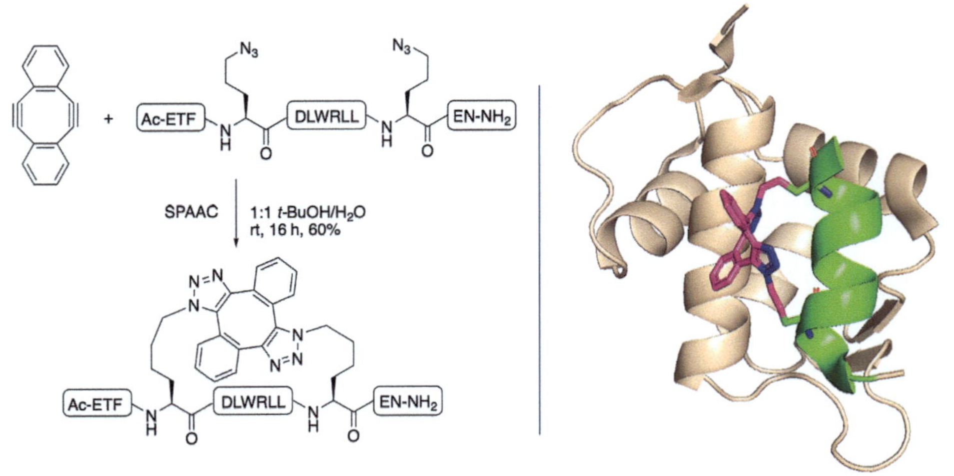

Figure 6.7 *Left: double-SPAAC stapling reaction of a p53-derived diazidopeptide; right: X-ray crystallographic structure of stapled peptide bound to MDM2 at 1.9 Å resolution (PDB ID: 5AFG); the stapled peptide is highlighted in green for the α-helical conformation and in magenta for the chemical staple.*

was pioneered in the nineties by Miller, Blackwell, and Grubbs through the application of ruthenium-catalyzed RCM to peptide side chains [7, 52] and later expanded through seminal contributions from the Verdine and Walensky groups, establishing hydrocarbon-stapled peptides as a powerful tool in the context of peptidomimetic design [5, 8, 11, 53].

The principle of all-hydrocarbon stapling involves the incorporation of olefin-bearing non-proteinogenic amino acids, typically α,α-disubstituted, at defined positions within a peptide sequence, followed by intramolecular RCM to generate a hydrocarbon macrocycle. Although monosubstituted alkenyl amino acids have been used, α,α-disubstituted residues remain preferred due to their strong helix-inducing properties [54]. The most common staple geometry bridges residues at the i to $i+4$ positions, corresponding to one turn of an α-helix, although other staples, such as i to $i+3$ and i to $i+7$ staples [55] and more complex chemical structures, such as doubly stapled peptides and stitched peptides, have been reported [56]. The successful application of RCM on solid support using α,α-disubstituted amino acids with varying olefin lengths and stereochemistry allows for the efficient synthesis of stapled peptides prior to resin cleavage, and in early reports revealed that i to $i+7$ staples often confer superior helicity and proteolytic stability as compared to i to $i+4$ analogues [8]. Thus, the compatibility of Grubbs' catalysts with solid-phase RCM facilitates the generation of stapled peptides across diverse sequences, including *N*-terminally tagged molecules.

Hydrocarbon-stapled peptides exhibit markedly improved properties relative to their linear counterparts, including enhanced α-helicity, increased resistance to proteolytic degradation, and often improved cellular uptake *via* endocytic pathways. These features have enabled the successful targeting of challenging PPIs traditionally considered "undruggable." Some remarkable examples include the already cited stapled analogues of the BID BH3 domain that induce apoptosis in leukemia cells [11], as well as stapled helical peptides acting as dual inhibitors of the p53-MDM2/MDMX interaction that suppress tumor growth [57].

The clinical relevance of hydrocarbon stapling is represented by advanced drug candidates, such as ALRN-5281 [58], a long-acting growth hormone-releasing hormone agonist, that successfully completed Phase I clinical trials for orphan endocrine disorders, as well as the cited compound ALRN-6924, designed to disrupt p53 interactions with MDM2 and MDMX, that was advanced to Phase I trials for solid tumors [59] and Phase II trials for hematological malignancies [60].

An important evolution of RCM-based stapling is the "stitching" approach, a method of double stapling in which two hydrocarbon staples share a common residue. This chemical structure is based on bis-pentenylglycine, which provides two olefinic side chains from a single quaternary amino acid to produce "stitched" bicyclic peptides with enhanced rigidity and helical content. Recent studies applying this concept to β-catenin-binding motifs demonstrated that stitched peptides combining i to $i+4$ and i to $i+3$ staples (resulting in an overall i to $i+7$ staple) achieved superior helicity, binding affinity, and cellular activity as compared to singly stapled analogues [61]. Structural analysis confirmed that such bicyclic peptides adopt well-defined helical conformations upon target binding, reducing the entropic penalty associated with complex formation (Figure 6.8).

Overall, all-hydrocarbon stapling *via* RCM represents a chemically elegant approach for stabilizing α-helical peptides that proved a powerful strategy for improving the druggability of peptides in the context of PPI targeting.

6.3.6 Charged Stapling

In a recent report in the context of the synthesis of stapled peptides with innovative linkers, the α-helix stabilization was coupled with the introduction of a positive charge through a guanidinium moiety directly into a hydrocarbon staple [62]. This approach was inspired by the occurrence of arginine residues at protein–protein interfaces, thus the guanidinium staple was designed to act as a constrained arginine mimetic capable of establishing hydrogen-bonding, ion pairing, cation-π, and van der Waals interactions. Guanidinium stapling was efficiently achieved on solid support using two orthogonally protected lysine residues with Mtt and Alloc and a sequence of steps consisting of selective deprotection and use of Fmoc-NCS to produce the intermediate thiourea with the first Lys to achieve the guanidinium moiety with the second one, thus enabling the modular control of staple size and helicity (Figure 6.9).

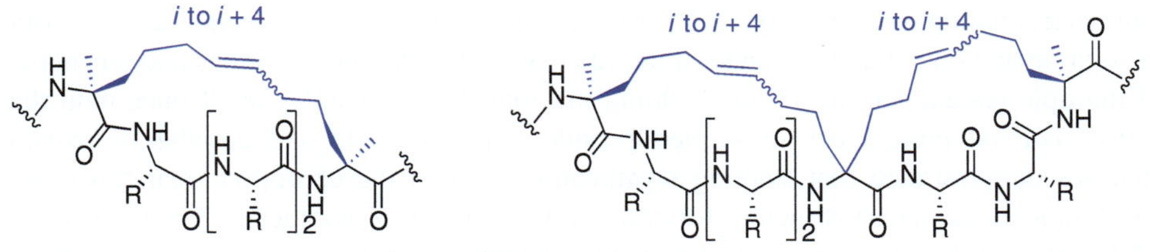

Figure 6.8 Chemical structure of cross-links used for peptide stapling (left) and stitching (right).

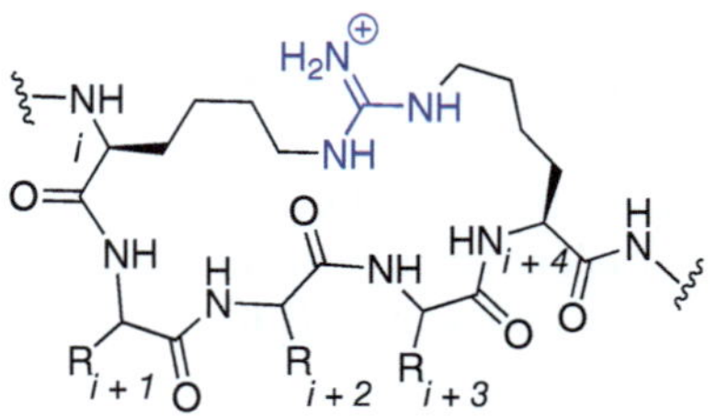

Figure 6.9 *Structure of a* i *to* i + *4 guanidinium stapled peptide.*

6.3.7 Peptide Stapling Through C—H Activation

Concerning peptide macrocyclization using C—H activation strategies, numerous reports have been published over last decade, as discussed in Chapter 5, especially in the context of transition metal-catalyzed C(sp^2)—H and C(sp^3)—H activation exploiting aromatic and aliphatic side chains, respectively, and using peptide backbone or additional chemical moieties as directing groups [63, 64]. Although still underrepresented, some reports accounted for specific peptide stapling resulting to the stabilization of secondary structures.

6.3.7.1 C(sp^2)—H Activation

A peptide-stapling strategy based on intramolecular direct arylation between a tryptophan residue and a phenylalanine-derived side chain was first described by James, demonstrating the feasibility of biaryl cross-links as conformational constraints in peptides [65]. Then, Lavilla developed a palladium-catalyzed intramolecular C—H arylation approach that enables the efficient macrocyclization of peptides containing tryptophan and iodinated phenylalanine or tyrosine residues. This methodology proved particularly effective for all-L peptide sequences in which the aromatic residues are positioned at i to $i+3$ or i to $i+4$ spacings, allowing the formation of well-defined biaryl-stapled macrocycles and demonstrating the robustness and generality of Pd-mediated C—H activation as a tool for constructing structurally constrained peptides using minimally modified proteinogenic amino acids [66]. Specifically, they introduced a novel peptide-stapling strategy based on the intramolecular Pd-catalyzed C—H activation to make direct carbon–carbon linkages between native aromatic amino acid side chains. The method exploits the innate reactivity of the tryptophan indole C2—H bond and an *m*-halogenated phenylalanine or tyrosine residue, as demonstrated by the development of stapled baratin and valorphin analogues enabling Trp–Phe or Trp–Tyr cross-linking, respectively, without the need for non-natural building blocks (Figure 6.10a). The reaction proved to proceed efficiently both in solution and on solid phase under mild Pd-catalyzed arylation conditions and amenable of multiple residue spacings, thus yielding macrocycles of tunable size. Concerning the conformational preferences of stapled peptides, spectroscopic and circular dichroism analyses demonstrated that stapling significantly reduces the conformational flexibility and promotes secondary-structure formation relative to linear analogues. Finally, stapled RGD- and NGR-containing peptides were developed and studied for their biological activity, demonstrating enhanced potency and a significant improvement of the resistance to proteolytic degradation (Figure 6.10b).

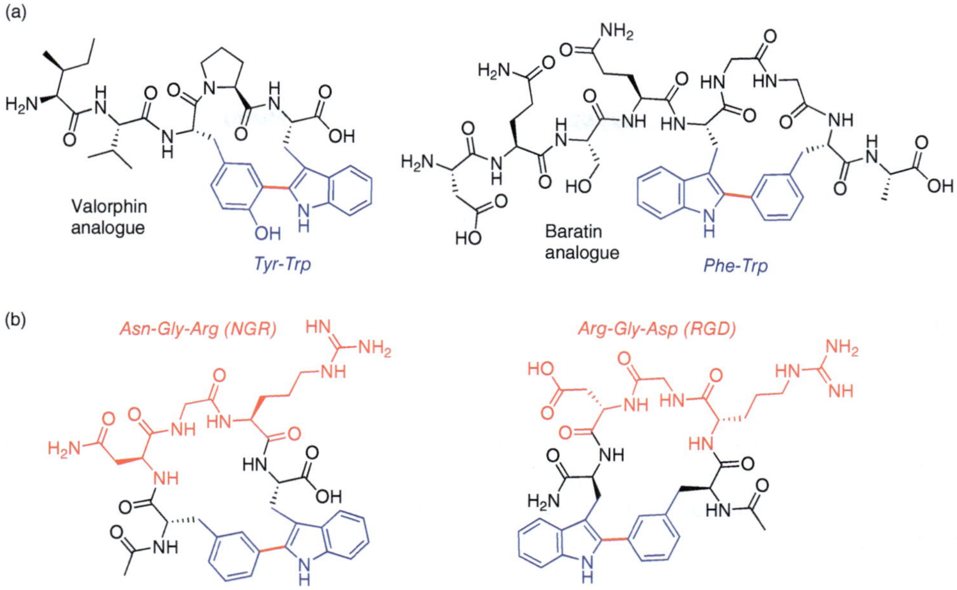

Figure 6.10 *(a) Structure of stapled valorphin and baratin analogues; (b) structure of stapled NGR and RGD peptides.*

6.3.7.2 $C(sp^3)$—H Activation

An efficient strategy for the site-selective arylation of C(sp^3)—H bonds of *N*-phthaloyl peptides using special peptide backbones was reported by Yu et al., thus overcoming the need of modification of building blocks before peptide synthesis that was considered a limitation for the use of C(sp^3)—H activation in peptidomimetic design [67]. Then, the Wang group applied this strategy to peptide stapling through Pd-catalyzed *N*-terminal of amino acid residue β-C(sp^3)—H arylation [68]. A major advance in the context of the synthesis of stapled peptides through SPPS was reported by Albericio et al., who described a conceptually innovative peptide-stapling strategy that substantially broadens the scope of available cross-linking chemistries by exploiting late-stage palladium-catalyzed C(sp^3)—H activation [69]. In contrast to most established stapling methods, which rely on prefunctionalized non-natural residues, this work introduced the first example of peptide macrocyclization *via* the selective activation of an aliphatic C(sp^3)—H bond, resulting in an all-hydrocarbon staple linkage formed through a C(sp^3)—C(sp^2) bond. In particular, such a stapling bond was formed between the β-C(sp^3)—H bond of an *N*-terminal alanine and the side chain of a phenylalanine at the meta position using an *m*-iodo-phenylalanine residue and using the peptide backbone as an intrinsic directing group, thus avoiding the need for auxiliary directing groups and enabling the late-stage modification of fully assembled peptides. The Pd-mediated β-C—H activation of a phthaloyl (Phth)-alanine was followed by intramolecular coupling with an iodophenylalanine residue to achieve efficient i to $i+3$ and i to $i+4$ macrocyclization with high diastereoselectivity, minimizing deiodination and homocoupling side reactions (Figure 6.11). The synthetic process carried out on solid phase demonstrated

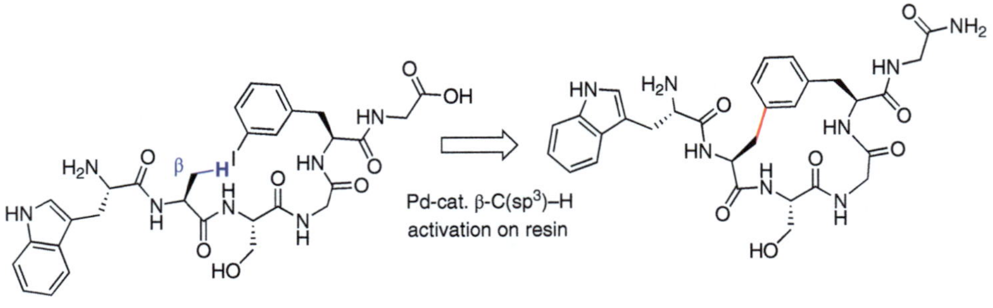

Figure 6.11 On-resin synthesis of stapled peptides through Pd-cat. $C(sp^3)$—H activation.

compatibility of the resin-bound peptide with the reaction conditions for the C—H functionalization and the possibility of removing the Phth protecting group after the C—H functionalization to continue the peptide synthesis on solid phase. The methodology showed broad functional-group tolerance, enabling the insertion of a wide range of proteinogenic amino acids under standard Fmoc protection, demonstrating chemoselectivity in the presence of competing aromatic C—H bonds. The structural flexibility was shown to strongly influence reactivity, with glycine-rich sequences favoring shorter staples.

6.3.8 Peptide Stapling *via* MCR

Multicomponent reactions enable peptide side-chain cross-linking that simultaneously stabilizes helical conformations and allows the increase of the chemical diversity by means of diversity-oriented derivatization [70]. In the context of the chemistry of peptide stapling, Rivera and Wessjohann introduced multicomponent stapling using single or double MCRs to directly bridge two side chains or connect them *via* a bifunctional linker at defined positions (i to $i+4$ or i to $i+7$). In the specific application of the Ugi MCR for the cyclization of peptides [71, 72] both single and double reactions were applied for the stapling procedure, incorporating additional exocyclic functionalities during ring closure and affording increased structural and functional diversification. In the single four-component MCR stapling of the resin-bound peptide, Lys and Asp/Glu side chains as amino and carboxylic components, respectively, were used by reaction with paraformaldehyde and an isocyanide (Figure 6.12).

Also, Dawson et al. reported an on-resin multicomponent stapling strategy enabling α-helical peptide stabilization, as demonstrated by circular dichroism and NMR, with simultaneous incorporation of diverse exocyclic lactam functionalities, including lipidic, anionic, saccharide, and labeling groups, representing the first example of such multifunctional peptide stapling [73].

In another report, Rivera and Wessjohann described the application of a three-component macrocyclization for late-stage *N*-terminus-side-chain peptide diversification. Specifically, the Petasis reaction (see Chapter 2), that consists of reacting an amine, an α-hydroxycarbonyl moiety, and a boronic acid to achieve polyfunctionalized amines, was used as an MCR chemistry for peptide stapling. Using dihydroxyacetone and aryl diboronic acids, peptides bearing two *N*-ε-Me-ornithines or two *N*-ε-Me-Lys were stapled

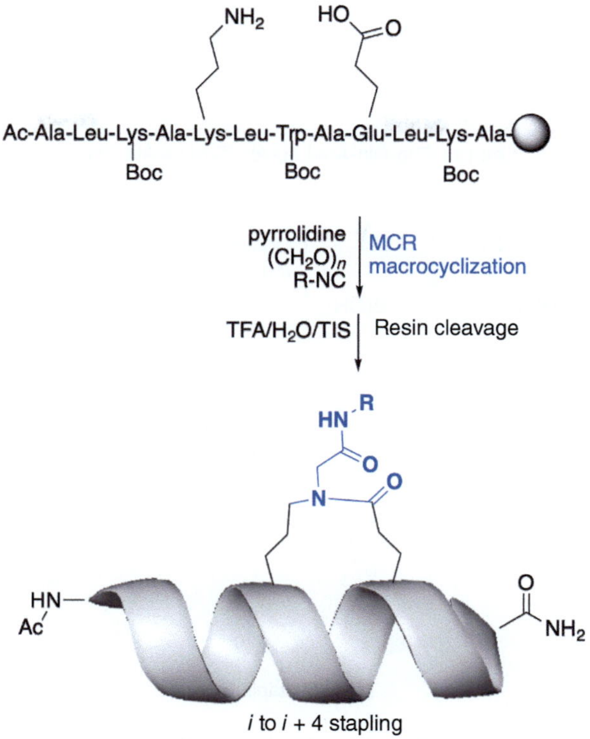

Figure 6.12 Application of the Ugi MCR for peptide stapling using paraformaldehyde and an isocyanide with lysine and aspartic or glutamic acid.

at i to $i+4$ or i to $i+7$ positions, respectively, yielding diverse macrocycles in good yields incorporating biologically relevant aldoses and ketoses (Figure 6.13) [74].

6.4 Therapeutic Applications

Secondary-structure mimetics are of primary relevance in the context of recent drug discovery programs encompassing PPIs, as these drug targets are characterized by well-defined structural motifs rather than narrow binding pockets. Among such interacting epitopes, the α-helix is particularly relevant as accounting for approximately 60% of secondary structures at the interaction interface of PPIs, as retrieved from the Protein Data Bank [75], and suggesting the development of α-helix mimetics as an attractive strategy for modulating PPIs.

In this context, peptide stapling has emerged as a powerful strategy to stabilize α-helical conformations and as a tool to develop peptidomimetic drug leads [25], taking advantage of macrocyclization chemistry, with all-hydrocarbon stapling being the most exploited approach, as described in previous paragraphs of this chapter. Beyond α-helices, β-hairpins also play critical roles in molecular recognition in biological contexts, although such folded epitopes have received less attention in stapling studies so far. Recent work has

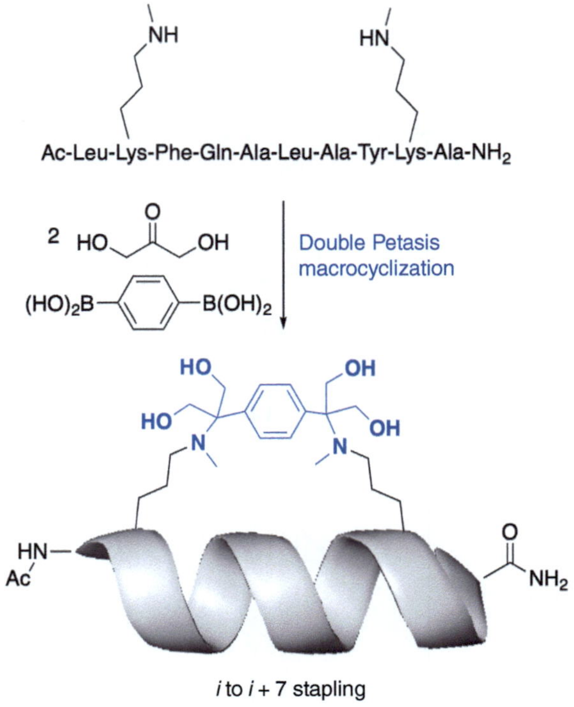

Figure 6.13 Representative application of the Petasis MCR for i to i + 7 peptide stapling.

demonstrated that constraining β-hairpin peptides through hydrocarbon stapling can influence their biological activity and structural stability. The systematic variation in staple position and length revealed that an optimal balance between conformational stability and flexibility is essential, particularly for antimicrobial peptides (AMPs), that exert their biological potential through interactions with membranes. In particular, single-stapled β-hairpins have shown enhanced potency against drug-resistant pathogens, whereas excessive rigidification through double stapling proved unfavorable [76]. Complementary advances have introduced novel stapling motifs specifically designed for antiparallel β-strands, yielding conformationally stable β-hairpin scaffolds showing proteolytic resistance and cellular penetration. Importantly, these constrained hairpins mimic the peptide motifs found at PPI interfaces, thus highlighting their potential as inhibitors of intracellular PPIs (Table 6.1, see also Chapter 10).

6.4.1 HIV Entry Inhibitor as a Representative Example of an α-Helix Mimetic

Despite major advances in antiretroviral therapy (ART), the development of effective treatments for human immunodeficiency virus type 1 (HIV-1) infection remains an unsolved challenge, and it is crucial to develop anti-HIV-1 agents that have a different mechanism from current therapies to overcome the limitations of the established antiretroviral drugs [87]. Specifically, the successful ART, although capable of suppressing viral loads to undetectable levels, cannot address the clearance of the HIV latent reservoir toward a complete

Table 6.1 *Representative bioactive stapled peptides. Data from [29].*

Secondary structure	Target class	Stapled peptide epitope	Primary molecular target	Disease area	Therapeutic strategy/rationale
α-Helix	p53 pathway (tumor suppressor regulation)	SAH-p53-8 and SAH-p53-14	MDM2/MDMX	Cancer	α-Helical p53 mimetics to disrupt p53-MDM2/MDMX PPIs and restoring p53 function [77].
α-Helix	p53 pathway targeted degradation	SPMI-HIF-derived stapled PROTACs	MDM2/MDMX	Cancer	Stapled α-helical ligands recruit E3 ligases to promote degradation of p53 suppressors [78].
α-Helix	Wnt/β-catenin signaling	StAx-3 and StAx-3-35R	β-Catenin	Cancer	Axin-derived α-helices competitively inhibit β-catenin PPIs, suppressing oncogenic Wnt signaling [79].
α-Helix	Wnt transcriptional co-activation	SAH-BCL9	BCL9-β-catenin	Cancer	Stabilized BCL9 α-helices disrupt transcriptional complexes downstream of β-catenin [80].
α-Helix	Apoptosis (BCL-2 family)	SAH-BH3-derived peptides	BCL-2/BCL-XL/MCL-1	Cancer	BH3 α-helix mimetics induce mitochondrial apoptosis [11].
α-Helix	Autophagy machinery	ATG5–ATG16L1 stapled peptides	ATG5/ATG16L1	Cancer	Disruption of α-helical PPIs impairs autophagosome formation [81].
α-Helix	Autophagy adaptors	Pen3-ortho and Pen8-ortho	LC3/GABARAP	Cancer	Amphipathic α-helices block ATG8-family protein interactions and cargo recruitment [82].
α-Helix	Viral capsid assembly	NYAD-1 and NYAD-203	HIV-1 capsid CTD	Infectious disease (HIV)	α-Helical stapled peptides disrupt capsid dimerization and viral assembly [83].
α-Helix	Bacterial transcription regulation	Stapled σ^{54} peptides	σ^{54} C-terminal domain	Infectious disease (bacteria)	Stapled α-helices penetrate Gram-negative bacteria and inhibit virulence gene transcription [84].
α-Helix	Nuclear receptor signaling	Stapled LXXLL-motif peptides	Estrogen receptor/other NRs	Cancer and metabolic diseases	Stabilized coactivator α-helices selectively inhibit nuclear receptor signaling [85].
β-Hairpin	Antimicrobial peptides	Stapled BTT3 analogues	Bacterial membranes	Infectious disease	Hydrocarbon-stapled β-hairpins improve antimicrobial potency and stability [76].
β-Hairpin	PPI inhibition	Mercaptoproline-based stapled antiparallel β-hairpins	Hairpin-type PPI interfaces	Cancer/intracellular targets	Stapling enforces β-hairpin geometry, stability, and mimics hairpin motifs found at PPIs [86].

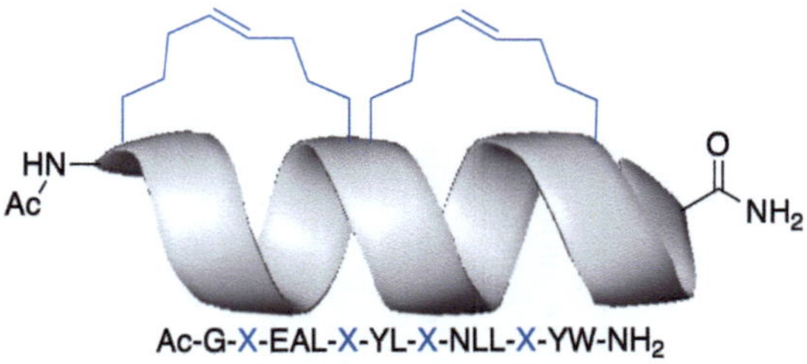

Figure 6.14 Structure of hydrocarbon double-stapled α-helical peptide D26 derived from the lentiviral LLP3 region of the HIV-1 gp41 (X amino acids refer to stapling positions at i *to* i + 4 *positions).*

eradication of the HIV-1/AIDS pandemic [88]. Moreover, the ART approach is challenged by the onset of viral resistance to several groups of currently available antivirals [89]. Virus inactivators represent a distinct class of anti-HIV agents that neutralize virions independently of intracellular replication, offering an attractive complement to conventional antiretroviral therapies. In this context, a recent study reported the rational design of D26, a hydrocarbon double-stapled α-helical peptide derived from the lentiviral lytic peptide-3 (LLP3) region of the HIV-1 gp41 (Figure 6.14). By applying all-hydrocarbon stapling, the authors successfully stabilized the native helical conformation of LLP3 with two independent hydrophobic cross-links with i to $i+4$ spacing, yielding a short bicyclic peptide capable of both inhibiting HIV-1 infection and directly inactivating cell-free virions [90].

D26 showed antiviral activity comparable to that of its linear precursor across diverse HIV-1 strains while displaying improved physicochemical and pharmacokinetic properties. Also, despite its peptidic nature, D26 showed detectable oral absorption, highlighting the pharmacological advantages imparted by hydrocarbon stapling. Thus, this work highlights hydrocarbon double stapling as a powerful tool for transforming lytic peptide motifs into drug-like antiviral agents and provides a convincing basis for further medicinal chemistry efforts to achieve durable viral suppression for the treatment of HIV-1 infection.

6.4.2 Representative Example of a β-Hairpin Antimicrobial Peptidomimetic

Recent peptide-stapling strategies have been explored to improve the clinical performance of AMPs, mainly focused on α-helical scaffolds, while β-hairpin AMPs remain underexplored. In a recent study, a systematic investigation of hydrocarbon stapling on β-hairpin AMPs was carried out through the design, synthesis, and evaluation of a focused peptide library based on a previously reported Gram-negative β-hairpin BTT3 AMP as the parent peptide [91]. Specifically, a series of 12 all-hydrocarbon stapled analogues were prepared *via* solid-phase peptide synthesis and RCM, varying staple position, number, and chain length [76] (Figure 6.15).

Structure–activity relationship analyses revealed a balance between antimicrobial potency, structural stabilization, and conformational flexibility. This study showed single-stapled peptides being more active than their double-stapled counterparts, highlighting the importance of preserving sufficient flexibility for establishing membrane interactions. Specifically, peptide 5(c6) was selected from the library of synthesized

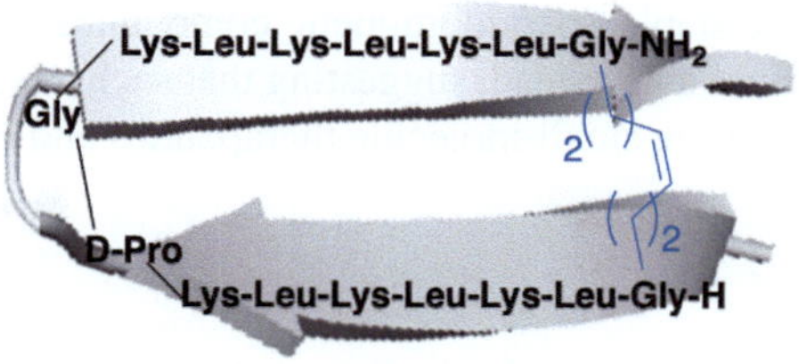

Figure 6.15 *Structure of stapled β-hairpin peptide BTT3.*

β-hairpin peptidomimetics, possessing a single six-carbon staple installed at the peptide termini, that exhibited high activity against colistin-resistant clinical isolates. Mechanistic studies to gain insight into the peptide binding to the bacterial membrane were carried out using imaging techniques and specifically using confocal microscopy to map the colocalization of FITC-labeled 5(c6) incubated with the bacterial membrane dye FM4-64 on *Escherichia coli*, showing the binding of FITC-labeled 5(c6) on the bacterial membrane, thus showing that 5(c6) is capable of translocating across bacterial membranes without inducing extensive membrane disruption, suggesting a non-lytic or minimally disruptive mode of action. These findings showed key design principles for stapled β-hairpin peptides and provide a valuable framework for the future development of antimicrobial agents targeting drug-resistant pathogens.

6.5 Conclusion

Stapled peptides represent an established approach to address highly validated yet historically undruggable PPI as therapeutic targets, and approved therapeutics are expected to emerge from this class in the future. This class of peptidomimetic compounds refers to the covalent tethering of two amino acid side chains within a peptide sequence to form a macrocycle to constrain the peptide backbone into a defined conformation, generally an α-helix. The systematic optimization of peptides into stapled peptides proved to deliver lead compounds with improved cellular potency, reduced off-target toxicity, and enhanced *in vivo* activity, even when starting from already optimized compounds, taking advantage of spectroscopic structural studies, such as circular dichroism spectroscopy, to assess secondary structure, such as α-helical and β-strand peptides with respect to unfolded peptides displaying random-coil conformations. Early chemical approaches to stabilize α-helical peptides were inspired by nature and mainly consisted of the formation of lactam bridges and disulfide bonds, which, although efficient in stabilizing helical structures, proved to be limited in chemical diversity. Further advances in chemical methods and strategic approaches enabled a wider toolbox for generating such peptidomimetics, classifying the applicable chemistry into one-component and two-component strategies, and synthetic approaches encompassing thioether linkages, RCM, Cu(I)-catalyzed azide–alkyne cycloaddition, Pd-catalyzed C—H activation, and other residue-specific pairing strategies employing linkers in a two-component approach. Numerous medicinal chemistry applications in diverse biomedical fields spanning from infectious disease to cancer demonstrate the fruitful approach of constraining bioactive peptide epitopes extracted from

PPI molecules into suitable stapled peptidomimetic compounds in the hit-to-lead process to transform a peptide hit into a drug lead, suggesting that such a peptidomimetic approach is valid to close the gap between small-molecule therapeutics and biologic drugs.

References

1. Ali AM, Atmaj J, Van Oosterwijk N, Groves MR, Dömling A. Stapled peptides inhibitors: a new window for target drug discovery. *Comput Struct Biotechnol J*. 2019;**17**:263–281.
2. Ebrahimi SB, Samanta D. Engineering protein-based therapeutics through structural and chemical design. *Nat Commun*. 2023;**14**:2411.
3. Robinson JA. Beta-hairpin peptidomimetics: design, structures and biological activities. *Acc Chem Res*. 2008;**41**:1278–1288.
4. Lenci E, Trabocchi A. Peptidomimetic toolbox for drug discovery. *Chem Soc Rev*. 2020;**49**:3262–3277.
5. Robertson NS, Spring DR. Using peptidomimetics and constrained peptides as valuable tools for inhibiting protein-protein interactions. *Molecules*. 2018;**23**:959.
6. Chu Q, Moellering RE, Hilinski GJ, Kim YW, Grossmann TN, Yeh JTH, Verdine GL. Towards understanding cell penetration by stapled peptides. *Med Chem Commun*. 2015;**6**:111–119.
7. Blackwell HE, Grubbs RH. Highly efficient synthesis of covalently cross-linked peptide helices by ring-closing metathesis. *Angew Chem Int Ed*. 1998;**37**:3281–3284.
8. Schafmeister CE, Po J, Verdine GL. An all-hydrocarbon cross-linking system for enhancing the helicity and metabolic stability of peptides. *J Am Chem Soc*. 2000;**122**:5891–5892.
9. Lau YH, de Andrade P, Wu Y, Spring DR. Peptide stapling techniques based on different macrocyclisation chemistries. *Chem Soc Rev*. 2015;**44**:91–102.
10. Bechtler C, Lamers C. Macrocyclization strategies for cyclic peptides and peptidomimetics. *RSC Med Chem*. 2021;**12**:1325–1351.
11. Walensky LD, Kung AL, Escher I, Malia TJ, Barbuto S, Wright RD, Wagner G, Verdine GL, Korsmeyer SJ. Activation of apoptosis in vivo by a hydrocarbon-stapled BH3 helix. *Science*. 2004;**305**:1466–1470.
12. Guerlavais V, Sawyer TK, Carvajal L, Chang YS, Graves B, Ren JG, Sutton D, Olson KA, Packman K, Darlak K, Elkin C, Feyfant E, Kesavan K, Gangurde P, Vassilev LT, Nash HM, Vukovic V, Aivado M, Annis DA. Discovery of Sulanemadlin (ALRN-6924), the first cell-permeating, stabilized α-helical peptide in clinical development. *J Med Chem*. 2023;**66**:9401–9417.
13. Merritt HI, Sawyer N, Arora PS. Bent into shape: folded peptides to mimic protein structure and modulate protein function. *Pept Sci*. 2020;**112**:e24145.
14. Makura Y, Ueda A, Kato T, Iyoshi A, Higuchi M, Doi M, Tanaka M. X-ray crystallographic structure of α-helical peptide stabilized by hydrocarbon stapling at i, $i+1$ positions. *Int J Mol Sci*. 2021;**22**:5364.
15. Sreerama N, Woody RW. Structural composition of betaI- and betaII-proteins. *Protein Sci*. 2003;**12**:384–388.
16. Kim D-H, Kang S-M. Stapled peptides: an innovative and ultimate future drug offering a highly powerful and potent therapeutic alternative. *Biomimetics*. 2024;**9**:537.

17. McWhinnie FS, Sepp K, Wilson C, Kunath T, Hupp TR, Baker TS, Houston DR, Hulme AN. Mono-substituted hydrocarbon diastereomer combinations reveal stapled peptides with high structural fidelity. *Chemistry*. 2018;**24**:2094–2097.
18. Chandramohan A, Josien H, Yuen TY, Duggal R, Spiegelberg D, Yan L, Juang YA, Ge L, Aronica PG, Kaan HYK, Lim YH, Peier A, Sherborne B, Hochman J, Lin S, Biswas K, Nestor M, Verma CS, Lane DP, Sawyer TK, Garbaccio R, Henry B, Kannan S, Brown CJ, Johannes CW, Partridge AW. Design-rules for stapled peptides with in vivo activity and their application to Mdm2/X antagonists. *Nat Commun*. 2024;**15**:489.
19. Kim YW, Grossmann TN, Verdine GL. Synthesis of all-hydrocarbon stapled α-helical peptides by ring-closing olefin metathesis. *Nat Protoc*. 2011;**6**:761–771.
20. de Araujo AD, Lim J, Wu KC, Hoang HN, Nguyen HT, Fairlie DP. Landscaping macrocyclic peptides: stapling hDM2-binding peptides for helicity, protein affinity, proteolytic stability and cell uptake. *RSC Chem Biol*. 2022;**3**:895–904.
21. Felix AM, Heimer EP, Wang CT, Lambros TJ, Fournier A, Mowles TF, Maines S, Campbell RM, Wegrzynski BB, Toome V et al. Synthesis, biological activity and conformational analysis of cyclic GRF analogs. *Int J Pept Protein Res*. 1988;**32**:441–454.
22. Feurstein C, Meyer V, Jung S. Structure-activity predictions from computational mining of protein databases to assist modular design of antimicrobial peptides. *Front Microbiol*. 2022;**13**:812903.
23. Moiola M, Memeo MG, Quadrelli P. Stapled peptides-a useful improvement for peptide-based drugs. *Molecules*. 2019;**24**:3654.
24. Speltz TE, Mayne CG, Fanning SW, Siddiqui Z, Tajkhorshid E, Greene GL, Moore TW. A "cross-stitched" peptide with improved helicity and proteolytic stability. *Org Biomol Chem*. 2018;**16**:3702–3706.
25. Zhang Y, Guo J, Cheng J, Zhang Z, Kang F, Wu X, Chu Q. High-throughput screening of stapled helical peptides in drug discovery. *J Med Chem*. 2023;**66**:95–106.
26. Bird GH, Madani N, Perry AF, Princiotto AM, Supko JG, He X, Gavathiotis E, Sodroski JG, Walensky LD. Hydrocarbon double-stapling remedies the proteolytic instability of a lengthy peptide therapeutic. *Proc Natl Acad Sci USA*. 2010;**107**:14093–14098.
27. Tian Y, Jiang Y, Li J, Wang D, Zhao H, Li Z. Effect of Stapling Architecture on physiochemical properties and cell permeability of stapled α-helical peptides: a comparative study. *ChemBioChem*. 2017;**18**:2087–2093.
28. Scholtz JM, Qian H, Robbins VH, Baldwin RL. The energetics of ion-pair and hydrogen-bonding interactions in a helical peptide. *Biochemistry*. 1993;**32**:9668–9676.
29. Li Y, Wu M, Fu Y, Xue J, Yuan F, Qu T, Rissanou AN, Wang Y, Li X, Hu H. Therapeutic stapled peptides: efficacy and molecular targets. *Pharmacol Res*. 2024;**203**:107137.
30. Li X, Chen S, Zhang WD, Hu HG. Stapled helical peptides bearing different anchoring residues. *Chem Rev*. 2020;**120**:10079–10144.
31. Woolley GA. Photocontrolling peptide alpha helices. *Acc Chem Res*. 2005;**38**: 486–493.
32. Wang Y, Chou DH. A thiol-Ene coupling approach to native peptide stapling and macrocyclization. *Angew Chem Int Ed Engl*. 2015;**54**:10931–10934.
33. Rojas AJ, Zhang C, Vinogradova EV, Buchwald NH, Reilly J, Pentelute BL, Buchwald SL. Divergent unprotected peptide macrocyclisation by palladium-mediated cysteine arylation. *Chem Sci*. 2017;**8**:4257–4263.

34. Spokoyny AM, Zou Y, Ling JJ, Yu H, Lin YS, Pentelute BL. A perfluoroaryl-cysteine S(N)Ar chemistry approach to unprotected peptide stapling. *J Am Chem Soc.* 2013;**135**:5946–5949.
35. Brown SP, Smith AB 3rd. Peptide/protein stapling and unstapling: introduction of s-tetrazine, photochemical release, and regeneration of the peptide/protein. *J Am Chem Soc.* 2015;**137**:4034–4037.
36. Phelan JC, Skelton NJ, Braisted AC, McDowell RS. A general method for constraining short peptides to an α-helical conformation. *J Am Chem Soc.* 1997;**119**:455–460.
37. Jackson DY, King DS, Chmielewski J, Singh S, Schultz PG. General approach to the synthesis of short .alpha.-helical peptides. *J Am Chem Soc.* 1991;**113**:9391–9392.
38. Shepherd NE, Hoang HN, Abbenante G, Fairlie DP. Single turn peptide alpha helices with exceptional stability in water. *J Am Chem Soc.* 2005;**127**:2974–2983.
39. Shepherd NE, Abbenante G, Fairlie DP. Consecutive cyclic pentapeptide modules form short alpha-helices that are very stable to water and denaturants. *Angew Chem Int Ed Engl.* 2004;**43**:2687–2690.
40. Harrison RS, Ruiz-Gómez G, Hill TA, Chow SY, Shepherd NE, Lohman RJ, Abbenante G, Hoang HN, Fairlie DP. Novel helix-constrained nociceptin derivatives are potent agonists and antagonists of ERK phosphorylation and thermal analgesia in mice. *J Med Chem.* 2010;**53**:8400–8408.
41. Khoo KK, Wilson MJ, Smith BJ, Zhang MM, Gulyas J, Yoshikami D, Rivier JE, Bulaj G, Norton RS. Lactam-stabilized helical analogues of the analgesic μ-conotoxin KIIIA. *J Med Chem.* 2011;**54**:7558–7566.
42. Brunel FM, Dawson PE. Synthesis of constrained helical peptides by thioether ligation: application to analogs of gp41. *Chem Commun.* 2005;2552–2554.
43. Brunel FM, Zwick MB, Cardoso RMF, Nelson JD, Wilson IA, Burton DR et al. Structure-function analysis of the epitope for 4E10, a broadly neutralizing human immunodeficiency virus type 1 antibody. *J Virol.* 2006;**80**:1680.
44. Timmerman P, Beld J, Puijk WC, Meloen RH. Rapid and quantitative cyclization of multiple peptide loops onto synthetic scaffolds for structural mimicry of protein surfaces. *ChemBioChem.* 2005;**6**:821–824.
45. Kolb HC, Finn MG, Sharpless KB. Click chemistry: diverse chemical function from a few good reactions. *Angew Chem Int Ed Engl.* 2001;**40**:2004–2021.
46. Tornøe CW, Christensen C, Meldal M. Peptidotriazoles on solid phase: [1,2,3]-triazoles by regiospecific copper(I)-catalyzed 1,3-dipolar cycloadditions of terminal alkynes to azides. *J Org Chem.* 2002;**67**:3057–3064.
47. Scrima M, Le Chevalier-Isaad A, Rovero P, Papini AM, Chorev M, D'Ursi AM. CuI-catalyzed azide–alkyne intramolecular i-to-($i+4$) side-chain-to-side-chain cyclization promotes the formation of helix-like secondary structures. *Eur J Org Chem.* 2010;**2010**:446–457.
48. Kawamoto SA, Coleska A, Ran X, Yi H, Yang CY, Wang S. Design of triazole-stapled BCL9 α-helical peptides to target the β-catenin/B-cell CLL/lymphoma 9 (BCL9) protein–protein interaction. *J Med Chem.* 2012;**55**:1137–1146.
49. Lau YH, Wu Y, de Andrade P, Galloway WR, Spring DR. A two-component 'double-click' approach to peptide stapling. *Nat Protoc.* 2015;**10**:585–594.
50. Lau YH, de Andrade P, Quah ST, Rossmann M, Laraia L, Skold N, Sum TJ, Rowling PJE, Joseph TL, Verma C, Hyvonen M, Itzhaki LS, Venkitaraman AR,

Brown CJ, Lane DP, Spring DR. Functionalised staple linkages for modulating the cellular activity of stapled peptides. *Chem Sci*. 2014;**5**:1804–1809.
51. Lau YH, Wu Y, Rossmann M, Tan BX, de Andrade P, Tan YS, Verma C, McKenzie GJ, Venkitaraman AR, Hyvönen M, Spring DR. Double strain-promoted macrocyclization for the rapid selection of cell-active stapled peptides. *Angew Chem Int Ed Engl*. 2015;**54**:15410–15413.
52. Miller SJ, Blackwell HE, Grubbs RH. Application of ring-closing metathesis to the synthesis of rigidified amino acids and peptides. *J Am Chem Soc*. 1996;**118**:9606–9614.
53. Bird GH, Bernal F, Pitter K, Walensky LD. Synthesis and biophysical characterization of stabilized alpha-helices of BCL-2 domains. *Methods Enzymol*. 2008;**446**:369–386.
54. Aillard B, Robertson NS, Baldwin AR, Robins S, Jamieson AG. Robust asymmetric synthesis of unnatural alkenyl amino acids for conformationally constrained α-helix peptides. *Org Biomol Chem*. 2014;**12**:8775–8782.
55. Kim YW, Kutchukian PS, Verdine GL. Introduction of all-hydrocarbon i, $i+3$ staples into alpha-helices via ring-closing olefin metathesis. *Org Lett*. 2010;**12**:3046–3049.
56. Hilinski GJ, Kim YW, Hong J, Kutchukian PS, Crenshaw CM, Berkovitch SS, Chang A, Ham S, Verdine GL. Stitched α-helical peptides via bis ring-closing metathesis. *J Am Chem Soc*. 2014;**136**:12314–12322.
57. Chang YS, Graves B, Guerlavais V, Tovar C, Packman K, To KH, Olson KA, Kesavan K, Gangurde P, Mukherjee A, Baker T, Darlak K, Elkin C, Filipovic Z, Qureshi FZ, Cai H, Berry P, Feyfant E, Shi XE, Horstick J, Annis DA, Manning AM, Fotouhi N, Nash H, Vassilev LT, Sawyer TK. Stapled α-helical peptide drug development: a potent dual inhibitor of MDM2 and MDMX for p53-dependent cancer therapy. *Proc Natl Acad Sci USA*. 2013;**110**:E3445–E3454.
58. Grigoryev Y. Stapled peptide to enter human testing, but affinity questions remain. *Nat Med*. 2013;**19**:120.
59. Meric-Bernstam F, Saleh MN, Infante JR, Goel S, Falchook GS, Shapiro G. Phase I trial of a novel stapled peptide ALRN-6924 disrupting MDMX- and MDM2-mediated inhibition of WT p53 in patients with solid tumors and lymphomas. *J Clin Oncol*. 2017;**35**(15-suppl):2505.
60. Payton M, Pinchasik D, Mehta A, Goel S, Zain JM, Sokol L, Jacobsen E, Patel MR, Horwitz SM, Meric-Bernstam F, Shustov A, Weinstock D, Aivado M, Annis DA. Phase 2a study of a novel stapled peptide ALRN-6924 disrupting MDMX- and MDM2-mediated inhibition of wild-type TP53 in patients with peripheral t-cell lymphoma. *Ann Oncol*. 2017;**28**:v355–v371.
61. Yeste-Vázquez A, Paulussen FM, Wendt M, Klintrot R, Schulte C, Wallraven K, van Gijzel L, Simeonov B, van der Gaag M, Gerber A, Maric HM, Hennig S, Grossmann TN. Structure-based design of bicyclic helical peptides that target the oncogene β-catenin. *Angew Chem Int Ed Engl*. 2024;**63**:e202411749.
62. Perdriau C, Luton A, Zimmeter K, Neuville M, Saragaglia C, Peluso-Iltis C, Osz J, Kauffmann B, Collie GW, Rochel N, Guichard G, Pasco M. Guanidinium-stapled helical peptides for targeting protein–protein interactions. *Angew Chem Int Ed Engl*. 2025;**64**:e202416348.
63. Rivera DG, Ojeda-Carralero GM, Reguera L, Van der Eycken EV. Peptide macrocyclization by transition metal catalysis. *Chem Soc Rev*. 2020;**49**:2039–2059.

64. Wang W, Lorion MM, Shah J, Kapdi AR, Ackermann L. Late-stage peptide diversification by position-selective C—H activation. *Angew Chem Int Ed Engl.* 2018;**57**:14700–14717.
65. Dong H, Limberakis C, Liras S, Price D, James K. Peptidic macrocyclization via palladium-catalyzed chemoselective indole C-2 arylation. *Chem Commun.* 2012;**48**:11644–11646.
66. Mendive-Tapia L, Preciado S, García J, Ramón R, Kielland N, Albericio F, Lavilla R. New peptide architectures through C—H activation stapling between tryptophan-phenylalanine/tyrosine residues. *Nat Commun.* 2015;**6**:7160.
67. Gong W, Zhang G, Liu T, Giri R, Yu JQ. Site-selective C(sp^3)—H functionalization of di-, tri-, and tetrapeptides at the *N*-terminus. *J Am Chem Soc.* 2014;**136**:16940–16946.
68. Tang J, He Y, Chen H, Sheng W, Wang H. Synthesis of bioactive and stabilized cyclic peptides by macrocyclization using C(sp^3)—H activation. *Chem Sci.* 2017;**8**:4565–4570.
69. Noisier AF, García J, Ionuţ IA, Albericio F. Stapled peptides by late-stage C(sp^3)—H activation. *Angew Chem Int Ed Engl.* 2017;**56**:314–318.
70. Reguera L, Rivera DG. Multicomponent reaction toolbox for peptide macrocyclization and stapling. *Chem Rev.* 2019;**119**:9836–9860.
71. Vasco AV, Pérez CS, Morales FE, Garay HE, Vasilev D, Gavín JA, Wessjohann LA, Rivera DG. Macrocyclization of peptide side chains by the Ugi reaction: achieving peptide folding and exocyclic *N*-functionalization in one shot. *J Org Chem.* 2015;**80**:6697–6707.
72. Ricardo MG, Morales FE, Garay H, Reyes O, Vasilev D, Wessjohann LA, Rivera DG. Bidirectional macrocyclization of peptides by double multicomponent reactions. *Org Biomol Chem.* 2015;**13**:438–446.
73. Assem N, Ferreira DJ, Wolan DW, Dawson PE. Acetone-linked peptides: a convergent approach for peptide macrocyclization and labeling. *Angew Chem Int Ed Engl.* 2015;**54**:8665–8668.
74. Ricardo MG, Llanes D, Wessjohann LA, Rivera DG. Introducing the petasis reaction for late-stage multicomponent diversification, labeling, and stapling of peptides. *Angew Chem Int Ed Engl.* 2019;**58**:2700–2704.
75. Bullock BN, Jochim AL, Arora PS. Assessing helical protein interfaces for inhibitor design. *J Am Chem Soc.* 2011;**133**:14220–14223.
76. Selvarajan V, Tram NDT, Xu J, Ngen STY, Koh JJ, Teo JWP, Yuen TY, Ee PLR. Stapled β-hairpin antimicrobial peptides with improved stability and activity against drug-resistant gram-negative bacteria. *J Med Chem.* 2023;**66**:8498–8509.
77. Bernal F, Wade M, Godes M, Davis TN, Whitehead DG, Kung AL, Wahl GM, Walensky LD. A stapled p53 helix overcomes HDMX-mediated suppression of p53. *Cancer Cell.* 2010;**18**:411–422.
78. Chen S, Li X, Li Y, Yuan X, Geng C, Gao S, Li J, Ma B, Wang Z, Lu W, Hu HG. Design of stapled peptide-based PROTACs for MDM2/MDMX atypical degradation and tumor suppression. *Theranostics.* 2022;**12**:6665–6681.
79. Grossmann TN, Yeh JT, Bowman BR, Chu Q, Moellering RE, Verdine GL. Inhibition of oncogenic Wnt signaling through direct targeting of β-catenin. *Proc Natl Acad Sci USA.* 2012;**109**(44):17942–17947.

80. Takada K, Zhu D, Bird GH, Sukhdeo K, Zhao JJ, Mani M, Lemieux M, Carrasco DE, Ryan J, Horst D, Fulciniti M, Munshi NC, Xu W, Kung AL, Shivdasani RA, Walensky LD, Carrasco DR. Targeted disruption of the BCL9/β-catenin complex inhibits oncogenic Wnt signaling. *Sci Transl Med*. 2012;**4**:148ra117.
81. Cui J, Ogasawara Y, Kurata I, Matoba K, Fujioka Y, Noda NN, Shibasaki M, Watanabe T. Targeting the ATG5-ATG16L1 protein-protein interaction with a hydrocarbon-stapled peptide derived from ATG16L1 for autophagy inhibition. *J Am Chem Soc*. 2022;**144**:17671–17679.
82. Brown H, Chung M, Üffing A, Batistatou N, Tsang T, Doskocil S, Mao W, Willbold D, Bast RC Jr, Lu Z, Weiergräber OH, Kritzer JA. Structure-based design of stapled peptides that bind GABARAP and inhibit autophagy. *J Am Chem Soc*. 2022;**144**:14687–14697.
83. Bhattacharya S, Zhang H, Cowburn D, Debnath AK. Novel structures of self-associating stapled peptides. *Biopolymers*. 2012;**97**:253–264.
84. Payne SR, Pau DI, Whiting AL, Kim YJ, Pharoah BM, Moi C, Boddy CN, Bernal F. Inhibition of bacterial gene transcription with an RpoN-based stapled peptide. *Cell Chem Biol*. 2018;**25**:1059–1066.e4.
85. Phillips C, Roberts LR, Schade M, Bazin R, Bent A, Davies NL, Moore R, Pannifer AD, Pickford AR, Prior SH, Read CM, Scott A, Brown DG, Xu B, Irving SL. Design and structure of stapled peptides binding to estrogen receptors. *J Am Chem Soc*. 2011;**133**:9696–9699.
86. Pace JR, Lampkin BJ, Abakah C, Moyer A, Miao J, Deprey K, Cerulli RA, Lin YS, Baleja JD, Baker D, Kritzer JA. Stapled β-hairpins featuring 4-mercaptoproline. *J Am Chem Soc*. 2021;**143**:15039–15044.
87. Simon V, Ho DD, Abdool KQ. HIV/AIDS epidemiology, pathogenesis, prevention, and treatment. *Lancet*. 2006;**368**:489–504.
88. Barton K, Winckelmann A, Palmer S. HIV-1 reservoirs during suppressive therapy. *Trends Microbiol*. 2016;**24**:345–355.
89. Wensing AM, Calvez V, Günthard HF, Johnson VA, Paredes R, Pillay D, Shafer RW, Richman DD. 2014 Update of the drug resistance mutations in HIV-1. *Top Antivir Med*. 2014;**22**:642–650.
90. Wang C, Zhang W, Xu L, Tu J, Su S, Li Q, Zhang T, Zheng L, Wang H, Zhuang X, Tang X, Yuan Y, Meng G, Lu L, Xiao J, Wang Q, Jiang S. Discovery of a double-stapled short peptide as a long-acting HIV-1 inactivator with potential for oral bioavailability. *J Med Chem*. 2024;**67**:9991–10004.
91. Tram NDT, Selvarajan V, Boags A, Mukherjee D, Marzinek JK, Cheng B, Jiang ZC, Goh P, Koh JJ, Teo JWP, Bond PJ, Ee PLR. Manipulating turn residues on de novo designed β-hairpin peptides for selectivity against drug-resistant bacteria. *Acta Biomater*. 2021;**135**:214–224.

7
Advances in Peptidomimetic Foldamers

7.1 Introduction

Secondary-structure mimetics are central to peptide-based drug discovery because they enable the reproduction of key recognition elements of proteins while overcoming the intrinsic limitations of native peptides, such as poor stability and bioavailability. Among these motifs, β-strands/β-sheets and α-helices dominate interactions between protein interfaces and between proteins with other biomolecules, thus motivating the development of foldamers and peptidomimetics that replicate their geometry and side-chain display.

Foldamers are a class of compounds within the toolbox of peptidomimetics that have been particularly instrumental in the context of oligomeric structure design for applications spanning from anticancer to antimicrobial compounds. Following a previous description on the basics of foldamers in the context of peptidomimetic compounds [1], foldamer chemistry has greatly modernized the field of biopolymer mimicry over the past two decades by demonstrating that well-defined and stable secondary structures can be generated from synthetic oligomeric backbones distinct from those of natural biomacromolecules [2]. These backbones may closely resemble natural peptides, such as aliphatic α-, β-, or γ-peptides and oligomers of *N*-alkyl glycines, also named peptoids, or possess significant structural differences, such as in m-phenylene ethynylene oligomers and aromatic oligoamides. A key feature of foldamers is their capacity to deviate from the structure of folded natural biopolymers while preserving the relationship between folding and function in the context of biological recognition processes. Consequently, foldamers are powerful sequence-based tools for addressing biological mechanisms with complementary approaches to conventional chemical biology and for developing innovative oligomers as diagnostic and therapeutic agents.

An important advance in the development of foldamers is the possibility of accessing the chemical space beyond traditional drug-likeness criteria, such as Lipinski's rules. This allows for the exploration of chemical structures and physicochemical properties inaccessible to small molecules or peptides. Nevertheless, the achievement of peptidomimetics of protein secondary structural elements responsible for key binding events is challenging, and only a limited number of foldamer-based systems have proved to reproduce such

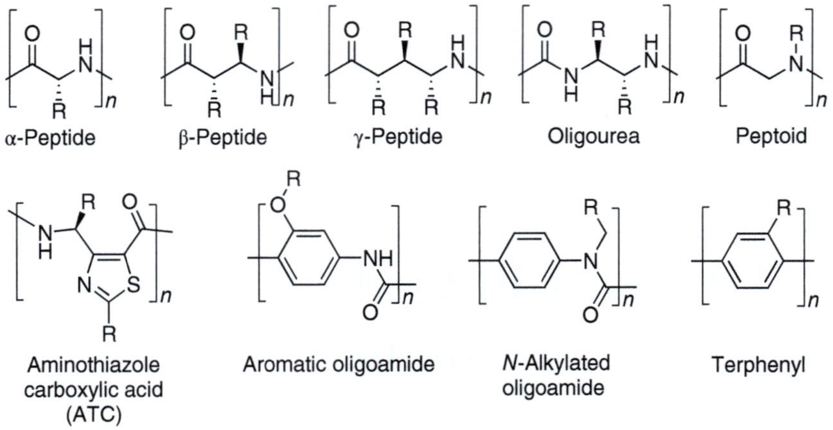

Figure 7.1 Representative aliphatic and aromatic peptidomimetic foldamers.

motifs with high affinity and specificity. Progress in this area requires a multidisciplinary approach to integrate highly modular synthetic strategies for higher-order foldamers, detailed structural information on foldamer–biomolecule complexes at the molecular level, and the integration of computational design methodologies.

Foldamers are commonly classified according to their backbone and can be broadly divided into aliphatic, aromatic, and hybrid structures. Fully aliphatic systems include α-, β-, and γ-peptides, peptoids, and peptidomimetic oligoureas that are conceived as peptide-based foldamers, whereas other foldamers are based on aromatic or heteroaromatic units such as oligoaryl- or heteroarylamides as non-peptide foldamers (Figure 7.1). Hybrid systems combining aliphatic and aromatic elements have been developed, too. Among the various secondary structures adopted by foldamers, although entries of β-strand and β-sheet-like structures are increasingly reported, helical conformations are the most prevalent, and extensive research into helix stabilization and helix mimicry has been documented in the field of peptidomimetics [3].

Concerning the development of foldamers as helix peptidomimetics, three general strategies have emerged for inducing helical structure, namely helical foldamers, helical surface mimetics, and helix stabilization approaches. Helical foldamers, including β-peptides and peptoids, are based on non-natural backbones and amino acid analogues capable of generating conformations that closely mimic the α-helix structures found in proteins. Such approaches have proven especially valuable for targeting protein–protein interactions (PPIs), which are generally mediated by extended helical interfaces [4].

From a biomedical perspective, peptide foldamers display a wide range of biological activities, including antimicrobial agents, cell-penetrating compounds, receptor ligands, and inhibition of PPIs, as reported [5–8]. β-Amino-acid-containing peptides are particularly attractive as oligomers for targeting PPIs implicated in disease, such as cancer, due to their relatively large interaction surfaces. Successful examples include the inhibition of the Bcl-2-BH3, p53-MDM2, and VEGF-VEGFR1 PPIs [9–12]. For more insight on the class of peptide and peptidomimetic foldamers and their biological applications, see [2, 13].

7.2 Secondary-Structure Mimetics

7.2.1 α-Helix Mimetics

The α-helix is the most extensively targeted secondary structure in foldamer and peptidomimetic research, given its prevalence in nature and major role in PPIs. Specifically, structural studies of helix-mediated peptide–protein contacts and PPIs revealed that approximately 60% of interactions are based on hot-spot residues situated on a single face of the helix, about one-third involve two faces, and about 10% require the contribution of all the conformational space occupied by the helix. This diversity has driven the development of a wide range of α-helix mimetics with varying levels of structural mimicry [14], generally reproducing the side-chain spacing at i, $i+4$, and $i+7$ positions but also encompassing the overall helix topology. In this context, two major design strategies have emerged for developing such peptidomimetics:

- non-helical scaffolds projecting substituents in an α-helix-like arrangement
- foldameric backbones adopting real helical conformations.

Non-helical scaffolds include rigid aromatic frameworks such as terphenyls and related heterocyclic systems, which mimic one or more faces of an α-helix, and have proved particularly effective for addressing PPIs. In parallel, foldamers derived from β-amino acids, mixed α/β-peptides, and peptoids have demonstrated intrinsic helical folding [15]. β-Peptides can form well-defined 12- and 14-helices, while α/β-peptides with appropriate residue arrangements closely mimic the geometry of α-helices. Peptoids, despite lacking backbone hydrogen-bond donors, can adopt stable helices driven by steric and side-chain effects.

7.2.2 β-Strand and β-Sheet Mimetics

Compared to α-helices, β-strand, and β-sheet mimetics are considerably less developed, as a consequence of the intrinsic properties of β-sheets requiring large, flat interaction surfaces stabilized by interstrand hydrogen bonds and the difficulty of avoiding the intrinsic propensity to aggregation and precipitation, which complicates both design and characterization. In natural peptides, β-sheets consist of aligned β-strands, each formed by extended peptide backbones with side chains alternating above and below the sheet plane. Parallel β-sheets are stabilized by 12-membered hydrogen-bond rings, whereas antiparallel β-sheets involve 10- and 14-membered rings.

Early *de novo* peptide-based foldamers capable of adopting sheet-like or hairpin conformations were reported for β- and γ-peptides, especially by the Gellman and Seebach groups. These studies established that noncanonical backbones could access mimetics of extended β-strand conformations, although the application of such peptide-based foldamers to aqueous environments has proven challenging in maintaining the secondary structural asset; thus, relatively few water-soluble β-sheet foldamers have been structurally characterized. In the context of peptide-based design, macrocyclization proved to stabilize β-strands and reduce aggregation, yielding improved β-hairpin and β-sheet stability in solution [16].

Also, the incorporation of unnatural turn-inducing units, such as β- or γ-amino acid turns, and the use of scaffolds reproducing hydrogen-bond templates have been envisaged as

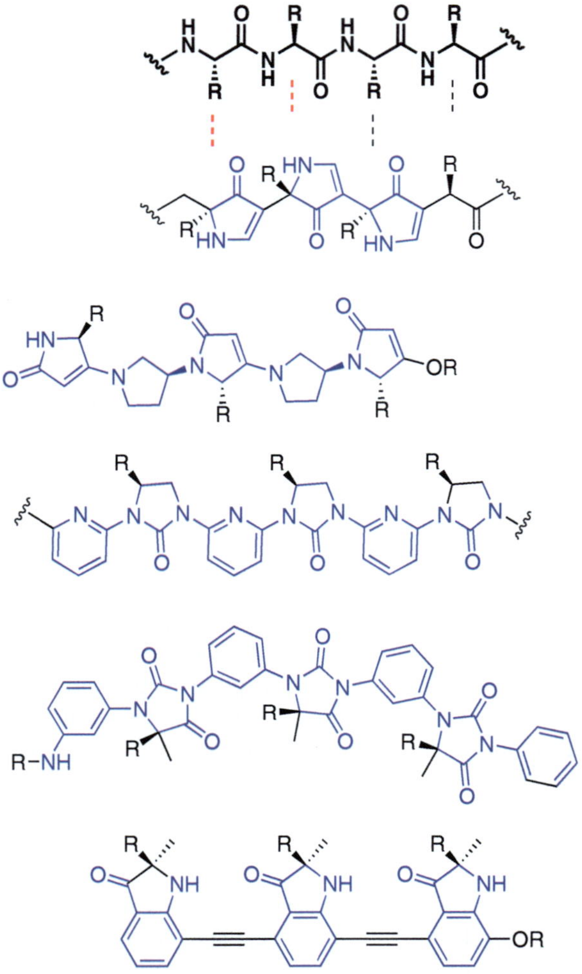

Figure 7.2 Examples of structural β-strands mimetics.

advanced tools for peptidomimetic design [17]. Specifically, an alternative strategy focused on structural β-strands mimetics is conceived on mimicking the spatial projection of key side chains at i, $i+2$, and $i+4$ positions rather than reproducing the full hydrogen-bonding network. Representative examples include oligopyrrolinones, pyrrolinone–pyrrolidine oligomers, indolin-3-ones, hydantoin oligomers, and aryl imidazolidinones (Figure 7.2) [18–22]. Although these non-peptide scaffolds have proven valuable for probing β-strand recognition motifs in protease inhibition and protein–protein interfaces, water-soluble variants suitable for functional studies in biological media are still underdeveloped.

7.3 Peptidomimetic Foldamers

As anticipated, the development of peptidomimetic foldamers focused on different structural arrangements based on resemblance to native peptide structure, and representative entries among foldamers' subclasses spanning from peptide-based oligomers to aromatic foldamers with non-peptide character and hybrid peptides displaying a combination of both features are herein discussed.

7.3.1 Peptide-Based Foldamers: Peptaibol and Aib-Oligomers

Apart from traditional peptide-derived foldamers encompassing α- to δ-unnatural amino acids, both linear and cyclic, other structural arrangements emerged over the years. For example, peptaibols and α-aminoisobutyric acid (Aib)-derived foldamers represent a distinctive class of helical foldamers based on oligomers of the achiral Aib [23]. Oligo-Aib peptides adopt 3_{10}- or α-helical conformations, existing as rapidly interconverting right- and left-handed forms, thus resulting in no defined helical screw sense in solution, as Aib is achiral [24], as confirmed by high-resolution crystallographic studies of model oligomeric Z-$(Aib)_{10}$-OH and by circular dichroism spectroscopy, the latter lacking the characteristic helical Cotton effects.

Given the biological relevance of helical structures in PPI and membrane interactions, significant efforts have been devoted to controlling helix handedness in Aib-based systems. In particular, oligomers derived from chiral cyclic α,α-disubstituted amino acids of varying ring size, such as 1-aminocyclobutanecarboxylic acid (ACBC), 1-aminocyclopentanecarboxylic acid (ACPC), and 1-aminocyclohexanecarboxylic acid (ACHC), have been studied [25]. These molecular scaffolds have been used for elucidating the capacity of peptaibols in disrupting membranes, similarly to the biological function of alamethicin, a channel-forming peptaibol antibiotic containing the non-proteinogenic Aib residue, suggesting that minimal terminal modifications, such as hydrophobic *C*-terminal substitutions, significantly modulate the membrane permeabilization and antimicrobial activity. More generally, the self-assembly properties of Aib-rich peptides in membranes have inspired the design of higher-order foldamers for the generation of a class of crystalline peptide-based materials named "foldecture" [26], mimetics of protein tertiary and quaternary organization.

7.3.2 Non-peptide α-Helix Mimetics

This subclass of peptidomimetics is mainly characterized by the structure of heterocyclic or benzamide foldamers, which constitute a group of foldamers designed to reproduce the spatial organization of side chains of α-helices. Complementary to peptide-based foldamers, these scaffolds typically incorporate aromatic rings within their backbones that are distinct from the characteristic backbone found in peptides and peptide-based foldamers, inducing strong geometric constraints and restricted rotational freedom, which leads to highly stable and predictable secondary structures, also taking advantage of noncovalent interactions such as π–π stacking and intramolecular hydrogen-bonding. Early work by Hamilton and coworkers introduced the basic concepts of aromatic oligoamide foldamers [27] through the combination of o-aminobenzoic acid with pyridine-2,6-dicarboxylic acid monomers, thus generating heteroaromatic oligomers capable of adopting helical conformations spontaneously. These helices are stabilized by well-defined intramolecular hydrogen-bonding networks between adjacent amide groups, giving helix structures similar to helicenes, demonstrating that such systems characterized by purely organic backbones could encode stable helical architectures without reliance on peptide hydrogen-bond patterns [28]. Subsequent studies expanded this approach to a wide range of oligoaryl- and

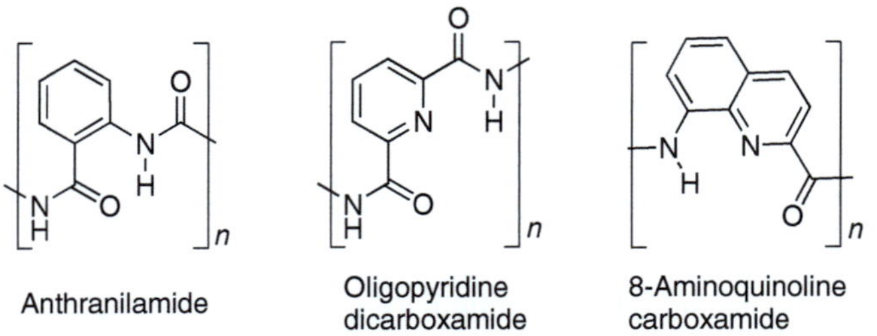

Figure 7.3 *Representative aromatic units constituting aromatic foldamers through hydrogen-bonding stabilization.*

heteroaryl-amide backbones, including oligomers of 4,6-dimethoxy-3-aminobenzoic acid, oligoquinolines, and mixed aromatic systems linked by hydrazone or urea functionalities. The shape of the helix foldamers can be modulated by a range of parameters, including the type of aromatic rings, the incorporation of hydrogen-bond donors or acceptors to constrain the backbone, and the aromatic heterocycles constituting the monomeric unit of the foldamer (e.g. benzene, pyridine, quinoline; see Figure 7.3) [29]. The introduction of chiral monomers also proved to induce a stereochemical selection along the foldamer chain [30].

Despite their potential as non-peptide helix mimetics, aromatic foldamers proved challenging in delivering bioactive compounds due to their hydrophobicity, impairing water solubility and bioactivity, and their synthetic complexity, as compared to peptide-derived foldamers, easily prepared through peptide synthesis. Nevertheless, some successful examples have been reported, such as the case of quinoline-based foldamers. Specifically, the first water-soluble helical quinoline oligoamides were obtained by adding multiple ammonium side chains in the backbone [31], showing enhanced helical stability in aqueous environments suitable for biological applications. Hybrid scaffolds combining aliphatic and aromatic building blocks have also been explored, offering original folding behaviors arising from the complementary properties of the monomeric units.

As anticipated, terphenyl-based scaffolds represent a milestone within the category of non-peptide α-helix mimetics (Figure 7.4) [32]. These rigid aromatic frameworks were designed to accommodate the substituents in positions corresponding to i, $i+3/4$, and $i+7$ residues along one face of an α-helix, thus mimicking the side-chain positioning involved in PPIs addressed by α-helices [33]. The approach proved successful in a number of reported examples, such as the inhibition of Bak-Bcl-xL [34]. Nevertheless, a significant shortcoming of such foldamer structures is their synthetic difficulty and poor aqueous solubility, which stimulated the search for alternative scaffolds with improved physico-chemical properties. One strategy involved replacing phenyl rings with heteroaromatic units to increase polarity and solubility, such as the application of terpyridyl scaffolds, in which phenyl rings are substituted with pyridine moieties (Figure 7.4). Additional heterocyclic frameworks incorporating oxazole or pyrazine rings further enhanced solubility through increased hydrogen-bonding capacity and hybrid 5-6-5 heterocyclic structures by replacing terminal six-membered rings with five-membered heterocycles such as imidazole and thiazole [35]. X-ray crystallographic studies confirmed that these scaffolds

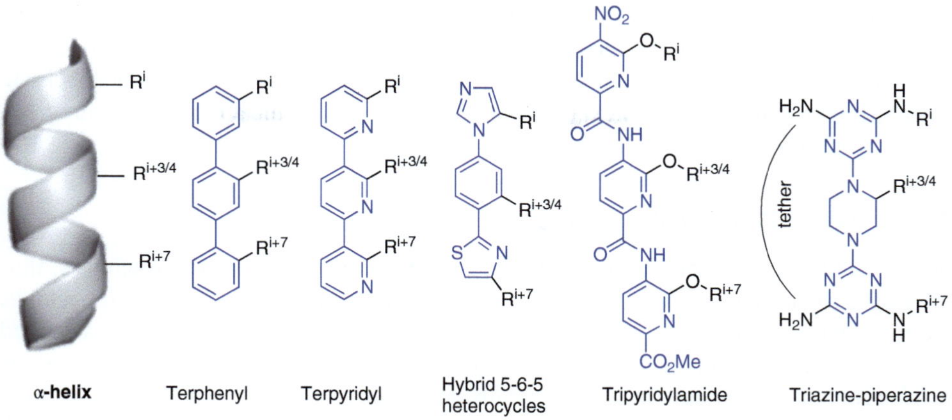

Figure 7.4 Structure of non-peptide α-helix foldamer mimetics with i, i + 3/4, and i + 7 side chains.

maintain α-helix-like side-chain projection (Figure 7.4). Other examples of heterocyclic terphenyl-like systems have been reported, such as the triazine–piperazine–triazine system in which the central phenyl ring was replaced by piperazine, displaying improved solubility and synthetic accessibility (Figure 7.4). These compounds were synthesized using one-bead-one-compound methodologies on solid-phase, thus enabling high-throughput screening applications [33]. Biomedical applications of non-peptide foldamers to modulate PPIs have been reported, such as tetra- and tripyridylamide derivatives as inhibitors of islet amyloid polypeptide (IAPP) fibrillation in lipid-dependent manners that protect insulin-secreting cells from amyloid toxicity. Similar strategies were studied in the context of Alzheimer's disease, where oligopyridylamide libraries yielded compounds capable of binding to amyloid-β peptides, thus impairing the formation of fibrils and overall reducing neurotoxicity in cellular models. Mechanistic studies revealed that subtle changes in charge and substitution pattern lead to distinct modes of interaction, either stabilizing helical conformations or capping aggregation-prone β-strand regions.

Also, oligopyridylamide foldamers proved to inhibit RNA–protein interactions. Specifically, in the context of HIV, such mimetics successfully disrupted the TAR RNA–TAT protein complex, which is recognized as a critical step in viral replication, demonstrating that α-helix mimetics can be adapted to target RNA interfaces, a traditionally challenging class of biomolecular interactions.

N-Alkylated oligobenzamides have also been developed as alternative α-helix mimetics with enhanced synthetic accessibility. Libraries generated *via* microwave-assisted solid-phase synthesis yielded compounds capable of disrupting p53-HDM2 and Mcl-1-NOXA-B interactions, thus promoting apoptosis in cancer cells [36]. Finally, dynamic combinatorial chemistry has emerged as a powerful approach for discovering α-helix mimetics. Specifically, by using oligoamide templates bearing aldehyde groups at positions corresponding to key helical residues, reversible libraries of imine- and hydrazone-based mimetics were generated in the presence of target proteins. Structural and spectroscopic studies confirmed that the selected compounds reproduce α-helical side-chain arrangements.

Overall, non-peptide α-helix mimetics based on heterocyclic and benzamide foldamers represent a versatile platform for modulating biomolecular interactions. Their structural modularity and analogy to biomolecules enable their application as building blocks to foldamer approaches in the context of peptidomimetics for chemical biology and drug discovery applications.

7.3.3 Hybrid Peptide Foldamers

Several studies on β-peptides demonstrated their intrinsic resistance to proteolysis, thus offering clear advantages for therapeutic applications. Among this class of compounds, the most extensively studied β-peptide secondary structure is the 14-helix, characterized by backbone hydrogen bonds to form 14-membered pseudorings and approximately three residues per turn. They have been successfully employed to target p53-MDM2 and MDMX interactions, to inhibit HIV gp41-mediated fusion, and to activate GLP-1 receptors with sub-micromolar affinities [37]. The shape of the helix proved dependent on the choice of β-amino acids, as the 14-helix is strongly favored in β^3-peptides and further stabilized by constrained residues such as ACHC [38], whereas the incorporation of five-membered cyclic residues such as ACPC promotes the formation of a tighter 12-helix. Moreover, alternating β^2/β^3 patterns or alternating chirality can lead to other helical forms, including 10/12 helices [39].

As a natural evolution in the study of synthetic peptide-based oligomeric compounds, hybrid foldamers have been conceived as synthetic oligomers combining the structural features of natural peptides with noncanonical building blocks, enabling access to conformational space beyond that of α-peptides while retaining elements of biological recognition. Specifically, hybrid foldamers proved to implement conformational control, functional mimicry, and enhanced stability by integrating β-, γ-, or other non-proteinogenic residues and by replacing peptide bonds with isosteric units [40–42]. Concerning the molecular scaffolds embedded in hybrid foldamers, these mixed oligomers comprise α/β-peptides, β-peptides, peptoids, and triazole-based systems.

These systems have become central to foldamer research, particularly in the context of PPI modulation, antimicrobial activity, and improved pharmacokinetic properties. Importantly, hybrid foldamers have demonstrated that the biological function does not require strict adherence to natural backbone chemistry, whereas the selective incorporation of unnatural elements can preserve or even enhance activity while improving pharmacological properties.

7.3.3.1 α/β-Peptide Foldamers

Among hybrid foldamers, α/β-peptides represent one of the most extensively studied and structurally well-characterized classes, which are synthesized by the systematic incorporation of β-amino acids into α-peptide backbones, thus modifying backbone geometry, hydrogen-bonding patterns, and side-chain orientation. Early studies demonstrated that α/β-peptides adopt well-defined secondary structures, particularly helices, when appropriate stereochemistry and sequence design using α- and β-amino acid building blocks are applied. In fact, the ability of α/β-peptides to form helices depends critically on specific dihedral angles of rotatable bonds in the backbone. In this context, specific repeating patterns, such as ααβ, αααβ, and ααβαααβ, have been shown to support stable helices, and among these, the ααβαααβ arrangement is particularly relevant as it introduces minimal distortion of the canonical α-helix backbone. In this arrangement, β-residues are position on one face of the helix, whereas the native α-residue side chains remain on the opposite face for molecular recognition. The incorporation of cyclic β-amino acids significantly enhances helix stability and reduces backbone flexibility, thus promoting predictable folding and stabilization of helices even in polar solvents. In particular, ACPC and ACHC proved especially valuable due to their synthetic accessibility and strong conformational constraints [43]. Specifically, the use of ACPC residues of both *trans* and *cis* stereochemistry enabled the exploration of large conformational space while maintaining side-chain functionality. *Trans*-ACPC substitutions stabilized α-helix-like conformations, whereas *cis*-ACPC residues introduced noncanonical backbone geometries that enabled new protein–ligand interactions, demonstrating that hybrid foldamers are not merely mimics of natural structures but can outperform native peptides through novel conformational solutions.

Concerning their use in drug discovery programs, α/β-peptides have been widely applied to inhibit α-helix-mediated PPIs. Relevant examples in the context of cancer research include the inhibition of the BH3-Bcl-2, p53-MDM2, and VEGF–VEGFR1 interactions, demonstrating the utility of such peptidomimetic foldamers in targeting extended helical interfaces rather than minimal hot spots. Also, the relevance of hybrid peptide foldamers was found essential in the case of the Bcl-xL-Bak PPI, as pure β-peptides adopting either 12- or 14-helical conformations proved insufficient to reproduce the binding affinity of the native α-helix, whereas alternating α,β-peptides stabilized by cyclic β-residues achieved micromolar inhibition, demonstrating the capability of such molecular frameworks to balance shape complementarity and affinity. Further optimization using mixed α/β+α chimeras improved the binding affinity by retaining key α-helical regions while enhancing the proteolytic stability [44] (Figure 7.5).

A design paradigm proposed in the context of hybrid foldamer research is the systematic "foldamerization," in which native bioactive peptides are scanned with constrained β-residues to identify positions tolerating substitution [11]. This approach has been particularly successful in the context of p53-MDM2/X inhibition as a key cancer target.

7.3.3.2 *Hybrid β-Sheet-Forming Peptides*

The strategy of developing hybrid peptidomimetic foldamers has been applied to β-sheet-forming peptides, too. For example, in the antiangiogenic peptide anginex [45],

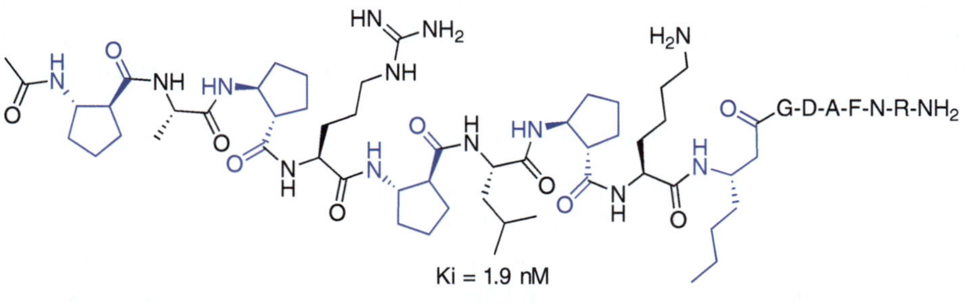

Figure 7.5 *Structure of chimeric (α/β + α)-peptide exhibiting inhibitory properties against Bcl-xL-Bak interaction; the cyclic β-amino acid scaffold* trans-ACPC *is highlighted in blue.*

the systematic α- to β^3-substitutions within the β-sheet core reduced the intrinsic folding propensity, although preserving the inducible β-sheet formation in the presence of interacting partners, such as lipid micelles, and the biological activity, specifically in the inhibition of endothelial cell proliferation, highlighting how hybrid foldamers can fine-tune folding equilibria, mitigate pathological aggregation, and improve enzymatic stability without abolishing the biological function.

7.3.3.3 Peptoids as Hybrid Foldamers

Peptoids (poly-*N*-substituted glycines) occupy a unique position among hybrid foldamers and have been early studied in the context of peptidomimetic foldamers [1]. The peculiar feature of this class of compounds resides in the structural arrangement of the oligomer, consisting of a backbone resembling that of α-peptides and of side chains being attached to the amide nitrogen rather than the α-carbon, thus eliminating backbone chirality and hydrogen-bond donation [46]. Interestingly, despite lacking canonical hydrogen-bond networks, peptoids can adopt stable helical conformations, and the folding is driven primarily by steric effects, *cis–trans* amide isomerization, and side-chain interactions. Indeed, the helicity can be controlled through the incorporation of chiral, bulky side chains at selected positions, compensating for the achirality of the backbone. A milestone example of this subclass of peptidomimetic foldamers is a 36-mer water-soluble peptoid containing repeating chiral side chains, which adopts a stable right-handed helix with exclusively *cis* amide bonds, demonstrating the remarkable stability of peptoid helices and emphasizing their robustness as foldamer scaffolds [47]. Concerning their applications in drug discovery, peptoids have been explored as α-helix mimetics in diverse biological contexts, including as antigens targeting VEGFR2 *in vitro* and *in vivo* [48, 49]. Their resistance to proteolysis makes them attractive for therapeutic development. Synthetic methodology has historically limited peptoid scalability and sequence definition both in solid-phase peptide synthesis and the sub-monomer approach. The latter, although providing excellent sequence control, requires large reagent excesses and is typically restricted to milligram scales. Nevertheless, recent advances using iterative exponential growth (IEG) strategies have been proposed to overcome these limitations (Figure 7.6), enabling the synthesis of sequence-defined peptoids on multi-gram scales with precise control over length, sequence, and stereochemistry using an orthogonally protected peptoid monomer with the secondary amine protected with a Cbz group and the carboxylic acid protected as a tert-butyl ester and an iterative cycle consisting of orthogonal deprotections using acid and hydrogenation, followed by amide coupling [50]. In this application, preformed peptoid monomers bearing different side-chain isosteres are required in the IEG synthetic cycle.

7.3.3.4 Triazolamers as Hybrid Foldamers

"Triazolamers" represent a distinct class of hybrid foldamers in which amide bonds are replaced or alternated with 1,2,3-triazole units formed *via* copper-catalyzed azide-alkyne cycloaddition (CuAAC) [51]. The oligomeric structure is characterized by repeating triazole units connected by single-carbon linkers accommodating side-chain isosteres in well-defined stereochemistry, adopting extended zigzag conformations mimetic of peptide β-strands. The resulting 1,4-disubstituted triazoles are metabolically inert and possess

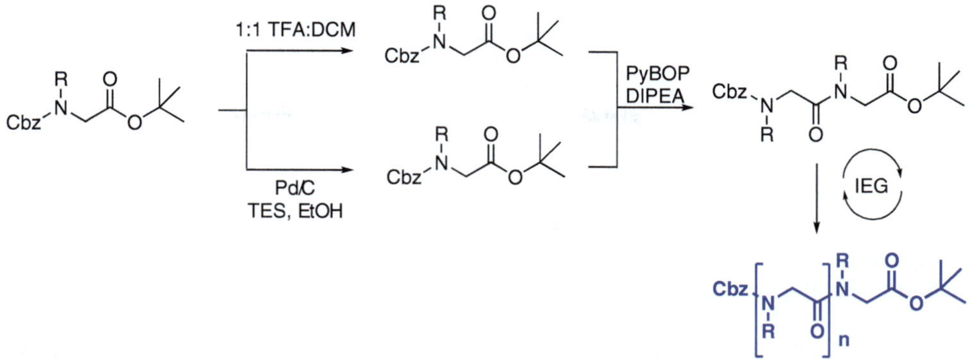

Figure 7.6 Peptoid synthesis through the IEG cycle involving orthogonal deprotections to expose reactive functional groups, followed by a coupling reaction.

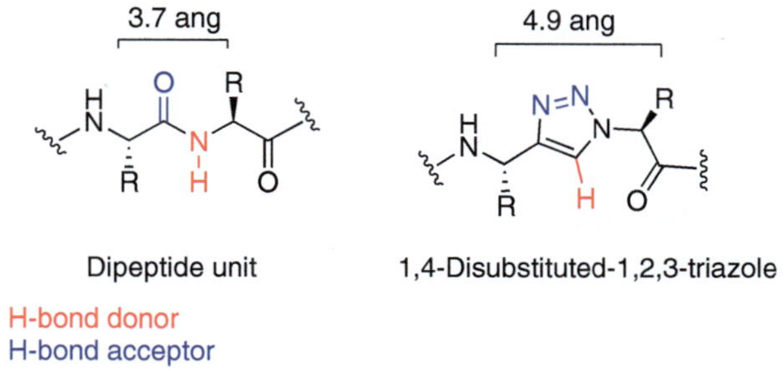

Figure 7.7 1,4-Disubstituted 1,2,3-triazole as a peptide bond isostere.

physicochemical properties sufficiently similar to *trans*-amide bonds to function as amide isosteres (Figure 7.7) [52–56], as also described in a previous book chapter on this topic [1].

The CuAAC reaction can be carried out either in solution or on solid support to afford 1,4-disubstituted triazoles. Optimal results are achieved using a sequential protocol when triflyl azide and Cu(II) are employed for the diazo transfer; in this approach, the copper is subsequently reduced in situ by ascorbate to initiate the CuAAC reaction (Figure 7.8) [57].

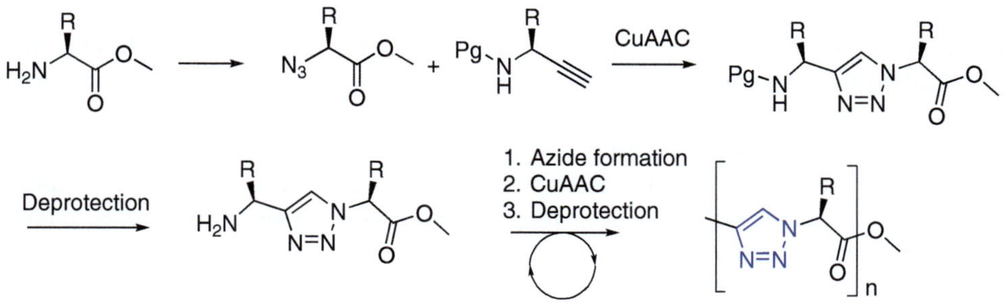

Figure 7.8 Solution-phase synthesis of triazolamers.

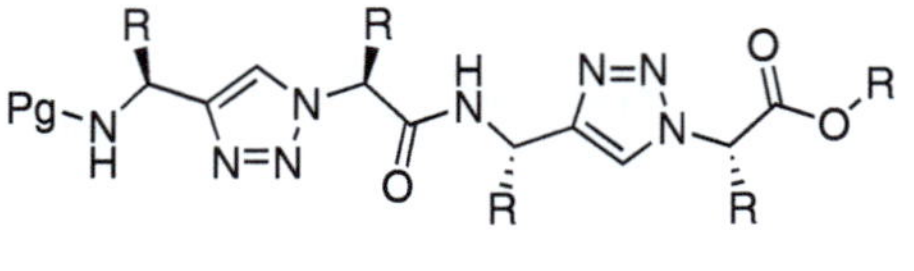

Figure 7.9 Structure of a peptidotriazole foldamer.

Peptidotriazole foldamers, in which peptide and triazole units alternate, combine features of both peptides and triazolamers (Figure 7.9). These hybrids can be synthesized *via* building-block or sub-monomer strategies, offering flexibility and modularity, and generally are developed using amino acid-derived azido acids and propargylamines. These systems exhibit diverse folding behaviors, including compact conformations stabilized by hydrophobic side chains. Molecular dynamics simulations and experimental studies revealed that both homo- and heterochiral oligomers can fold into well-defined structures in organic solvents and water. Click-polyaddition strategies enable the construction of higher-order oligomers. The robustness, modularity, and metabolic stability of triazole-containing backbones make them attractive alternatives to traditional peptide scaffolds.

7.4 Application of Secondary-Structure Mimetic Foldamers in Drug Discovery

The predictability of secondary-structure formation, sequence design, and modularity of foldamers have enabled remarkable progress in biological applications over the past two decades. Secondary-structure mimetics are now being applied to target membranes, protein surfaces, and nucleic acids, with α-helical mimetics being particularly dominant in PPI inhibition. Nevertheless, while α-helix mimetics dominate current applications, advances in β-strand design and aggregation control are gradually delivering novel entries in this subtype of foldamers, and the continued integration of high-resolution structural characterization and computational modeling is essential for translating such molecular designs into robust therapeutic agents.

Foldamer and peptidomimetic scaffolds have emerged as versatile tools in chemical biology and drug discovery, enabling the rational design of molecules that mimic the key structural and functional features of natural peptides. By reproducing the secondary structure of target proteins, foldamers allow researchers to modulate PPIs and other biological processes with high specificity. Specifically, foldamer applications can be classified according to their therapeutic context. Concerning design principles and strategies have focused on mimicking the spatial orientation of key residues, which is critical for engaging protein surfaces, and on engineering the native peptide backbone to stabilize helical or extended conformations while enhancing proteolytic resistance. The synthetic accessibility is being addressed in design strategies with preference toward modular approaches to favor the development of libraries for subsequent screening activities.

In antimicrobial research, helix- and strand-mimetic peptidomimetics such as β-peptides, peptoids, peptides, and oligo-acyl-lysines mimic antimicrobial peptides (AMPs) and disrupt bacterial membranes. Other mimetics interfere with biofilm formation, quorum sensing, or lipopolysaccharide binding, thus providing activity against both Gram-positive and Gram-negative pathogens. β-Sheet-inspired compounds have also been used to modulate aggregation while retaining biological activity, as demonstrated in modified angiogenesis inhibitors.

In oncology, α-helix mimetics have been extensively applied to disrupt PPIs involving Bcl-2 family proteins, p53-MDM2, and related pathways. The combination of high-resolution structural data and compatibility with solid-phase synthesis has enabled the optimization of potency and selectivity. Recent α-, β-, and γ-peptidomimetics demonstrate how hybrid backbones can match native peptide activity while offering superior stability. Concerning cell uptake, which is particularly relevant for intracellular targets, terphenyl and oligobenzamide foldamers demonstrated intrinsic permeability ideal for mimicking α-helices in the context of cancer PPI inhibition, including p53-MDM2 and BCL-2 family proteins [58]. Also, hybrid triazine–piperazine–triazine foldamers and oligooxopiperazines displayed α-helix mimetic capabilities for enhanced MCL-1 binding and for targeting p300-HIF-1α and other transcription factor PPIs, respectively. Foldamers with higher peptide character, such as β-peptides, peptoids, and hybrid α-, β-, and γ-peptides, proved ideal for antimicrobial activity and BCL-2 mimicry, as well as intracellular PPI modulation.

7.4.1 Cancer-Targeting Foldamers

PPIs play a central role in cancer biology, regulating cell proliferation, apoptosis, and transcriptional programs. Targeting protein–protein interfaces has been demonstrated to be a powerful therapeutic strategy in cancer drug discovery, given their central role in cancer biology, regulating cell proliferation, apoptosis, and transcriptional programs. Nevertheless, this is particularly challenging using small molecules due to the large, flat nature of protein surfaces. In this context, foldamers provide a well-conceived solution by mimicking the α-helical motifs commonly found at PPI interfaces. In this context, two major strategies have been developed:

- helix-guided foldamers, characterized by artificial backbones possessing predictable helical moieties closely resembling natural α-peptides, such as β-peptides forming 14-helices, α/β-heterogeneous oligoamides, and foldamers incorporating γ- or cyclic residues, allowing accurate placement of key residues at positions corresponding to i, $i+4$, and $i+7$ on the helical surface.
- minimalist topographical mimetics, consisting of rigid, extended scaffolds that reproduce the spatial orientation of key side chains and differ significantly from the natural peptide backbone, such as terphenyl, triazine–piperazine–triazine, oligooxopiperazine, and oligopyridylamide scaffolds.

Both approaches have enabled the targeting of therapeutically relevant PPIs, including p53-MDM2, BCL-2 family, and p300-HIF-1α, as discussed below [10, 59, 60].

7.4.1.1 Targeting the Antiapoptotic BCL-2 Family Proteins

The BCL-2 family regulates apoptosis through pro-survival and pro-apoptotic proteins [61]. BH3-only proteins contain α-helices that bind to antiapoptotic members like BCL-xL and

MCL-1. Foldamer mimetics of the BH3 α-helix have been designed to induce apoptosis selectively in cancer cells by inhibiting antiapoptotic BCL-2 proteins. Challenges in this context include the need to mimic the length of the BH3 helix, which shows four consecutive turns, and the presentation of hydrophobic and polar residues on opposite faces of the helix. The application of foldamer strategies has included:

- small non-helical scaffolds with appended side chains that replicate the hydrophobic and conserved carboxylate residues of BH3 peptides [62];
- terphenyl and triazine–piperazine–triazine scaffolds targeting BCL-xL or MCL-1;
- multi-faced helix mimetics displaying side chains on more than one face of the helix to improve binding specificity.

Some terphenyl-based compounds demonstrate up to 100-fold selectivity over unrelated PPIs, illustrating the selectivity that can be achieved with foldamer design.

7.4.1.2 Foldamers Targeting the p53-MDM2 Interaction

Restoring wild-type p53 activity in tumors overexpressing MDM2 is a key anticancer strategy [63]. The p53 activation domain forms an α-helix that binds MDM2 via three hydrophobic residues (Phe19, Trp23, and Leu26) [64]. In this context, foldamer-based strategies have focused on:

- stabilizing the helical conformation using α,α-dialkylated residues, side-chain cross-links, and hydrogen-bond surrogates [65–67];
- non-peptide scaffolds that reproduce the spatial orientation of the p53 hot-spot residues, including terphenyls, oligopyridylamides, oligobenzamides, and oligooxopiperazines;
- dual-faced helix mimetics designed to mimic more than one face of the helix, improving interaction with MDM2 and related targets.

Several compounds bearing terphenyl and oligobenzamide scaffolds proved to block the p53-MDM2 interaction at sub-micromolar concentrations, demonstrating in-cell activity (Figure 7.10).

7.4.1.3 Foldamers Targeting the p300-HIF-1α Interaction

The hypoxia-inducible factor 1α (HIF-1α) regulates tumor adaptation to low oxygen conditions by recruiting coactivators such as p300 *via* helical transactivation domains. In this context, disruption of such PPI interaction has been addressed with various approaches, including foldamers as follows (Figure 7.11):

- oligooxopiperazine derivatives, which mimic side chains of key hydrophobic residues in the third helix of HIF-1α, with lead compounds capable of binding the p300 protein with a Kd of 533 nM, modulating downstream gene expression in lung cancer cells [68];
- 3-*O*-alkylated oligobenzamides reproducing the i, $i+4$, and $i+7$ positions of HIF-1α helices, inhibiting the peptide–protein interaction in vitro [69];
- hybrid molecules combining foldamer mimetics with segments of the native peptide to enhance specificity.

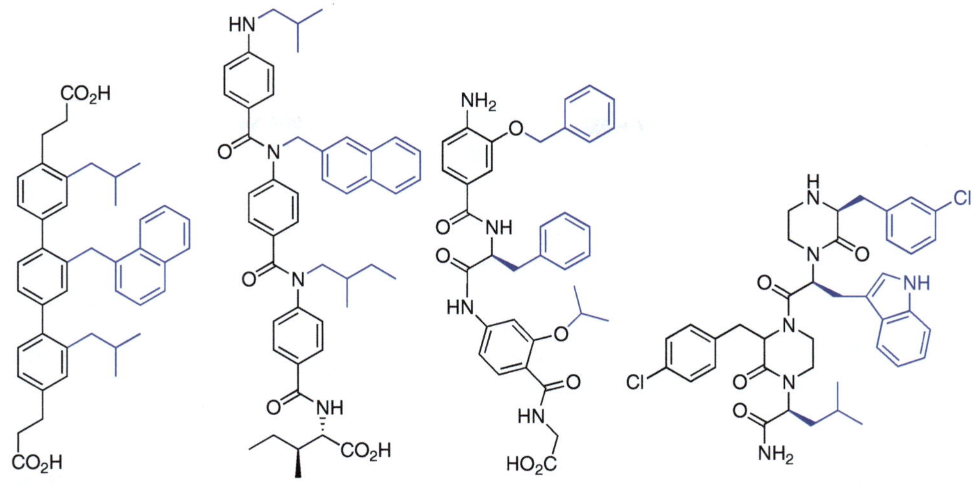

Terphenyl **Oligobenzamide** **Oligooxopiperazine**

Figure 7.10 *Inhibitors of the p53-hDM2 interaction based on extended and rigid α-helix mimetic scaffolds.*

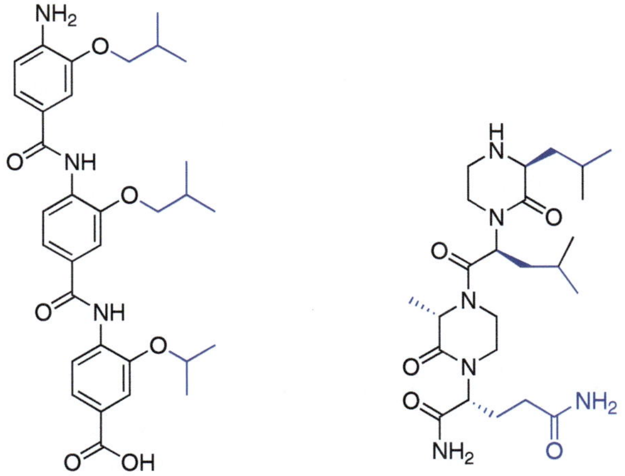

Oligobenzamide **Oligooxopiperazine**

Figure 7.11 *Structure of 3-O-alkylated oligobenzamide (left) and oligooxopiperazine (right) derivatives designed to mimic the HIF-1α third helix.*

7.4.2 Antimicrobial Foldamers

Membrane-active AMPs serve as natural defense mechanisms against bacteria. Synthetic foldamers that mimic AMPs have been developed to overcome limitations such as proteolytic degradation, toxicity, and bacterial resistance [70], generally displaying cationic charge to promote selective interaction with negatively charged bacterial membranes, amphiphilicity to arrange hydrophobic and cationic residues on opposite surfaces of a helix

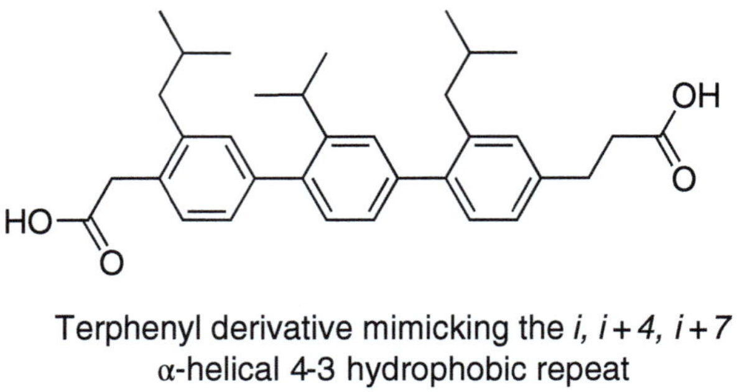

Figure 7.12 *Structure of a terphenyl foldamer antagonist for the disruption of the gp41 core, mimicking the α-helical 4-3 hydrophobic repeat regions.*

to induce membrane disruption, and proteolytic stability through unnatural backbones or cyclization [5]. To meet all these requirements for developing antimicrobial peptidomimetics capable of forming globally amphiphilic helices, which mimic the helical structure of AMPs, several foldamers have been developed, such as β-peptides forming 10-, 12-, or 14-helices [71], peptoids, *N,N′*-linked aliphatic oligoureas, oligo(urea/amide) hybrids, and sulfono-γ-AApeptides [72].

AMP-mimicking foldamers proved to exert antibacterial effects primarily by disrupting bacterial membranes, leading to cell lysis. In particular, they have demonstrated activity against both Gram-positive and Gram-negative bacteria, as well as biofilm inhibition, where foldamers prevent bacterial community formation by interfering with quorum sensing, lipopolysaccharide binding, targeting LPS on Gram-negative bacteria, and mitigating endotoxin activity and cell wall synthesis inhibition, modeled after natural antibiotics such as teixobactin.

7.4.3 Antiviral Foldamers

Foldamers have also been explored as viral fusion inhibitors, although these approaches are less developed compared to cancer and antimicrobial applications. For example, in the case of HIV-1, several strategies aim to inhibit viral fusion by targeting helical domains involved in membrane interaction, including D-peptide inhibitors identified via mirror-image phage display [73], β^3-peptide analogues designed to mimic hydrophobic pocket-binding domains [74], and terphenyl scaffolds presenting hydrophobic side chains to mimic heptad repeats in viral fusion proteins in the HR1 and HR2 regions (Figure 7.12) [75]. While potency remains below that of clinically used antivirals, these studies illustrate the potential of foldamers to selectively interfere with viral protein interactions and entry mechanisms.

7.5 Conclusion

Foldamer-based peptidomimetics represent a powerful and modular approach to targeting biological interfaces. The identification of foldamers that selectively recognize biopolymer targets is achieved by precise control over monomer sequence, secondary structure, and an expanded toolbox of building blocks. By combining secondary-structure mimicry with backbone engineering, foldamers provide predictable, tunable, and biocompatible molecular scaffolds capable of modulating PPIs, disrupting membranes, and interfering with viral mechanisms. Nevertheless, precise mimicry of protein secondary structures involved in binding remains challenging, and only limited foldamer-based strategies are routinely available. Progress requires modular chemistries, higher-order molecular platforms, the structural characterization at the molecular level of foldamer–biomolecule complexes, and advanced computational design. Foldamer chemistry, including protease-resistant backbones that fold in water, has been advanced, enabling various applications ranging from PPI inhibition to antimicrobial and hormone mimetics. In particular, parallel to aliphatic α/β-peptides, which dominate the panorama of peptidomimetic foldamers, emerging backbones have been proposed to broaden the therapeutic potential. Also, the continued exploration of hybrid foldamer structures, along with structural validation and in-cell studies, is expected to expand their therapeutic applications further. Concerning the *in vivo* evaluation of foldamers as candidate drugs, pharmacokinetic studies remain scarce but represent the next critical frontier in drug discovery using foldamer peptidomimetics.

References

1. Trabocchi A, Guarna A. *Peptidomimetics in Organic and Medicinal Chemistry : The Art of Transforming Peptides in Drugs*. Chichester, West Sussex, United Kingdom: John Wiley & Sons Inc., 2014.
2. Pasco M, Dolain C, Guichard G. Foldamers in medicinal chemistry. In: *Comprehensive Supramolecular Chemistry II*. Elsevier, 2017.
3. Guarracino DA, Riordan JA, Barreto GM, Oldfield AL, Kouba CM, Agrinsoni D. Macrocyclic control in helix mimetics. *Chem Rev*. 2019;**119**:9915–9949.
4. Guarracino DA, Bullock BN, Arora PS. Mini review: protein-protein interactions in transcription: a fertile ground for helix mimetics. *Biopolymers*. 2011;**95**:1–7.
5. Tew GN, Scott RW, Klein ML, Degrado WF. De novo design of antimicrobial polymers, foldamers, and small molecules: from discovery to practical applications. *Acc Chem Res*. 2010;**43**:30–39.
6. Potocky TB, Silvius J, Menon AK, Gellman SH. HeLa cell entry by guanidinium-rich beta-peptides: importance of specific cation-cell surface interactions. *ChemBioChem*. 2007;**8**:917–926.
7. Johnson LM, Barrick S, Hager MV, McFedries A, Homan EA, Rabaglia ME, Keller MP, Attie AD, Saghatelian A, Bisello A, Gellman SH. A potent a/b-peptide analogue of GLP-1 with prolonged action in vivo. *J Am Chem Soc*. 2014;**136**:12848–12851.
8. Wójcik P, Berlicki Ł. Peptide-based inhibitors of protein–protein interactions. *Bioorg Med Chem Lett*. 2016;**26**:707–713.
9. Lee EF, Smith BJ, Horne WS, Mayer KN, Evangelista M, Colman PM, Gellman SH, Fairlie WD. Structural basis of Bcl-x_L recognition by a BH3-mimetic a/b-peptide generated by sequence-based design. *ChemBioChem*. 2011;**12**:2025–2032.
10. Harker EA, Daniels DS, Guarracino DA, Schepartz A. Beta-peptides with improved affinity for hDM2 and hDMX. *Bioorg Med Chem*. 2009;**17**:2038–2046.

11. Fortuna P, Twarda-Clapa A, Skalniak L, Ożga K, Holak TA, Berlicki Ł. Systematic 'foldamerization' of peptide inhibiting p53-MDM2/X interactions by the incorporation of trans- or cis-2-aminocyclopentanecarboxylic acid residues. *Eur J Med Chem.* 2020;**208**:112814.
12. Haase HS, Peterson-Kaufman KJ, Lan Levengood SK, Checco JW, Murphy W, Gellman SH. Extending foldamer design beyond a-helix mimicry: a/b-peptide inhibitors of vascular endothelial growth factor signaling. *J Am Chem Soc.* 2012;**134**:7652–7655.
13. Swenson CS, Mandava G, Thomas DM, Moellering RE. Tackling undruggable targets with designer peptidomimetics and synthetic biologics. *Chem Rev.* 2024;**124**:13020–13093.
14. Bullock BN, Jochim AL, Arora PS. Assessing helical protein interfaces for inhibitor design. *J Am Chem Soc.* 2011;**133**:14220–14223.
15. Chongsiriwatana NP, Patch JA, Czyzewski AM, Dohm MT, Ivankin A, Gidalevitz D, Zuckermann RN, Barron AE. Peptoids that mimic the structure, function, and mechanism of helical antimicrobial peptides. *Proc Natl Acad Sci USA.* 2008;**105**:2794–2799.
16. Robinson JA. Beta-hairpin peptidomimetics: design, structures and biological activities. *Acc Chem Res.* 2008;**41**:1278–1288.
17. Nowick JS, Chung DM, Maitra K, Maitra S, Stigers KD, Sun Y. An unnatural amino acid that mimics a tripeptide β-strand and forms β-sheetlike hydrogen-bonded dimers. *J Am Chem Soc.* 2000;**122**:7654–7661.
18. Smith AB 3rd, Charnley AK, Hirschmann R. Pyrrolinone-based peptidomimetics. Let the enzyme or receptor be the judge. *Acc Chem Res.* 2011;**44**:180–193.
19. Raghuraman A, Ko E, Perez LM, Ioerger TR, Burgess K. Pyrrolinone-pyrrolidine oligomers as universal peptidomimetics. *J Am Chem Soc.* 2011;**133**:12350–12353.
20. Wyrembak PN, Hamilton AD. Alkyne-linked 2,2-disubstituted-indolin-3-one oligomers as extended beta-strand mimetics. *J Am Chem Soc.* 2009;**131**:4566–4567.
21. Jamieson AG, Russell D, Hamilton AD. A 1,3-phenyl-linked hydantoin oligomer scaffold as a β-strand mimetic. *Chem Commun.* 2012;**48**:3709–3711.
22. Sutherell CL, Thompson S, Scott RT, Hamilton AD. Aryl-linked imidazolidin-2-ones as non-peptidic β-strand mimetics. *Chem Commun.* 2012;**48**:9834–9836.
23. Das S, Ben Haj Salah K, Djibo M, Inguimbert N. Peptaibols as a model for the insertions of chemical modifications. *Arch Biochem Biophys.* 2018;**658**:16–30.
24. Gessmann R, Brückner H, Petratos K. The crystal structure of Z-$(Aib)_{10}$-OH at 0.65 Å resolution: three complete turns of 3_{10}-helix. *J Pept Sci.* 2016;**22**:76–81.
25. Tanda K, Eto R, Kato K, Oba M, Ueda A, Suemune H, Doi M, Demizu Y, Kurihara M, Tanaka M. Peptide foldamers composed of six-membered ring α,α- disubstituted α-amino acids with two changeable chiral acetal moieties. *Tetrahedron.* 2015;**71**:3909–3914.
26. Yoo SH, Lee HS. Foldectures: 3D molecular architectures from self-assembly of peptide foldamers. *Acc Chem Res.* 2017;**50**:832–841.
27. Hamuro Y, Geib SJ, Hamilton AD. Novel molecular scaffolds: formation of helical secondary structure in a family of oligoanthranilamides. *Angew Chem Int Ed.* 1994;**33**:446–448.
28. Koert U, Harding MM, Lehn JM. DNH deoxyribonucleohelicates: self assembly of oligonucleosidic double-helical metal complexes. *Nature.* 1990;**346**:339–342.

29. Odriozola I, Kyritsakas N, Lehn JM. Structural codons: linearity/helicity interconversion by pyridine/pyrimidine exchange in molecular strands. *Chem Commun.* 2004;62–63.
30. Dolain C, Jiang H, Léger JM, Guionneau P, Huc I. Chiral induction in quinoline-derived oligoamide foldamers: assignment of helical handedness and role of steric effects. *J Am Chem Soc.* 2005;**127**:12943–12951.
31. Gillies ER, Deiss F, Staedel C, Schmitter JM, Huc I. Development and biological assessment of fully water-soluble helical aromatic amide foldamers. *Angew Chem Int Ed Engl.* 2007;**46**:4081–4084.
32. Saraogi I, Hamilton AD. alpha-Helix mimetics as inhibitors of protein–protein interactions. *Biochem Soc Trans.* 2008;**36**:1414–1417.
33. Algar S, Martín-Martínez M, González-Muñiz R. Evolution in non-peptide α-helix mimetics on the road to effective protein–protein interaction modulators. *Eur J Med Chem.* 2021;**211**:113015.
34. Ernst JT, Becerril J, Park HS, Yin H, Hamilton AD. Design and application of an alpha-helix-mimetic scaffold based on an oligoamide-foldamer strategy: antagonism of the Bak BH3/Bcl-xL complex. *Angew Chem Int Ed Engl.* 2003;**42**:535–539.
35. Cummings CG, Ross NT, Katt WP, Hamilton AD. Synthesis and biological evaluation of a 5-6-5 imidazole-phenyl-thiazole based alpha-helix mimetic. *Org Lett.* 2009;**11**:25–28.
36. Barnard A, Miles JA, Burslem GM, Barker AM, Wilson AJ. Multivalent helix mimetics for PPI-inhibition. *Org Biomol Chem.* 2015;**13**:258–264.
37. Cabrele C, Martinek TA, Reiser O, Berlicki Ł. Peptides containing β-amino acid patterns: challenges and successes in medicinal chemistry. *J Med Chem.* 2014;**57**:9718–9739.
38. Appella DH, Christianson LA, Karle IL, Powell DR, Gellman SH. Synthesis and characterization of trans-2-aminocyclohexanecarboxylic acid oligomers: an unnatural helical secondary structure and implications for β-peptide tertiary structure. *J Am Chem Soc.* 1999;**121**:6206–6212.
39. Rueping M, Schreiber JV, Lelais G, Jaun B, Seebach D. Mixed β2/β3-hexapeptides and β2/β3-nonapeptides folding to (P)-helices with alternating twelve- and ten-membered hydrogen-bonded rings. *Helv Chim Acta.* 2002;**85**:2577–2593.
40. Appella DH, Christianson LA, Klein DA, Richards MR, Powell DR, Gellman SH. Synthesis and structural characterization ofhelix-forming b-peptides: trans-2-aminocyclopentanecarbox-ylic acid oligomers. *J Am Chem Soc.* 1999;**121**:7574–7581.
41. Sharma GV, Jayaprakash P, Narsimulu K, Ravi Sankar A, Ravinder Reddy K, Radha Krishna P, Kunwar AC. A left-handed 9-helix in gamma-peptides: synthesis and conformational studies of oligomers with dipeptide repeats of C-linked carbo-gamma4-amino acids and gamma-aminobutyric acid. *Angew Chem Int Ed Engl.* 2006;**45**:2944–2947.
42. Arndt H-D, Ziemer B, Koert U. Folding propensity of cyclo-hexylether-d-peptides. *Org Lett.* 2004;**6**:3269–3272.
43. Choi SH, Guzei IA, Spencer LC, Gellman SH. Crystallographiccharacterization of helical secondary structures in a/b-peptides with 1:1 residue alternation. *J Am Chem Soc.* 2008;**130**:6544–6550.

44. Sadowsky JD, Schmitt MA, Lee HS, Umezawa N, Wang S, Tomita Y, Gellman SH. Chimeric (α/β + α)-peptide ligands for the BH3-recognition cleft of Bcl-xL: critical role of the molecular scaffold in protein surface recognition. *J Am Chem Soc*. 2005;**127**:11966–11968.
45. Mayo KH, Ilyina E. A folding pathway for betapep-4 peptide 33mer: from unfolded monomers and beta-sheet sandwich dimers to well-structured tetramers. *Protein Sci*. 1998;**7**:358–368.
46. Kirshenbaum K, Barron AE, Goldsmith RA, Armand P, Bradley EK, Truong KT, Dill KA, Cohen FE, Zuckermann RN. Sequence-specific polypeptoids: a diverse family of heteropolymers with stable secondary structure. *Proc Natl Acad Sci USA*. 1998 Apr;**95**:4303–4308.
47. Sanborn TJ, Wu CW, Zuckermann RN, Barron AE. Extreme stability of helices formed by water-soluble poly-*N*-substituted glycines (polypeptoids) with alpha-chiral side chains. *Biopolymers*. 2002;**63**:12–20.
48. Merritt HI, Sawyer N, Arora PS. Bent into shape: folded peptides to mimic protein structure and modulate protein function. *Pept Sci*. 2020;**112**:e24145.
49. Udugamasooriya DG, Dineen SP, Brekken RA, Kodadek T. A peptoid "antibody surrogate" that antagonizes VEGF receptor 2 activity. *J Am Chem Soc*. 2008;**130**: 5744–5752.
50. Clapperton AM, Hood C, Tran H. Sequence-defined peptoids via iterative exponential growth. *Chem Sci*. 2025;**16**:9638–9647.
51. Schröder DC, Kracker O, Fröhr T, Góra J, Jewginski M, Nieß A, Antes I, Latajka R, Marion A, Sewald N. 1,4-Disubstituted 1*H*-1,2,3-triazole containing peptidotriazolamers: a new class of peptidomimetics with interesting foldamer properties. *Front Chem*. 2019;**7**:155.
52. Brik A, Alexandratos J, Lin YC, Elder JH, Olson AJ, Wlodawer A, Goodsell DS, Wong CH. 1,2,3-triazole as a peptide surrogate in the rapid synthesis of HIV-1 protease inhibitors. *ChemBioChem*. 2005;**6**:1167–1169.
53. Lee T, Cho M, Ko SY, Youn HJ, Baek DJ, Cho WJ, Kang CY, Kim S. Synthesis and evaluation of 1,2,3-triazole containing analogues of the immunostimulant alpha-GalCer. *J Med Chem*. 2007;**50**:585–589.
54. Appendino G, Bacchiega S, Minassi A, Cascio MG, De Petrocellis L, Di Marzo V. The 1,2,3-triazole ring as a peptido- and olefinomimetic element: discovery of click vanilloids and cannabinoids. *Angew Chem Int Ed Engl*. 2007;**46**:9312–9315.
55. Angell YL, Burgess K. Peptidomimetics via copper-catalyzed azide-alkyne cycloadditions. *Chem Soc Rev*. 2007;**36**:1674–1689.
56. Bock VD, Speijer D, Hiemstra H, van Maarseveen JH. 1,2,3-Triazoles as peptide bond isosteres: synthesis and biological evaluation of cyclotetrapeptidemimics. *Org Biomol Chem*. 2007;**5**:971.
57. Angelo NG, Arora PS. Nonpeptidic foldamers from amino acids: synthesis and characterization of 1,3-substituted triazole oligomers. *J Am Chem Soc*. 2005;**127**: 17134–17135.
58. Campbell F, Plante JP, Edwards TA, Warriner SL, Wilson AJ. *N*-alkylated oligoamide alpha-helical proteomimetics. *Org Biomol Chem*. 2010;**8**:2344–2351.

59. Checco JW, Kreitler DF, Thomas NC, Belair DG, Rettko NJ, Murphy WL, Forest KT, Gellman SH. Targeting diverse protein–protein interaction interfaces with α/β-peptides derived from the Z-domain scaffold. *Proc Natl Acad Sci USA*. 2015;**112**:4552–4557.
60. Burslem GM, Kyle HF, Breeze AL, Edwards TA, Nelson A, Warriner SL, Wilson AJ. Towards "bionic" proteins: replacement of continuous sequences from HIF-1α with proteomimetics to create functional p300 binding HIF-1α mimics. *Chem Commun*. 2016;**52**:5421–5424.
61. Juin P, Geneste O, Gautier F, Depil S, Campone M. Decoding and unlocking the BCL-2 dependency of cancer cells. *Nat Rev Cancer*. 2013;**13**:455–465.
62. Gloaguen C, Voisin-Chiret AS, Sopkova-de Oliveira Santos J, Fogha J, Gautier F, De Giorgi M, Burzicki G, Perato S, Pétigny-Lechartier C, Simonin-Le Jeune K, Brotin E, Goux D, N'Diaye M, Lambert B, Louis MH, Ligat L, Lopez F, Juin P, Bureau R, Rault S, Poulain L. First evidence that oligopyridines, α-helix foldamers, inhibit Mcl-1 and sensitize ovarian carcinoma cells to Bcl-xL-targeting strategies. *J Med Chem*. 2015;**58**:1644–1668.
63. Lane DP, Hupp TR. Drug discovery and p53. *Drug Discov Today*. 2003;**8**:347–355.
64. Kussie PH, Gorina S, Marechal V, Elenbaas B, Moreau J, Levine AJ, Pavletich NP. Structure of the MDM2 oncoprotein bound to the p53 tumor suppressor transactivation domain. *Science*. 1996;**274**:948–953.
65. Banerjee R, Basu G, Chène P, Roy S. Aib-based peptide backbone as scaffolds for helical peptide mimics. *J Pept Res*. 2002;**60**:88–94.
66. Bernal F, Wade M, Godes M, Davis TN, Whitehead DG, Kung AL, Wahl GM, Walensky LD. A stapled p53 helix overcomes HDMX-mediated suppression of p53. *Cancer Cell*. 2010;**18**:411–422.
67. Henchey LK, Porter JR, Ghosh I, Arora PS. High specificity in protein recognition by hydrogen-bond-surrogate α-helices: selective inhibition of the p53/MDM2 complex. *ChemBioChem*. 2010;**11**:2104–2107.
68. Tosovská P, Arora PS. Oligooxopiperazines as nonpeptidic alpha-helix mimetics. *Org Lett*. 2010;**12**:1588–1591.
69. Burslem GM, Kyle HF, Breeze AL, Edwards TA, Nelson A, Warriner SL, Wilson AJ. Small-molecule proteomimetic inhibitors of the HIF-1α-p300 protein–protein interaction. *ChemBioChem*. 2014;**15**:1083–1087.
70. Marr AK, Gooderham WJ, Hancock RE. Antibacterial peptides for therapeutic use: obstacles and realistic outlook. *Curr Opin Pharmacol*. 2006;**6**:468–472.
71. Arvidsson PI, Ryder NS, Weiss HM, Gross G, Kretz O, Woessner R, Seebach D. Antibiotic and hemolytic activity of a beta2/beta3 peptide capable of folding into a 12/10-helical secondary structure. *ChemBioChem*. 2003;**4**:1345–1347.
72. She F, Nimmagadda A, Teng P, Su M, Zuo X, Cai J. Helical 1:1 α/sulfono-γ-AA heterogeneous peptides with antibacterial activity. *Biomacromolecules*. 2016;**17**:1854–1859.
73. Welch BD, VanDemark AP, Heroux A, Hill CP, Kay MS. Potent D-peptide inhibitors of HIV-1 entry. *Proc Natl Acad Sci USA*. 2007;**104**:16828–16833.
74. Stephens OM, Kim S, Welch BD, Hodsdon ME, Kay MS, Schepartz A. Inhibiting HIV fusion with a beta-peptide foldamer. *J Am Chem Soc*. 2005;**127**:13126–13127.
75. Ernst JT, Kutzki O, Debnath AK, Jiang S, Lu H, Hamilton AD. Design of a protein surface antagonist based on alpha-helix mimicry: inhibition of gp41 assembly and viral fusion. *Angew Chem Int Ed Engl*. 2002;**41**:278–281.

Part III

Applications and Case Studies in Drug Discovery

8

Peptidomimetic SARS-CoV-2 Main Protease Inhibitors

8.1 Introduction

The pandemic coronavirus disease 2019 (COVID-19) has represented a very significant threat to global health, triggered by the acute respiratory syndrome coronavirus 2 (SARS-CoV-2) infection. The virus was isolated in December 2019 in Wuhan, China, showing a similarity with other coronaviruses such as SARS-CoV and MERS-CoV (Middle East Respiratory Syndrome coronavirus). On January 30, 2020, a health emergency was declared by the World Health Organization (WHO), and in March 2020, the pandemic state was announced, resulting in about 777 million infections and more than 7 million deaths worldwide [1].

Since then, efforts in scientific research have been made at a global level to evaluate existing therapeutics for treating COVID-19 [2–4] and to develop new therapies [5–7] and vaccines [8] capable of treating the SARS-CoV-2 infection. These efforts resulted in the development of a range of vaccines and antivirals, and in May 2023 the WHO declared COVID-19 no longer a global health emergency. Although the introduction of vaccines allowed for reducing significantly serious and critical cases among people infected, immunosuppressed or unvaccinated subjects still represent a vulnerable target, and new variants of SARS-CoV-2 that continue to develop are an issue [9]. Therefore, the achievement of therapeutic approaches is required to complete vaccine development. A range of therapies were made available for the treatment of COVID-19 [10], including antivirals, inflammation inhibitors, antirheumatic drugs, plasma, and therapeutic antibodies.

Major efforts to block viral replication focus on fundamental infection-related events aimed mainly at blocking the Spike-ACE2 interaction and inhibiting viral proteins required for replication, such as the Mpro protease and RNA-dependent RNA polymerase (RdRp). As a result, a wide number of inhibitors of Mpro and RdRp proteins have been identified, and three direct-acting antivirals have been approved so far, such as Paxlovid, targeting Mpro with nirmatrelvir as a protease inhibitor, and both molnupiravir and remdesivir as inhibitors of RdRp.

8.2 Structure and Replication Cycle of SARS-CoV-2

SARS-COV-2 operates through a complex infection mechanism consisting of several protein–protein interactions based on large and extended interaction surfaces [11–13]. Coronaviruses are RNA viruses belonging to the *Coronaviridae* family, with SARS-CoV-2 possessing a genomic identity rate of approximately 82% with other identified coronaviruses, such as SARS-CoV and MERS-CoV. The genome is composed of two main open reading frames (ORFs), namely ORF1a and ORF1b. Also, there are other ORFs that encode structural proteins, such as the Spike glycoprotein (S), the membrane (M), envelope (E), and nucleocapsid proteins (N) [14]. Finally, there are 16 nonstructural proteins (NSPs) that have an important role in infection and viral replication [15–18]. The entry of the virus into the host cell is mediated by the Spike (S) glycoprotein, which forms a homotrimer assembly in its active form, with each monomer being composed of two subunits:

- S1 contains a receptor-binding domain (RBD) for binding to the angiotensin-converting enzyme 2 (ACE2) receptor, which can be found in the "open" or "closed" conformation, and specifically for binding ACE2 with high affinity only in the "open" conformation (Figure 8.1).
- S2 mediates the fusion of the virus to human cells and its entry through a fusion peptide, heptad repeat 1 and 2 (HR1 and HR2), transmembrane, and cytoplasmic domains, which mediate the fusion of the viral particles to the host cell.

When the Spike protein of SARS-CoV-2 engages the ACE2 receptor, the complex undergoes a proteolytic cleavage operated by transmembrane serine protease TMPRSS2, leading to the formation of an S2′ site capable of interacting with ACE2 and mediating the beginning of the cell fusion process [19–22]. Once the viral genome has entered the cell, it starts replicating and undergoes all the steps for the synthesis of the viral proteins.

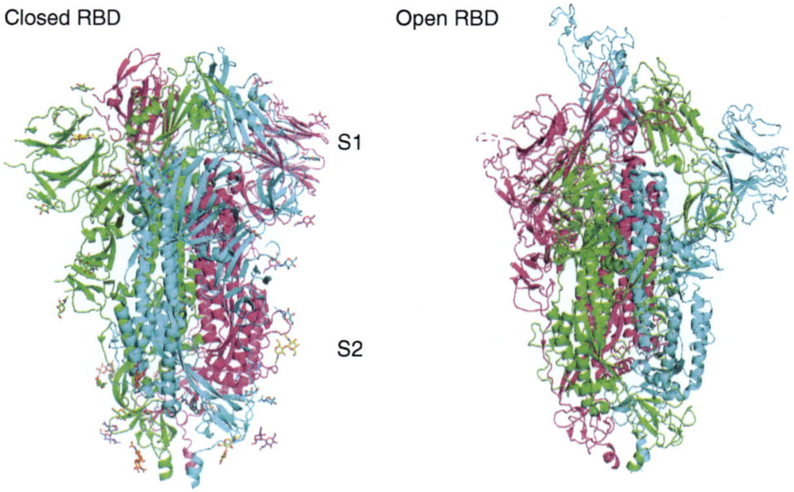

Figure 8.1 Structure of the Spike protein of SARS-CoV-2 with the S1 portion in closed (PDB: 6VXX) or open conformation (PDB: 7FB1).

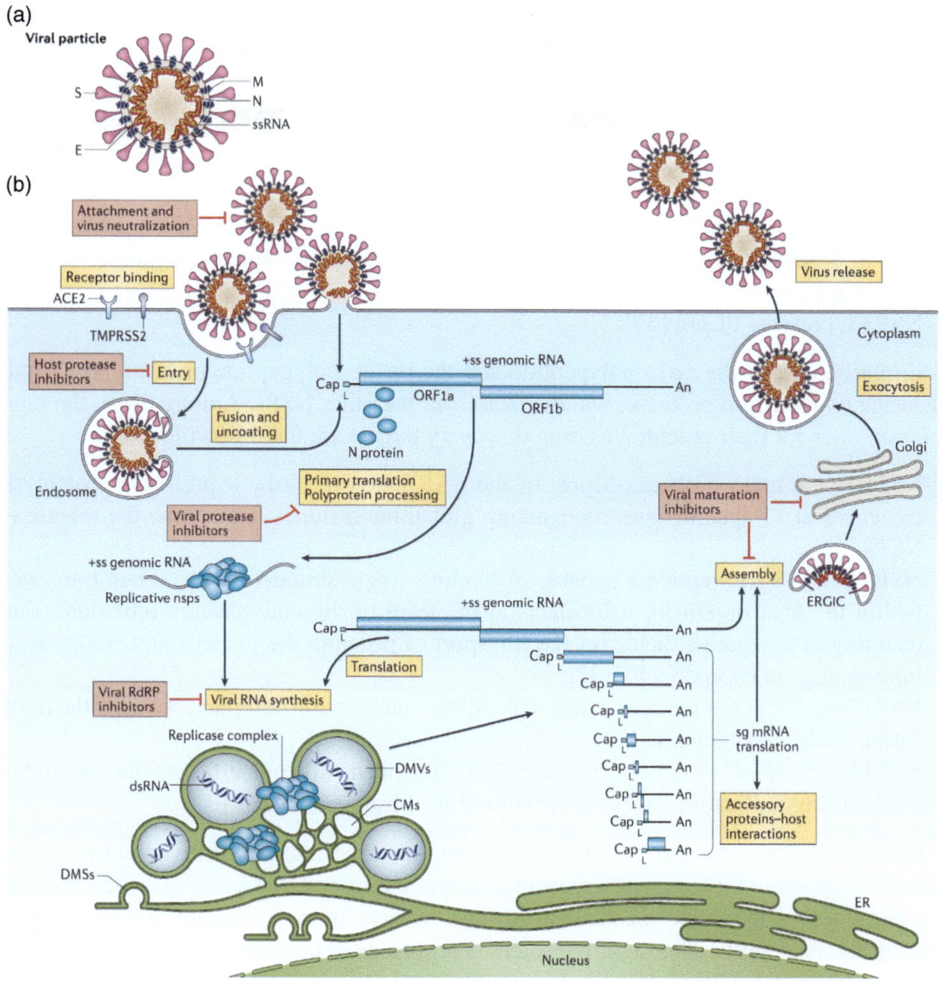

Figure 8.2 *Schematic representation of the SARS-CoV-2 replication cycle: a) structure of coronavirus virion; b) coronavirus life cycle. Source: [11] / with permission of Springer Nature.*

Finally, the viral particles are assembled and released to continue viral replication and spread (Figure 8.2) [23].

8.2.1 The Replication Cycle: A Focus on Nonstructural Proteins

As anticipated, the genome encodes 16 NSPs [24]. These proteins are translated from ORF1a and ORF1ab to encode NSP1 to NSP11 and NSP12 to NSP16, respectively [25–27]. Following the translation of the polypeptides generated by the two ORFs, proteolytic cleavage occurs by two viral proteases, namely the main protease (Mpro, also called 3-chymotrypsin-like protease, 3CLpro) and the papain-like protease (PLpro). Once released after proteolytic cleavage, these NSPs perform various biological functions starting from viral replication. The protease PLpro performs the proteolytic cleavage on the polypeptide pp1a that is translated from ORF1a, generating the release of NSP1, NSP2, and NSP3. These proteins have different roles:

- NSP1 allows for the inhibition of the translation of host proteins, interacting directly with the pocket of the 40S ribosome; furthermore, this bond leads to the lack of production of factors for the innate immune response [28].
- NSP2 interacts with Prohibitin 1 (PHB1) and Prohibitin 2 (PHB2), which mediate biological processes related to the morphology of mitochondria and the regulation of transcription factors [29].
- NSP3 represents PLpro [30].

Similarly, part of the pp1a polypeptide and the pp1ab polypeptide undergo proteolytic cleavage by the Mpro protease, which releases all the other NSPs. Among these, the most relevant ones for their potential as drug discovery targets are the following [31]:

- NSP5 is the main protease, Mpro, of the SARS-CoV-2 virus. It performs proteolytic cleavages at 11 specific sites recognizing glutamine residues to promote the release of NSPs [32].
- NSP9 is a dimeric replicase capable of binding single-stranded RNA. When translated within the host organism, it localizes at the level of the endoplasmic reticulum, thus resulting in a negative factor for the transport of p65 into the nucleus and acting as an immunological escape factor [33].
- NSP12 is an RdRp at the core of the replication-transcription complex, favoring the replication of the viral genome [34].
- NSP13, a SARS-CoV-2 helicase, is an ATP-dependent RdRp that favors the opening of RNA during replication and transcription of the viral genome [35].

The NSPs of the SARS-CoV-2 virus have multiple roles and are involved in many viral processes. Many of these play their role in viral replication and transcription; others mediate the immunological escape mechanism toward the host immune system. Therefore, preventing the proteolytic cleavage by proteases represents a key step in finding therapeutic strategies against SARS-CoV-2 viral infection.

8.3 Approved Vaccines and Antiviral Treatments

COVID-19 vaccines represented a fundamental progress in global public health, effectively reducing severe illness, hospitalization, and mortality rates. Developed through mRNA, viral vector, and protein-based platforms, they stimulate adaptive immune responses. Accordingly, widespread vaccination contributed to driving immunity, limiting viral transmission, and mitigating socioeconomic disruption, emphasizing their critical role in pandemic control.

Also, to fight SARS-CoV-2 infection, approaches involving the use of antiviral therapies and monoclonal antibodies have also been taken into consideration. These drugs act in various ways, including inhibiting the entry of the virus into the cell, inhibiting the fusion of the viral membrane with the cellular membrane and endocytosis, and blocking the activity of the Mpro protease or RdRp [10, 36].

Approved monoclonal antibodies, such as early bamlanvinimab and sotrovimab, regdanvimab, and tixagevimab/cilgavimab, were capable to block the viral entry into human cells by targeting the Spike protein and blocking the virus from attaching to ACE2 receptors [37]. The use of these antibodies in high-risk patients has been shown to result in lower hospitalizations and deaths associated with COVID-19 as compared to a placebo group. This therapy also accelerated the decline in SARS-CoV-2 viral load, both in high-risk patients and in moderate- or mild-risk patients when the two antibodies were administered concomitantly.

In this context, inhibitors of Mpro and RdRp proteins have been developed for the treatment of COVID-19, and three direct-acting antivirals have been approved, namely Paxlovid, targeting Mpro with PF-07321332 (nirmatrelvir) as a protease inhibitor, and both molnupiravir and remdesivir as inhibitors of the RdRp. Remdesivir is a repurposed antiviral drug conceived as an adenosine analogue metabolized to remdesivir triphosphate as the active metabolite. This molecule is incorporated by RNA polymerase into nascent viral RNA, thus competing with the natural substrate, and leads to early chain termination during replication and subsequent inhibition of viral proliferation [38]. Although remdesivir was recommended only in specific conditions, it may have an effect in improving the clinical situation in the early stages of the disease. It reduces mortality in the early phase of infection, the need for respiratory assistance, and invasive mechanical ventilation among patients affected by COVID-19 [39]. Molnupiravir [40] is an oral antiviral prodrug of *N*-hydroxycytidine that inhibits SARS-CoV-2 replication by inducing viral RNA mutagenesis. Clinical trials demonstrated reduced hospitalization and mortality in high-risk patients with early COVID-19, though efficacy proved to be lower than Paxlovid's as a protease inhibitor. Nevertheless, it remains an important option for antiviral treatment. Accordingly, a strategy to fast-forward the generation of a therapeutic targeting SARS-CoV-2 infection was identified at Pfizer following previous research on the development of an inhibitor of the key viral protease Mpro of the related coronavirus SARS-CoV-1, resulting in the development of Paxlovid as an antiviral therapeutic agent consisting of two compounds: PF-07321332, an oral covalent inhibitor of the major SARS-CoV-2 Mpro protease, capable of binding to the catalytic residue Cys145, and ritonavir, an HIV-1 protease inhibitor. Ritonavir was taken into account for inhibiting the metabolism of PF-07321332, allowing the reduction of the administered dose of the protease inhibitor.

8.4 Therapeutic Targets and Approaches

Therapeutic approaches that focus on coronaviruses are diverse and can be grouped based on the targets selected and the phases of the SARS-CoV-2 life cycle [41]:

- Inhibition of viral entry through chemical compounds or monoclonal antibodies. SARS-CoV-2 uses the Spike S protein on its surface to attach to the host cell through a protein–protein interaction with the ACE2 receptor.

This interaction can be a target to identify an appropriate treatment [42]. In particular, the RBD domain of the S protein is a key target for antibody-mediated binding disruption. Alternatively, excess soluble ACE2 was proposed as a solution to support viral neutralization by competitively binding to SARS-CoV-2 [43].

- Blockade of viral entry and internalization through inhibition of TMPRSS2, as it promotes the viral entry into the cell through proteolytic cleavage and activation of the viral envelope glycoproteins.
- Inhibition of viral endocytosis, which, in the case of SARS-CoV-2, can be mediated by clathrin.

The kinase AAK1 controls clathrin-mediated endocytosis, and a potential kinase inhibitor is the already reported baricitinib [44].

- Neutralization of viral multiplication through small molecules and antibodies that can act on functional proteins responsible for SARS-CoV-2 replication. Potential targets are represented by viral proteases, polymerases, and helicases.

The already mentioned remdesivir acts against RdRp [45]. The major protease Mpro is a very important target that will be discussed further, as it is responsible for the synthesis of NSP.

- Inhibition of viral particle packaging and structural proteins, namely the envelope protein E, the membrane protein M and the nucleocapsid protein N.

These proteins have been envisaged as ideal targets due to their essential role in viral proliferation and their structural differences from host proteins, thus minimizing potential side effects from drugs targeting such proteins. They are also involved in suppressing the host immune system, thus giving the virus an advantage over the host [46, 47]. The N protein encapsulates the genomic RNA of the virus during its life cycle in order to protect the genome and allow it to penetrate the host cell. This indicates that the N protein is essential for viral replication, especially in the early phase. The N protein works to blocking RNA from silencing and suppressing siRNA-mediated RNA interference; thus, siRNA-based therapies can be exploited to target the translation of E, M, and N proteins and inhibit viral replication [48].

8.5 Structure and Function of the Main Protease (Mpro)

The two viral proteases particularly relevant in SARS-CoV-2 viral replication are the PLPro and the Mpro (or 3CLPro). These proteases have been recognized as excellent targets for drug development. The main functions of these proteases are to perform proteolytic cuts of two polyproteins (pp1a and pp1ab) that derive from the translation of the ORFs present in the SARS-CoV-2 genome. The Mpro has received particular attention from the scientific community as a therapeutic target, as it performs an important proteolytic cut inside the host cell, releasing most of the NSPs encoded within the viral genome. The structure of the protease has been characterized in detail through crystallographic analyses (Figure 8.3), which revealed the protease consisting of two protomers, A and B, each protomer containing three domains called I, II, and III. In addition to the three domains mentioned above, the presence of seven amino acid residues, the *N*-finger, which mediate the dimerization of the protomers, also emerged. It has been observed that the removal of the *N*-finger sequence negatively modulates the activity of the protease itself [31, 49].

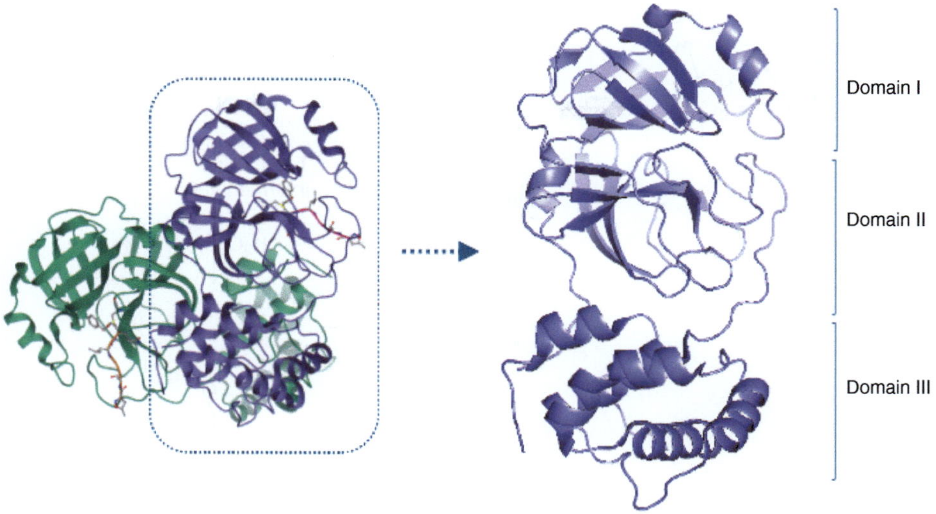

Figure 8.3 *Crystal structure of Mpro (PDB: 6LU7): left: dimeric structure; right: highlight of domains I, II, and III.*

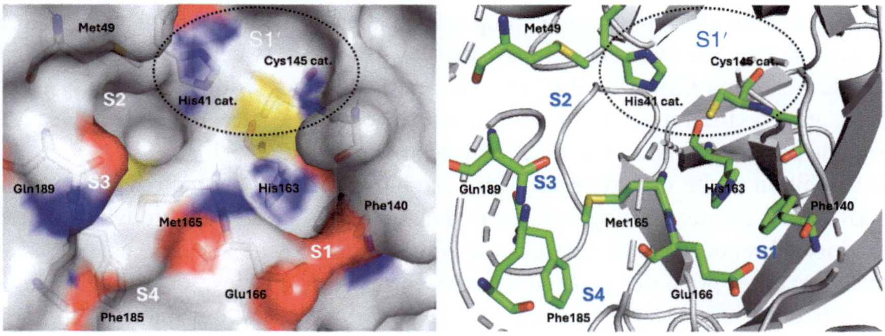

Figure 8.4 *X-ray crystallographic structure of the Mpro binding site highlighting key subsites and the catalytic dyad consisting of His41 and Cys145 (PDB: 6M0K). Adapted from [50].*

The catalytic site of Mpro fits between the beta barrel structures of domain I (chymotrypsin-like domain) and domain II (picornavirus 3C protease-like domain). Domain III is composed of 5 globular helices and a loop that connects both domains I and II through a salt bridge that occurs between residues Arg 4 and Arg 298 (Figure 8.4) [51]. At the *N*-terminal end of the B protomer, Serine 1 interacts with residues Phe140 and Glu166 of the A protomer, resulting in importance for the formation and stabilization of the S1 binding pocket during the dimerization of Mpro. In the binding site, a highly conserved catalytic dyad, histidine–cysteine, is present. In particular, Cys145 has a nucleophilic role in the proteolysis step, while His41 acts as a base catalyst [52].

Based on the structure of Mpro and the autoproteolytic cleavage mechanism, a nucleophilic-like reaction mechanism for the cleavage of polyproteins by Mpro has been proposed [53]. As described in Figure 8.5, the thiol of Cys145 is deprotonated by the

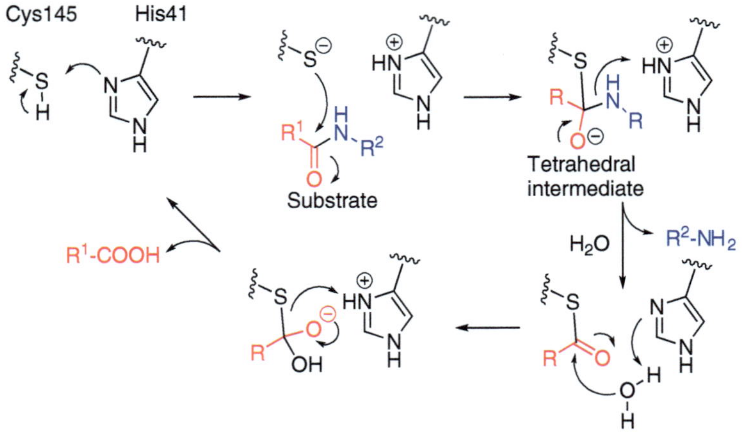

Figure 8.5 *Catalytic mechanism of the Cys-His dyad of Mpro.*

basicity of His41, which is spatially close, and the formed sulfide anion performs a nucleophilic attack on the carbonyl group of the amide bond of the viral polyprotein. Thus, a new peptide with a terminal amino group is formed, and His41 returns to its deprotonated form. Specifically, the thioester complex that has formed between the carbonyl group of the amide bond and Cys145 is hydrolyzed, and the peptide with the carboxyl group is released, and Cys145 returns to its initial conformation. In this catalytic mechanism, after each single proteolysis, the catalytic dyad is recovered.

The inhibition of the viral protease Mpro can block many of the biological functions of the SARS-CoV-2 virus, starting from the replication of the viral RNA through its transcription and its subsequent proliferation. Analyzing the amino acid sequence encoding Mpro, compared to other proteases isolated in other coronavirus species, a very high conservation rate is observed. The binding site appears to be one of the most conserved portions of the protease structure, with a mutation rate of 0.1% [54]. The substrate of Mpro, from the *N*-terminus to the *C*-terminus, is composed of seven amino acids numbered, according to Schechter-Berger nomenclature [55], as P4–P3–P2–P1–P1′, and the cleavage site is found between the P1 and P1′ positions [56]. In particular, a Gln residue is always required in the P1 position [50]. The substrate binding site, as described above, is located between domains I and II. It can in turn be divided into various subsites, which perform different functions. The binding pocket that interacts with the majority of the substrate is composed of two hidden sites (S1 and S2), a surface subsite with hydrophobic character (S4), and various extended subsites that are accessible to the solvent (S3, S5, and S1). The catalytic cysteine-histidine dyad is located in the S1′ subsite. As reported in Figure 8.4 and Table 8.1, each subsite contributes to the specificity of the binding pocket. The substrate positions also have different specificities; P1, P2, and P1′ determine the substrate specificity toward Mpro, while P4 and P3 and P3′ stabilize the binding.

The high conservation rate of Mpro, combined with the fact that humans do not possess a homologous protease, makes it a very important therapeutic target for the development of antiviral drugs. The main inhibitors of Mpro are peptidomimetic in nature.

Table 8.1 *Summary of Mpro subsites.*

Subsite	Residues	Specificity
S1	Phe140, Gly143, Ser144, and Cys145, His163, Glu166, and His172	Gln
S2	Thr25, His41, and Cys145	Leu, Phe, Met, and Val
S3	His41, Met49, and Met165	None
S4	Met165 and Glu166	Hydrophobic amino acids

8.6 Peptidomimetic Mpro Inhibitors

Peptidomimetic covalent inhibitors are developed by structurally mimicking the natural substrate of the SARS-CoV-2 Mpro. Initially, this type of peptidomimetic inhibitor was conceived to have an inhibitory effect against the 3Cpro or Mpro proteases of other viruses such as rhinovirus or MERS and SARS-CoV-1, and some peptidomimetic compounds active against the latter also showed some activity against SARS-CoV-2 [57]. The consensus sequence of protease substrate that is the basis for peptide-derived peptidomimetic design is characterized by the presence of glutamine at the P1 and leucine and valine at the P2 and P3 positions, respectively. This information was combined with a structure-based drug discovery approach to design compounds optimized for SARS-CoV-2 Mpro on the basis of the 3D structure of the binding pocket.

Peptidomimetic inhibitors of cysteine proteases can be divided generally into two major groups, namely covalent and non-covalent inhibitors. Depending on the chemical nature of the warhead positioned at the P1′ site, these inhibitors typically incorporate functional groups such as aldehydes, sulfonates, nitriles, ketones, and Michael acceptors (Figure 8.6). Each warhead is able to establish different bonds with the catalytic residues, based on the electronic properties, the substituent groups, and the steric effects. From a purely mechanical point of view, covalent inhibitors interact with Mpro in two steps: the inhibitor approaches the binding site of the protease to form a non-covalent complex until the warhead group is in proximity to the catalytic dyad (Cys145/His41). Then, the warhead group interacts with Cys145, forming a covalent bond. If the warhead group does not establish covalent bonds, the molecule is classified as a non-covalent inhibitor. If the

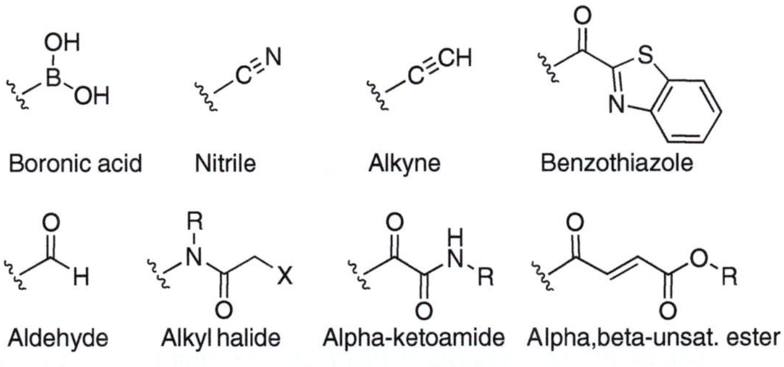

Figure 8.6 *Representative electrophilic warheads used for targeting cysteine proteases. Source: [58] / MDPI / CC BY 4.0.*

interactions between the warhead group and the catalytic dyad remain the same as when the inhibitor is in the binding pocket, it is defined as a non-covalent inhibitor [59].

In recent years, extensive efforts have been directed toward designing peptidomimetic inhibitors bearing aldehyde warheads, with structural modifications introduced at the P1, P2, P3, and P4 positions to optimize activity. Among the various classes of inhibitors, those designed on the Mpro hydrolysis substrate are the most extensively reported, demonstrating potent inhibitory effects against Mpro, with several compounds achieving nanomolar activity. Nevertheless, the peptide character and the presence of reactive electrophilic warheads present in peptidomimetic compounds were associated with bioavailability and safety issues. For example, nirmatrelvir required coadministration with ritonavir as a boosting agent to improve oral bioavailability. Non-covalent small-molecule inhibitors possess minor peptide character and establish their activity through weak reversible interactions with amino acid residues at S1, S2, and S4 subsites of Mpro, showing safety and oral bioavailability properties significantly improved, as in the case of S-217622 and WPV01, which were accepted for clinical trials.

8.6.1 Peptidomimetic Inhibitors with a Ketone Warhead

Peptidomimetic inhibitors that present the ketone warhead group showed the highest inhibitory potency toward Mpro. In the literature, there are many peptidomimetic antivirals that are based on the presence of a ketone group as a warhead. Hoffman et al. from Pfizer proposed PF-00835231 (lufotrelvir) [60] as a reversible inhibitor of Mpro that presents a hydroxymethylketone group as a warhead (Figure 8.7, top left). This antiviral was proposed as a drug during SARS-CoV infection in 2002–2003. To improve the solubility of PF-00835231, a phosphate group was added, forming the prodrug

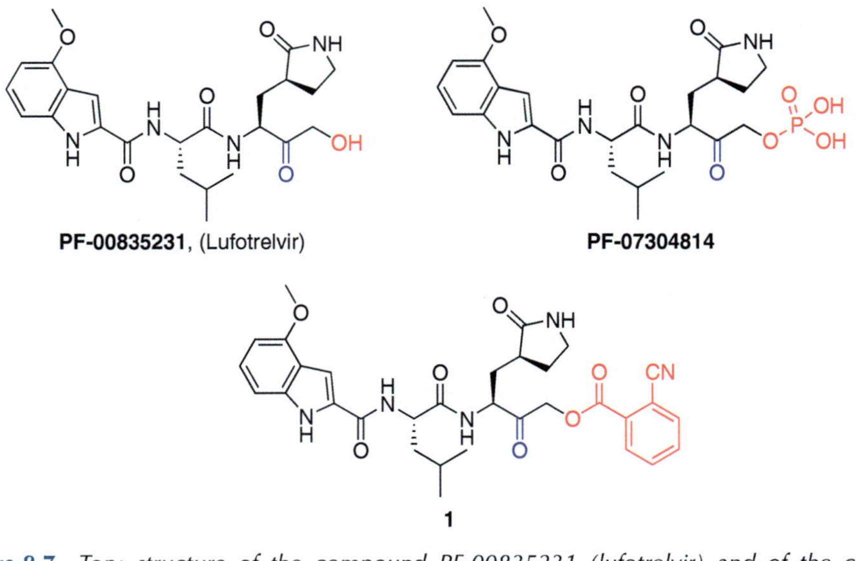

*Figure 8.7 Top: structure of the compound PF-00835231 (lufotrelvir) and of the corresponding prodrug PF-07304814; bottom: structure of **1** containing the acyloxymethyl ketone warhead.*

PF-07304814 (Figure 8.7, top right). In tissues, the phosphate group is eliminated by alkaline phosphatase enzymes, releasing PF-00835231, which showed good inhibitory activity with an IC_{50} = 0.27 nM. By modifying the warhead group, peptidomimetic inhibitors have been developed that feature the acyloxymethylketone group. In this case, the interaction with Cys145 becomes irreversible covalent due to the release of the acyl group. Hoffmann et al. [60] developed compound **1** (Figure 8.7, bottom), which has an IC_{50} =17 nM. The acyloxymethylketone group can have two different interaction mechanisms. The cyanobenzoate group is used both as an excellent leaving group and to improve the electron density of the compound.

Hayashi et al. [61] developed several benzothiazole-ketone-containing peptidomimetics with low IC_{50} inhibition values toward Mpro in the micromolar range. Structural biology studies showed the benzothiazole fitting very well into the S1' subsite (Figure 8.8). Following a SAR analysis of these compounds and structural studies [62], compound YH-53 was taken into account as a compound with a benzothiazole-ketone warhead for its drug development potential against SARS-CoV-2. Specifically, the carbonyl group in position P1 fits into the S1 subsite; the carbonyl and amino groups of pyrrolidin-2-one mediate hydrogen bonds with His163 and Glu166, and the 4-methoxy-indole group at position P3 proved to increase the inhibitory activity.

8.6.2 Peptidomimetic Inhibitors with an α-Ketoamide Warhead

Using α-ketoamides as a warhead group instead of ketones represents an advantage because it provides an additional hydrogen-bonding acceptor. Hilgenfeld et al. have designed and synthesized many broad-spectrum peptidomimetics, diversified in the substituent at position P2 [63]. The study highlighted the binding mode of compound **2**, possessing an IC_{50} = 0.24 μM. In particular, through the crystallographic structures, it was possible to define the mode of interaction between Cys145 and the α-ketoamide group: the hydroxyl group is stabilized through hydrogen bonding with His41, and the oxygen atom of the amide group forms a hydrogen bond with Gly143, stabilizing the drug inside the binding pocket of Mpro (Figure 8.9).

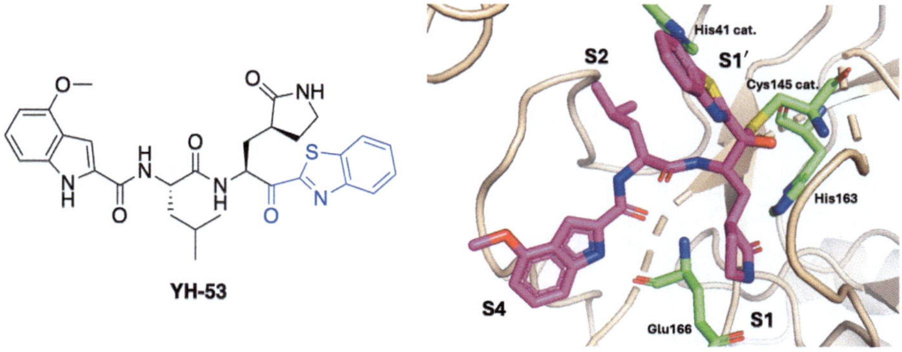

Figure 8.8 *Left: structure of compound YH-53; 3D close-up of X-ray. Crystallographic structure of YH-53 in the active site of Mpro (PDB:7E18).*

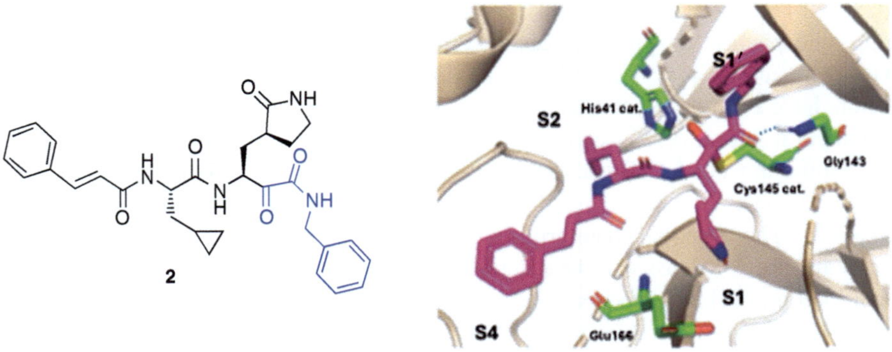

Figure 8.9 Left: structure of compound **2**; right: X-ray crystal model of the binding to the active site (PDB:5N5O).

8.6.3 Peptidomimetic Inhibitors with an Aldehyde Warhead

Mpro peptidomimetic inhibitors that have aldehyde warheads are able to covalently bind to Cys145 and to establish hydrogen bonds with the main chain of Cys145 [64]. These compounds are derived conceptually from other protease inhibitors, such as telaprevir or boceprevir, showing typical side-chain isosteres at P1, P2, and P3 positions for optimal interaction with the corresponding enzyme's subsites. For example, Qiao et al. inserted an aldehyde group at the P1 position, a bicyclic proline motif in the P2 position, and aryl groups at the P3 position [65]. Among the developed inhibitors, compound **3** showed an IC_{50} = 7.6 nM (Figure 8.10). X-ray crystallographic studies highlighted the aldehyde

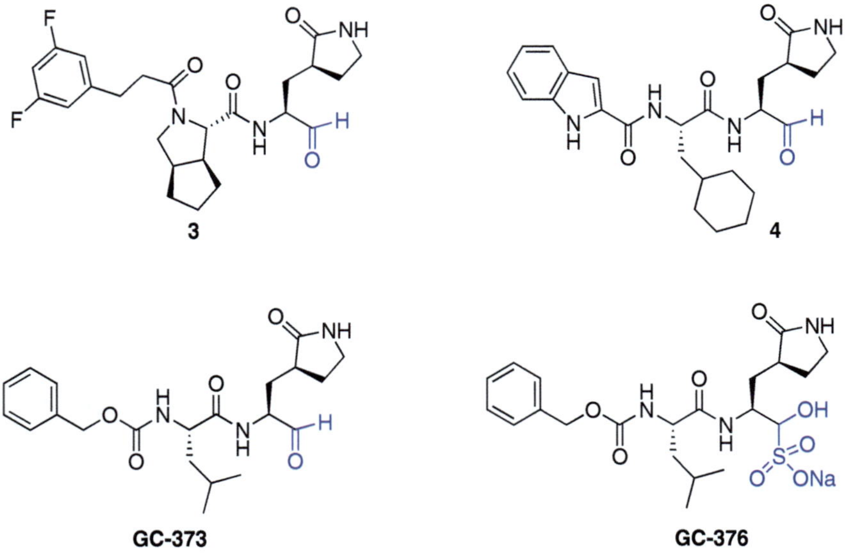

Figure 8.10 Covalent peptidomimetic inhibitors of Mpro bearing the free aldehyde warhead or a sodium bisulfite as a prodrug (GC-376).

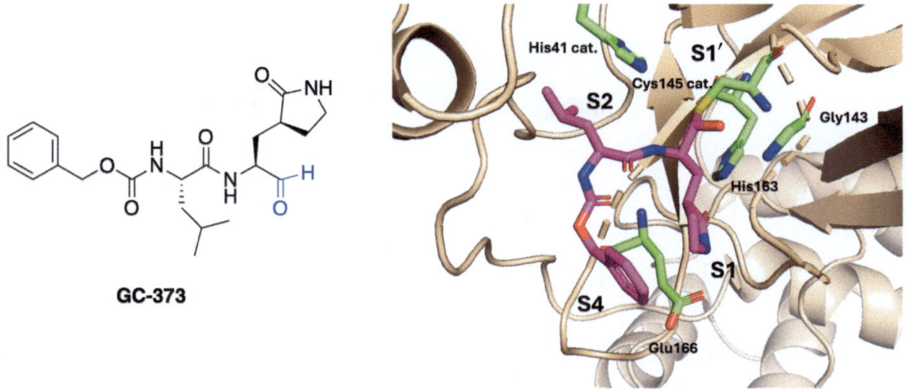

Figure 8.11 *Left: structure of GC-373; right: X-ray crystallographic structure of the inhibitor binding to the active site (PDB:6WTJ).*

carbonyl forming a covalent bond with Cys145, the lactam at the P1 position interacting with the S1 subsite and its nitrogen and oxygen heteroatoms forming hydrogen bonds with the residues of the subsite (Figure 8.10). Dai et al. developed the peptidomimetic inhibitor **4** using an aldehyde as a warhead group and a (*S*)-γ-lactam as a substituent in position P1. The lactam and the aldehyde present the usual interactions as described above. In position P2, there is a cyclohexyl group that inserts into the subsite S2, giving rise to many hydrophobic interactions. The indole group present in position P3 is exposed to the solvent and also to the subsites S3 and S4 [50]. Some peptidomimetic inhibitors that have aldehydes as warheads can be developed as bisulfite adducts, which turn out to be prodrugs. The introduction of the bisulfite group has the sole purpose of increasing the aqueous solubility of the drugs and improving the pharmacokinetics. As reported by Vandyck and Deval [66], the peptidomimetic compound GC-376 is a sodium bisulfite prodrug of the aldehyde GC-373 (Figure 8.10). GC-373 has shown good antiviral activity *in vivo*; however, inhibitors with aldehyde warheads can be toxic, as they can form Schiff bases and DNA adducts that induce oxidative stress and inflammation (Figure 8.11). Furthermore, GC-376 and GC-373 have been found to be active against the human proteins Cathepsin L and B proteases, representing an issue for clinical development.

8.6.4 Peptidomimetic Inhibitors with Michael-Accepting Warheads

Peptidomimetic inhibitors that have a Michael-accepting warhead are derived from the drug rupintrivir, which entered clinical trials for the treatment of the common cold [67], and several peptidomimetic inhibitors have been proposed accordingly. In particular, N3 has been recognized as a broad-spectrum inhibitor of Mpro. Jin et al. [49] revealed the structure of N3 with Mpro, showing how the beta carbon of the unsaturated ester forms a covalent bond with the thiol of Cys145, indicating that a Michael addition occurs. Furthermore, the lactam occupies the S1 subsite, and the isobutyl and methyl groups of Leu and Ala, respectively, interact with the S2 and S4 subsites, forming hydrophobic interactions, whereas the isopropyl group of Val increases the solvent exposure area (Figure 8.12). The high reactivity and covalent binding *via* a Michael addition have been recently validated, and some

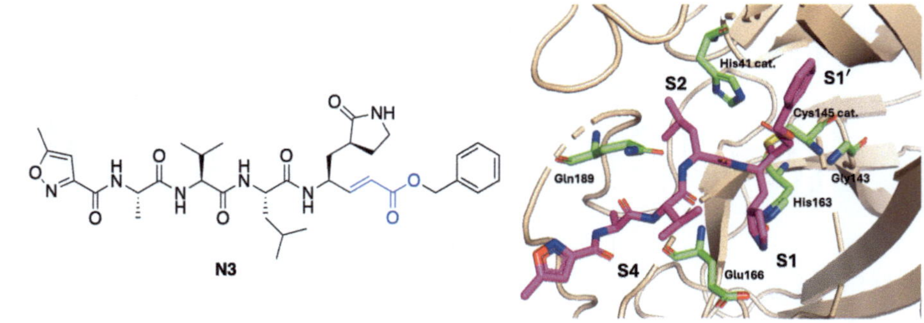

Figure 8.12 Left: structure of N3; right: X-ray crystallographic structure of the ligand binding to the active site (PDB:6LU7). Source: Adapted from [68].

drugs have been approved by the FDA. However, further evaluation of risks and side effects needs to be assessed, as the ethyl ester group used as a Michael acceptor can be hydrolyzed by carboxylesterases present in plasma, thus essentially eliminating the selected warhead group.

8.6.5 Peptidomimetic Inhibitors with a Nitrile Warhead

The nitrile warhead group has been widely used for the drug design of reversible covalent peptidomimetic inhibitors. Although the nitrile group is relatively inert, its electrophilicity can be increased if it has adjacent electron-withdrawing groups. The proposed mechanism for the binding of the nitrile group to the thiol of Cys145 is an electrophilic–nucleophilic reaction that ends with the formation of a thioimidate covalent bond [69]. Representative peptidomimetic inhibitors possessing the nitrile as a warhead group are reported as follows:

- nirmatrelvir (PF-07321332), the first peptidomimetic inhibitor approved by the FDA and EMA for Mpro;
- simnotrelvir (SIM-0417), a peptidomimetic inhibitor that has been approved by the Chinese National Medical Products Administration (NMPA).

Some studies performed on the P1 portion have shown that the 2-piperidone group increases the affinity of the substrate in the Mpro inhibition mechanism [70]. Pomotrelvir (PBI-0451) has the 2-piperidone group that fits into the S1 subsite and also establishes a network of multiple hydrogen bonds with the C=O and N—H groups found in the amide group of the protease backbone [71].

8.7 The Road to Approval of Paxlovid as an Mpro Inhibitor Drug

In 2002, Hoffman et al. from Pfizer developed PF-00835231 (Figure 8.13), an inhibitor of the Mpro of the SARS-CoV virus [72]. It was initially designed as a SARS-CoV-1 Mpro inhibitor and was found to bind tightly to the SARS-CoV-2 Mpro active site, too, through a reversible covalent interaction with Cys145 of the catalytic dyad with His41 [60]. The modification of the peptide character of PF-00835231 was envisaged with the aim of improving

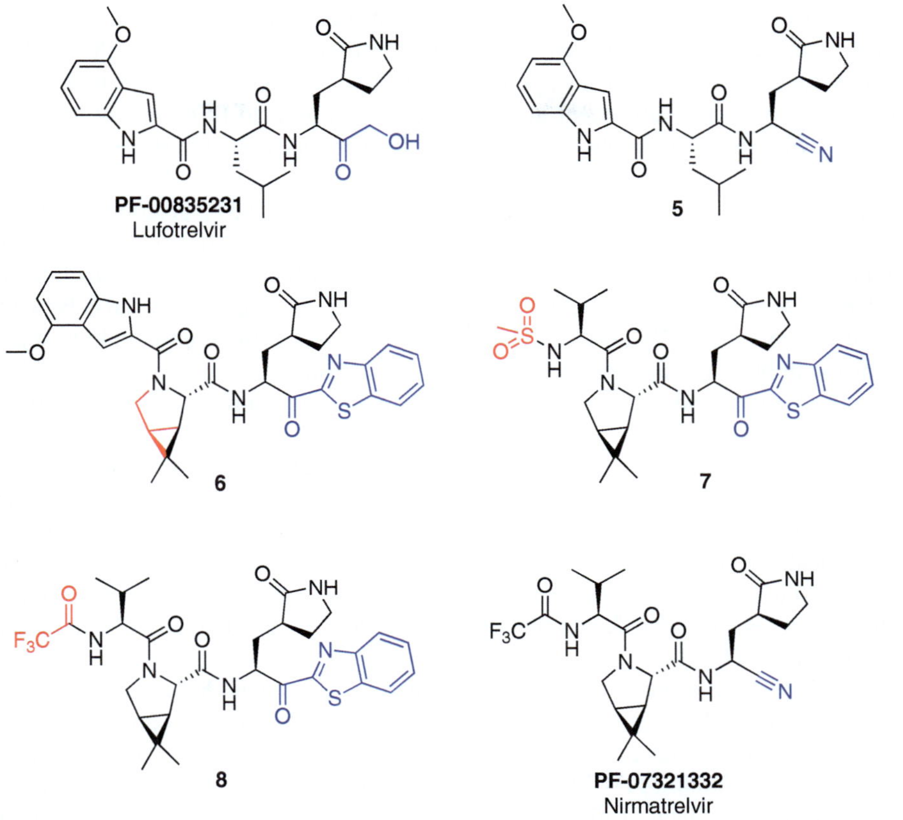

Figure 8.13 *Structure of peptidomimetic Mpro inhibitors from lufotrelvir to nirmatrelvir.*

small-molecule permeability through reduction of the number of hydrogen-bond donors (HBDs) [73, 74].

Following enzymatic assays, its efficacy against the Mpro of SARS-CoV-2 was demonstrated, as the structure of the protease showed a percentage of sequence homology of 100%. Nevertheless, PF-00835231 (lufotrelvir) demonstrated low passive permeability and low oral absorption. Therefore, Owen et al. at Pfizer removed the hydroxymethylketone group and planned to work in parallel with two distinct functional groups, namely the nitrile and the benzothiazol-2-yl ketone. They also modified the P3 portion, inserting branched acyclic substituents for improving the interactions with the S3 subsite of Mpro [75]. The P1 lactam was maintained during the drug optimization process for the key role as a Gln mimetic at the P1′ cleavage site possessing a single HBD, which is a highly specific recognition element for Mpro [76].

The introduction of the nitrile group greatly improved oral absorption; however, compound **5** showed a lower IC_{50} than PF-00835231 (Figure 8.13). Owen et al. envisaged some modifications to the P2 portion, introducing 6,6-dimethyl-3-azabicyclohexane as a cyclic leucine mimetic. In fact, this substitution eliminated some hydrogen bonds that formed between the amide backbone of compound **5** in the P2 and P3 positions. Compound **6**

(Figure 8.13) was developed first, which features the benzothiazole ketone as the warhead group. Compound **6** showed reduced inhibitory activity as compared to PF-00835231, possibly due to the absence of any hydrogen bonds from the ligand backbone to the Gln189 residue. However, the substituent at the P2 position proved to fit optimally into the S2 subsite, giving rise to hydrophobic interactions with residues Met49, Met169, His41, and Gln189.

To improve interactions at the S3 subsite, the cyclic group in the P3 position was replaced in favor of branched acyclic groups, capable of bringing hydrogen bonds to Gln189. Compound **7** (Figure 8.13) presented a methanesulfonamide, which improved interactions with residues present in the S3 subsite, showing a better IC_{50} than compounds **4** and **6**. By screening substituents in the P3 position, Owen et al. identified trifluoroacetamide as a group useful for improving the pharmacokinetics and increasing the stability of the inhibitor, resulting in compound **8** (Figure 8.13) with a significantly improved IC_{50} as compared to compound **7**.

Finally, the replacement of the benzothiazole-ketone warhead with the nitrile group led to the development of PF-07321332 (Figure 8.13), which showed better inhibitory activity than the other tested compounds. Peptidomimetic PF-07321332 was selected as the final clinical candidate because of the potent inhibitory profile toward SARS-CoV-2 Mpro (K_i = 3.11 nM) with remarkable antiviral activity in VeroE6 cells (IC_{50} = 74.5 nM). The binding mode is similar to **7**, and the nitrile at P1′ forms a reversible thioimidate adduct with the catalytic Cys145 (Figure 8.14).

PF-07321332 was selected as a drug candidate over compound **8**, based on the ease of synthetic scale-up. Furthermore, Owen et al. focused on enzymes capable of degrading PF-07321332. Metabolic studies showed that the oxidoreductase enzyme CYP3A4 was able to selectively inhibit PF-07321332 at a rate of approximately 82%. Further studies on the oxidative metabolism of PF-07321332 confirmed that CYP3A4 was the predominant metabolic enzyme. These data presented an excellent therapeutic opportunity, as the coadministration of PF-07321332 and a selective CYP3A4 inhibitor, ritonavir (RTV), significantly increased the therapeutic concentrations of the SARS-CoV-2 Mpro inhibitor. PF-07321332, successively named nirmatrevir, and ritonavir, administered together,

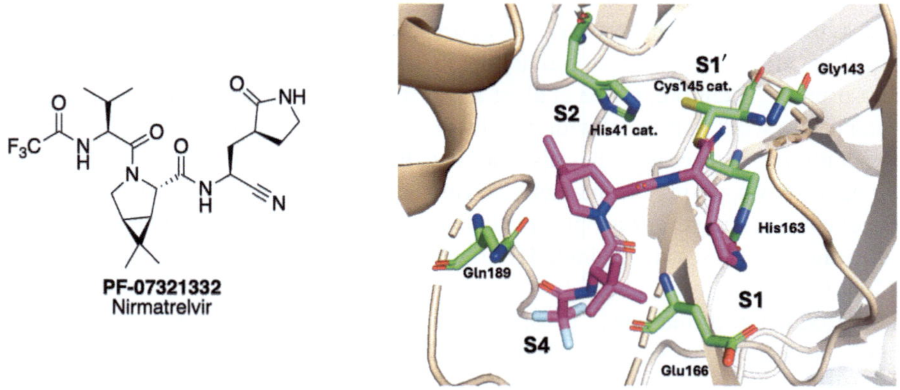

Figure 8.14 Left: structure of PF-07321332; right: X-ray crystallographic structure of SARS-CoV-2 main protease in complex with protease inhibitor PF-07321332 (PDB:7VH8).

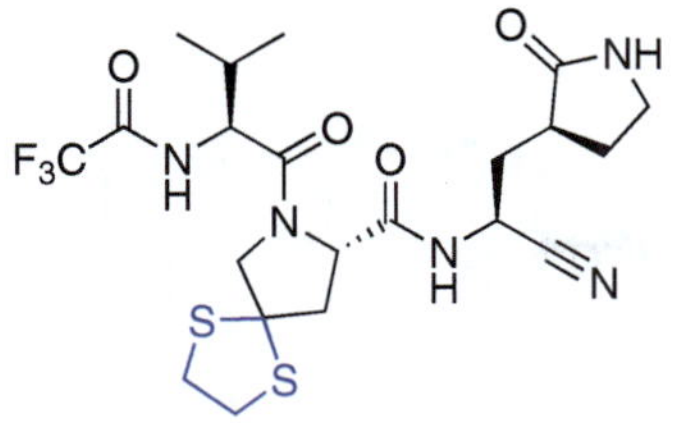

Simnotrelvir

Figure 8.15 Structure of Simnotrelvir.

constitute the SARS-CoV-2 antiviral Paxlovid, developed by the company Pfizer. In January 2022, the EMA authorized the conditional use of Paxlovid for the treatment of COVID-19. Also, it was confirmed that the antiviral activity of Paxlovid is not compromised by SARS-CoV-2 variants [77].

Simnotrelvir [78] is a next-generation oral Mpro inhibitor developed for COVID-19 therapy (Figure 8.15). When boosted with ritonavir, it demonstrated potent antiviral efficacy across circulating SARS-CoV-2 variants, including strains resistant to earlier antivirals. Clinical trials indicated that Simnotrelvir accelerates viral clearance, shortens hospitalization, and improves time to clinical recovery as compared with standard care, also showing a favorable safety profile, a high barrier to resistance, and favorable pharmacokinetic stability supporting once-daily dosing.

8.8 PROTACs in Antiviral Drug Discovery Through Target Degradation

Proteolysis Targeting Chimera (PROTAC) represents an innovative technology for protein degradation [79]. These are heterobifunctional compounds possessing two active ligands connected by a linker. One of the ligands is specific toward the protein of interest, while the other binds specifically to the E3 protein ligase, which mediates the ubiquitination and subsequent degradation of the protein of interest. PROTACs are favorable over traditional inhibitors for their property to degrade proteins, the possibility of using weak ligands, better selectivity toward targets, and the ability to overcome drug resistance. The first PROTAC developed for the treatment of infectious diseases was proposed to degrade the NS3/4A protease of the hepatitis C virus [80]. Alugubelli et al. [81] developed some of the very first PROTACs capable of degrading the Mpro of SARS-CoV-2. Specifically, starting from boceprevir, in 2020 they synthesized compound MPI8, which was the warhead selected for developing PROTACs, identified through a synthesized library of PROTACs presenting ligands for Mpro and derivatives of thalidomide-4-OH as ligands for CRBN (cereblon), a substrate for the E3 ligase. Among the various PROTACs developed, only MPD2 proved to function as a potent degrader of Mpro, with a DC_{50} of 296 nM (Figure 8.16). To elucidate the degradation mechanism of MPD2, Alugubelli et al. developed competition assays, indicating that both the specific ligand for CRBN and MPI8 are essential in the binding phase of the proteins of interest and in the degradation of Mpro.

Trabocchi, D'Onofrio, Ciofi-Baffoni et al. [82] developed other PROTACs based on the drug pomalidomide as a ligand for the CRBN receptor and a derivative of GC-373 selected

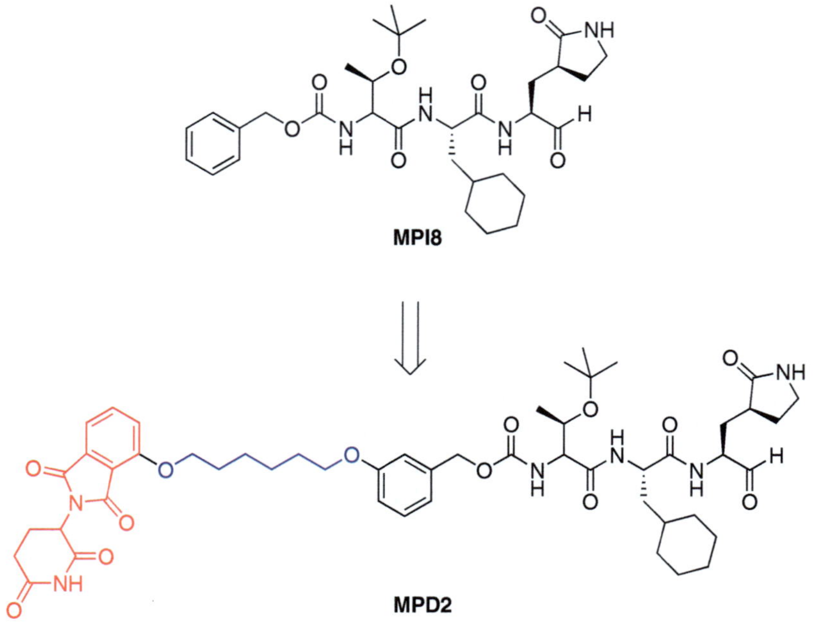

Figure 8.16 Structure of Mpro inhibitor MPI8 and of the corresponding PROTAC, MPD2.

as a specific ligand for Mpro. In particular, the warhead aldehyde group was replaced with an α,β-unsaturated methyl ester, resulting in the generation of two PROTACs (Figure 8.17).

The inhibitory activity of PROTAC1 and PROTAC2 was evaluated through enzymatic inhibition assays, and the collected data highlighted a very different activity for both PROTACs, as PROTAC1 showed an IC50 of 1.35 μM, while PROTAC2 did not show any significant activity. The binding model of PROTAC1 to Mpro was assessed by NMR spectroscopy and X-ray crystallographic studies. The ligand of Mpro forms a reversible covalent bond with Cys145, forms hydrogen bonds with residues His41, Phe140, and Glu166, and hydrophobic interactions with the nonpolar subsites. The reported studies showed that the PROTAC approach can be a new strategy for drug discovery of novel drugs for the SARS-CoV-2 virus. Furthermore, the PROTAC strategy is able to overcome drug resistance that is developing in COVID-19 patients.

8.9 Conclusion

The pandemic disease COVID-19, caused by severe acute respiratory SARS-CoV-2, represented a major threat to global health in 2020 and the subsequent two-three years. Although the introduction of vaccines allowed for achieving effective protection against infection, severe illness, hospitalization, and death from COVID-19, therapeutic approaches were required to complete vaccine development, especially for immunosuppressed or unvaccinated subjects. In this context, the protease Mpro was soon identified as an ideal drug target

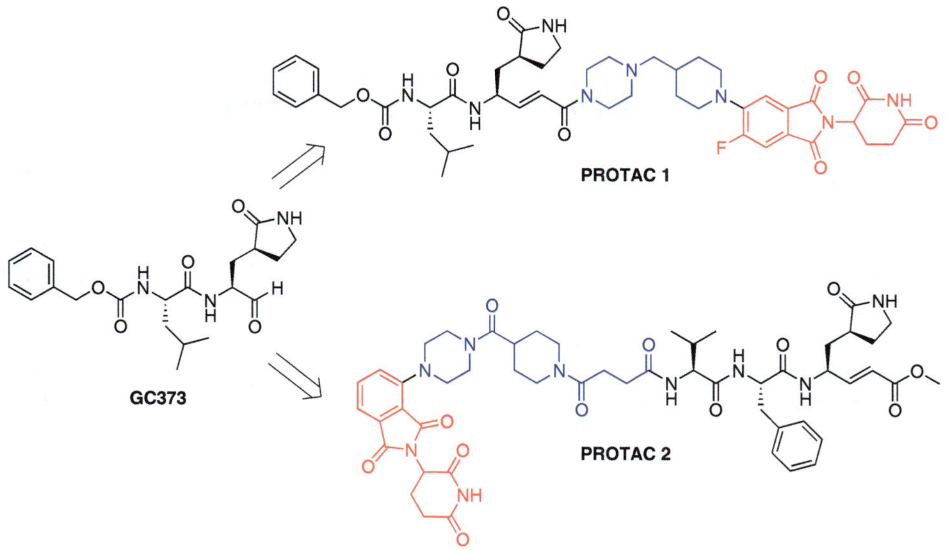

Figure 8.17 *Structure of the compound GC-373 and structures of PROTAC1 and PROTAC2.*

among coronaviruses, as it plays a key role in viral maturation by processing polyproteins that are translated by viral RNA.

Indeed, since the breakthrough use of peptidomimetics in the development of HIV protease inhibitors during the 1990s, the design of new antiviral agents has been often based on this strategy to target proteases critical in numerous viral infections. Accordingly, the design process followed the typical peptidomimetic approach, introducing localized changes to the native peptide, such as altering the peptide bond or incorporating specific chemical groups to refine local and overall molecular properties. Notably, the incorporation of covalent reactive groups ("warheads") proved effective in engaging essential residues at the protease active site that drive substrate cleavage. Following this approach, a wide number of compounds were reported over the last few years as both covalent and non-covalent inhibitors, and this successful approach was demonstrated by the approval of Paxlovid for clinical use, consisting of targeting PF-07321332 (nirmatrelvir) as an MPro inhibitor containing a nitrile group as the warhead. Research in this area remains active, with a focus on developing novel inhibitors and degraders, particularly addressing new genetic variants of SARS-CoV-2 that continue to emerge.

References

1. WHO. *Coronavirus Disease COVID-19 Pandemic*. WHO. https://www.who.int/emergencies/diseases/novel-coronavirus-2019.
2. RECOVERY Collaborative Group. Aspirin in patients admitted to hospital with COVID-19 (RECOVERY): a randomised, controlled, open-label, platform trial. *Lancet*. 2022;**399**:143–151.

3. RECOVERY Collaborative Group. Dexamethasone in hospitalized patients with COVID-19. *N Engl J Med*. 2020;**384**:693–704.
4. RECOVERY Collaborative Group. Effect of hydroxychloroquine in hospitalized patients with COVID-19. *New Engl J Med*. 2020;**383**:2030–2040.
5. Xiu S, Dick A, Ju H, Mirzaie S, Abdi F, Cocklin S, Zhan P, Liu X. Inhibitors of SARS-CoV-2 entry: current and future opportunities. *J Med Chem*. 2020;**63**: 12256–12274.
6. Gil C, Ginex T, Maestro I, Nozal V, Barrado-Gil L, Cuesta-Geijo MÁ, Urquiza J, Ramírez D, Alonso C, Campillo NE, Martinez A. COVID-19: drug targets and potential treatments. *J Med Chem*. 2020;**63**:12359–12386.
7. Choudhry N, Zhao X, Xu D, Zanin M, Chen W, Yang Z, Chen J. Chinese therapeutic strategy for fighting COVID-19 and potential small-molecule inhibitors against severe acute respiratory syndrome coronavirus 2 (SARS-CoV-2). *J Med Chem*. 2020;**63**:13205–13227.
8. Baden LR, El Sahly HM, Essink B, Kotloff K, Frey S, Novak R, Diemert D, Spector SA, Rouphael N, Creech CB, J MG, Khetan S, Segall N, Solis J, Brosz A, Fierro C, Schwartz H, Neuzil K, Corey L, Gilbert P, Janes H, Follmann D, Marovich M, Mascola J, COVE Study Group. Efficacy and safety of the mRNA-1273 SARS-CoV-2 vaccine. *N Engl J Med*. 2021;**384**:403–416.
9. Callaway E. Beyond Omicron: what's next for COVID's viral evolution. *Nature*. 2021;**600**:204–207.
10. Robinson PC, Liew DFL, Tanner HL, Grainger JR, Dwek RA, Reisler RB, Steinman L, Feldmann M, Ho LP, Hussell T, Moss P, Richards D, Zitzmann N. COVID-19 therapeutics: challenges and directions for the future. *Proc Natl Acad Sci USA*. 2022;**119**:e2119893119.
11. V'kovski P, Kratzel A, Steiner S, Stalder H, Thiel V. Coronavirus biology and replication: implications for SARS-CoV-2. *Nat Rev Microbiol*. 2021;**19**(3):155–170.
12. Harrison AG, Lin T, Wang P. Mechanisms of SARS-CoV-2 transmission and pathogenesis. *Trends Immunol*. 2020;**41**:1100–1115.
13. Panahi Y, Gorabi AM, Talaei S, Beiraghdar F, Akbarzadeh A, Tarhriz V, Mellatyar H. An overview on the treatments and prevention against COVID-19. *Virol J*. 2023;**20**:23.
14. Wang M, Zhang N. Focus on the 2019 novel coronavirus (SARS-CoV-2). *Future Microbiol*. 2020;**15**:905–918.
15. Shereen MA, Khan S, Kazmi A, Bashir N, Siddique R. COVID-19 infection: origin, transmission, and characteristics of human coronaviruses. *J Adv Res*. 2020;**24**:91–98.
16. Kadam SB, Sukhramani GS, Bishnoi P, Pable AA, Barvkar VT. SARS-CoV-2, the pandemic coronavirus: molecular and structural insights. *J Basic Microbiol*. 2021;**61**:180–202.
17. Wu CR, Yin WC, Jiang Y, Xu HE. Structure genomics of SARS-CoV-2 and its Omicron variant: drug design templates for COVID-19. *Acta Pharmacol Sin*. 2022;**43**: 3021–3033.
18. Naqvi AAT, Fatima K, Mohammad T, Fatima U, Singh IK, Singh A, Atif SM, Hariprasad G, Hasan GM, Hassan MI. Insights into SARS-CoV-2 genome, structure, evolution, pathogenesis and therapies: structural genomics approach. *Biochim Biophys Acta Mol basis Dis*. 2020;**1866**:165878.

19. Hoffmann M, Kleine-Weber H, Schroeder S, Krüger N, Herrler T, Erichsen S, Schiergens TS, Herrler G, Wu NH, Nitsche A, Müller MA, Drosten C, Pöhlmann S. SARS-CoV-2 cell entry depends on ACE2 and TMPRSS2 and is blocked by a clinically proven protease inhibitor. *Cell*. 2020;**181**:271–280.e8.
20. Jackson CB, Farzan M, Chen B, Choe H. Mechanisms of SARS-CoV-2 entry into cells. *Nat Rev Mol Cell Biol*. 2022;**23**:3–20.
21. Scialo F, Daniele A, Amato F, Pastore L, Matera MG, Cazzola M, Castaldo G, Bianco A. ACE2: the major cell entry receptor for SARS-CoV-2. *Lung*. 2020;**198**:867–877.
22. Shulla A, Heald-Sargent T, Subramanya G, Zhao J, Perlman S, Gallagher T. A transmembrane serine protease is linked to the severe acute respiratory syndrome coronavirus receptor and activates virus entry. *J Virol*. 2011;**85**:873–882.
23. Cevik M, Kuppalli K, Kindrachuk J, Peiris M. Virology, transmission, and pathogenesis of SARS-CoV-2. *BMJ*. 2020;**371**:m3862.
24. Bai C, Zhong Q, Gao GF. Overview of SARS-CoV-2 genome-encoded proteins. *Sci China Life Sci*. 2022;**65**:280–294.
25. Tam D, Lorenzo-Leal AC, Hernández LR, Bach H. Targeting SARS-CoV-2 non-structural proteins. *Int J Mol Sci*. 2023;**24**:13002.
26. Thoms M, Buschauer R, Ameismeier M, Koepke L, Denk T, Hirschenberger M, Kratzat H, Hayn M, Mackens-Kiani T, Cheng J, Straub JH, Stürzel CM, Fröhlich T, Berninghausen O, Becker T, Kirchhoff F, Sparrer KMJ, Beckmann R. Structural basis for translational shutdown and immune evasion by the Nsp1 protein of SARS-CoV-2. *Science*. 2020;**369**:1249–1255.
27. Yadav R, Chaudhary JK, Jain N, Chaudhary PK, Khanra S, Dhamija P, Sharma A, Kumar A, Handu S. Role of structural and non-structural proteins and therapeutic targets of SARSCoV-2 for COVID-19. *Cells*. 2021;**10**:821.
28. Schubert K, Karousis ED, Jomaa A, Scaiola A, Echeverria B, Gurzeler LA, Leibundgut M, Thiel V, Mühlemann O, Ban N. SARS-CoV-2 Nsp1 binds the ribosomal mRNA channel to inhibit translation. *Nat Struct Mol Biol*. 2020;**27**:959–966.
29. Gamble SC, Chotai D, Odontiadis M, Dart DA, Brooke GN, Powell SM, Reebye V, Varela-Carver A, Kawano Y, Waxman J, Bevan CL. Prohibitin, a protein downregulated by androgens, represses androgen receptor activity. *Oncogene*. 2007;**26**:1757–1768.
30. Calleja DJ, Lessene G, Komander D. Inhibitors of SARS-CoV-2 PLpro. *Front Chem*. 2022;**10**:876212.
31. Hu Q, Xiong Y, Zhu GH, Zhang YN, Zhang YW, Huang P, Ge GB. The SARS-CoV-2 main protease (Mpro): structure, function, and emerging therapies for COVID-19. *MedComm*. 2020;**2022**(3):e151.
32. Roe MK, Junod NA, Young AR, Beachboard DC, Stobart CC. Targeting novel structural and functional features of coronavirus protease nsp5 (Mpro, Mpro) in the age of COVID-19. *J Gen Virol*. 2021;**102**:001558.
33. Makiyama K, Hazawa M, Kobayashi A, Lim K, Voon DC, Wong RW. NSP9 of SARS-CoV-2 attenuates nuclear transport by hampering nucleoporin 62 dynamics and functions in host cells. *Biochem Biophys Res Commun*. 2022;**586**:137–142.
34. Hillen HS, Kokic G, Farnung L, Dienemann C, Tegunov D, Cramer P. Structure of replicating SARS-CoV-2 polymerase. *Nature*. 2020;**584**:154–156.

35. Chen J, Malone B, Llewellyn E, Grasso M, Shelton PMM, Olinares PDB, Maruthi K, Eng ET, Vatandaslar H, Chait BT, Kapoor TM, Darst SA, Campbell EA. Structural basis for helicase-polymerase coupling in the SARS-CoV-2 replication-transcription complex. *Cell*. 2020;**182**:1560–1573.e13.
36. Sanders JM, Monogue ML, Jodlowski TZ, Cutrell JB. Pharmacologic treatments for coronavirus disease 2019 (COVID-19): a review. *JAMA*. 2020;**323**:1824–1836.
37. Hwang YC, Lu RM, Su SC, Chiang PY, Ko SH, Ke FY, Liang KH, Hsieh TY, Wu HC. Monoclonal antibodies for COVID-19 therapy and SARS-CoV-2 detection. *J Biomed Sci*. 2022;**29**:1.
38. Singh AK, Singh A, Singh R, Misra A. Remdesivir in COVID-19: a critical review of pharmacology, pre-clinical and clinical studies. *Diabetes Metab Syndr*. 2020; **14**(4):641–648.
39. Angamo MT, Mohammed MA, Peterson GM. Efficacy and safety of remdesivir in hospitalised COVID-19 patients: a systematic review and meta-analysis. *Infection*. 2022;**50**:27–41.
40. Jayk Bernal A, Gomes da Silva MM, Musungaie DB, Kovalchuk E, Gonzalez A, Delos Reyes V, Martín-Quirós A, Caraco Y, Williams-Diaz A, Brown ML, Du J, Pedley A, Assaid C, Strizki J, Grobler JA, Shamsuddin HH, Tipping R, Wan H, Paschke A, Butterton JR, Johnson MG, De Anda C, MOVe-OUT Study Group. Molnupiravir for oral treatment of Covid-19 in nonhospitalized patients. *N Engl J Med*. 2022;**386**:509–520.
41. Anand U, Jakhmola S, Indari O, Jha HC, Chen ZS, Tripathi V, Pérez de la Lastra JM. Potential therapeutic targets and vaccine development for SARS-CoV-2/COVID-19 pandemic management: a review on the recent update. *Front Immunol*. 2021;**12**:658519.
42. Prajapat M, Sarma P, Shekhar N, Avti P, Sinha S, Kaur H, Kumar S, Bhattacharyya A, Kumar H, Bansal S, Medhi B. Drug targets for corona virus: a systematic review. *Indian J Pharm*. 2020;**52**:56–65.
43. Batlle D, Wysocki J, Satchell K. Soluble angiotensin-converting enzyme 2: a potential approach for coronavirus infection therapy? *Clin Sci (Lond)*. 2020;**134**:543–545.
44. Baglivo M, Baronio M, Natalini G, Beccari T, Chiurazzi P, Fulcheri E, Petralia PP, Michelini S, Fiorentini G, Miggiano GA, Morresi A, Tonini G, Bertelli M. Natural small molecules as inhibitors of coronavirus lipid-dependent attachment to host cells: a possible strategy for reducing SARS-COV-2 infectivity? *Acta Biomed*. 2020;**91**:161–164.
45. Wang M, Cao R, Zhang L, Yang X, Liu J, Xu M, Shi Z, Hu Z, Zhong W, Xiao G. Remdesivir and chloroquine effectively inhibit the recently emerged novel coronavirus (2019-nCoV) in vitro. *Cell Res*. 2020;**30**:269–271.
46. Borgio JF, Alsuwat HS, Al Otaibi WM, Ibrahim AM, Almandil NB, Al Asoom LI, Salahuddin M, Kamaraj B, AbdulAzeez S. State-of-the-art tools unveil potent drug targets amongst clinically approved drugs to inhibit helicase in SARS-CoV-2. *Arch Med Sci*. 2020;**16**:508–518.
47. Cherian R, Chandra B, Tung ML, Vuylsteke A. COVID-19 conundrum: clinical phenotyping based on pathophysiology as a promising approach to guide therapy in a novel illness. *Eur Respir J*. 2020;**56**:2002135.
48. Mu J, Xu J, Zhang L, Shu T, Wu D, Huang M, Ren Y, Li X, Geng Q, Xu Y, Qiu Y, Zhou X. SARS-CoV-2-encoded nucleocapsid protein acts as a viral suppressor of RNA interference in cells. *Sci China Life Sci*. 2020;**63**:1413–1416.

49. Jin Z, Du X, Xu Y, Deng Y, Liu M, Zhao Y, Zhang B, Li X, Zhang L, Peng C, Duan Y, Yu J, Wang L, Yang K, Liu F, Jiang R, Yang X, You T, Liu X, Yang X, Bai F, Liu H, Liu X, Guddat LW, Xu W, Xiao G, Qin C, Shi Z, Jiang H, Rao Z, Yang H. Structure of Mpro from SARS-CoV-2 and discovery of its inhibitors. *Nature*. 2020;**582**:289–293.
50. Dai W, Zhang B, Jiang XM, Su H, Li J, Zhao Y, Xie X, Jin Z, Peng J, Liu F, Li C, Li Y, Bai F, Wang H, Cheng X, Cen X, Hu S, Yang X, Wang J, Liu X, Xiao G, Jiang H, Rao Z, Zhang LK, Xu Y, Yang H, Liu H. Structure-based design of antiviral drug candidates targeting the SARS-CoV-2 main protease. *Science*. 2020;**368**:1331–1335.
51. Wei P, Fan K, Chen H, Ma L, Huang C, Tan L, Xi D, Li C, Liu Y, Cao A, Lai L. The N-terminal octapeptide acts as a dimerization inhibitor of SARS coronavirus 3C-like proteinase. *Biochem Biophys Res Commun*. 2006;**339**:865–872.
52. Zhang L, Lin D, Sun X, Curth U, Drosten C, Sauerhering L, Becker S, Rox K, Hilgenfeld R. Crystal structure of SARS-CoV-2 main protease provides a basis for design of improved α-ketoamide inhibitors. *Science*. 2020;**368**:409–412.
53. Huang C, Wei P, Fan K, Liu Y, Lai L. 3C-like proteinase from SARS coronavirus catalyzes substrate hydrolysis by a general base mechanism. *Biochemistry*. 2004;**43**(15):4568–4574.
54. Gao K, Wang R, Chen J, Tepe JJ, Huang F, Wei GW. Perspectives on SARS-CoV-2 main protease inhibitors. *J Med Chem*. 2021;**64**:16922–16955.
55. Schechter I, Berger A. On the size of the active site in proteases. I. Papain. *Biochem Biophys Res Commun*. 1967;**27**:157–162.
56. Hsu MF, Kuo CJ, Chang KT, Chang HC, Chou CC, Ko TP, Shr HL, Chang GG, Wang AH, Liang PH. Mechanism of the maturation process of SARS-CoV 3CL protease. *J Biol Chem*. 2005;**280**:31257–31266.
57. Vandyck K, Deval J. Considerations for the discovery and development of 3-chymotrypsin-like cysteine protease inhibitors targeting SARS-CoV-2 infection. *Curr Opin Virol*. 2021;**49**:36–40.
58. Borges PHO, Ferreira SB, Silva FP Jr. Recent advances on targeting proteases for antiviral development. *Viruses*. 2024;**16**:366.
59. De Cesco S, Kurian J, Dufresne C, Mittermaier AK, Moitessier N. Covalent inhibitors design and discovery. *Eur J Med Chem*. 2017;**138**:96–114.
60. Hoffman RL, Kania RS, Brothers MA, Davies JF, Ferre RA, Gajiwala KS, He M, Hogan RJ, Kozminski K, Li LY, Lockner JW, Lou J, Marra MT, Mitchell LJ Jr, Murray BW, Nieman JA, Noell S, Planken SP, Rowe T, Ryan K, Smith GJ 3rd, Solowiej JE, Steppan CM, Taggart B. Discovery of ketone-based covalent inhibitors of coronavirus 3CL proteases for the potential therapeutic treatment of COVID-19. *J Med Chem*. 2020;**63**:12725–12747.
61. Thanigaimalai P, Konno S, Yamamoto T, Koiwai Y, Taguchi A, Takayama K, Yakushiji F, Akaji K, Chen SE, Naser-Tavakolian A, Schön A, Freire E, Hayashi Y. Development of potent dipeptide-type SARS-CoV 3CL protease inhibitors with novel P3 scaffolds: design, synthesis, biological evaluation, and docking studies. *Eur J Med Chem*. 2013;**68**:372–384.
62. Konno S, Kobayashi K, Senda M, Funai Y, Seki Y, Tamai I, Schäkel L, Sakata K, Pillaiyar T, Taguchi A, Taniguchi A, Gütschow M, Müller CE, Takeuchi K, Hirohama M, Kawaguchi A, Kojima M, Senda T, Shirasaka Y, Kamitani W, Hayashi Y.

3CL protease inhibitors with an electrophilic arylketone moiety as anti-SARS-CoV-2 agents. *J Med Chem*. 2022;**65**:2926–2939.

63. Zhang L, Lin D, Kusov Y, Nian Y, Ma Q, Wang J, von Brunn A, Leyssen P, Lanko K, Neyts J, de Wilde A, Snijder EJ, Liu H, Hilgenfeld R. α-Ketoamides as broad-spectrum inhibitors of coronavirus and enterovirus replication: structure-based design, synthesis, and activity assessment. *J Med Chem*. 2020;**63**:4562–4578.
64. Shanghai Institute of Materia Medica, Chinese Academy of Sciences (2021). Pharmaceutical use of aldehyde-based compound, WO2021151265.
65. Qiao J, Li YS, Zeng R, Liu FL, Luo RH, Huang C, Wang YF, Zhang J, Quan B, Shen C, Mao X, Liu X, Sun W, Yang W, Ni X, Wang K, Xu L, Duan ZL, Zou QC, Zhang HL, Qu W, Long YH, Li MH, Yang RC, Liu X, You J, Zhou Y, Yao R, Li WP, Liu JM, Chen P, Liu Y, Lin GF, Yang X, Zou J, Li L, Hu Y, Lu GW, Li WM, Wei YQ, Zheng YT, Lei J, Yang S. SARS-CoV-2 Mpro inhibitors with antiviral activity in a transgenic mouse model. *Science*. 2021;**371**:1374–1378.
66. Vandyck K, Deval J. Considerations for the discovery and development of 3chymotrypsin like cysteine protease inhibitors targeting SARS CoV 2 infection. *Curr Opin Virol*. 2021;**49**:36–40.
67. Matthews DA, Dragovich PS, Webber SE, Fuhrman SA, Patick AK, Zalman LS, Hendrickson TF, Love RA, Prins TJ, Marakovits JT, Zhou R, Tikhe J, Ford CE, Meador JW, Ferre RA, Brown EL, Binford SL, Brothers MA, DeLisle DM, Worland ST. Structure-assisted design of mechanism-based irreversible inhibitors of human rhinovirus 3C protease with potent antiviral activity against multiple rhinovirus serotypes. *Proc Natl Acad Sci USA*. 1999;**96**:11000–11007.
68. Chen R, Gao Y, Liu H, Li H, Chen W, Ma J. Advances in research on 3C-like protease (Mpro) inhibitors against SARS-CoV-2 since 2020. *RSC Med Chem*. 2022;**14**:9–21.
69. Tian L, Qiang T, Yang X, Gao Y, Zhai X, Kang K, Du C, Lu Q, Gao H, Zhang D, Xie X, Liang C. Development of de-novo coronavirus 3-chymotrypsin-like protease (Mpro) inhibitors since COVID-19 outbreak: a strategy to tackle challenges of persistent virus infection. *Eur J Med Chem*. 2024;**264**:115979.
70. Enanta Pharmaceuticals, Inc. (2022). Functionalized peptides for inhibition of coronavirus 3C or 3C-like protease and for treatment and prevention of coronavirus infection, WO2022036018.
71. Tong X, Keung W, Arnold LD, Stevens LJ, Pruijssers AJ, Kook S, Lopatin U, Denison M, Kwong AD. Evaluation of in vitro antiviral activity of SARS-CoV-2 Mpro inhibitor pomotrelvir and cross-resistance to nirmatrelvir resistance substitutions. *Antimicrob Agents Chemother*. 2023;**67**:e0084023.
72. Tuttle JB, Allais C, Allerton CMN, Anderson AS, Arcari JT, Aschenbrenner LM, Avery M, Bellenger J, Berritt S, Boras B, Boscoe BP, Buzon LM, Cardin RD, Carlo AA, Coffman KJ, Dantonio A, Di L, Eng H, Farley KA, Ferre RA, Gajiwala KS, Gibson SA, Greasley SE, Hurst BL, Kadar EP, Kalgutkar AS, Lachapelle EA, Lanyon LF, Lee J, Lee J, Lian Y, Liu W, Martínez-Alsina LA, Mason SW, Noell S, Novak J, Obach RS, Ogilvie K, O'Neil SV, Ostner G, Owen DR, Patel NC, Pettersson M, Singh RS, Rai DK, Reese MR, Sakata S, Sammons MF, Sathish JG, Sharma R, Steppan CM, Stewart A, Updyke L, Verhoest PR, Wei L, Wright SW, Yang E, Yang Q, Zhu Y. Discovery of nirmatrelvir (PF-07321332): a potent, orally active inhibitor of the severe

acute respiratory syndrome coronavirus 2 (SARS CoV-2) main protease. *J Med Chem.* 2025;**68**:7003–7030.

73. Lipinski CA, Lombardo F, Dominy BW, Feeney PJ. Experimental and computational approaches to estimate solubility and permeability in drug discovery and development settings. *Adv Drug Deliv Rev.* 2001;**46**:3–26.
74. Varamini P, Toth I. Recent advances in oral delivery of peptide hormones. *Expert Opin Drug Deliv.* 2016;**13**:507–522.
75. Owen DR, Allerton CMN, Anderson AS, Aschenbrenner L, Avery M, Berritt S, Boras B, Cardin RD, Carlo A, Coffman KJ, Dantonio A, Di L, Eng H, Ferre R, Gajiwala KS, Gibson SA, Greasley SE, Hurst BL, Kadar EP, Kalgutkar AS, Lee JC, Lee J, Liu W, Mason SW, Noell S, Novak JJ, Obach RS, Ogilvie K, Patel NC, Pettersson M, Rai DK, Reese MR, Sammons MF, Sathish JG, Singh RSP, Steppan CM, Stewart AE, Tuttle JB, Updyke L, Verhoest PR, Wei L, Yang Q, Zhu Y. An oral SARS-CoV-2 Mpro inhibitor clinical candidate for the treatment of COVID-19. *Science.* 2021;**374**:1586–1593.
76. Miczi M, Golda M, Kunkli B, Nagy T, Tőzsér J, Mótyán JA. Identification of host cellular protein substrates of SARS-COV-2 main protease. *Int J Mol Sci.* 2020;**21**:9523.
77. Ullrich S, Ekanayake KB, Otting G, Nitsche C. Main protease mutants of SARS-CoV-2 variants remain susceptible to nirmatrelvir. *Bioorg Med Chem Lett.* 2022;**62**:128629.
78. Zhao L, Li C, Wang M, Zhou M, Jiang L, Zhang W, Yu J, Wang W, Zhou K, Pan K, Lam H-Y, Hung IF-N, Chan K-H, Liu L, Wang F, Zhao X, Chen Y. Potent antiviral activity of simnotrelvir against key epidemic SARS-CoV-2 variants with a high resistance barrier. *Antimicrob Agents Chemother.* 2025;**69**:e0155624.
79. Békés M, Langley DR, Crews CM. PROTAC targeted protein degraders: the past is prologue. *Nat Rev Drug Discov.* 2022;**21**:181–200.
80. de Wispelaere M, Du G, Donovan KA, Zhang T, Eleuteri NA, Yuan JC, Kalabathula J, Nowak RP, Fischer ES, Gray NS, Yang PL. Small molecule degraders of the hepatitis C virus protease reduce susceptibility to resistance mutations. *Nat Commun.* 2019;**10**:3468.
81. Alugubelli YR, Xiao J, Khatua K, Kumar S, Ma Y, Ma XR, Vulupala VR, Atla SR, Blankenship L, Coleman D, Neuman BW, Liu WR, Xu S. Discovery of first-in-class PROTAC degraders of SARS-CoV-2 main protease. *J Med Chem.* 2024; **67**:6495–6507.
82. Grifagni D, Lenci E, De Santis A, Orsetti A, Barracchia CG, Tedesco F, Bellini Puglielli R, Lucarelli F, Lauriola A, Assfalg M, Cantini F, Calderone V, Guardavaccaro D, Trabocchi A, D'Onofrio M, Ciofi-Baffoni S. Development of a GC-376 based peptidomimetic PROTAC as a degrader of 3-chymotrypsin-like protease of SARS-CoV-2. *ACS Med Chem Lett.* 2024;**15**:250–257.

9

Antimicrobial Peptidomimetics

9.1 Introduction

It is well accepted that antibiotics have revolutionized modern medicine over the last century, drastically reducing mortality from bacterial infections and enabling medical procedures such as surgery, organ transplantation, and cancer chemotherapy. Nevertheless, the global escalation of antimicrobial resistance (AMR) has been recognized as one of the most critical public health risks of the 21st century [1]. According to the World Health Organization, bacterial infections resistant to conventional antibiotics account for more than 700,000 deaths annually, and studies have estimated that 10 million deaths per year may potentially occur by 2050 if new solutions toward AMR are not implemented. This is particularly relevant as an impact on the economy and society, with cumulative global losses estimated to exceed USD 100 trillion over the next three decades [2].

Such resistance phenomena arise when microorganisms, particularly bacteria, develop mechanisms capable of evading the inhibitory effects of antimicrobial agents [3]. Drug-resistant infections compromise treatment efficacy, thus prolonging hospitalization, increasing morbidity and mortality, and enhancing the risk of epidemics [4]. The emergence of multidrug-resistant pathogens is represented by the occurrence of the so-called *Enterococcus faecium*, *Staphylococcus aureus*, *Klebsiella pneumoniae*, *Acinetobacter baumannii*, *Pseudomonas aeruginosa*, and *Enterobacter* spp. (ESKAPE) pathogens, as these organisms have evolved to "escape" nearly all available antibiotic classes. Resistance phenomena can arise through multiple mechanisms [5, 6], including:

- self-mutation or modification
- intercellular resistance gene transfer
- decreased or changed membrane permeability
- hydrolase or inactivated enzyme action
- target protection protein
- active efflux system

Also, adaptive resistance involving metabolic changes, cell morphological alterations, and biofilm formation diminishes the effectiveness of antimicrobial therapy. The reasons for the occurrence of such resistance phenomena are mainly due to the abuse of antibiotics in clinical, agricultural, and community settings, which have accelerated these evolutionary processes, and to the limited efforts toward the discovery of new antibiotic classes over the last few decades. Thus, there is an urgent need for novel therapeutic strategies capable of overcoming resistance [7]. In this context, AMPs have attracted considerable attention due to their broad-spectrum activity and reduced propensity for resistance development [8, 9]. AMPs represent a promising approach against AMR for improved efficacy in the post-antibiotic era [10].

9.2 Bacterial Resistance Mechanisms

AMR mechanisms are categorized into three groups, namely inherent, acquired, and adaptive [6, 11]. Inherent resistance is based on the structural characteristics that impair the function of antibiotics, such as the outer membrane of Gram-negative bacteria that limits drug entry [12]. Acquired resistance results from genetic alterations, such as point mutations or the transfer of resistance genes, that can lead to the production of enzymes capable of degrading antibiotics, alter drug targets, or enhance drug efflux. The third form is adaptive resistance, which is a reversible physiological response to environmental stress, such as sub-inhibitory antibiotic concentrations, thus enhancing bacterial tolerance without permanent genetic change.

At the molecular level, bacteria use a wide range of resistance strategies to neutralize antimicrobial agents (Figure 9.1) [13]. The principal mechanisms include:

- *Enzymatic degradation*: bacteria produce enzymes, such as β-lactamases and aminoglycoside-modifying enzymes, that inactivate antibiotics before they reach their targets;
- *Target modification*: structural alterations of antibiotic-binding sites, such as ribosomal methylation or DNA gyrase mutations, reduce antibiotic affinity and efficacy;
- *Reduced membrane permeability*: in Gram-negative bacteria, the reduced expression of the outer membrane porins and changes in lipopolysaccharide (LPS) composition limit the antibiotic penetration;
- *Active efflux pumps*: membrane-associated transporters, such as AcrAB-TolC, actively extrude antibiotics out of the cell, thus preventing their accumulation at bactericidal concentrations;
- *Biofilm formation*: bacteria exhibit increased tolerance through reduced metabolic activity and the formation of physical barriers to drug diffusion.

Resistance mechanisms are often displayed within a single pathogen to produce multidrug-resistant phenotypes. For example, *A. baumannii* and *K. pneumoniae* may exhibit resistance machanisms involving efflux pump activation combined to target modification and reduced membrane permeability [14, 15]. The clinical consequences are an increased incidence of infections that are resistant to first-line antibiotics.

Traditional antibiotics typically act on a single molecular target, such as those involving cell-wall biosynthesis, protein biosynthesis, or DNA replication. Consequently, the acquisition of a resistance gene can make these agents ineffective. In contrast, AMPs and peptidomimetics display multiple modes of action, including membrane disruption and interaction with intracellular targets, thus reducing the likelihood of resistance development [16]. Nevertheless, resistance to AMPs can occur through changes in membrane charge or composition or through the expression of proteases that degrade peptides, although requiring a significant fitness cost on the bacteria.

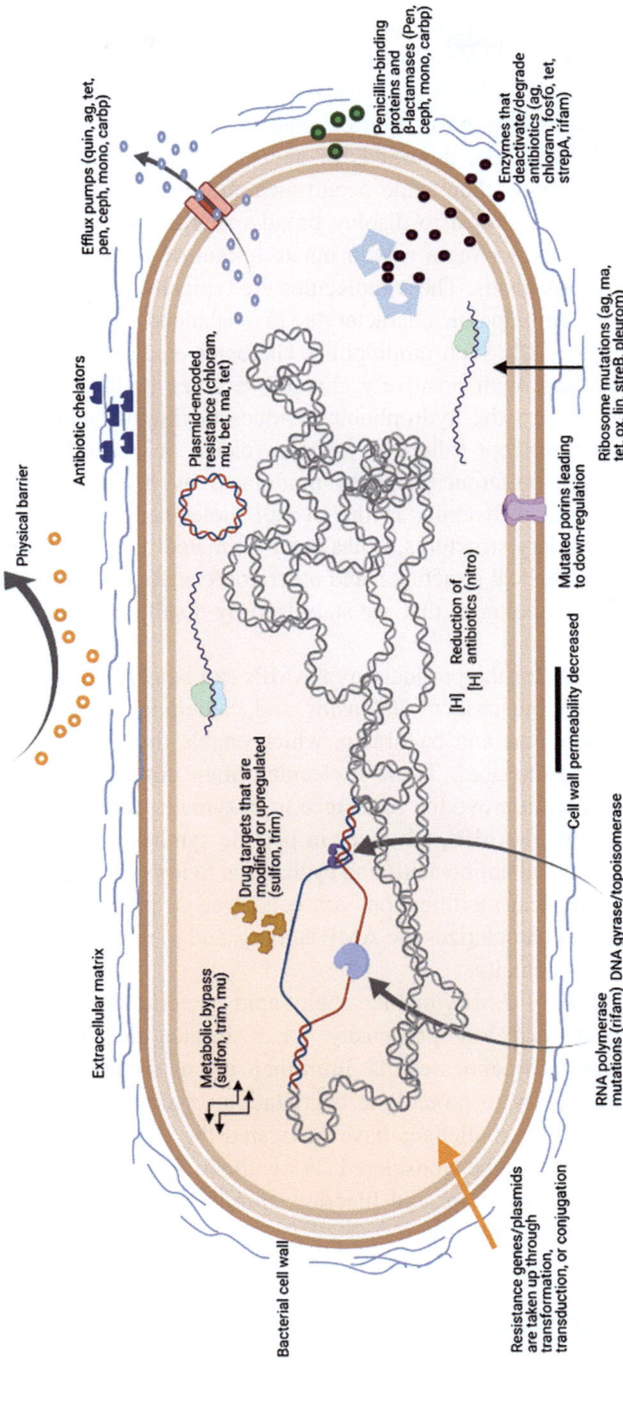

Figure 9.1 *Bacterial mechanisms of resistance to antibiotics. Source: [13] / The Royal Society of Chemistry / CC BY 3.0.*

9.3 Definition of Antimicrobial Peptides

AMPs represent one of the most promising compound classes for developing next-generation therapeutics against AMR [17]. Following the FDA approval of daptomycin as the first AMP in 2003 [18], AMPs have become a recognized therapeutic approach, and increasing efforts have been devoted to their characterization and therapeutic development. Accordingly, thousands of natural and synthetic AMPs have been identified to date, highlighting their wide occurrence and evolutionary conservation. In particular, AMPs have been shown to display broad-spectrum activity against bacteria, fungi, viruses, and parasites, having a role in innate immunity and as a primary defense tool in a wide range of organisms. These molecules are typically short to medium in size and possess cationic and amphipathic character due to a balanced ratio of hydrophobic and hydrophilic moieties [19, 20]. Such amphiphilic character enables selective interaction with microbial membranes, with positively charged residues binding to the negatively charged bacterial surface and the hydrophobic residues facilitating membrane insertion and disruption. AMPs can adopt α-helical, β-sheet, folded, or extended conformations, and these conformational preferences can be modulated by pH or lipid composition, showing significant chemical diversity in the overall molecular shape. Concerning the occurrence of such secondary structures, it has been estimated that about 60% of known AMPs are organized in α-helical structures, and other conformational preferences account for β-sheet or mixed α/β topologies that are stabilized by disulfide bonds or head-to-tail cyclization.

The biological processes for the production of AMPs can be divided into those encompassing ribosomes, as in the case of defensins and cathelicidins, and non-ribosomal pathways, such as polymyxins and bacitracin, which enable the assembly of AMPs by non-ribosomal peptide synthetases. These molecules often contain D-amino acids and thiazole derivatives, which improve the resistance to enzymatic degradation. Concerning the development of modified AMPs, advances in peptide synthesis have allowed for the incorporation of noncanonical amino acids and cyclization to improve peptide stability and bioavailability. Such chemical modifications are conceived to preserve the cationic and hydrophobic balance that characterizes the AMP activity, and to reduce the degradation by proteases and the immunogenicity.

The clinical advantages of AMPs include their rapid bactericidal activity, often with broad-spectrum efficacy, and low propensity for resistance development. However, natural AMPs show several weaknesses that limit their use as therapeutics, such as poor *in vivo* stability, susceptibility to proteolytic degradation, cytotoxicity to host cells, and high production costs. These challenges have motivated the development of antimicrobial peptidomimetics that can be considered as synthetic or semisynthetic analogues capable of reproducing the structure and function of AMPs while exhibiting improved pharmacokinetics and reduced toxicity.

9.4 Classification of Antimicrobial Peptides

AMPs are characterized by diversity in source, structure, biosynthetic pathway, and biological function, and different classifications can be outlined based on these features [8]. Comprehensive databases such as APD3 and DRAMP have categorized thousands of AMP sequences, structures, and activities, thus enabling data-driven design [21]. As of November 2025, the DRAMP database recorded 14,214 natural and synthetic AMPs, including 96 drug candidates at preclinical or clinical stages. Multiple complementary classifications have been proposed to categorize AMPs, namely by source organism, structural features, biosynthetic mechanism, mechanism of action, and amino acid composition.

Classification by source. AMPs are ubiquitous in nature and have been isolated from diverse organisms:

- bacterial peptides (bacteriocins), as narrow-spectrum AMPs produced by bacteria to inhibit competitors;
- fungal peptides, including peptaibols and other peptide antibiotics;
- plant-derived peptides such as thionins, defensins, and lipid transfer proteins, which protect against microbial invasion;
- insect peptides, including defensins, cecropins, and attacins, playing key roles in arthropod innate immunity;
- amphibian peptides, such as magainins, dermaseptins, and temporins, secreted on skin surfaces to provide broad-spectrum antimicrobial defense;
- vertebrate peptides, primarily defensins and cathelicidins, are key to epithelial and systemic immune protection.

Classification by structure. AMPs are most commonly grouped according to secondary structural motifs (Figure 9.2), which characterize their physicochemical behavior and mode of membrane interaction [22], overall influencing the antimicrobial potency, target specificity, and resistance to enzymatic degradation. The four major structural families are:

- α-helical peptides (A), such as magainins and LL-37, that are linear, amphipathic, and typically lacking disulfide bonds;

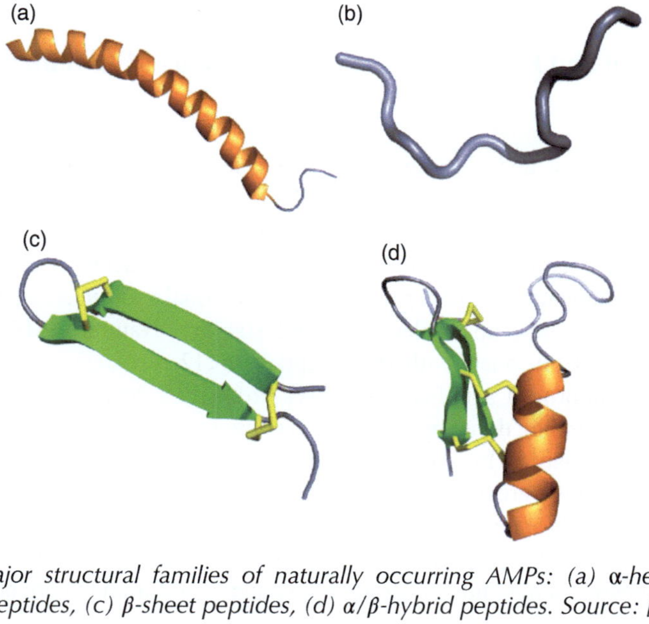

Figure 9.2 *Major structural families of naturally occurring AMPs: (a) α-helical peptides, (b) extended peptides, (c) β-sheet peptides, (d) α/β-hybrid peptides. Source: [22] / with permission of Elsevier.*

- extended peptides (B), such as indolicidin [23], that are rich in proline, tryptophan, or arginine, adopting flexible conformations;
- β-sheet peptides (C), such as defensins and protegrins, that are stabilized by intramolecular disulfide bridges, providing protease resistance;
- α/β-hybrid peptides (D), such as human β-defensins, that contain both helical and β-structured elements.

Classification by biosynthetic mechanism. As anticipated, AMPs can be divided into ribosomally synthesized peptides (RiPPs), such as lantibiotics, such as nisin and microcins, that are gene-encoded and often posttranslationally modified; and non-RiPPs, produced by non-ribosomal peptide synthetases without mRNA templates, often incorporating noncanonical amino acids and unusual modifications that enhance stability and activity.

Classification by function. AMPs are grouped according to their biological activity, namely antibacterial (polymyxin B, active against Gram-negative bacteria), antifungal (histatins, targeting *Candida* spp.), antiviral (LL-37, which disrupts viral envelopes), antiparasitic (dermaseptins), and immunomodulatory activity, that influence cytokine production and immune cell recruitment.

Classification by mechanism. AMPs can be divided into membrane-disruptive peptides, which destabilize microbial membranes, and intracellular-targeting peptides, which cross membranes to inhibit nucleic acid or protein synthesis.

Classification by amino acid composition. AMPs can further be classified depending on the peptide character, such as:

- cationic peptides, enriched in lysine and arginine, capable of promoting electrostatic interactions with bacterial membranes;
- anionic peptides, which are less common, often functioning *via* metal ion chelation;
- amphipathic peptides, characterized by isolated hydrophobic and hydrophilic residues that enable membrane insertion and disruption.

9.5 Structural and Physicochemical Determinants of Antimicrobial Peptides

The biological activity and selectivity of AMPs and their peptidomimetic analogues are basically governed by their structural and physicochemical properties, specifically facilitating interactions with microbial membranes and intracellular targets. These properties are net charge, hydrophobicity, amphipathic character, secondary structure, peptide length, aggregation behavior, and posttranslational modification, all responsible for antimicrobial potency, host selectivity, and stability. Natural AMPs are typically constituted by 10–50 residues, with short peptides (<12 residues) generally exhibiting reduced potency, improved stability, and lower production costs and longer sequences providing multiple binding domains, although displaying a higher tendency to aggregation and enzymatic degradation. In this view, modifications such as truncation, dimerization, or cyclization strategies are used to balance potency and stability in both natural and synthetic analogues. An interesting feature of AMPs is the conformational modification upon interaction with the target. Specifically, AMPs typically display remarkable structural flexibility and possess a random coil structure in aqueous solution but are able to establish ordered conformations upon membrane contact. In fact, secondary structure facilitates the interaction between AMPs and the microbial membrane, being responsible for membrane orientation, insertion depth, and mechanism of action. The major classes of AMPs are those reported in par. 9.2 and Figure 9.2, represented by α-helical peptides, β-sheet peptides, α/β hybrid peptides, and extended peptides. Concerning the features displayed by modified AMPs, the structural stabilization through cyclization, hydrocarbon stapling, or disulfide bonds enhances the proteolytic resistance and the thermal stability required for therapeutic application. An important insight about the 3D structure of AMPs is characterized by the arrangement of hydrophobic and hydrophilic domains that constitute the so-called amphipathic character as a distinctive hallmark of AMPs. Such a dual feature is essential on one side for the selective targeting of negatively charged bacterial membranes, taking advantage of cationic residues like arginine, lysine, and histidine, that promote the electrostatic attraction to anionic surfaces rich in phosphatidylglycerol and cardiolipin, without affecting the zwitterionic membranes of mammalian cells, whereas, on the other side, hydrophobic residues like leucine, isoleucine, valine, phenylalanine, and tryptophan interact with the lipid bilayer. It has been estimated that the optimal hydrophobicity window accounts for 40–60% in hydrophobic content, and this is crucial for optimal activity and safety, as AMPs possessing high hydrophobicity exhibit hemolytic activity and aggregation, whereas insufficient hydrophobicity compromises the membrane affinity. Such physicochemical properties linked to the amphipathic side-chain distribution, which is essential for the activity, have been referred to as the "hydrophobic moment" as a critical parameter for balancing antimicrobial activity and host-cell safety [13, 24].

The hydrophilic character is closely linked to the "cationic character" as another key structural feature of most natural AMPs, which involves the net positive charge of these molecules, ideally ranging between +2 and +9 at physiological pH. This cationic character enables the charge interaction of AMPs with bacterial membranes, which are typically more negatively charged than their eukaryotic counterparts. In this context, arginine residues, possessing guanidine on the side chain, provide stronger binding than lysine through bidentate hydrogen-bonding interactions, although excessive positive charge can promote nonspecific binding to host cells or serum proteins. The overall charge distribution depends on the isoelectric point (pI) of the molecules, affecting solubility and aggregation, as peptides with high pI values (>10) tend to aggregate under physiological ionic strength, whereas lower pI values may reduce the positive charge content necessary for complementary charge interactions to bacterial membranes. Also, thermal and pH stability can influence the pharmacological performance in the context of balanced hydrophobic/hydrophilic content and rigidity. For example, Du et al. reported that the incorporation of β- and γ-residues or *N*-alkylated backbones enhances the hydrophobic/hydrophilic balance and rigidity, thus improving serum stability without compromising the antimicrobial potency [25].

Finally, balanced AMP aggregation is a key property that can either enhance the biological activity by promoting oligomeric pore formation or reduce the bioavailability through precipitation. Aggregation propensity depends on charge density, aromaticity, and hydrophobic content. In this view, the modification of AMPs through the incorporation of D-amino acids, peptoid backbones, or noncanonical amino acids can reduce the tendency to aggregation.

In view of addressing possible modes to improve the function of AMPs, *C*-terminal amidation has been demonstrated as a useful approach to increase the net positive charge to stabilize α-helical conformations. Also, *N*-terminal acetylation and lipidation proved to enhance membrane affinity and binding, thus prolonging the half-life. Similarly, the introduction of noncanonical amino acids or backbone modifications can reduce conformational flexibility and proteolysis, allowing for the introduction of novel functional groups for targeting.

9.6 Mechanism of Action of Antimicrobial Peptides

AMPs operate through mechanisms distinct from conventional antibiotics, typically combining membrane disruption, intracellular targeting, and immunomodulation, thus reducing the possibility of resistance development (Figure 9.3) [26–28]. Understanding

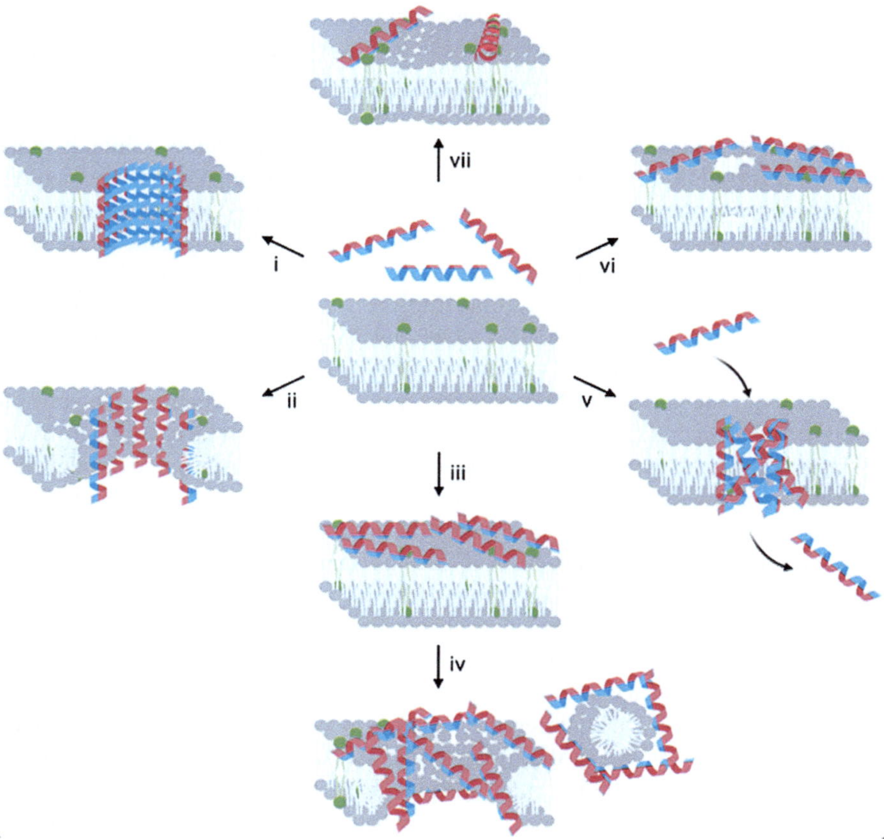

Figure 9.3 Membrane disruption mechanisms displayed by AMPs: the membrane is shown in gray and negatively charged phospholipids are highlighted in green; the amphipathic nature of helical peptides is represented by the blue color indicating the surface with positively charged residues and the red color indicating the surface with hydrophobic amino acids: (i) barrel-stave model, (ii) toroidal model peptides, (iii) carpet model, (iv) detergent model, (v) aggregate model, (vi) electroporation model, and (vii) phospholipid clustering. Source: [13] / The Royal Society of Chemistry / CC BY 3.0.

these mechanisms is essential for the rational design of synthetic peptidomimetics that replicate natural AMP activity while improving pharmacological stability and safety [29].

The most extensively studied mode of action of AMPs involves the interaction with microbial membranes. Cationic peptides are electrostatically attracted to negatively charged components of bacterial envelopes, such as LPSs and anionic phospholipids like phosphatidylglycerol and cardiolipin [30]. This binding facilitates peptide insertion into membranes, thus leading to membrane permeabilization and cell death. These processes lead to rapid microbial lysis and are difficult for bacteria to counteract, as they involve physical perturbation rather than a specific biochemical target. Several biophysical models have been reported to describe AMP-induced membrane disruption:

- *Barrel-stave model*: amphipathic α-helices insert perpendicularly into the lipid bilayer, forming transmembrane channels whose hydrophilic faces line the pore lumen [31]. Peptides such as alamethicin and pardaxin exemplify this model;
- *Toroidal pore model*: peptides induce the continuous bending of the lipid monolayers, creating pores lined with both peptides and lipid headgroups [32]. This mechanism is common in magainins and protegrins;
- *Carpet model*: peptides orient parallel to the membrane surface, forming a carpet-like layer that destabilizes lipid packing and causes membrane disintegration in a detergent-like fashion. LL-37 and cecropins are typical examples;
- *Detergent model*: as the peptide to lipid ratio increases, this can lead to membrane dissolution;
- *Aggregate model*: at high peptide concentrations, AMPs can aggregate into supramolecular complexes and delaminate bilayers, dissolving membranes into peptide-lipid micelles;
- *Electroporation model*: peptide accumulation on the membrane surface causes an electric potential buildup, resulting in a membrane disruption.

Another key mechanism of action of numerous AMPs is their ability to penetrate the cytoplasm and to reach intracellular targets through several modes of action, including:

- inhibition of nucleic acid synthesis that prevents replication through the binding of AMP to bacterial DNA and RNA;
- inhibition of protein synthesis that interferes with ribosomal function or the translation process;
- enzyme inhibition that blocks the activity of enzymes essential for cell-wall biosynthesis or central metabolism [33];
- metal ion chelation and efflux pump inhibition, forming complexes with divalent cations (Mg^{2+} and Fe^{2+}) or disrupting efflux systems, thus impairing bacterial homeostasis.

A dual mode of action consisting of membrane disruption and intracellular inhibition significantly reduces the tendency to develop bacterial resistance by targeting more than one essential process at the same time. In addition, it has been discovered that AMPs possess immunomodulatory properties that enhance host defense. Specifically, AMPs have a role in several processes, such as: 1) the recruitment of immune cells to infection sites like neutrophils, macrophages, and dendritic cells; 2) the regulation of cytokine networks; 3) the promotion of phagocytosis and clearance of pathogens; 4) the activation of wound healing and angiogenesis by stimulating the proliferation of epithelial cells. The human cathelicidin

LL-37 is an example of this multifunctionality, acting as both a bactericidal peptide and a chemotactic agent that modulates Toll-like receptor signaling.

Overall, AMPs and their synthetic analogues maintain broad-spectrum efficacy due to their multimodal mechanisms. Nevertheless, some adaptive mechanisms to the effects of AMPs have been reported, including the alteration of membrane charge and the activation of efflux pumps.

9.7 Representative Examples of Antimicrobial Peptides

Numerous compounds that have advanced through preclinical and clinical development demonstrated the chemical diversity and biological versatility of AMPs and their peptidomimetics. These molecules highlight both the structural variety and the therapeutic potential of AMPs.

9.7.1 Natural and Semisynthetic Antimicrobial Peptides

Magainins are natural compounds isolated from the skin of *Xenopus laevis* that are considered the earliest discovered eukaryotic AMPs [34]. In particular, magainin-2 proved to show rapid and nonspecific bacteriolysis through a toroidal pore mechanism, displaying low cytotoxicity toward mammalian cells due to its amphipathic character. Although the clinical formulation "pexiganan" demonstrated only modest efficacy in treating diabetic foot ulcers, its development demonstrated the possibility of developing AMPs as therapeutics [35]. Another example of natural AMPs is represented by LL-37 [36] as an entry to a multifunctional compound capable of disrupting bacterial membranes, neutralizing LPS, and modulating cytokine signaling pathways. Structure–activity studies on this molecule showed that the truncation of its *N*-terminal hydrophobic domain significantly reduces the antibacterial potency, thus highlighting the essential role of amphipathic helicity for the biological activity. Among β-sheet peptides, defensins represent the prototypical disulfide-stabilized AMPs. In particular, human α- and β-defensins exhibit broad-spectrum antimicrobial activity and function in combination with lysozyme. Concerning the properties displayed by structured peptides, β-sheet motifs confer remarkable thermostability and proteolytic resistance, and such properties have inspired the design of cyclic β-hairpin mimetics (Table 9.1).

9.7.2 Clinically Approved and Advanced Antimicrobial Peptides

Polymyxins, and particularly polymyxin B and polymyxin E (also named colistin), remain among the few AMP-derived antibiotics approved for clinical use against multidrug-resistant Gram-negative pathogens (Figure 9.4). These molecules are characterized by a

Table 9.1 *Representative natural antimicrobial peptides. Data from [22].*

Peptide	Sequence	Charge
Magainin-2 (amphibian)	GIGKFLHSAKKFGKAFVGEIMNS	+3
LL-37 (human)	LLGDFFRKSKEKIGKEFKRIVQRIKDFLRNLVPRTES	+6
Indolicidin (bovine)	ILPWKWPWWPWRR-NH_2	+4
HBD-1 (human)	DHYNCVSSGGQCLYSACPIFTKIQGTCYRGKAKCCK	+4

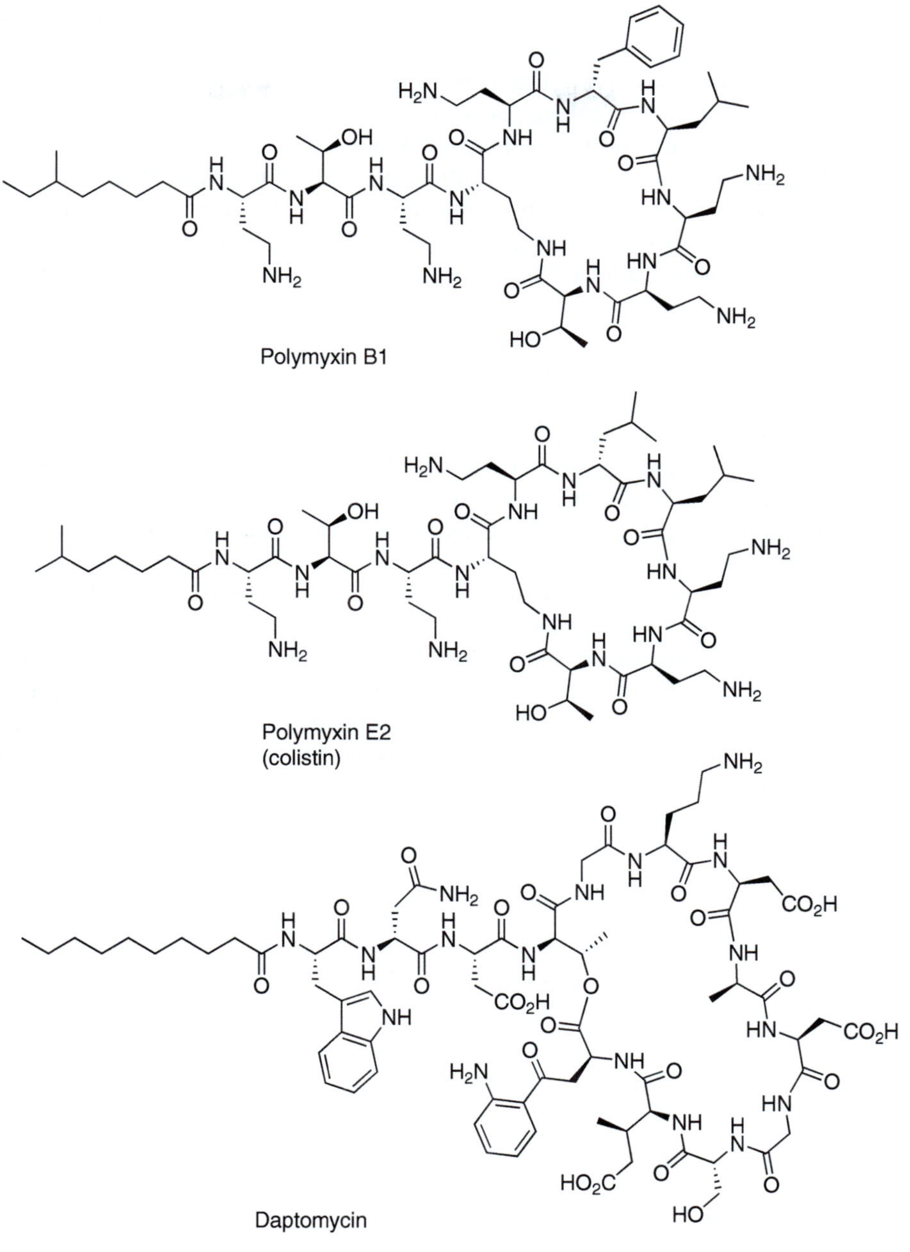

Figure 9.4 *Representative clinically approved antimicrobial peptides.*

cyclic lipopeptide scaffold that contains a cationic diaminobutyric acid-rich ring linked to a fatty acid tail, enabling strong electrostatic and hydrophobic interactions with bacterial membranes. Although displaying remarkable antimicrobial potency, their application is rather limited due to nephrotoxicity and neurotoxicity. Thus, next-generation analogues such as NAB739 and SPR741 [37, 38], characterized by modified lipid tails, have been developed to retain antimicrobial efficacy while reducing toxicity. Daptomycin is a cyclic lipopeptide produced by *Streptomyces roseosporus* that represents another example of a clinically approved AMP [18]. Daptomycin exerts its bactericidal activity against Gram-positive bacteria through calcium-dependent membrane insertion and rapid depolarization. Its approval in 2003 for the treatment of *S. aureus* bacteremia and endocarditis is considered a major milestone in the therapeutic application of natural peptide antibiotics. Although resistance *via* cell-membrane remodeling and phospholipid redistribution has occasionally emerged, such cases remain rare, highlighting the potential of combination therapy approaches to maintain the efficacy in the antimicrobial therapeutic treatment.

9.8 Antimicrobial Peptidomimetics

Antimicrobial peptidomimetics are synthetic or semisynthetic compounds engineered to mimic the structure and function of natural AMPs while overcoming their inherent limitations, such as susceptibility to proteolysis, high production costs, and potential cytotoxicity. These molecules retain broad-spectrum antimicrobial activity while offering enhanced stability, bioavailability, and potency.

Peptidomimetics encompass a variety of structural classes that include small-molecule analogs, possessing lower molecular weight, improved stability, and easier chemical synthesis; oligomers, consisting of short polymeric chains that mimic AMP's amphipathic and cationic characters; and noncanonical amino acids that are incorporated into peptide sequences to fine-tune charge, hydrophobicity, conformational rigidity, and resistance to enzymatic degradation. Small-molecule peptidomimetics are particularly promising due to ease of synthesis and optimization and high stability. Several candidates have progressed to clinical trials, including CSA-13, PMX30063 (brilacidin), and LTX-109 (Figure 9.5).

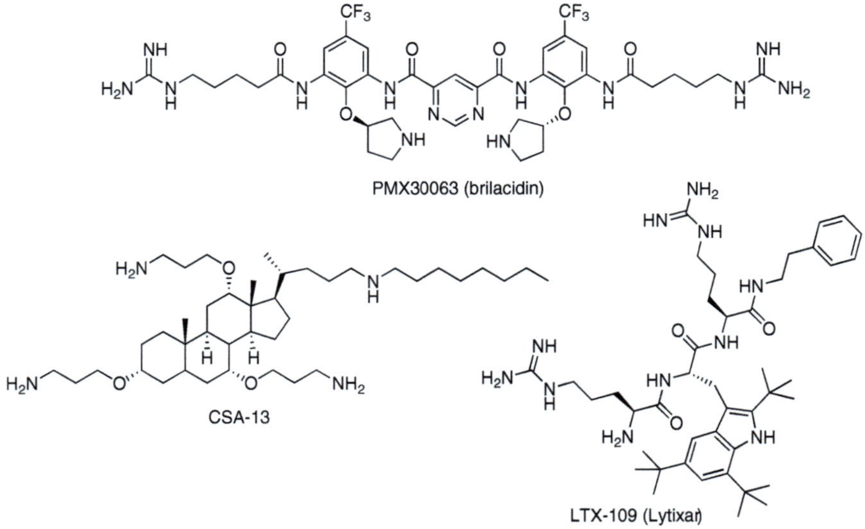

Figure 9.5 Structure of small-molecule antimicrobial peptidomimetics in clinical trials.

CSA-13 is a small-molecule cationic mimetic exhibiting broad-spectrum bactericidal activity, PMX30063 has been designed for enhanced stability and membrane-targeting potency, and LTX-109 demonstrated topical efficacy against skin infections, reproducing the amphipathic structure to mimic natural AMPs.

Advances in peptide synthesis have greatly expanded the toolbox for generating peptidomimetics with optimized properties. Solid-phase peptide synthesis using the Fmoc strategy is applied for the stepwise assembly of AMP on an insoluble resin. Although costly for large peptides, automated flow-based SPPS may allow for the rapid production of long and complex sequences. The incorporation of noncanonical amino acids allows for introducing novel side chains, thus enabling control over charge distribution, hydrophobicity, and structural stability. Such modifications improve proteolytic resistance, membrane interaction, and selectivity. Examples include D-amino acids, fluorinated residues, and bulky hydrophobic groups.

As a complementary approach, cell-free protein synthesis takes advantage of lysates from microorganisms, such as *E. coli* or *Bacillus subtilis*, with transcriptional and translational machinery to produce peptides rapidly. Although limited by available translation systems, this approach facilitates the incorporation of noncanonical amino acids, reduces cytotoxicity, and enables the scalable production of otherwise toxic sequences.

Finally, genetic code expansion is an approach that allows for the site-specific insertion of noncanonical amino acids into RiPPs by reassigning codons and introducing orthogonal tRNA/aminoacyl-tRNA synthetase pairs. This enables custom chemical functionalities that are challenging to achieve via traditional methods, offering unparalleled site-specificity although requiring complex genetic engineering.

9.8.1 Types of Antimicrobial Peptidomimetics

Antimicrobial peptidomimetics are designed to reproduce the amphipathic topology and cationic charge distribution characteristic of AMPs but differ in backbone chemistry, rigidity, and molecular weight. Current classifications account for several types of compounds, including small-molecule peptidomimetics, foldamers, oligopeptides, peptoids, stapled peptides, and hybrid or conjugate systems.

Small-molecule peptidomimetics are low-molecular-weight compounds designed to mimic the structural and amphipathic features of AMPs while maintaining the resistance to enzymatic degradation [39]. They act mainly through membrane disruption or inhibition of intracellular processes, as in the case of peptide AMPs, but by adding chemical stability, ease of synthesis, and oral bioavailability in their peptidomimetic design. Early examples of antimicrobial peptidomimetics are the ceragenins, which are cholic-acid-based amphiphiles that mimic the facial amphipathic character of AMPs. CSA-13 (Figure 9.5) and related derivatives display broad-spectrum activity against both Gram-positive and Gram-negative bacteria, including methicillin-resistant *S. aureus* and *P. aeruginosa*. Their mechanism of action involves membrane depolarization similarly to that of natural AMPs. Ceragenins, first described by Savage and coworkers [40], replicate the cationic and hydrophobic topology of AMPs and are chemically stable and resistant to proteolysis.

Another family of small-molecule mimetics comprises arylamide oligomers, designed to reproduce the amphipathic pattern of α-helical AMPs using aromatic scaffolds. Among these molecules, PMX-30063 (brilacidin), developed by DeGrado and collaborators

(Figure 9.5), demonstrated strong antistaphylococcal activity and has reached advanced clinical trials for skin and soft-tissue infections, showing comparable efficacy to daptomycin and improved stability. These arylamide molecules achieve stability through intramolecular π-stacking interactions and can be fine-tuned by substituent modification. In general, the major group of peptidomimetics in this field is represented by *foldamers*, which include synthetic oligomers designed to adopt well-defined secondary structures such as α-helices or β-sheets, reminiscent of natural AMPs [41]. They combine structural rigidity with tunable amphipathic and cationic character. β-Peptide foldamers are generated through a sequence of β-amino acids and mimic the α-helical conformation of AMPs by forming stable 14- or 12-helices that resist proteolytic degradation and display potent antimicrobial activity. Helical oligopeptide foldamers have been developed with selective activity against Gram-positive bacteria and reduced cytotoxicity. Numerous studies have demonstrated the antimicrobial potential of peptide foldamers. Linear aromatic oligomers with amphipathic character and potent antimicrobial activity were reported by the groups of DeGrado and Tew [42–44], while Huc and coworkers designed the first fully water-soluble helical quinoline oligoamides by attaching ammonium side chains along the backbone [45], showing that the helical stability of these molecules increases in polar media as compared to organic solvents. Heterogeneous backbones have also been explored with abiotic foldamers to form hybrid scaffolds exhibiting unique folding behavior due to the absence of common features between aliphatic and aromatic building blocks. In this context, synthetic foldamers serve as stable analogues that retain amphiphilic character and mimic the mechanisms of AMPs [46]. These include β-peptide 14-, 12-, and 10/12-helices [47, 48], peptoids [49], *N,N′*-linked aliphatic oligoureas, oligo(urea/amide) hybrids, and sulfono-γ-AApeptides [50]. Such structures are designed to form globally amphiphilic helices by clustering hydrophobic and cationic residues on the helix surface, reproducing AMP topology (Figure 9.6).

Cationic peptidomimetic oligomers represent another category of synthetic AMP mimetics consisting of short sequences engineered to enhance the antimicrobial activity and pharmacological stability through structural modifications such as incorporation of D-amino acids, cyclization, and *N*-terminal acetylation. D-Amino acid substitution preserves antimicrobial potency and reduces proteolytic susceptibility. Cyclic oligopeptides derived from natural AMPs exhibit improved membrane binding and serum stability. Cationic polymers and dendrimers, including poly(ethylene imine), poly(lysine), and poly(amidoamine) derivatives, act as high-molecular-weight mimics capable of multivalent electrostatic interaction with bacterial membranes. Their activity can be optimized by balancing hydrophobicity and charge to minimize mammalian cytotoxicity.

Peptoids have been exploited for the development of antimicrobial peptidomimetics, too. In fact, being characterized by an oligomeric structure composed of *N*-substituted glycines, these molecules have been developed as sequence-specific analogues capable of maintaining the amphipathic and cationic properties of peptides while providing high resistance to proteolysis [51]. In peptoids, the increased conformational flexibility is provided by the side chains being attached to the amide nitrogen rather than the α-carbon, thus removing backbone hydrogen bonds. Despite their flexibility, these molecules can adopt polyproline-type helical structures and amphipathic arrangements analogous to α-helices. Interestingly, the absence of natural amide hydrogens contributes to enhanced

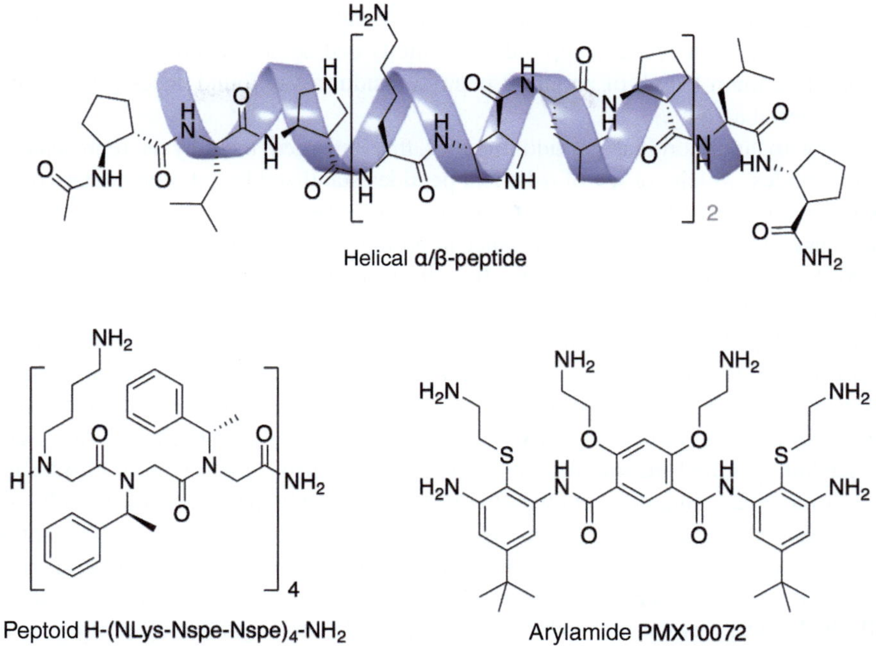

Figure 9.6 *Representative antimicrobial foldamers.*

serum stability, and recent studies have explored hybrid peptoid-peptide conjugates that combine rigidity with biological recognition [52].

Cationic peptoids showed potent activity against multidrug-resistant *A. baumannii* and *P. aeruginosa* with low hemolytic activity [49]. Also, peptoid libraries have been designed for broad-spectrum antimicrobial action with tunable hydrophobicity and minimal toxicity.

In a work by Patch and Barron [53], a series of peptoid oligomers as helical, cationic, and facially amphipathic mimics of the magainin-2 amide were developed and studied for their antibacterial potency (Figure 9.7). Conformational studies and biological assays demonstrated that certain peptoids display characteristic helical circular dichroism conformation and exhibit selective (nonhemolytic) and potent antibacterial activity against both Gram-positive and Gram-negative bacteria. Interestingly, peptoids showing weak

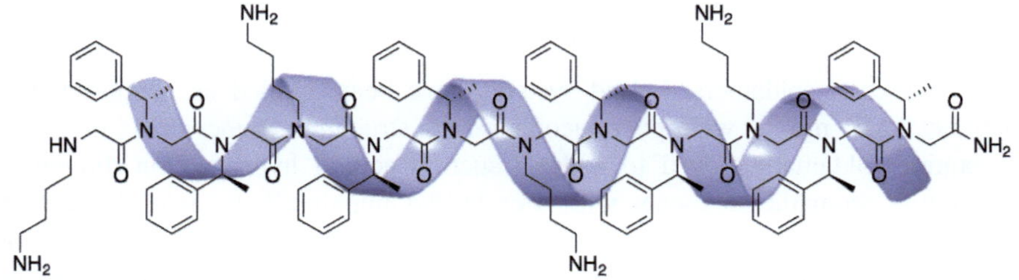

Figure 9.7 *Structure of peptoid analogue of the natural antimicrobial peptide magainin-2 amide.*

helical character proved ineffective as antibiotics, and similarly to the natural magainin peptides, a correlation between peptoid lipophilicity and hemolytic propensity was found. Finally, a minimum length of about 12 peptoid residues was found necessary to achieve the antibacterial activity.

An entry to antimicrobial peptidomimetics that emerged recently is represented by stapled peptides, which consist of α-helical peptides stabilized by intramolecular covalent cross-links, commonly hydrocarbon staples, that preserve secondary structure and amphipathic character while improving resistance to proteolysis and enhancing cell permeability [54, 55]. Hydrocarbon staples are typically introduced through olefin metathesis between α,α-disubstituted nonnatural amino acids at i to $i+4$ or $i+7$ relative positions. The concept, initially developed for peptide-based inhibitors of protein–protein interactions, has been successfully extended to AMPs. In general, cross-linking methods to achieve stapled peptides include ring-closing metathesis using Grubbs catalysts to form hydrocarbon staples, amide bond formation between Lys/Asp or Lys/Glu side chains, disulfide or thioether linkages, and triazole formation *via* click chemistry. Each type of cross-linking approach modulates in a different way the rigidity, hydrophobicity, and overall charge of the peptide, thus allowing for fine-tuning of activity and selectivity. Also, computational and molecular dynamics approaches now inform the rational positioning of staples to maximize membrane interaction and amphipathic balance. Stapling enhances α-helicity by 20–40%, increases thermal stability by up to 40 °C, and significantly prolongs serum half-life. Membrane binding affinity and lytic activity generally increase with helical stabilization provided that the staple does not interfere with amphipathic orientation.

Several remarkable stapled AMPs have been described in the literature (Figure 9.8) [56], including:

- magainin-2 analogues [57] bearing hydrocarbon stapling side chains for helix stabilization. The helical structure of the peptides was found to be strongly stabilized and their antimicrobial activity enhanced. Interestingly, peptide stapling between the first and fifth positions from the *N*-terminus showed higher antimicrobial activity as compared to magainin-2 against both Gram-positive and Gram-negative bacteria without exerting significant hemolytic activity;
- stapled aurein 1.2 derivatives retained the natural peptide's α-helical content in physiological salt and serum, maintaining potent antimicrobial activity [58];
- stapled LL-37 fragments, produced by i to $i+4$ or i to $i+7$ hydrocarbon cross-links, exhibited improved stability and antibacterial efficacy *via* effective membrane perturbation and anti-inflammatory efficacy through compact LPS binding [59];
- double-stapled β-peptide foldamers achieved exceptional helicity and salt tolerance, expanding the structural toolkit beyond α-amino acids.

In the context of foldamers, hybrid systems have been applied as peptidomimetics to combine AMP motifs with pharmacophoric or structural frameworks. In particular, arylamide foldamers proved to address such structural hybridization through the combination of aromatic backbones with peptide-like amphiphilicity, as in PMX-30063, which was advanced to clinical trials for the treatment of acute bacterial skin infections. Peptidomimetic hybrids containing α-, β-, and γ-amino acids or peptoid monomers demonstrated enhanced structural diversity and therapeutic potential [52].

AMP-antibiotic hybrid conjugates exploit synergistic mechanisms, as represented by vancomycin-arginine conjugates that proved to maintain the therapeutic activity

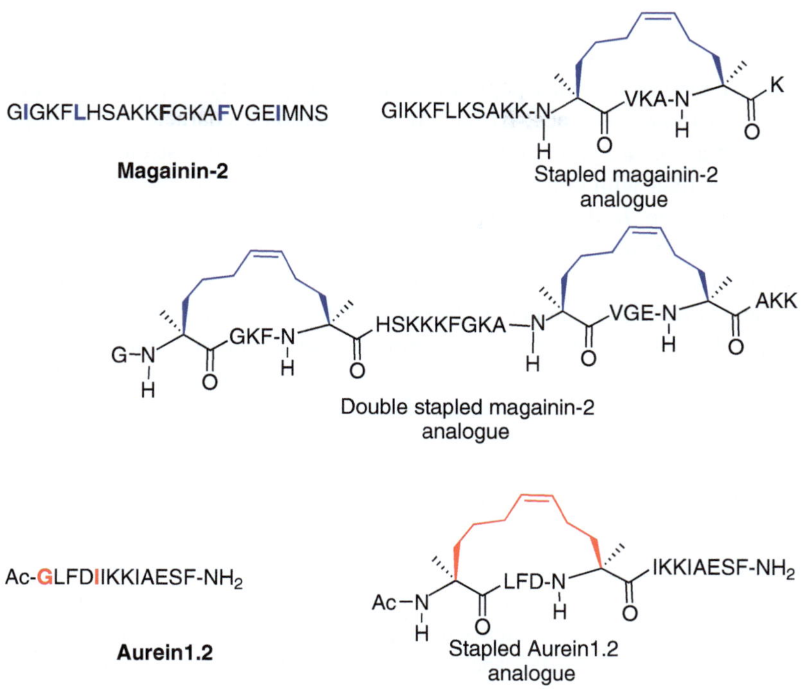

Figure 9.8 *Representative stapled and double-stapled antimicrobial peptidomimetics.*

against resistant *Enterococcus species* by coupling cell-wall targeting to membrane permeabilization [60]. Other examples of AMP-antibiotic conjugates include rifampicin or chloramphenicol to facilitate the intracellular delivery of such traditional drugs [61]. Another hybrid solution encompassing AMPs is represented by lipidated peptides, such as lipo-LL37 analogues, that demonstrated improved serum stability and biofilm penetration.

Synthetic polymeric AMP mimetics achieved *via* ring-opening polymerization, mimic the activity of natural AMPs showing rapid bactericidal action and low mammalian toxicity. Also, the incorporation of noncanonical amino acids represents an additional strategy for modulating the properties of AMP mimetics. Substitution with D-amino acids, *N*-methyl residues, or fluorinated side chains enhances the resistance to proteolysis and improves the bioavailability. For instance, LTX-109, a tripeptide containing an unnatural tryptophan residue, showed robust activity against *E. coli* and *S. aureus* and has advanced to Phase II clinical trials. Overall, these chemical modifications expand the diversity of peptidomimetic scaffolds, enabling the fine tuning of conformation and amphipathic character, resulting to improved membrane interaction.

9.9 Conclusion

AMPs and their peptidomimetic analogues represent a promising approach against multidrug-resistant pathogens, including Gram-negative bacteria and members of the ESKAPE group. Natural AMPs provide an established therapeutic strategy for broad-spectrum antimicrobial action through different effects, such as membrane disruption, intracellular targeting and immunomodulation. Despite their promise, the clinical translation of AMPs remains still challenging due to 1) high production costs, particularly for long or complex peptides; 2) the susceptibility of peptides to proteolytic degradation and rapid clearance; 3) the excessive cationic character or hydrophobicity, which can increase host cytotoxicity. While the occurrence of resistance mechanisms is generally slower than with conventional antibiotics, bacterial adaptations can reduce the efficacy of AMPs, too. In this context, antimicrobial peptidomimetics encompass a structurally diverse group of synthetic molecules that replicate the amphipathic and cationic features of natural AMPs while addressing their inherent limitations. From small-molecule ceragenins and arylamide foldamers to β-peptide foldamers, peptoids, stapled helices, and hybrid conjugates, these systems share a common design approach, aiming to preserve the essential physicochemical features responsible for antimicrobial activity while enhancing the stability and therapeutic performance. These molecules are conceived using several strategies, such as backbone modifications, cyclization, stapling, use of noncanonical amino acids, and foldamer design. Specifically, several peptidomimetic classes, including small molecules, foldamers, oligopeptides, peptoids, and stapled peptides, demonstrate the versatility of synthetic approaches, allowing for modulating the amphipathic and cationic character and reproducing the secondary structure to target specific pathogens.

Combination therapies with conventional antibiotics to exploit synergistic effects, targeted and responsive delivery systems to improve bioavailability, and studies to reduce off-target effects are among the therapeutic approaches that are being studied. Indeed, computational chemistry, and high-throughput screening are being applied more often to accelerate sequence optimization and preclinical evaluation. More generally, these molecules hold the potential to add innovation to anti-infective therapy and provide novel compounds as new drug leads for combating multidrug-resistant bacterial infections.

References

1. Ho CS, Wong CTH, Aung TT, Lakshminarayanan R, Mehta JS, Rauz S, McNally A, Kintses B, Peacock SJ, de la Fuente-Nunez C, Hancock REW, Ting DSJ. Antimicrobial resistance: a concise update. *Lancet Microbe*. 2025;**6**:100947.
2. de Kraker ME, Stewardson AJ, Harbarth S. Will 10 million people die a year due to antimicrobial resistance by s2050? *PLoS Med*. 2016;**13**:e1002184.
3. Wright GD. Mechanisms of resistance to antibiotics. *Curr Opin Chem Biol*. 2003;**7**:563–569.
4. Tacconelli E, Carrara E, Savoldi A, Harbarth S, Mendelson M, Monnet DL, Pulcini C, Kahlmeter G, Kluytmans J, Carmeli Y, Ouellette M, Outterson K, Patel J, Cavaleri M, Cox EM, Houchens CR, Grayson ML, Hansen P, Singh N, Theuretzbacher U, Magrini N, WHO Pathogens Priority List Working Group. Discovery, research, and development of new antibiotics: the WHO priority list of antibiotic-resistant bacteria and tuberculosis. *Lancet Infect Dis*. 2018;**18**:318–327.
5. Blair JM, Webber MA, Baylay AJ, Ogbolu DO, Piddock LJ. Molecular mechanisms of antibiotic resistance. *Nat Rev Microbiol*. 2015;**13**:42–51.
6. Darby EM, Trampari E, Siasat P, Gaya MS, Alav I, Webber MA, Blair JMA. Molecular mechanisms of antibiotic resistance revisited. *Nat Rev Microbiol*. 2023;**21**:280–295.

7. Brown ED, Wright GD. Antibacterial drug discovery in the resistance era. *Nature*. 2016 Jan;**21**(529):336–343.
8. del Olmo M, Andreu C. Current status of the application of antimicrobial peptides and their conjugated derivatives. *Molecules*. 2025;**30**:3070.
9. Marciano CL, Félix de Lima JV, Couto Rosa MS, do Nascimento RA, AdO F, ICd S, Chrysostomo-Massaro TN, NGd R-G, Cabral H. A comprehensive overview of antimicrobial peptides: broad-spectrum activity, computational approaches, and applications. *Antibiotics*. 2025;**14**:1115.
10. Wang Y, Song M, Chang W. Antimicrobial peptides and proteins against drug-resistant pathogens. *Cell Surf*. 2024;**12**:100135.
11. Su M, Su Y. Recent advances in amphipathic peptidomimetics as antimicrobial agents to combat drug resistance. *Molecules*. 2024;**29**:2492.
12. Nikaido H. Outer membrane barrier as a mechanism of antimicrobial resistance. *Antimicrob Agents Chemother*. 2003;**47**:1–4.
13. Gan BH, Gaynord J, Rowe SM, Deingruber T, Spring DR. The multifaceted nature of antimicrobial peptides: current synthetic chemistry approaches and future directions. *Chem Soc Rev*. 2021;**50**:7820–7880.
14. Kyriakidis I, Vasileiou E, Pana ZD, Tragiannidis A. *Acinetobacter baumannii* antibiotic resistance mechanisms. *Pathogens*. 2021;**10**:373.
15. Kidd TJ, Mills G, Sá-Pessoa J, Dumigan A, Frank CG, Insua JL, Ingram R, Hobley L, Bengoechea JA. A *Klebsiella pneumoniae* antibiotic resistance mechanism that subdues host defences and promotes virulence. *EMBO Mol Med*. 2017;**9**:430–447.
16. Zhang QY, Yan ZB, Meng YM, Hong XY, Shao G, Ma JJ, Cheng XR, Liu J, Kang J, Fu CY. Antimicrobial peptides: mechanism of action, activity and clinical potential. *Mil Med Res*. 2021 Sep;**9**(8):48.
17. Oliveira Júnior NG, Souza CM, Buccini DF, Cardoso MH, Franco OL. Antimicrobial peptides: structure, functions and translational applications. *Nat Rev Microbiol*. 2025;**23**:687–700.
18. Heidary M, Khosravi AD, Khoshnood S, Nasiri MJ, Soleimani S, Goudarzi M. Daptomycin. *J Antimicrob Chemother*. 2018;**73**:1–11.
19. Ageitos JM, Sánchez-Pérez A, Calo-Mata P, Villa TG. Antimicrobial peptides (AMPs): ancient compounds that represent novel weapons in the fight against bacteria. *Biochem Pharmacol*. 2017;**133**:117–138.
20. Lazzaro BP, Zasloff M, Rolff J. Antimicrobial peptides: application informed by evolution. *Science*. 2020;**368**:eaau5480.
21. Wang G, Li X, Wang Z. APD3: the antimicrobial peptide database as a tool for research and education. *Nucleic Acids Res*. 2016;**44**:D1087–D1093.
22. Koehbach J, Craik DJ. The vast structural diversity of antimicrobial peptides. *Trends Pharmacol Sci*. 2019;**40**:517–528.
23. Friedrich CL, Rozek A, Patrzykat A, Hancock RE. Structure and mechanism of action of an indolicidin peptide derivative with improved activity against gram-positive bacteria. *J Biol Chem*. 2001;**276**:24015–24022.
24. Johnson TS, Bourdine AA, Deber CM. Hydrophobic moment drives penetration of bacterial membranes by transmembrane peptides. *J Biol Chem*. 2023;**299**:105266.

25. Du Y, Li L, Zheng Y, Liu J, Gong J, Qiu Z, Li Y, Qiao J, Huo YX. Incorporation of non-canonical amino acids into antimicrobial peptides: advances, challenges, and perspectives. *Appl Environ Microbiol.* 2022;**88**:e0161722.
26. Zasloff M. Antimicrobial peptides of multicellular organisms. *Nature.* 2002;**415**: 389–395.
27. Hancock REW, Sahl HG. Antimicrobial and host-defense peptides as new anti-infective therapeutic strategies. *Nat Biotechnol.* 2006;**24**:1551–1557.
28. Yeaman MR, Yount NY. Mechanisms of antimicrobial peptide action and resistance. *Pharmacol Rev.* 2003;**55**:27–55.
29. Shai Y. Mode of action of membrane active antimicrobial peptides. *Biopolymers.* 2002;**66**:236–248.
30. Brogden KA. Antimicrobial peptides: pore formers or metabolic inhibitors in bacteria? *Nat Rev Microbiol.* 2005;**3**:238–250.
31. Matsuzaki K. Control of cell selectivity of antimicrobial peptides. *Biochim Biophys Acta.* 1999;**1462**:1–10.
32. Huang HW. Action of antimicrobial peptides: two-state model. *Biochemistry.* 2000;**39**:8347–8352.
33. Mahlapuu M, Håkansson J, Ringstad L, Björn C. Antimicrobial peptides: an emerging category of therapeutic agents. *Front Cell Infect Microbiol.* 2016;**6**:194.
34. McMillan KAM, Coombs MRP. Review: examining the natural role of amphibian antimicrobial peptide magainin. *Molecules.* 2020;**25**:5436.
35. Flamm RK, Rhomberg PR, Simpson KM, Farrell DJ, Sader HS, Jones RN. in vitro spectrum of pexiganan activity when tested against pathogens from diabetic foot infections and with selected resistance mechanisms. *Antimicrob Agents Chemother.* 2015;**59**:1751–1754.
36. Burton MF, Steel PG. The chemistry and biology of LL-37. *Nat Prod Rep.* 2009;**26**:1572–1584.
37. Wesseling CMJ, Martin NI. Synergy by perturbing the Gram-negative outer membrane: opening the door for Gram-positive specific antibiotics. *ACS Infect Dis.* 2022;**8**:1731–1757.
38. Vaara M. Polymyxin derivatives that sensitize Gram-negative bacteria to other antibiotics. *Molecules.* 2019;**24**:249.
39. Fjell CD, Hiss JA, Hancock RE, Schneider G. Designing antimicrobial peptides: form follows function. *Nat Rev Drug Discov.* 2012;**11**:37–51.
40. Lai XZ, Feng Y, Pollard J, Chin JN, Rybak MJ, Bucki R, Epand RF, Epand RM, Savage PB. Ceragenins: cholic acid-based mimics of antimicrobial peptides. *Acc Chem Res.* 2008;**41**:1233–1240.
41. Horne WS, Gellman SH. Foldamers with heterogeneous backbones. *Acc Chem Res.* 2008;**41**:1399–1408.
42. Hamuro Y, Schneider JP, DeGrado WF. De novo design of antibacterial bpeptides. *J Am Chem Soc.* 1999;**121**:12200–12201.
43. Tew GN, Liu D, Chen B, Doerksen RJ, Kaplan J, Carroll PJ, Klein ML, DeGrado WF. De novo design of biomimetic antimicrobial polymers. *Proc Natl Acad Sci USA.* 2002;**99**:5110–5114.

44. Tew GN, Scott RW, Klein ML, DeGrado WF. De novo design of antimicrobial polymers, foldamers, and small molecules: from discovery to practical applications. *Acc Chem Res*. 2010;**43**:30–39.
45. Gillies ER, Deiss F, Staedel C, Schmitter J-M, Huc I. Development and biological assessment of fully water-soluble helical aromatic amide foldamers. *Angew Chem Int Ed Engl*. 2007;**46**:4081–4084.
46. Scott RW, DeGrado WF, Tew GN. De novo designed synthetic mimics of antimicrobial peptides. *Curr Opin Biotechnol*. 2008;**19**:620–627.
47. Schmitt MA, Weisblum B, Gellman SH. Unexpected relationships between structure and function in alpha,beta-peptides: antimicrobial foldamers with heterogeneous backbones. *J Am Chem Soc*. 2004;**126**:6848–6849.
48. Arvidsson PI, Ryder NS, Weiss HM, Gross G, Kretz O, Woessner R, Seebach D. Antibiotic and hemolytic activity of a beta2/beta3 peptide capable of folding into a 12/10-helical secondary structure. *ChemBioChem*. 2003;**4**:1345–1347.
49. Chongsiriwatana NP, Patch JA, Czyzewski AM, Dohm MT, Ivankin A, Gidalevitz D, Zuckermann RN, Barron AE. Peptoids that mimic the structure, function, and mechanism of helical antimicrobial peptides. *Proc Natl Acad Sci USA*. 2008;**105**: 2794–2799.
50. She F, Nimmagadda A, Teng P, Su M, Zuo X, Cai J. Helical 1:1 α/sulfono-γ-AA heterogeneous peptides with antibacterial activity. *Biomacromolecules*. 2016;**17**:1854–1859.
51. Zuckermann RN, Kodadek T. Peptoids as potential therapeutics. *Curr Opin Mol Ther*. 2009;**11**:299–307.
52. Olsen CA. Peptoid-peptide hybrid backbone architectures. *ChemBioChem*. 2010;**11**: 152–160.
53. Patch JA, Barron AE. Helical peptoid mimics of magainin-2 amide. *J Am Chem Soc*. 2003;**125**:12092–12093.
54. Schafmeister CE, Po J, Verdine GL. Stapled α-helical peptides. *J Am Chem Soc*. 2000;**122**:5891–5892.
55. Walensky LD, Bird GH. Hydrocarbon-stapled peptides: principles and applications. *J Med Chem*. 2014;**57**:6275–6288.
56. Li Y, Wu M, Fu Y, Xue J, Yuan F, Qu T, Rissanou AN, Wang Y, Li X, Hu H. Therapeutic stapled peptides: efficacy and molecular targets. *Pharmacol Res*. 2024;**203**:107137.
57. Hirano M, Saito C, Yokoo H, Goto C, Kawano R, Misawa T, Demizu Y. Development of antimicrobial stapled peptides based on magainin 2 sequence. *Molecules*. 2021;**26**(2):444.
58. Zheng M, Wang R, Chen S, Zou Y, Yan L, Zhao L, Li X. Design, synthesis and antifungal activity of stapled aurein1.2 peptides. *Antibiotics*. 2021;**10**:956.
59. Zhang Y, Zheng M, Wang Z, Liu Z, Chen S, Li X, Shi Y, Hu H. Discovery of novel antibacterial agent for the infected wound treatment: all-hydrocarbon stapling optimization of LL-37. *Theranostics*. 2024;**14**:1181–1194.
60. Ruczyński J, Prochera K, Kaźmierczak N, Kosznik-Kwaśnicka K, Piechowicz L, Mucha P, Rekowski P. New conjugates of vancomycin with cell-penetrating peptides – synthesis, antimicrobial activity, cytotoxicity, and BBB permeability studies. *Molecules*. 2024;**29**:5519.
61. Gray DA, Wenzel M. Multitarget approaches against multiresistant superbugs. *ACS Infect Dis*. 2020;**6**:1346–1365.

10

Principles and Case Studies of Peptidomimetic Inhibitors of PPI

10.1 Introduction

As introduced in Section 1.9.1 of Chapter 1, protein–protein interactions (PPIs) are important mediators of biological processes and play a critical role in the progression of numerous disease states [1]. Approximately 650,000 disease-relevant PPIs have been reported within the human interactome [2, 3], most of these interactions still being underexplored. PPIs are typically dynamic and occur over relatively large protein–protein contact surface areas, ranging from approximately 1,500 to 3,000 $Å^2$. This is considerably larger than the surface area required for inhibition by small molecules, which generally accounts for 300–1,000 $Å^2$ [4, 5]. In addition, PPI interfaces are often flat and lack of well-defined binding pockets, contributing to their established classification as "undruggable." Nevertheless, recent advances in peptide discovery have renewed their attention, offering a wide range of structurally diverse sequences. Thus, around 70 peptides have been marketed, and several other clinically developing peptides have been recently reported and classified by structure, therapeutic targets, and design strategies [6]. Despite their large size, not all residues within a PPI interface contribute equally to the binding. In fact, it has been discovered that a limited number of amino acids, generally termed "hot spots," are responsible for the majority of binding affinity and specificity within PPI [7]. These hot spots are typically rich in hydrophobic residues, and as a consequence of protein folding, they are often distributed within the protein surface rather than being adjacent in the primary sequence. The presence of such hot spots enables the rational design of inhibitors capable of disrupting PPIs by targeting these interactive regions [8, 9]. Hot spots are frequently associated with defined protein secondary-structural motifs, such as α-helices, β-sheets, and β-turns. Among these, α-helices have received particular attention, as they account for approximately 50–60% of all secondary structures found at protein–protein interfaces and mediate a significant amount of therapeutically relevant PPIs. In many cases, hot spot residues are found on a single face of the helix; thus, helix mimicry has emerged as a promising strategy for the development of potent PPI inhibitors, and in this context peptidomimetics capable of mimicking one or multiple faces of an α-helix have been successfully reported [10]. Although β-strands and β-sheets play fundamental roles at protein–protein interfaces, too, their exploitation as PPI modulators remains limited. Structural analyses

Table 10.1 Main advantages and disadvantages of small molecules, peptides, and antibodies as PPI inhibitors.

PPI inhibitor class	Advantages	Disadvantages
Small molecules	Ability to penetrate the cell membrane, potential for oral administration	Off-target effects and associated side effects
Peptides	High target specificity, high binding affinity	Short half-life, poor oral bioavailability, physicochemical instability, and poor solubility
Antibodies	High target specificity, high efficiency	Immunogenicity-related side effects, generally limited to extracellular targets due to high MW

of the Protein Data Bank have revealed that β-strands at PPI interfaces can interact in diverse ways, involving side-chain recognition and backbone hydrogen-bonding interactions [11]. In this context, several scaffolds designed to mimic β-structure interactions have been developed, thus highlighting their potential as alternative PPI inhibition strategies [12].

Advances in structural biology, molecular modeling, and screening technologies over the past decade have significantly expanded the drug discovery background for PPIs, transforming them into a promising class of therapeutic targets. Despite the challenging nature of PPIs for conventional small molecule approaches, advanced technologies such as fragment-based drug discovery, computer-aided drug design, virtual screening, DNA-encoded libraries, and allosteric site targeting have enabled the systematic identification, optimization, and development of PPI inhibitors.

PPI inhibitors are broadly classified into three categories: small molecules, antibodies, and peptides or their derivatives, including peptidomimetics [9]. The principal advantages and disadvantages associated with small molecules, peptides, and antibodies as PPI inhibitors are summarized in Table 10.1.

Small molecules often exhibit limited selectivity for biological macromolecules, which can result in off-target effects and toxicity [13]. Antibodies can effectively compete with PPIs, but due to their high molecular weight, their use is generally restricted to extracellular targets, and the risk of immune responses is an issue. Peptidomimetics, conceived as peptide-like molecules, are designed to incorporate amino acid analogues and other chemical entities designed to reproduce the three-dimensional structure of the interacting peptide epitope [14]. By mimicking secondary-structural elements such as α-helices and β-sheets at PPI interfaces, peptidomimetics can disrupt PPIs with high specificity and potency. Their molecular weight lies between that of small molecules and antibodies, thus addressing the necessary target specificity and synthetic accessibility, although their clinical application is often limited by poor serum stability due to proteolytic degradation. Accordingly, several successful applications of this concept have been reported in the context of PPI inhibition, and some comprehensive reviews are recommended for further insight [10, 15, 16].

10.2 Design of Peptidomimetic PPI Inhibitors

The achievement of molecules capable of mimicking peptides in their bioactive conformations remains a central goal in the development of PPI inhibitors. Advances in PPI-focused technologies have enabled the precise arrangement of side-chain functionalities analogous to peptide secondary structures, giving rise to the development of peptidomimetics targeting PPIs [17]. In this context, peptidomimetic design is largely informed by structural biology and computational analysis for the identification of interacting residues in PPIs. Crystal structures of protein complexes, alanine scanning, and *in silico* methods are used to map interaction interfaces and to identify "hot spots" as key binding determinants [18]. Experimental techniques such as surface plasmon resonance (SPR) and isothermal titration calorimetry (ITC) further contribute to identifying minimal binding epitopes to be used as templates for iterative structural optimization and biological evaluation.

Chemical optimization strategies of hit peptides generally aim to stabilize the bioactive conformation, enhance inhibitory potency, and improve overall physicochemical properties. These include backbone and side-chain modifications, amide-bond isosteres, the incorporation of D-amino acids, and conformational constraints. Several classes of peptidomimetics have demonstrated efficacy in modulating PPIs, including stapled peptides, foldamers, and non-peptide small molecules. Low-molecular-weight peptide analogs often serve as starting points, as they preserve a peptide-like character and enable the straightforward modification by introducing unnatural amino acids or alternative functional groups. These compounds can adopt linear, cyclic, or branched conformations, thus enabling fine-tuning over structure and activity. In a parallel approach, non-peptide scaffolds have been developed to replace the peptide backbone while maintaining the spatial orientation of key side chains. Non-peptide inhibitors are generally developed either through hit-to-lead optimization of peptide precursors or through virtual and high-throughput screening strategies.

With reference to protein secondary structures as key peptide epitopes targeting PPIs, given the relevance of α-helices in PPIs, these structures have been mainly studied in peptidomimetic design. The structural feature of α-helical interfaces, characterized by side chains oriented on the same face of the helix at positions i, $i + 4$, and $i + 7$ [19], have inspired the development of rigid scaffolds capable of reproducing helical side-chain projection. Representative examples among peptidomimetic scaffolds of this series span from peptide-like scaffolds, like peptoids and peptide foldamers, to foldamers with reduced peptide character, such as terphenyl derivatives [20], and terephthalamides, which offer increased rigidity, improved aqueous solubility, and reduced synthetic simplicity as compared with terphenyl-based helix mimetics [21].

10.3 Peptidomimetics Targeting Cancer PPI

Studies on PPI networks have been particularly addressed in cancer research, giving a comprehensive view of cellular functions and biological processes and how disruptions in specific interactions contribute to cancer. Investigating the structural interfaces of cancer-related proteins and how they interact is essential for understanding their roles within these networks. Thus, the detailed structural investigation of protein interfaces is highly valuable for addressing fundamental questions about interaction mechanisms and disease-associated dysfunction for developing peptide, peptidomimetic, and small molecule inhibitors and modulators of PPI. Over the last two decades, major cancer-related PPIs have been investigated, and several peptidomimetic and small molecule inhibitors have been advanced to preclinical and clinical studies, accordingly (Table 10.2). An overview of the peptidomimetic approach for targeting RGD integrin as a PPI target has already been reported in the previous book title on peptidomimetics [23], and an example of other recent advances over the last decade is reported in the next paragraph with a case study on the p53-MDM2 PPI.

10.4 Case Study in Cancer: Targeting the p53-MDM2 PPI

The p53 gene (TP53) encodes a key tumor suppressor protein that plays a central role as a transcription factor regulating cell-cycle arrest, DNA repair, senescence, and apoptosis in response to cellular stress such as DNA damage or hypoxia [24]. Because of this essential role, p53 is often referred to as the "guardian of the genome." Wild-type p53 suppresses tumor formation, whereas mutant p53 frequently acquires tumor-promoting functions. Mutations in the TP53 gene are among the most common genetic alterations in human cancers and are strongly associated with tumor initiation, progression, treatment resistance, and poor clinical outcomes [25]. In addition to genetic mutation, functional inactivation of p53 can arise from aberrant PPIs, especially with MDM2 (mouse double minute 2). This (and human double minute 2, hDM2) is the primary negative regulator of p53, binding directly to the *N*-terminal transactivation domain of p53 and suppressing its transcriptional activity to promote its degradation [26]. MDM2 also functions as an E3 ubiquitin ligase, ubiquitinating p53 and targeting it for proteasomal degradation [27]. The p53-MDM2 PPI is strongly associated with oncogenesis, as the overexpression of MDM2 is observed in several cancer types, including osteosarcoma, lung cancer, and brain tumors, and is closely linked to p53 inactivation and tumor development. X-ray crystallography of the p53-MDM2 complex revealed that the p53 activation domain adopts an α-helical conformation upon binding. Three key hydrophobic residues, namely Phe19, Trp23, and Leu26, are situated on one face of this helix and interact with a hydrophobic cleft on MDM2, thus forming the core interaction interface (Figure 10.1) [28]. This well-defined binding motif has made the p53-MDM2 interaction a prototypical target for PPI-focused drug discovery and a standard system for α-helix mimicry [29].

Despite the challenges in developing PPI inhibitors due to flat interaction surfaces or intrinsic disorder, as particularly observed for transcription factors such as p53, significant advance has been made in targeting the p53-MDM2 interface. Accordingly, diverse classes of antagonists, ranging from low-molecular-weight small molecules to peptides and peptidomimetics, have been developed to disrupt this interaction [30], and several inhibitors have advanced into clinical trials [31], although some have been discontinued due to toxicity or limited efficacy. Indeed, disrupting the p53-MDM2 PPI to restore wild-type p53 activity remains one of the most intensively studied and promising strategies for anticancer therapy [16, 22, 32]. The different approaches followed for the development of inhibitors of the p53-MDM2 PPI have been grouped according to the peptidomimetic organization in

Table 10.2 *Major cancer-related PPIs with peptidomimetic inhibitors that have been reported in the literature or advanced into clinical/preclinical development. Data from [22].*

Cancer PPI target	PPI function in cancer	Peptide/peptidomimetic inhibitors	Small molecule inhibitors
p53-MDM2/MDMX	Negative regulation of p53 tumor suppression	ALRN-6924 (Sulanemadlin), stapled peptide in clinical trials	Nutlin-3, MDM2 antagonist; RG-7112, potent small molecule MDM2 inhibitor in clinical trials
Bcl-2 Family (Bcl-2-BAX/Bcl-XL)	Antiapoptotic interactions	Stapled peptides and peptidic peptidomimetics	Venetoclax, BH3 mimetic targeting Bcl-2, approved for hematologic cancers
Integrin–ECM (αvβ3/αvβ5)	Cell adhesion, angiogenesis	Cilengitide, cyclic RGD peptide targeting integrins in clinical trials	Non-peptide RGD peptidomimetics
RAS-GEF	Oncogenic signaling activation	LUNA18 (Paluratide), cyclic peptide pan-RAS inhibitor	Emerging small molecules modulating KRAS/effector interactions
Myc-Max	Transcriptional activation of oncogenic programs	Designed peptides/peptidomimetics	Small organic antagonists from screening libraries
Keap1-Nrf2	Oxidative stress response, cancer survival	Peptide inhibitors with enhanced affinity	Small molecules disrupting Keap1-Nrf2 PPI
PD-1/PD-L1	Immune checkpoint evasion	Peptide interface mimetics	CA-170, small molecule PD-1/PD-L1 modulator in clinical trials
14-3-3 PPIs	Signaling regulation	–	Stabilizers like Fusicoccin A derivatives
CD40–CD40L	Immune modulation in tumor microenvironment	–	Small molecule inhibitors derived from sulfonic acid scaffolds

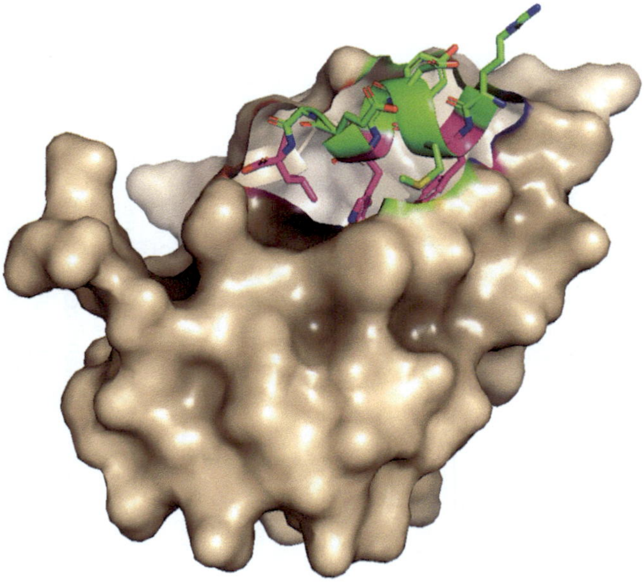

Figure 10.1 X-ray crystallographic structure of human MDM2 in complex with an optimized p53 peptide (PDB: 1T4F); the p53 peptide epitope is highlighted in green, with key hot-spot interacting residues Phe19, Trp23, and Leu26 highlighted in magenta.

Classes A–D [17], from strict structural similarity with the interacting peptide epitope to non-peptide small molecule ligands.

10.4.1 Stapled Peptides

Early efforts to inhibit the p53-MDM2 interaction focused on short peptides derived from the p53 transactivation domain. An important study showed that peptides encompassing the 19-26 sequence of p53 could effectively inhibit MDM2 when the α-helical conformation was stabilized, thus establishing helix mimicry as a key design principle [33]. This insight prompted extensive development of peptide-based p53 peptidomimetics of Class A type, including multiple side-chain cross-linking strategies designed to enhance helicity, binding affinity, and stability with respect to the parent peptide epitope [34]. Beyond α-peptides, salt-bridge-stabilized 14-helical β-peptides presenting hDM2-binding residues on one face of the helix were explored as alternative p53 mimetics [35]. Subsequent computer-aided optimization led to β-peptides with nanomolar binding affinity to both hDM2 and hDMX, which were achieved by replacing a central indole side chain with unnatural aromatic amino acids containing 3-trifluoromethylphenyl or 3,4-dichlorophenyl groups in the side chain [36]. In order to address limited cell permeability, hydrocarbon side-chain bridging was introduced as a helix stabilization approach. Specifically, β-peptides bearing i to $i + 4$ diether or hydrocarbon staples, particularly those bridged between positions 4 and 7 of the parent epitope, showed slightly improved MDM2 binding and significantly enhanced the cellular uptake as compared to linear analogs. Based on these results, α-helical stapled peptides derived directly from p53 became the dominant strategy for targeting the intracellular p53-MDM2 interface, and multiple stapling approaches were investigated, including lactamization, Diels-Alder cyclization, and Ugi multicomponent reactions beyond all-hydrocarbon stapling. Concerning the latter approach, Verdine's group designed 10 stapled variants of the 14–29 sequence of p53 using all-hydrocarbon staples [34], showing increased α-helicity, proteolytic stability, and improved membrane permeability. In particular, the hydrocarbon stapling between Ser20 and Pro27 in an i to $i + 7$ fashion proved optimal, identifying SAH-p53-8 as the lead compound with a K_d of 55 nM for the interaction with MDM2 and exhibiting a dose-dependent cytotoxicity in SJSA-1 osteosarcoma cells with an EC_{50} of 8.8 μM. Also, this molecule proved to cross cell membranes and induce apoptosis in tumor cells, thus acting as a dual MDM2/MDMX inhibitor.

Then, a major advance resulted from phage display, allowing for the identification of a peptide, pDI, that was ~300-fold more potent than the p53 14–29 epitope for disrupting the p53-MDM2/MDMX PPI, as reported by Chen et al. [37]. This peptide was optimized using all-hydrocarbon stapling to yield ATSP-7041, which displayed sevenfold higher α-helicity than pDI and ~50-fold greater inhibition of SJSA-1 cell growth as compared to the SAH-p53-8 stapled peptide [38]. The structure of ATSP-7041 was conceived, retaining the canonical p53 binding triad (Phe19, Trp23, and Leu26) as the hot spot for this PPI and incorporating Tyr22 as a fourth key residue. Subsequent modification of Leu26 to cyclobutylalanine and *C*-terminal extension with Ala-Ala produced a peptide with very high affinity for both MDM2 ($K_i = 0.9$ nM) and MDMX ($K_i = 7$ nM). In preclinical assays, ATSP-7041 demonstrated strong pharmacological activity *in vitro* and in p53-WT mouse xenograft models, although it showed several drawbacks related to solubility and synthetic scale-up due to the nonnatural cyclobutylalanine residue [38]. These limitations required further optimization, overall resulting in the achievement of the clinical candidate ALRN-6924 (sulanemadlin), developed by Aileron Therapeutics [39]. ALRN-6924 (Figure 10.2)

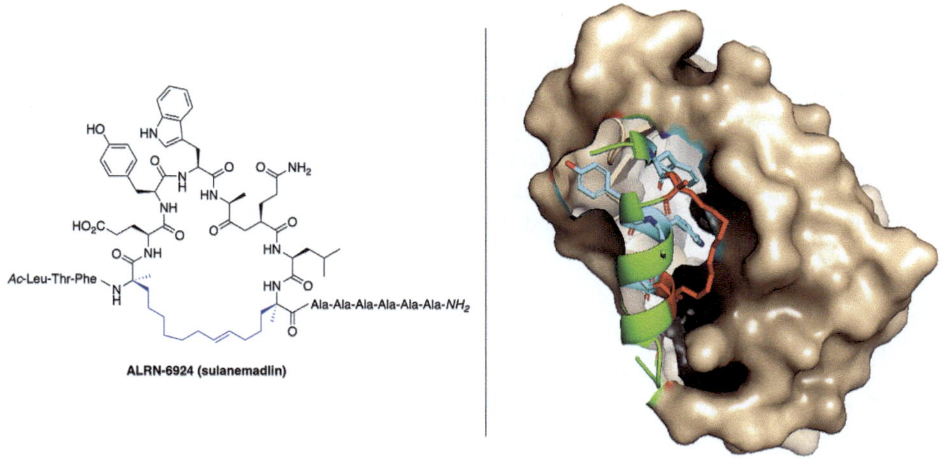

Figure 10.2 *Left: structure of ALRN-6924 (sulanemadlin); right: X-ray co-crystal structure of ALRN-6924 bound to MDMX (PDB: 8GJS). ALRN-6924 is shown as a green ribbon, the hydrocarbon staple in red, and MDMX as a surface rendering. The core hydrophobic triad of Phe19, Trp23, Leu26, and Tyr22 is shown in cyan.*

was designed as a 17-amino-acid stapled peptide that mimics the *N*-terminal α-helix of p53 and is the first dual MDM2/MDMX inhibitor to enter clinical trials [40]. ALRN-6924 binds both MDM2 and MDMX with high affinity, activates the p53 signaling pathway, induces p21-mediated cell-cycle arrest, and shows antitumor activity *in vitro* and *in vivo*. In preclinical models, it also functioned as a radioprotectant and reduced chemotherapy-induced toxicity [41]. Overall, stapled peptide technology, particularly the hydrocarbon staples, in conjunction with phage-display optimization, enabled the generation of potent and cell-permeable p53 mimetics that overcome the limitations of small molecule inhibitors and culminated in the clinical advancement of ALRN-6924.

10.4.2 Foldamers and Structural Mimetics

Foldamers have emerged as powerful Class B mimetics for reproducing α-helical epitopes involved in PPIs. Their modularity, structural tunability, and resistance to proteolysis make them attractive scaffolds for medicinal chemistry applications. In general, beyond individual scaffold optimization, foldamer research has diversified into two complementary design strategies for α-helix mimicry. The first focuses on helicity as a design principle relying on unnatural backbones with predictable folding behavior and similarity to α-peptides. The second adopts a different approach, focusing on the spatial projection of side chains using rigid, extended backbones that are structurally distinct from peptides and are often described as "topographical mimics." Both strategies have yielded promising PPI inhibitors across multiple target classes.

Early foldamer-based approaches to p53-MDM2 inhibition focused on β-peptides engineered to mimic the α-helical transactivation domain of p53. In a seminal study, the key hydrophobic hot spot residues Phe19, Trp23, and Leu26 were displayed on one face of a 14-helical β-peptide, with helix stabilization achieved through an electrostatic macrodipole strategy (Figure 10.3) [35, 42]. These constructs showed binding affinity for MDM2 in the micromolar range, thus validating the concept that nonnatural helical backbones could reproduce the essential recognition elements of the p53 helix. However, subsequent investigations revealed that many β-peptides exhibited limited binding potency, suggesting that the 14-helix geometry does not optimally match the spatial arrangement of residues found in the native p53-MDM2 interface.

To improve the binding profile, several structural refinements were explored, including the incorporation of nonnatural side chains, which highlighted the importance of fine-tuning side-chain orientation and physicochemical properties [36]. In parallel,

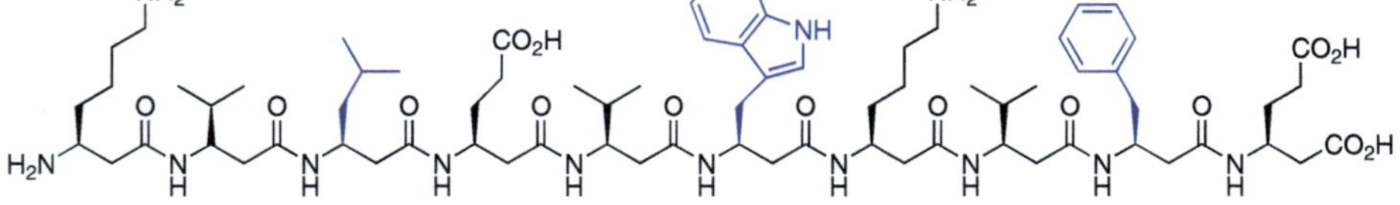

χ-Peptide foldamer χ53-1

Figure 10.3 Structure of β-peptide foldamer β53-1 (the interacting side chains are highlighted in blue).

strategies to overcome poor cellular uptake, a common limitation found in peptide-based foldamers, were developed, including conjugation to cell-penetrating peptides [43] and the introduction of conformational constraints such as side-chain-to-side-chain cross-links and β-homoarginine residues, which enhanced both rigidity and membrane permeability [44].

The toolbox of helical foldamers includes not only β-peptides but also heterogeneous α/β-oligoamides that combine α- and β-amino acids in defined patterns to stabilize helical conformations while preserving the native α-amino acid hot spots required for binding. Mixed ααβαααβ and αααβ sequences, for example have been shown to reproduce the binding surfaces of α-helical inhibitors targeting gp41 and BCL-xL, respectively. Accordingly, α/β-peptide foldamers targeting MDM2 showed improved affinity and conformational preorganization relative to earlier β-peptide designs [45]. In contrast to helix-stabilized systems, achiral peptoids, which are characterized by backbone flexibility and the absence of hydrogen-bonding constraints, displayed only moderate inhibition of the p53-MDM2 interaction, thus emphasizing the importance of structural arrangement in foldamer-based PPI inhibitors [46].

In parallel with foldamer development, some of the most potent p53-MDM2 inhibitors reported to date consisted of Class C structural mimetics based on rigid, non-peptide scaffolds. Specifically, extended aromatic scaffolds such as terphenyls, aromatic oligoamides [47], oligopyridines [48], and oligooxopiperazines [49] have been developed as non-peptide α-helix mimics capable of inhibiting PPIs, such as p53-MDM2 (Figure 10.4). Early work on trisubstituted terphenyls demonstrated that substitution at the ortho position could mimic the i, $i+4$, and $i+7$ arrangement of an α-helix, enabling the selective binding of MDM2 *in vitro*. Several terphenyl derivatives inhibited the p53-MDM2 interaction at sub-micromolar concentrations and showed selectivity over related PPIs such as BCL-xL/Bak and BCL-2/Bak. Interestingly, small structural changes, such as switching

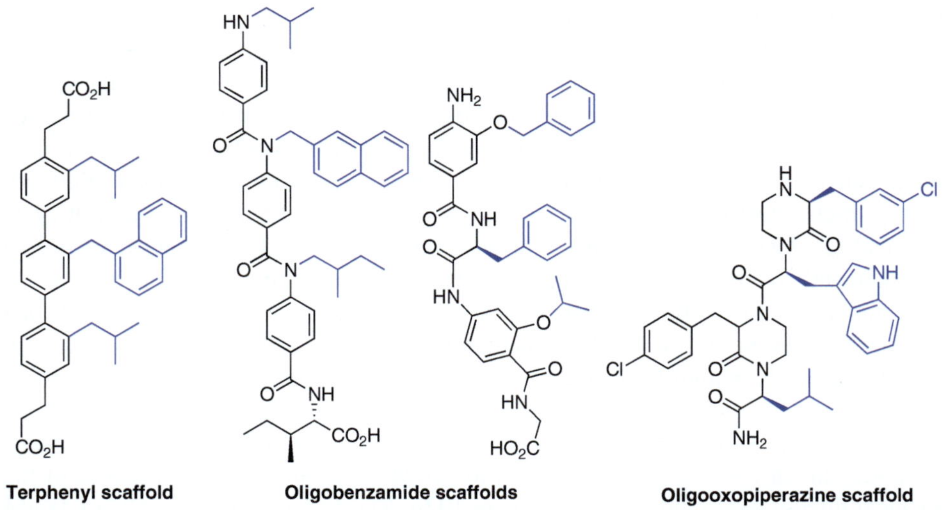

Figure 10.4 *Peptidomimetic foldamers as inhibitors of the p53-hDM2 interaction based on extended and rigid α-helix mimetic scaffolds (the interacting side-chain isosteres are highlighted in blue).*

between 1- and 2-naphthyl substituents, could reverse selectivity profiles, highlighting the sensitivity of topographical mimics to subtle geometric differences. Further optimization led to heteroaromatic scaffolds, including oligopyridylamides and oligobenzamides, which offered improved solubility, modular synthesis, and compatibility with solid-phase methods. Focused libraries of 2-*O*- and 3-*O*-alkylated and *N*-alkylated oligobenzamides yielded low-micromolar p53-MDM2 inhibitors, including membrane-permeable compounds capable of disrupting the interaction in cells. Hybrid oligoamides incorporating proteinogenic α-amino acids allowed for fine control over selectivity between MDM2 and related targets such as MCL-1, with stereochemistry playing a pivotal role. Aliphatic oligooxopiperazines were also explored as alternative α-helix mimetic frameworks.

Additional non-peptide systems, covalently constrained oligobenzamide helix mimetics, pyrrolopyrimidines, phenyl-piperazine-triazines, pyrazines, picolinamides, and pyridazines, have further expanded the toolbox of Class C peptidomimetic inhibitors. Some of these compounds not only disrupted the p53-MDM2 interaction but also interfered with p53-MDMX binding, leading to p53 stabilization and apoptosis in cancer cells. Hydrogen-bond-guided mimetics, while generally less potent, also demonstrated *in vitro* inhibitory activity. Overall, foldamers and structural mimetics have expanded the class of PPI inhibitors by demonstrating that complex α-helical recognition motifs can be effectively reproduced using both peptide-inspired and synthetic scaffolds.

10.4.3 Small Molecule Peptidomimetics

Class D mimetics comprise small molecule and peptidomimetic inhibitors designed to disrupt the p53-MDM2 PPI by directly occupying the p53-binding pocket of MDM2. Unlike foldamers or extended topographical mimics, these compounds are based on compact, rigid scaffolds optimized through classical medicinal chemistry and structure-based design. Relevant examples of Class D inhibitors are the Nutlin family (Figure 10.5, top), discovered by Hoffmann-La Roche by high-throughput screening of synthetic compound libraries using SPR assays [50, 51]. Nutlins are tetrasubstituted imidazoline derivatives that competitively inhibit p53 binding by mimicking the spatial arrangement of the critical α-helical side chains. Crystallographic studies revealed that Nutlins occupy the three hydrophobic sub-pockets of MDM2 corresponding to Phe19, Trp23, and Leu26 of p53, thus effectively blocking the interaction. These compounds exhibit nanomolar inhibitory potency, high selectivity, and robust activation of wild-type p53, leading to cell-cycle arrest and apoptosis in a p53-dependent manner. Importantly, Nutlins demonstrated *in vivo* antitumor activity in human xenograft models, thus establishing proof of concept for small molecule PPI inhibition. Structure-based optimization of the Nutlin scaffold led to clinically relevant derivatives. In particular, the discovery of RG7112 was reported in 2013 (Figure 10.5, bottom left), and it was developed by modifying Nutlin-3 to improve the pharmacokinetic and stability profiles. Key changes included the replacement of the 4-methoxy group with a *tert*-butyl substituent and optimization of the piperazine moiety. The co-crystal structure of RG7112 bound to MDM2 confirmed the maintenance of the canonical three-pocket binding mode (Figure 10.6). RG7112 exhibited an IC_{50} of 18 nM and became the first MDM2 inhibitor to enter clinical trials for advanced solid tumors, where it activated wild-type p53 in tumors with MDM2 amplification or overexpression. Further refinement produced RG7388 (Figure 10.5, bottom right), which incorporates

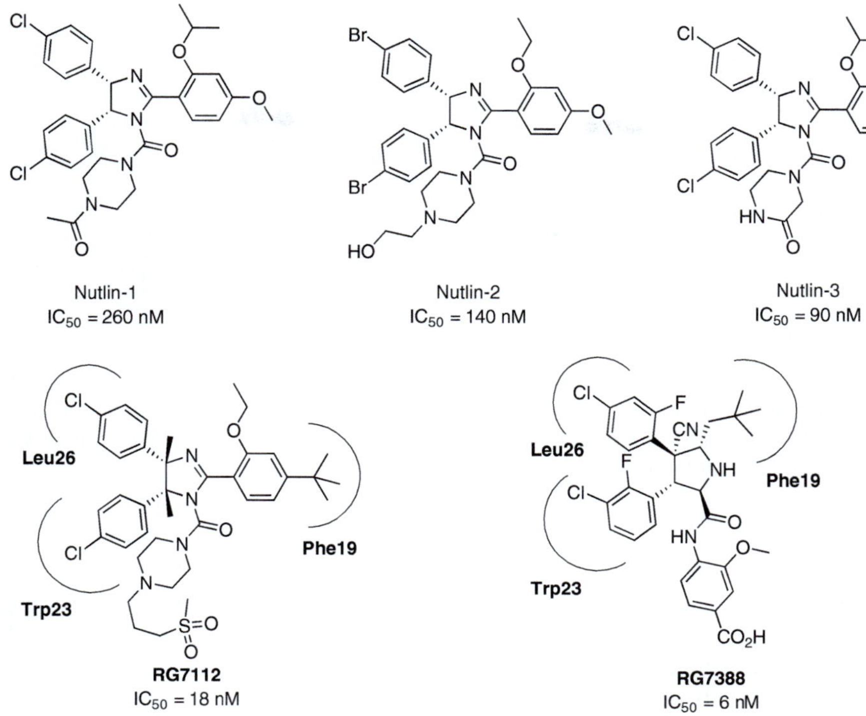

Figure 10.5 *Top: chemical structure of Nutlins; bottom: chemical structures of RG7112 and RG7388 and their indicated binding mode.*

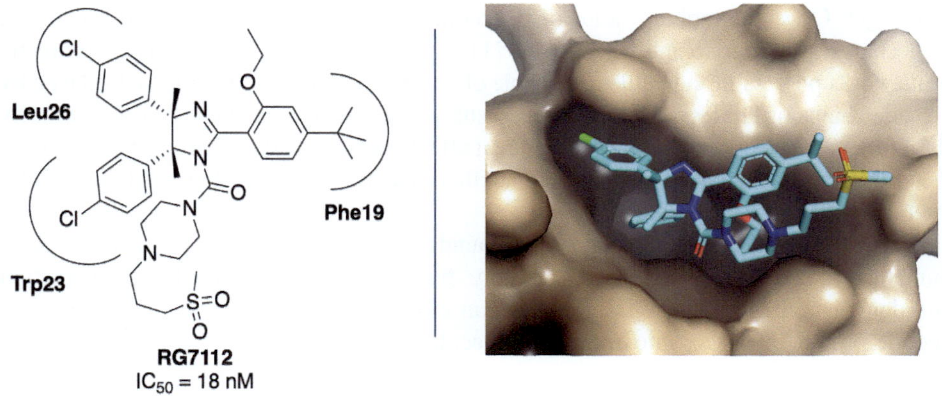

Figure 10.6 *Left: chemical structure of Class D peptidomimetic RG7112; right: X-ray crystallographic structure of RG7112 (cyan) bound in the p53 binding pocket of MDM2 (PDB:4IPF).*

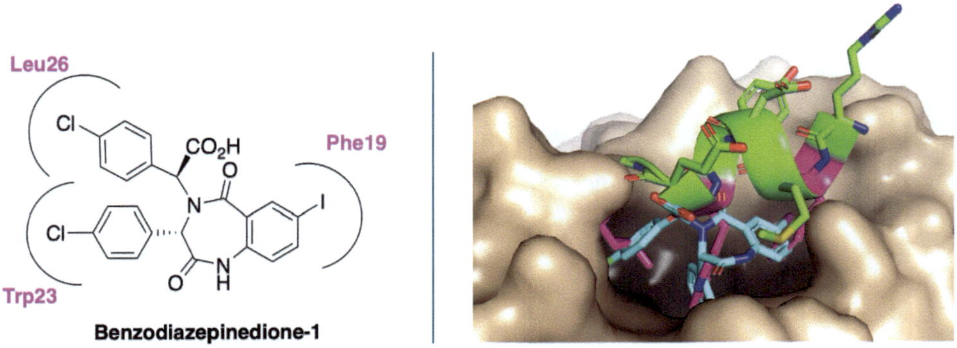

Figure 10.7 Left: structure of Class D peptidomimetic benzodiazepinedione-1; right: X-ray crystallographic structure of benzodiazepinedione-1 (cyan) bound in the p53 binding pocket of MDM2 (PDB 1T4E) and superimposed on the helix p53 epitope shown in green, with key hot spot residues highlighted in magenta (PDB 1T4F).

a more sterically demanding benzene derivative while retaining oxygen-containing substituents. This compound showed enhanced potency with an IC_{50} of 6 nM, as well as improved bioavailability and favorable metabolic stability, thus making it a useful chemical probe for studying the p53-MDM2 PPI.

In parallel with Nutlins, other molecular scaffolds have been explored. In particular, benzodiazepinediones, developed by Johnson & Johnson Pharmaceuticals, represent another class of peptidomimetic inhibitors that bind the p53 pocket of MDM2 and reproduce the spatial orientation of α-helical side chains. These molecules inhibit the p53-MDM2 interaction *in vitro*, further validating the feasibility of small molecule approaches for this target. Crystallographic studies of benzodiazepinediones bound to MDM2 highlight a mechanism similar to nutlins and RG7112 that consists of applying rigid aromatic frameworks possessing substituents capable of reproducing those interactions displayed by the side chains of Phe19, Trp23, and Leu26 sub-pockets with high shape complementarity (Figure 10.7). The optimization of substituent effects, was studied to significantly influence the affinity and selectivity by addressing auxiliary cavities or establishing additional polar interactions.

The success of Nutlins and related compounds has positioned Class D small molecule mimetics as a clinically advanced strategy for targeting the p53-MDM2 interaction and demonstrates that Class D mimetics can effectively translate structural features of peptide-protein interfaces into drug-like molecules.

10.5 Case Study in Infectious Diseases: Targeting the ACE2–SARS-CoV-2 Spike Protein PPI

The pandemic disease COVID-19 [52, 53] caused by severe acute respiratory syndrome coronavirus 2 (SARS-CoV-2) was responsible for over 779 million infections and 7 million deaths worldwide, as of July 2026 [54]. Although the introduction of vaccines allowed to reduce critical cases among people actually infected to 0.1%, immunosuppressed or unvaccinated subjects still represented a vulnerable target, and concern was raised about new genetic variants of SARS-CoV-2 that continued to emerge [55]. Therefore, research toward the implementation of therapeutic approaches was envisaged to complete vaccine development. Overall, for the treatment of COVID-19, a range of established therapies included antivirals, inflammation inhibitors, antirheumatic drugs, plasma, and therapeutic antibodies [56]. Major efforts to block viral replication focused on key infection-related events, specifically aimed at blocking the Spike-ACE2 interaction and inhibiting viral proteins required for replication, such as the Mpro (or 3CLpro) protease and RNA-dependent RNA polymerase (RdRp). As a result, several small molecule inhibitors of Mpro and RdRp proteins were identified, and three direct-acting antivirals were approved, such as Paxlovid, targeting Mpro with PF-07321332 (then named nirmatrelvir) as a protease inhibitor, and molnupiravir and remdesivir as inhibitors of RdRp.

Among the various potential drug targets against SARS-CoV-2 infection [57, 58], the Spike protein was soon identified as an important target for virus neutralization, given its crucial role in the entry of the virus into the host cell [59]. Thus, the PPI occurring between the S1 subunit of the spike and the active site of the ACE2 receptor was envision as a target for identifying an effective treatment [60] and the inhibition of virus binding to the host cell receptor and subsequent internalization as a strategy to block viral entry through chemical compounds or monoclonal antibodies, accordingly. Neutralization mechanisms involve the competition for the ACE2 binding site, steric hindrance on the ACE2 binding site by adjacent sites, or the inhibition of structural rearrangements required for fusion, and numerous small molecules and biologics capable of interacting with the Spike protein have been identified, accordingly. Also, nanobodies capable of neutralizing the virus, variants of soluble ACE2, and rationally designed peptides and miniproteins have also been identified [61]. Cryo-EM and X-ray crystallographic studies carried out soon after the COVID-19 outbreak revealed the structural details of the interaction between Spike RBD (receptor-binding domain) and ACE2 [60], revealing a large and flat interaction surface not favorable to the design of small molecules to inhibit PPIs. The structure also revealed the α-helix 1 moiety of ACE2 interacting with the RBD of the Spike protein and consisting of multiple hot spots (Figure 10.8).

In this context, peptides were conceived to offer an accessible solution to disrupt PPIs by binding to regions containing multiple contact hot spots [62]. Accordingly, Pentelute et al. [63, 64] hypothesized that inhibiting the interaction between ACE2 and Spike *via* peptide-based ligands could prevent the viral entry into human cells, thus providing a novel therapeutic opportunity. Accordingly, the group developed a synthetic and screening workflow to rapidly discover peptides capable of binding to the SARS-CoV-2 Spike RBD. Based on the structure of the Spike RBD-ACE2 complex, molecular dynamics simulations, and automated fast-flow peptide synthesis, they identified a 23-amino acid peptide (SBP1) interacting with the RBD with nanomolar affinity (K_d = 47 nM), as evaluated through bio-layer interferometry. SBP1 was reported as a first-in-class peptide Spike RBD ligand with potential inhibition of viral entry into human cells, although no antiviral activity was reported. The SBP1 completely proteinogenic sequence IEEQAKTFLDKFNHEAEDLFYQS was chosen based on the sequence of α-helix 1 and is not expected to be immunogenic.

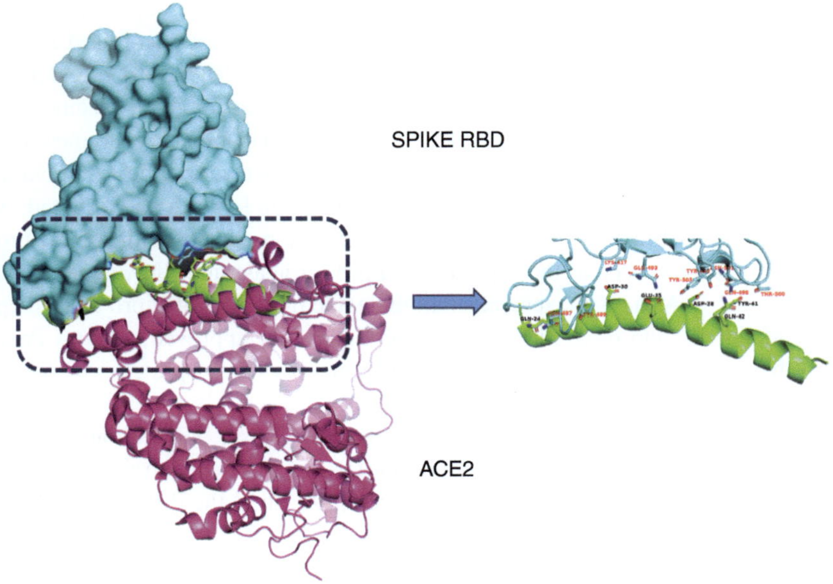

Figure 10.8 *Left: Spike RBD-ACE2 protein–protein interaction between Spike RBD (cyan) and ACE2 (magenta), highlighting the ACE2 α-helix 1 in green (PDB: 6M0J); right: close-up of ACE2 α-helix 1 interacting with Spike RBD. The amino acid residues important for the interaction are shown in red and black for the Spike RBD and α-helix 1, respectively.*

Successively, Sharma et al. [65] reported the rational design, synthesis, and biological evaluation of a series of small peptides capable of inhibiting the Spike RBD-ACE2 interaction, based on the ACE2 sequence, specifically α-helices 1, 3, and 11. Two peptides, SAP1 and SAP6, were identified to inhibit SARS-CoV-2 infection, demonstrating that it is possible to block viral infection by inhibiting the Spike-ACE2 interaction using peptide-based inhibitors. SAP6, containing the conserved EDLFYQ sequence as the epitope targeting the Spike RBD, highlighted the importance of the *N*-terminal α-helix 1 of ACE2 in the interaction with the viral protein.

Another bioactive peptide targeting the Spike RBD-ACE2 PPI was identified by the Karoyan group. The sequence was conceived as a mimetic of the *N*-terminal α-helix 1, containing most of the residues in contact with the Spike RBD and showing a high propensity to form a helical structure in aqueous solution. This peptide proved to bind to the Spike RBD with high affinity, with an IC_{50} in the nM range [66]. Almost all the synthesized peptides showed an α-helical structure in the CD spectra. Specifically, an analysis of the role of amino acid residues in the sequence for stabilizing the α-helix 1 conformation was reported, with the aim to carry out sequence optimization to achieve α-helix 1 analogs with a high propensity to fold and to generate ordered structures. In this context, the introduction of a proline residue revealed a strongly destabilizing effect on the helical conformation, whereas the introduction of homo-tyrosine led only to a slight decrease in the helical content. Subsequently, by combining the properties of hit peptides from the initial screening, two additional peptides were synthesized, incorporating homo-tyrosine. These peptides showed high efficiency in blocking SARS-CoV-2 infection at micromolar concentrations in VeroE6 cells and lung cells, with IC_{50} values in the nM range. The best peptides were then analyzed for their binding to Spike RBD using bio-layer interferometry, showing that the most effective peptides interacted with an estimated K_d of less than 1 nM and demonstrating an absence of toxicity to cells.

Other examples of ACE2-based peptides to inhibit Spike-ACE2 interaction included miniproteins, as reported by Baker's lab [67], as compounds possessing advantages over monoclonal antibodies for ease of administration due to higher stability and reduced size, being potentially more effective, and facilitating the synthetic scale-up in terms of application and costs, although not mitigating the risk of potential immunogenicity, which in this case was not observed.

10.5.1 Stapled Peptides

In the context of addressing the development of small molecule therapeutics as a suitable chemotherapeutic approach for the treatment of COVID-19 infections, the design of more stable and convenient therapeutics consisting of stapled peptides was reported [68]. Specifically, starting from the 23-residue SBP1 peptide developed from the α-helix 1 sequence, which showed high affinity but no reported antiviral activity data [63], Debnath's group hypothesized that SBP1 was not stable against proteolysis and, more importantly, that it did not have the correct secondary structure required for binding to the RBD of SARS-CoV-2. Based on α-helix 1 and its hot spots, the group designed stapled peptides as improved molecules capable of preventing SARS-CoV-2 from binding to ACE2. As discussed in Chapter 6, the stapling of peptide sequences increases the helicity of small peptides in solution, improves resistance to proteolytic digestion, increases potency, and often improves cell penetration [69, 70]. In fact, α-helix 1 is expected to lose its helical conformation in solution, thus losing the orientation of all or most of the amino acid residues involved in the interaction. Taking this into account, Debnath and coworkers used a well-established stapling technique to force the peptide into a helical conformation. Three of the four synthesized stapled peptides (namely compounds **1**, **2**, and **4**, Figure 10.9) showed strong inhibitory activity against SARS-CoV-2 in pseudovirus and viral infectivity assays *in vitro*. Furthermore, peptides with antiviral activity against SARS-CoV-2 exhibited a high proportion of helicity (60–94%), demonstrating these molecules to possess therapeutic potential for COVID-19 by targeting the Spike protein of SARS-CoV-2, preventing the virus from entering the cell and causing the infection.

Other examples of stapled peptides were reported from Jamieson's group [71], who synthesized a series of peptidomimetics based on α-helix 1, too. Similar to Debnath's group, they produced a series of peptides constrained into a helical conformation and tested their efficacy in inhibiting the Spike RBD-ACE2 PPI and in preventing viral infection. Accordingly, they introduced a chemical staple at each available position on the side of the α-helix not involved in the interaction. Specifically, the staples were incorporated at positions i to $i+4$ or i to $i+7$, thus forcing the peptide to make one or two helical turns, respectively. Although these stapled peptides established the helical structure in solution, as confirmed by circular dichroism, none of them showed activity in neutralization assays. Furthermore, the binding to the Spike RBD was not demonstrated by either immunofluorescence or *in vitro* fluorescence polarization assays. Nevertheless, pseudovirus assays in

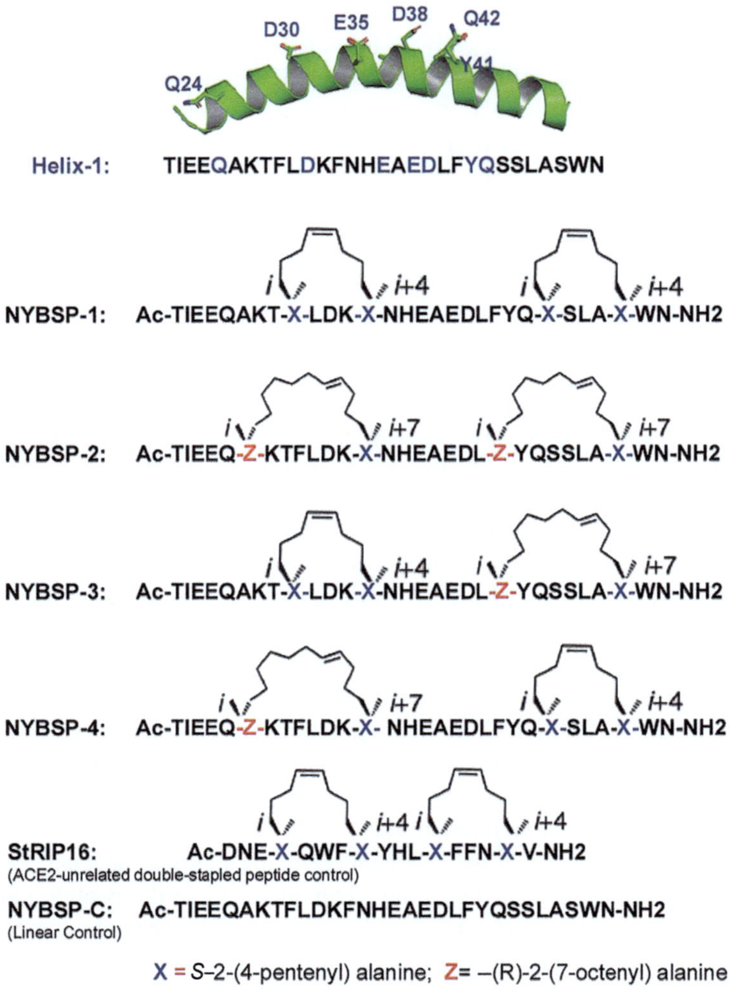

Figure 10.9 Double-stapled peptides as anti-SARS-CoV-2 inhibitors. The linear peptide, NYBSP-C, and an ACE2-unrelated double-stapled peptide were used as controls. Source: Reproduced under CC BY 4.0 License from Ref. [68].

several reports have shown that many double-stapled peptides inhibit viral entry into cells, suggesting that double-stapled peptides are a useful approach to inducing helicity.

Our contribution in this field consisted of the design and synthesis of a series of peptide derivatives based on the ACE2 α-helix 1 epitope to disrupt the SARS-CoV-2 Spike RBD-ACE2 interaction [72]. Specifically, i to $i+4$ stapling strategies were applied to stabilize the secondary helical structure, leading to the identification of a 9-mer RCM-stapled peptide capable of competing with recombinant ACE2 for Spike RBD binding in the micromolar range (Figure 10.10). Competitive ELISA assays and circular dichroism analyses demonstrated that ring-closing metathesis more effectively stabilized α-helicity than triazole stapling achieved *via* click chemistry, assisted by the incorporation

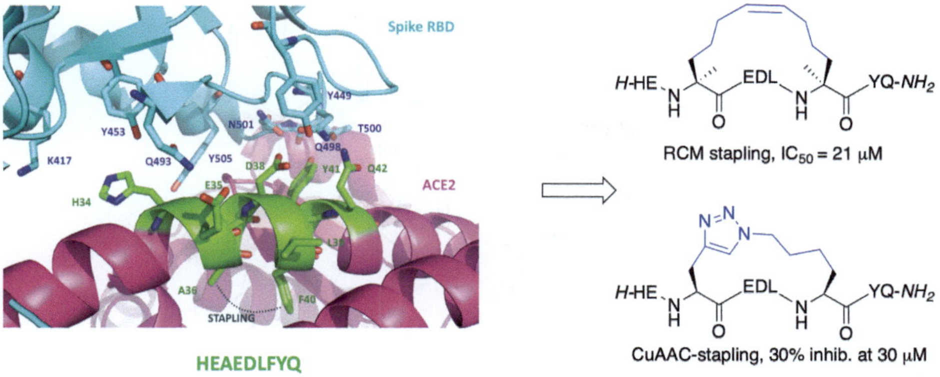

Figure 10.10 *Left: magnification of RBD (cyan)-ACE2 (purple) interaction, highlighting the selected ACE2 α-helix 1 epitope (green, PDB: 6M0J), the key interacting amino acids, and the selected amino acids for the* i *to* i *+4 stapling region; right: RCM- and CuAAC-based stapled peptides developed from the HEAEDLFYQ peptide epitope. Source: [72] / Elsevier / CC BY 4.0.*

of quaternary amino acids. Molecular dynamics simulations indicated that the stapled peptide exhibited increased structural rigidity, enhanced affinity for the Spike RBD, and steric interference with ACE2 binding relative to the parent epitope. LC–MS/MS plasma stability assays further showed improved stability at 48 hours as compared with the native sequence. Overall, these findings supported the development of helix-based peptidomimetics targeting the Spike RBD-ACE2 PPI.

The application of peptidomimetic concepts to targeting ACE2-spike PPI in SARS-CoV-2 drug discovery demonstrates the soundness of this approach also in the field of viral diseases, and peptidomimetics targeting the Spike RBD-ACE2 PPI are still an active area of research. Recent advances include computationally designed peptides with measurable Spike RBD binding, early peptidomimetics with limited inhibitory activity, and newer optimized peptides that showed efficacy against multiple viral variants in biological assays. Despite these developments, converting the binding affinity into strong, broad antiviral activity across variants remains a significant challenge.

10.5.2 Macrocyclic Peptides

Other approaches encompassing the application of macrocyclic peptides targeting the viral Spike protein through its RBD were reported by Zanardi et al. [73]. Based on prior *in silico* studies, polymyxin B (Figure 10.11) was identified as a promising scaffold capable of binding two distinct hot spot regions (BS1 and BS2) at the Spike RBD-ACE2 interface. Accordingly, these regions were defined using X-ray structures and computational alanine scanning [74]. Using the polymyxin B core, a series of novel peptidomimetics, including monomers and heterodimers, were designed to simultaneously target both binding sites. SPR assays showed that selected monomers and heterodimers interacted with the Spike RBD with micromolar affinity, with heterodimers displaying stronger binding than monomers. While these molecules did not fully prevent infection by live SARS-CoV-2, a heterodimer produced a modest but measurable inhibition of viral entry in

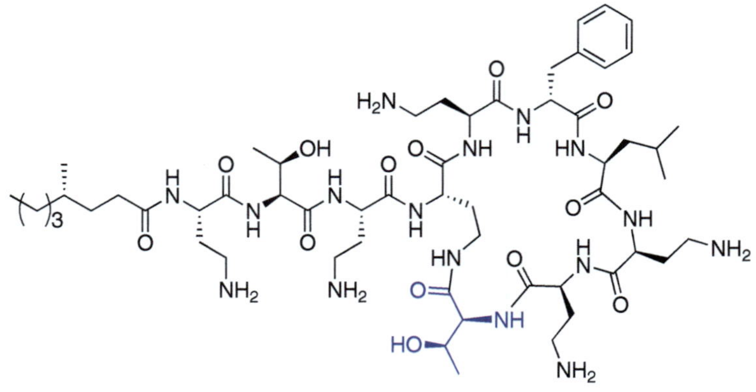

Figure 10.11 *Structure of macrocyclic peptide polymyxin B, highlighting the Thr residue used for replacement with 3-azido-alanine for CuAAC bioconjugation with BS1 monomer.*

ACE2-expressing cell lines, thus validating previous modeling predictions. These findings provided the first proof-of-feasibility for medium-sized heterodimeric macrocyclic peptide constructs targeting the Spike RBD.

10.5.3 Foldamers

Foldamer structures were also reported as a therapeutic strategy against SARS-CoV-2, focusing on inhibiting the interaction between the viral Spike protein and the human ACE2 receptor. Specifically, helical peptide foldamers were developed as peptidomimetics of the ACE2 α1 helix, using *trans*-2-aminocyclopentanecarboxylic acid (*trans*-ACPC) residues to induce and stabilize helical conformations [75]. Chimeric α,β-peptides incorporating cyclic β-amino acids such as *trans*-ACPC are well known for their predictable folding and enhanced structural rigidity [76]. These features make them attractive scaffolds for developing PPI inhibitors. Guided by these principles, a computer-aided design approach was used to generate and optimize helical foldamers capable of binding the SARS-CoV-2 Spike RBD and inhibiting the interaction with ACE2. Circular dichroism and NMR confirmed that the resulting peptides adopted stable helical conformations in solution. Specifically, a sequential three-step approach was followed, aimed at reaching hit peptide foldamer ligands:

- Design of helical peptide fragments from fragments of spike-binding proteins from ACE2, LCB1 and LCB3.
- Incorporation of constrained β-amino acid (*trans*-ACPC) using sequences ααβαααβ and αααααβ.
- Optimization of the hybrid peptide foldamer through sequence optimization based on an iterative, computer-aided, and experimentally validated iterative approach.

Initial designs focused on "foldamerizing" the ACE2 α1 helix by inserting *trans*-ACPC residues in defined sequence patterns (ααβαααβ and αααααβ). Although these peptides were conformationally stable, they showed limited improvement in the RBD binding. Consequently, a revised strategy was adopted based on "foldamerized" variants of the

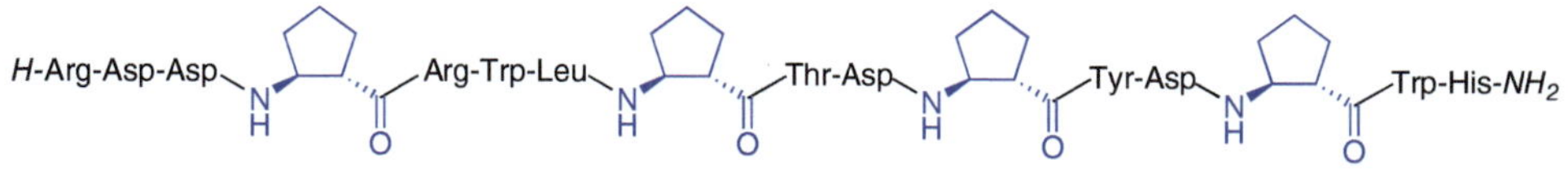

Figure 10.12 Structure of lead hybrid peptide foldamer showing best inhibition toward Spike RBD-ACE2 PPI.

computationally designed spike binders LCB1 and LCB3, which are known to bind the Spike RBD protein with high affinity [77]. Four iterative rounds of optimization were performed starting from a "foldamerized" LCB3 fragment. Early modifications identified a molecule in which two tryptophan residues significantly enhanced the binding by anchoring the peptide into hydrophobic clefts on the Spike RBD. Subsequent substitutions revealed that replacing a tryptophan with an arginine enabled additional hydrogen bonding to key RBD residues, yielding a molecule with improved affinity. Further targeted mutations at selected positions led to another potent compound, with a K_d of 1.1 μM. Overall, this stepwise optimization culminated in an 18-amino-acid helical foldamer with high affinity for the SARS-CoV-2 Spike RBD protein ($K_d = 650$ nM) and effective inhibition of the Spike-ACE2 interaction with an IC_{50} of 1.3 μM (Figure 10.12). Computational energy calculations and molecular dynamics simulations corroborated the experimental results, thus demonstrating increasingly stronger interactions and enhanced complex stability. Overall, these findings highlighted helical peptide foldamers as robust, tunable scaffolds for targeting viral PPIs.

10.5.4 Small Molecule Peptidomimetics

Concerning the identification of small molecules targeting the Spike-ACE2 interaction, by using advanced screening technologies, including virtual screening approaches, a range of potential Spike protein-ACE2 PPI inhibitors have been identified among natural products, including flavonoids, terpenoids, phenols, and alkaloids, guiding efficient discovery from natural sources and supporting the development of broad-spectrum anti-coronavirus therapeutics [78, 79]. In this context, effective direct-acting antivirals remain necessary to complement preventive strategies. In a report by our group, constrained peptidomimetics were designed to inhibit viral entry by mimicking a short epitope of the ACE2 α1 helix [80]. Based on prior literature identifying the ACE2 sequence EDLFYQ as a minimal spike-binding epitope [65], this region was selected for further optimization. To improve the conformational rigidity and drug-like properties, the Leu-Phe segment, not directly interacting with the Spike RBD, was replaced with constrained dipeptide isosteres. Thus, a small library of Fmoc-protected piperazinone and diazepanone dipeptide isosteres was synthesized efficiently using iminium chemistry and incorporated into the ACE2-derived hexapeptide. The resulting peptidomimetics inhibited the Spike RBD-ACE2 PPI in the micromolar range, demonstrating that conformational constraint enhanced PPI inhibition as compared with the parent peptide (Figure 10.13). In addition, the presence of a *C*-terminal glutamine, shared by both the ACE2 epitope and the SARS-CoV-2 main protease (Mpro or 3CLpro) recognition motif, enabled evaluation of dual antiviral activity. Remarkably, in a fluorimetric enzyme inhibition kinetic assay, two constrained peptidomimetics showed significant inhibition of Mpro activity, whereas the native peptide was inactive, thus

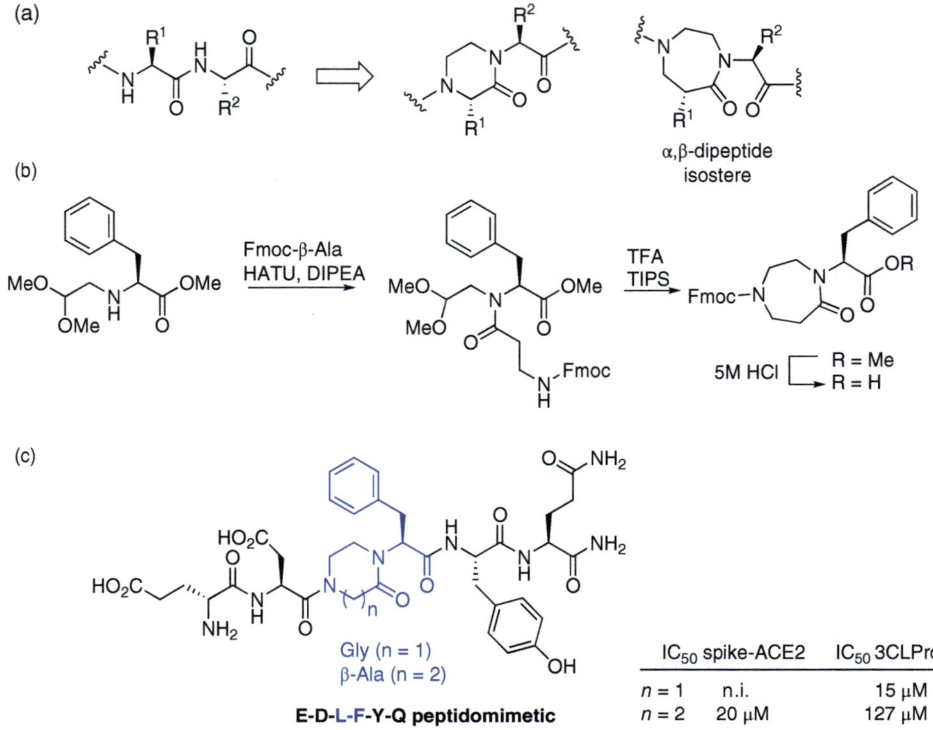

	IC_{50} spike-ACE2	IC_{50} 3CLPro
n = 1	n.i.	15 μM
n = 2	20 μM	127 μM

Figure 10.13 Small molecule peptidomimetic targeting spike-ACE2 PPI. (a) Structure of piperazinone and diazepinone as dipeptide isosteres; (b) synthesis of diazepinone-derived dipeptide isostere; (c) structure of constrained EDLFYQ peptidomimetic.

highlighting constrained dipeptide isosteres as key structural elements for improving peptide-based PPI inhibitors and supporting their potential as multitarget antivirals against coronavirus infections.

10.6 Conclusion

PPIs represent an important and challenging target class in contemporary drug discovery due to their central role in regulating cellular networks and in diseases such as cancer and viral infections. However, the large and often flat nature of PPI interfaces has historically limited the inhibition potency of traditional small molecule inhibitors. Although significant progress has modified the opinion of PPIs as undruggable targets over the past two decades, it is now clear that no single therapeutic approach can universally address their complexity.

In this context, peptidomimetics occupy a valuable position between small molecules and biologics. By mimicking key secondary-structural elements and key hot spot residues, peptidomimetics can achieve high specificity and affinity while retaining design flexibility typical of small molecules and better than biologics. The successful targeting of α-helical

PPIs, such as Bcl-2 family interactions, the p53-MDM2 interface in cancer and the Spike RBD-ACE2 PPI in the context of infectious diseases, has validated this strategy and established key design principles. Nevertheless, despite these advances, the clinical translation of peptidomimetic PPI inhibitors remains limited due to the presence of persistent challenges-for those molecules displaying a relevant peptide character, particularly poor metabolic stability and limited oral bioavailability. While several chemical strategies in the context of peptidomimetic chemistry, such as cyclization, backbone modification, use of nonnatural amino acids and of conformational constraint have improved the stability and potency, such modifications often introduce increased synthetic complexity. As a result, peptidomimetic optimization remains highly target-dependent and typically requires iterative refinement.

Concerning secondary-structure mimicking, peptidomimetic design has typically focused on α-helical interfaces due to their prevalence at PPIs. In contrast, β-strand- and β-sheet-mediated interactions, also displaying biological significance, remain underexplored due to the greater complexity of mimicking their structural asset with new chemical entities. Thus, expanding the peptidomimetic toolbox of modular scaffolds capable of mimicking diverse β-mediated interactions represents an important opportunity for future innovation. Nevertheless, advances in structural biology and computational modeling are facilitating new design strategies in this field. Specifically, high-resolution structural data and molecular dynamics calculations, supported by machine learning-assisted design, enable the identification of hot spots and the optimization of structural requirements for addressing binding interfaces. Complementary biophysical techniques further support the transformation of peptide leads into drug-like mimetics.

Overall, peptidomimetics constitute a key strategy for PPI modulation. Although challenges concerning hit-to-lead studies and formulation still persist, the integration of chemical innovation with computational and biological data is advancing the knowledge and the peptidomimetic toolbox in this field, thus enabling the ability to achieve successful results toward PPI modulators as candidate drugs in diverse biomedical areas.

References

1. Zhang G, Andersen J, Gerona-Navarro G. Peptidomimetics targeting protein–protein interactions for therapeutic development. *Protein Pept Lett*. 2018;**25**:1076–1089.
2. Bonetta L. Protein–protein interactions: interactome under construction. *Nature*. 2010;**468**:851–854.
3. Gonzalez MW, Kann MG. Chapter 4: Protein interactions and disease. *PLoS Comput Biol*. 2012;**8**:e1002819.
4. Jones S, Thornton JM. Principles of protein–protein interactions. *Proc Natl Acad Sci USA*. 1996;**93**:13–20.
5. Wells JA, McClendon CL. Reaching for high-hanging fruit in drug discovery at protein–protein interfaces. *Nature*. 2007;**450**:1001–1009.
6. Cabri W, Cantelmi P, Corbisiero D, Fantoni T, Ferrazzano L, Martelli G, Mattellone A, Tolomelli A. Therapeutic peptides targeting PPI in clinical development: overview, mechanism of action and perspectives. *Front Mol Biosci*. 2021;**8**:697586.

7. Clackson T, Wells JA. A hot-spot of binding energy in a hormone–receptor interface. *Science*. 1995;**267**:383–386.
8. Geppert T, Hoy B, Wessler S, Schneider G. Context-based identification of protein–protein interfaces and "hot-spot" residues. *Chem Biol*. 2011;**18**:344–353.
9. Lu H, Zhou Q, He J, Jiang Z, Peng C, Tong R, Shi J. Recent advances in the development of protein-protein interactions modulators: mechanisms and clinical trials. *Signal Transduct Target Ther*. 2020;**5**:213.
10. Robertson NS, Spring DR. Using peptidomimetics and constrained peptides as valuable tools for inhibiting protein–protein interactions. *Molecules*. 2018;**23**:959.
11. Watkins AM, Arora PS. Anatomy of β-strands at protein–protein interfaces. *ACS Chem Biol*. 2014;**9**:1747–1754.
12. Robinson JA. Beta-hairpin peptidomimetics: design, structures and biological activities. *Acc Chem Res*. 2008;**41**:1278–1288.
13. Lin A, Giuliano CJ, Palladino A, John KM, Abramowicz C, Yuan ML, Sausville EL, Lukow DA, Liu L, Chait AR, Galluzzo ZC, Tucker C, Sheltzer JM. Off-target toxicity is a common mechanism of action of cancer drugs undergoing clinical trials. *Sci Transl Med*. 2019;**11**:eaaw8412.
14. Lenci E, Trabocchi A. Peptidomimetic toolbox for drug discovery. *Chem Soc Rev*. 2020;**49**:3262–3277.
15. Lloyd Mabonga L, Kappo AP. Peptidomimetics: a synthetic tool for inhibiting protein–protein interactions in cancer. *Int J Pept Res Ther*. 2020;**26**:225–241.
16. Wang J, Zheng P, Yu J, Yang X, Zhang J. Rational design of small-sized peptidomimetic inhibitors disrupting protein–protein interaction. *RSC Med Chem*. 2024; **15**:2212–2225.
17. Pelay-Gimeno M, Glas A, Koch O, Grossmann TN. Structure-based design of inhibitors of protein–protein interactions: mimicking peptide binding epitopes. *Angew Chem Int Ed Engl*. 2015;**54**:8896–8927.
18. Ibarra AA, Bartlett GJ, Hegedüs Z, Dutt S, Hobor F, Horner KA, Hetherington K, Spence K, Nelson A, Edwards TA, Woolfson DN, Sessions RB, Wilson AJ. Predicting and experimentally validating hot-spot residues at protein–protein interfaces. *ACS Chem Biol*. 2019;**14**:2252–2263.
19. Khoo KK, Wilson MJ, Smith BJ, Zhang MM, Gulyas J, Yoshikami D, Rivier JE, Bulaj G, Norton RS. Lactam-stabilized helical analogues of the analgesic μ-conotoxin KIIIA. *J Med Chem*. 2011;**54**:7558–7566.
20. Yin H, Lee GI, Sedey KA, Kutzki O, Park HS, Orner BP, Ernst JT, Wang HG, Sebti SM, Hamilton AD. Terphenyl-based Bak BH3 alpha-helical proteomimetics as low-molecular-weight antagonists of Bcl-xL. *J Am Chem Soc*. 2005;**127**: 10191–10196.
21. Yin H, Hamilton AD. Terephthalamide derivatives as mimetics of the helical region of Bak peptide target Bcl-xL protein. *Bioorg Med Chem Lett*. 2004;**14**:1375–1379.
22. Xie X, Yu T, Li X, Zhang N, Foster LJ, Peng C, Huang W, He G. Recent advances in targeting the "undruggable" proteins: from drug discovery to clinical trials. *Signal Transduct Target Ther*. 2023;**8**:335.
23. Trabocchi A, Guarna A. Case study 2: peptidomimetic ligands for αvβ3 integrin. In: *Peptidomimetics in Organic and Medicinal Chemistry*. Chichester: John Wiley & Sons, Ltd, 2014.

24. Hernández Borrero LJ, El-Deiry WS. Tumor suppressor p53: biology, signaling pathways, and therapeutic targeting. *Biochim Biophys Acta Rev Cancer*. 2021;**1876**: 188556.
25. Chène P. Inhibiting the p53-MDM2 interaction: an important target for cancer therapy. *Nat Rev Cancer*. 2003;**3**:102–109.
26. Burgess A, Chia KM, Haupt S, Thomas D, Haupt Y, Lim E. Clinical overview of MDM2/X-targeted therapies. *Front Oncol*. 2016;**6**:7.
27. Moll UM, Petrenko O. The MDM2-p53 interaction. *Mol Cancer Res*. 2003;**1**: 1001–1008.
28. Kussie PH, Gorina S, Marechal V, Elenbaas B, Moreau J, Levine AJ, Pavletich NP. Structure of the MDM2 oncoprotein bound to the p53 tumor suppressor transactivation domain. *Science*. 1996;**274**:948–953.
29. Pazgier M, Liu M, Zou G, Yuan W, Li C, Li C, Li J, Monbo J, Zella D, Tarasov SG, Lu W. Structural basis for high-affinity peptide inhibition of p53 interactions with MDM2 and MDMX. *Proc Natl Acad Sci USA*. 2009;**106**:4665–4670.
30. Huang J. Current developments of targeting the p53 signaling pathway for cancer treatment. *Pharmacol Ther*. 2021;**220**:107720.
31. Zhu H, Gao H, Ji Y, Zhou Q, Du Z, Tian L, Jiang Y, Yao K, Zhou Z. Targeting p53-MDM2 interaction by small-molecule inhibitors: learning from MDM2 inhibitors in clinical trials. *J Hematol Oncol*. 2022;**15**:91.
32. Wu D, Li Y, Zheng L, Xiao H, Ouyang L, Wang G, Sun Q. Small molecules targeting protein–protein interactions for cancer therapy. *Acta Pharm Sin B*. 2023;**13**: 4060–4088.
33. García-Echeverría C, Chène P, Blommers MJ, Furet P. Discovery of potent antagonists of the interaction between human double minute 2 and tumor suppressor p53. *J Med Chem*. 2000;**43**:3205–3208.
34. Bernal F, Tyler AF, Korsmeyer SJ, Walensky LD, Verdine GL. Reactivation of the p53 tumor suppressor pathway by a stapled p53 peptide. *J Am Chem Soc*. 2007;**129**: 2456–2457.
35. Kritzer JA, Hodsdon ME, Schepartz A. Solution structure of a beta-peptide ligand for hDM2. *J Am Chem Soc*. 2005;**127**:4118–4119.
36. Michel J, Harker EA, Tirado-Rives J, Jorgensen WL, Schepartz A. In Silico Improvement of beta3-peptide inhibitors of p53 × hDM2 and p53 × hDMX. *J Am Chem Soc*. 2009;**131**:6356–6357.
37. Phan J, Li Z, Kasprzak A, Li B, Sebti S, Guida W, Schönbrunn E, Chen J. Structure-based design of high affinity peptides inhibiting the interaction of p53 with MDM2 and MDMX. *J Biol Chem*. 2010;**285**:2174–2183.
38. Chang YS, Graves B, Guerlavais V, Tovar C, Packman K, To KH, Olson KA, Kesavan K, Gangurde P, Mukherjee A, Baker T, Darlak K, Elkin C, Filipovic Z, Qureshi FZ, Cai H, Berry P, Feyfant E, Shi XE, Horstick J, Annis DA, Manning AM, Fotouhi N, Nash H, Vassilev LT, Sawyer TK. Stapled α-helical peptide drug development: a potent dual inhibitor of MDM2 and MDMX for p53-dependent cancer therapy. *Proc Natl Acad Sci USA*. 2013;**110**:E3445–E3454.
39. Guerlavais V, Sawyer TK, Carvajal L, Chang YS, Graves B, Ren JG, Sutton D, Olson KA, Packman K, Darlak K, Elkin C, Feyfant E, Kesavan K, Gangurde P, Vassilev LT, Nash HM, Vukovic V, Aivado M, Annis DA. Discovery of sulanemadlin

(ALRN-6924), the first cell-permeating, stabilized α-helical peptide in clinical development. *J Med Chem.* 2023;**66**:9401–9417.
40. Zhou X, Singh M, Sanz Santos G, Guerlavais V, Carvajal LA, Aivado M, Zhan Y, Oliveira MMS, Westerberg LS, Annis DA, Johnsen JI, Selivanova G. Pharmacologic activation of p53 triggers viral mimicry response thereby abolishing tumor immune evasion and promoting antitumor immunity. *Cancer Discov.* 2021;**11**:3090–3105.
41. Carvajal LA, Neriah DB, Senecal A, Benard L, Thiruthuvanathan V, Yatsenko T, Narayanagari SR, Wheat JC, Todorova TI, Mitchell K, Kenworthy C, Guerlavais V, Annis DA, Bartholdy B, Will B, Anampa JD, Mantzaris I, Aivado M, Singer RH, Coleman RA, Verma A, Steidl U. Dual inhibition of MDMX and MDM2 as a therapeutic strategy in leukemia. *Sci Transl Med.* 2018;**10**:eaao3003.
42. Kritzer JA, Stephens OM, Guarracino DA, Reznik SK, Schepartz A. β-Peptides as inhibitors of protein–protein interactions. *Bioorg Med Chem.* 2005;**13**:11–16.
43. Hintersteiner M, Kimmerlin T, Garavel G, Schindler T, Bauer R, Meisner NC, Seifert JM, Uhl V, Auer M. A highly potent and cellularly active β-peptidic inhibitor of the p53/hDM2 interaction. *ChemBioChem.* 2009;**10**:994–998.
44. Bautista AD, Appelbaum JS, Craig CJ, Michel J, Schepartz A. Bridged β3-peptide inhibitors of p53-hDM2 complexation–correlation between affinity and cell permeability. *J Am Chem Soc.* 2010;**132**:2904–2906.
45. Patgiri A, Joy ST, Arora PS. Nucleation effects in peptide foldamers. *J Am Chem Soc.* 2012;**134**:11495–11502.
46. Hara T, Durell SR, Myers MC, Appella DH. Probing the structural requirements of peptoids that inhibit HDM2-p53 interactions. *J Am Chem Soc.* 2006;**128**:1995–2004.
47. Barnard A, Long K, Martin HL, Miles JA, Edwards TA, Tomlinson DC, Macdonald A, Wilson AJ. Selective and potent proteomimetic inhibitors of intracellular protein-protein interactions. *Angew Chem Int Ed Engl.* 2015;**54**:2960–2965.
48. Gloaguen C, Voisin-Chiret AS, Sopkova-de Oliveira Santos J, Fogha J, Gautier F, De Giorgi M, Burzicki G, Perato S, Pétigny-Lechartier C, Simonin-Le Jeune K, Brotin E, Goux D, N'Diaye M, Lambert B, Louis MH, Ligat L, Lopez F, Juin P, Bureau R, Rault S, Poulain L. First evidence that oligopyridines, α-helix foldamers, inhibit Mcl-1 and sensitize ovarian carcinoma cells to Bcl-xL-targeting strategies. *J Med Chem.* 2015;**58**:1644–1668.
49. Lao BB, Drew K, Guarracino DA, Brewer TF, Heindel DW, Bonneau R, Arora PS. Rational design of topographical helix mimics as potent inhibitors of protein–protein interactions. *J Am Chem Soc.* 2014;**136**:7877–7888.
50. Vassilev LT, Vu BT, Graves B, Carvajal D, Podlaski F, Filipovic Z, Kong N, Kammlott U, Lukacs C, Klein C, Fotouhi N, Liu EA. in vivo activation of the p53 pathway by small-molecule antagonists of MDM2. *Science.* 2004;**303**:844–848.
51. Grasberger BL, Lu T, Schubert C, Parks DJ, Carver TE, Koblish HK, Cummings MD, LaFrance LV, Milkiewicz KL, Calvo RR, Maguire D, Lattanze J, Franks CF, Zhao S, Ramachandren K, Bylebyl GR, Zhang M, Manthey CL, Petrella EC, Pantoliano MW, Deckman IC, Spurlino JC, Maroney AC, Tomczuk BE, Molloy CJ, Bone RF. Discovery and cocrystal structure of benzodiazepinedione HDM2 antagonists that activate p53 in cells. *J Med Chem.* 2005;**48**:909–912.
52. Zhou P, Yang XL, Wang XG, Hu B, Zhang L, Zhang W, Si HR, Zhu Y, Li B, Huang CL, Chen HD, Chen J, Luo Y, Guo H, Jiang RD, Liu MQ, Chen Y, Shen XR, Wang X,

Zheng XS, Zhao K, Chen QJ, Deng F, Liu LL, Yan B, Zhan FX, Wang YY, Xiao GF, Shi ZL. A pneumonia outbreak associated with a new coronavirus of probable bat origin. *Nature*. 2020;**579**:270–273.

53. Wu F, Zhao S, Yu B, Chen YM, Wang W, Song ZG, Hu Y, Tao ZW, Tian JH, Pei YY, Yuan ML, Zhang YL, Dai FH, Liu Y, Wang QM, Zheng JJ, Xu L, Holmes EC, Zhang YZ. A new coronavirus associated with human respiratory disease in China. *Nature*. 2020;**579**:265–269.
54. WHO. Coronavirus disease (COVID-19) pandemic.
55. Callaway E. Beyond omicron: what's next for COVID's viral evolution. *Nature*. 2021;**600**:204–207.
56. Robinson PC, Liew DFL, Tanner HL, Grainger JR, Dwek RA, Reisler RB, Steinman L, Feldmann M, Ho LP, Hussell T, Moss P, Richards D, Zitzmann N. COVID-19 therapeutics: challenges and directions for the future. *Proc Natl Acad Sci USA*. 2022;**119**:e2119893119.
57. Saxena A. Drug targets for COVID-19 therapeutics: ongoing global efforts. *J Biosci*. 2020;**45**:87.
58. Liu C, Zhou Q, Li Y, Garner LV, Watkins SP, Carter LJ, Smoot J, Gregg AC, Daniels AD, Jervey S, Albaiu D. Research and development on therapeutic agents and vaccines for COVID-19 and related human coronavirus diseases. *ACS Cent Sci*. 2020;**6**:315–331.
59. Noy-Porat T, Makdasi E, Alcalay R, Mechaly A, Levy Y, Bercovich-Kinori A, Zauberman A, Tamir H, Yahalom-Ronen Y, Israeli M, Epstein E, Achdout H, Melamed S, Chitlaru T, Weiss S, Peretz E, Rosen O, Paran N, Yitzhaki S, Shapira SC, Israely T, Mazor O, Rosenfeld R. A panel of human neutralizing mAbs targeting SARS-CoV-2 spike at multiple epitopes. *Nat Commun*. 2020;**11**:4303.
60. Wrapp D, Wang N, Corbett KS, Goldsmith JA, Hsieh CL, Abiona O, Graham BS, McLellan JS. Cryo-EM structure of the 2019-nCoV spike in the prefusion conformation. *Science*. 2020;**367**:1260–1263.
61. Pomplun S. Targeting the SARS-CoV-2-spike protein: from antibodies to miniproteins and peptides. *RSC Med Chem*. 2020;**12**:197–202.
62. Josephson K, Ricardo A, Szostak JW. mRNA display: from basic principles to macrocycle drug discovery. *Drug Discov Today*. 2014;**19**:388–399.
63. Zhang G, Pomplun S, Loftis AR, Tan X, Loas A, Pentelute BL (2020). Investigation of ACE2 N-terminal fragments binding to SARS-CoV-2 Spike RBD. bioRxiv 2020.03.19.999318.
64. Pomplun S, Jbara M, Quartararo AJ, Zhang G, Brown JS, Lee YC, Ye X, Hanna S, Pentelute BL. De novo discovery of high-affinity peptide binders for the SARS-CoV-2 spike protein. *ACS Cent Sci*. 2021;**7**:156–163.
65. Larue RC, Xing E, Kenney AD, Zhang Y, Tuazon JA, Li J, Yount JS, Li PK, Sharma A. Rationally designed ACE2-derived peptides inhibit SARS-CoV-2. *Bioconjug Chem*. 2021;**32**:215–223.
66. Karoyan P, Vieillard V, Gómez-Morales L, Odile E, Guihot A, Luyt CE, Denis A, Grondin P, Lequin O. Human ACE2 peptide-mimics block SARS-CoV-2 pulmonary cells infection. *Commun Biol*. 2021;**4**:197.
67. Cao L, Goreshnik I, Coventry B, Case JB, Miller L, Kozodoy L, Chen RE, Carter L, Walls AC, Park YJ, Strauch EM, Stewart L, Diamond MS, Veesler D,

Baker D. De novo design of picomolar SARS-CoV-2 miniprotein inhibitors. *Science*. 2020;**370**:426–431.
68. Curreli F, Victor SMB, Ahmed S, Drelich A, Tong X, Tseng CK, Hillyer CD, Debnath AK. Stapled peptides based on human angiotensin-converting enzyme 2 (ACE2) potently inhibit SARS-CoV-2 infection in vitro. *MBio*. 2020;**11**:e02451–e024520.
69. Cromm PM, Spiegel J, Grossmann TN. Hydrocarbon stapled peptides as modulators of biological function. *ACS Chem Biol*. 2015;**10**:1362–1375.
70. Verdine GL, Hilinski GJ. Stapled peptides for intracellular drug targets. *Methods Enzymol*. 2012;**503**:3–33.
71. Morgan DC, Morris C, Mahindra A, Blair CM, Tejeda G, Herbert I, Turnbull ML, Lieber G, Willett BJ, Logan N, Smith B, Tobin AB, Bhella D, Baillie G, Jamieson AG. Stapled ACE2 peptidomimetics designed to target the SARS-CoV-2 spike protein do not prevent virus internalization. *Pept Sci*. 2021;**113**:e24217.
72. Calugi L, Sautariello G, Lenci E, Mattei ML, Coppa C, Cini N, Contini A, Trabocchi A. Identification of a short ACE2-derived stapled peptide targeting the SARS-CoV-2 spike protein. *Eur J Med Chem*. 2023;**249**:115118.
73. Bugatti K, Sartori A, Battistini L, Coppa C, Vanhulle E, Noppen S, Provinciael B, Naesens L, Stevaert A, Contini A, Vermeire K, Zanardi F. Novel polymyxin-inspired peptidomimetics targeting the SARS-CoV-2 Spike:hACE2 interface. *Int J Mol Sci*. 2023;**24**:8765.
74. Maffucci I, Contini A. In silico drug repurposing for SARS-CoV-2 main proteinase and spike proteins. *J Proteome Res*. 2020;**19**:4637–4648.
75. Marković V, Shaik JB, Ożga K, Ciesiołkiewicz A, Lizandra Perez J, Rudzińska-Szostak E, Berlicki Ł. Peptide foldamer-based inhibitors of the SARS-CoV-2 S protein-human ACE2 interaction. *J Enzyme Inhib Med Chem*. 2023;**38**:2244693.
76. Pilsl LKA, Reiser O. a/b-peptide foldamers: state of the art. *Amino Acids*. 2011;**41**: 709–718.
77. Wu L, Chen Q, Liu K, Wang J, Han P, Zhang Y, Hu Y, Meng Y, Pan X, Qiao C, Tian S, Du P, Song H, Shi W, Qi J, Wang H-W, Yan J, Gao GF, Wang Q. Broad host range of SARS-CoV-2 and the molecular basis for SARS-CoV-2 binding to cat ACE2. *Cell Discov*. 2020;**6**:68.
78. Ma LL, Liu HM, Liu XM, Yuan XY, Xu C, Wang F, Lin JZ, Xu RC, Zhang DK. Screening S protein – ACE2 blockers from natural products: strategies and advances in the discovery of potential inhibitors of COVID-19. *Eur J Med Chem*. 2021;**226**:113857.
79. Dong M, Galvan Achi JM, Du R, Rong L, Cui Q. Development of SARS-CoV-2 entry antivirals. *Cell Insight*. 2024;**3**:100144.
80. Tedesco F, Calugi L, Lenci E, Trabocchi A. Peptidomimetic small-molecule inhibitors of 3CLPro activity and spike-ACE2 interaction: toward dual-action molecules against coronavirus infections. *J Org Chem*. 2022;**87**:12041–12051.

11

Peptidomimetics Targeting the Alzheimer's Disease

11.1 Introduction

The first study describing what would later be recognized as Alzheimer's disease (AD) was published more than a century ago by the German physician Alois Alzheimer. The patient Auguste D began to suffer from dementia at the age of 51, with symptoms that did not correspond to any disease known at the time. Six years after the onset of symptoms like memory problems, paranoia, behavior, and language problems, she died of septicemia. During the autopsy, Alzheimer found an atrophic brain and abnormal protein deposits, later recognized as senile plaques and neurofibrillary tangles (NFTs).

According to the World Health Organization (WHO) [1], about 57 million individuals worldwide were living with dementia as of 2021. Annually, nearly 10 million new cases are diagnosed, and dementia ranks as the seventh leading cause of death globally and of disability and dependency among older adults. Dementia arises from a range of diseases that impair brain function, and AD is the most prevalent form, accounting for approximately 60–70% of cases, thus being recognized as a global health priority [2]. AD is a neurodegenerative disease affecting mostly older adults, which is characterized by gradual memory loss and generalized impairment of cognitive functions [3, 4]. It is the most common form of dementia, with a similar incidence of the disease throughout the world population [5] that increases exponentially with age [6, 7] and with about twice the number of diagnosed cases occurring in women [8].

The cost of healthcare expenditure generated by the management of the disease is expected to increase, confirming the significant socioeconomic impact of AD [9]. Accordingly, in 2010, the total global cost of dementia was estimated at $604 billion, and in 2019, at $1.3 trillion [10]. Remarkably, around 50% of this cost was associated with the care provided by informal carers, such as family members. In addition to advanced age, other risk factors have been identified (Table 11.1), such as genetic factors, environmental exposures, smoking, alcoholism, dietary habits, and cardiovascular factors in addition to aging, which suggest that AD has a multifactorial etiology [24].

Table 11.1 *Major risk factors for Alzheimer's disease.*

Category	Causes/risk factors	Description	References
Genetic factors	APP, PSEN1, PSEN2, and ApoE ε4	Mutations in APP, presenilin 1 and 2 genes cause familial early-onset AD; ApoE ε4 allele increases risk of late-onset AD	11–13
Age	Advanced age	Greatest risk factor; incidence increases exponentially with age	[14, 15]
Environmental factors	Exposure to toxins and pollution	Possible contribution from environmental toxins, though evidence is inconclusive	[16, 17]
Cardiovascular health	Hypertension, diabetes, obesity, and smoking	Vascular health influences AD risk; comorbidities exacerbate cognitive decline	[18, 19]
Lifestyle factors	Diet, physical inactivity, and cognitive engagement	Healthy diet, exercise, and mental activity may reduce risk	[20, 21]
Other medical conditions	Traumatic brain injury (TBI) and depression	TBI and mood disorders linked to increased AD risk	[22, 23]

Based on the recent 2024 Lancet Commission report [25], 14 modifiable risk factors have been identified, spanning early, mid, and late life, that can potentially prevent or delay up to 45% of dementia cases. These include addressing education, hearing, high blood pressure, smoking, obesity, depression, physical activity, diabetes, alcohol, head injuries, air pollution, social contact, vision loss, and high cholesterol. This is considered a milestone in the context of dementia prevention, intervention and care, and it has been envisaged that addressing these factors, particularly in midlife, is critical, as they account for nearly half of all global dementia cases. The 14 modifiable risk factors for dementia are herein listed as follows:

1. *Less education in early life*: lower educational attainment in childhood.
2. *Hearing impairment*: untreated midlife hearing loss.
3. *Hypertension*: high blood pressure in midlife.
4. *Smoking*: current tobacco use.
5. *Obesity*: high body mass index.
6. *Depression*: untreated, chronic depression.
7. *Physical inactivity*: lack of regular exercise.
8. *Diabetes*: type 2 diabetes mellitus.
9. *Excessive alcohol consumption*: high, harmful alcohol intake.
10. *Head injury*: traumatic brain injury.
11. *Air pollution*: long-term exposure.
12. *Infrequent social contact*: social isolation and loneliness.
13. *Vision loss*: untreated vision impairment.
14. *High LDL cholesterol*: high level of "bad" cholesterol in midlife.

AD is traditionally classified in two groups based on the age of onset:

- early-onset AD (EOAD), affecting people under the age of 65 (5–10% of all cases);
- late-onset AD (LOAD), occurring after the age of 65, and documented as the most common form.

Furthermore, three different variants of EOAD and LOAD are recognized, namely autosomal dominant, familial, and sporadic. Currently, the three genes associated with early-onset AD are:

- APP (amyloid precursor protein), which is found on chromosome 21;
- presenilin 1 (PSEN1), which is found on chromosome 14;
- presenilin 2 (PSEN2) present on chromosome 1.

As for late-onset AD, the gene for apolipoprotein (ApoE) on chromosome 19 appears to be involved, particularly in the E4 allele. All forms of AD are defined by the same pathological characteristics. Specifically, AD is a neurodegenerative disease characterized by the loss of multiple cognitive functions, including memory loss, especially short-term, difficulty with language, impaired judgment, confusion, and challenges with visual and spatial relations. In addition to cognitive symptoms, and during the moderate-to-advanced stages of the disease (also termed "mild cognitive impairments," MCI), behavioral and personality changes occur, such as mood swings, social withdrawal, repetitive behaviors, getting lost or disoriented, especially in familiar areas, changes in sleep patterns, paranoia, and impulsive behavior. In the advanced stage of the disease, complete spatial–temporal and sensorial disorientation occurs, with total impairment of long- and short-term memory. The person will need complete support, especially following the complications of dehydration and malnutrition that can occur in this stage of the disease.

The speed of progression of AD and the life expectancy of a person affected by AD are highly variable for each individual. About 1 in 10 people suffer from more than one type of dementia at the same time. This is called mixed dementia, the most common form of which is the onset of AD together with vascular dementia, caused by problems with blood flow to the brain. On average, people affected by AD live for 8–10 years after diagnosis, and death generally occurs due to the onset of complications. Specifically, the progression of AD can be summarized in three critical neurochemical stages [26]:

- *Prodromal or preclinical stage* is the initial stage characterized by elevated levels of amyloid-β (Aβ) 1–42 in the brain, which may result from a shift in APP processing from the α-secretase to the β/γ-secretase pathway or from impaired clearance of Aβ peptides across the blood–brain barrier (BBB). Potential compensatory mechanisms include microglial activation that supports neuroprotection and the removal of Aβ 1–42 into amyloid plaques. Despite such neuropathological changes, cognitive function remains intact during this phase, as concentrations of the more neurotoxic oligomeric Aβ 1–42 species are still low.
- *Cognitive dysfunction phase* is the stage where compensatory neuroprotective mechanisms become insufficient, and oligomeric Aβ 1–42 accumulates. These molecules bind to neuronal membrane proteins, disrupt intracellular signaling, and are internalized *via* receptor-mediated endocytosis. The resultant mitochondrial dysfunction leads to

the production of reactive oxygen species (ROS), which further impair cellular signaling and activate kinases that phosphorylate tau protein. The subsequent synaptic dysfunction and kinase dysregulation cause the emerging cognitive impairments observed during this phase.
- *-Neurodegenerative phase* is characterized by a shift from protective to proinflammatory neuroinflammatory responses with increased production of ROS and reactive nitrogen species (RNS), leading to extensive neuronal damage and activation of apoptotic or necrotic pathways. Alternatively, a complete loss of microglial function may occur, resulting in nonfunctional glial cells. The progressive formation of NFTs ultimately leads to widespread neuronal death through apoptosis or necrosis.

AD affects neurons in several regions of the brain, particularly in the cerebral cortex where thought and language occur, the basal ganglia, and the hippocampus, essential for memory storage [27]. The two recognized pathological hallmarks of AD are:

- extracellular Aβ deposits in the cerebral cortex, also referred to as senile plaques;
- intracellular NFTs consisting of paired helical filaments in the subcortical gray matter.

Senile plaques or amyloid plaques are made up of accumulations of the Aβ peptide, while NFTs are generated by accumulation of hyperphosphorylated tau protein (Figure 11.1). Aβ deposition and NFTs lead to the loss of synapses and neurons, resulting in atrophy of the above-cited affected areas of the brain.

The mechanism by which Aβ peptides and NFTs cause such damage is still not fully understood. There are numerous hypotheses, among which the most accredited is the amyloid one.

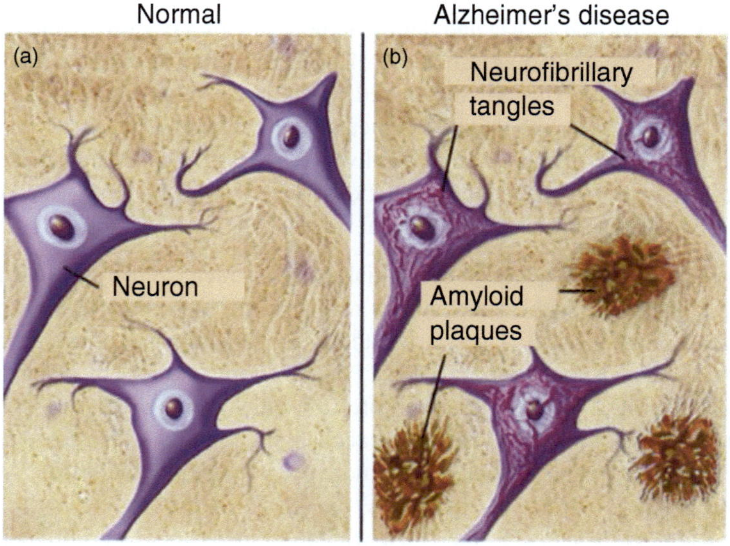

Figure 11.1 *Amyloid plaques and neurofibrillary tangles as major pathological hallmarks of AD that are absent in healthy brain tissue; (a) normal; (b) Alzheimer's disease. Source: [28] / Lidsen Publishing Inc. / CC BY 4.0.*

11.2 The Amyloid Hypothesis

In mid-1980s, the amyloid hypothesis was proposed for the AD, and it has been one of the most accredited for decades [29]. It suggests that the deposition of Aβ protein is central to the pathogenesis of the disease. The hypothesis is based on two key observations:

- The identification of Aβ as the primary protein of senile plaques.
- The identification of several mutations in familial AD that lead to the accumulation of Aβ.

The Aβ protein derives from the proteolysis of the APP. This is a transmembrane protein consisting of 770 amino acids [30] that is widely expressed in normal human tissues and shows particularly high levels in the brain [31], where it exerts functions for regulating the survival of neurons. It is postulated that the alteration of the metabolism of the APP protein is at the basis of the excessive formation of Aβ with consequent accumulation causing damage to neurons and their death. Usually, Aβ fragments are eliminated, whereas in AD, for unknown reasons, they accumulate, forming hard and insoluble plaques.

In the amyloid hypothesis the progressive accumulation of Aβ peptide fragments in the brain generates a series of events ultimately resulting to neuronal cell death, loss of neuronal synapses, and neurotransmitter deficits, all of which contribute to the clinical symptoms of dementia. These Aβ polypeptide fragments, consisting of 39–42 amino acids, accumulate in AD to form hard and insoluble plaques. The Aβ protein is derived from its precursor APP, following a series of sequential proteolytic cleavages. The APP is translationally localized in the endoplasmic reticulum and subsequently modified during the secretory pathway. During transit from the endoplasmic reticulum to the plasma membrane, the APP protein undergoes *N*- and *O*-glycosylation, sulfation, and phosphorylation [32, 33], enabling internalization into the membrane, thanks to the presence of a signal sequence at the *C*-terminal end, which constitutes the intracytoplasmic domain of approximately 100 amino acids with an amino acid sequence called Aβ. The three main enzymes that are involved in the metabolism of APP, namely α-, β-, and γ-secretase, act at the level of the amino acid sequence Aβ. Two proteolytic pathways for the cleavage of APP have been identified (Figure 11.2), the *non-amyloidogenic* pathway and the *amyloidogenic* one.

The *non-amyloidogenic pathway* refers to the cleavage of βAPP by α- and γ-secretase, without giving rise to Aβ. It begins with a proteolytic cleavage by α-secretase to generate the soluble alpha fragment (sAPPα) in the extracellular space and the *C*-terminal C83 peptide fragment into the cytoplasm, which in turn can be cleaved by γ-secretase, releasing the p3 peptide into the extracellular space and *C*-terminal fragments into the cytoplasm. The non-amyloidogenic pathway occurs predominantly on the cell surface where α-secretase is present, while the amyloidogenic pathway occurs in the organelles of the endocytic pathway where β- and γ-secretases operate [30].

The amyloidogenic pathway refers to the cleavage of APP at the *N*-terminal domain by BACE1, to form *C*-terminal fragments on the extracellular side that include the entire Aβ sequence, followed by cleavage by γ-secretase to give rise to Aβ [35, 36]. In particular, the amyloidogenic pathway is originated by β-secretase (also called β site-APP Cleaving Enzyme 1, BACE1) that cleaves APP, generating the soluble beta fragment (sAPPβ) in the extracellular space and the C99 *C*-terminal fragment in the cytoplasm. The latter is processed by γ-secretase, leading to the production of the APP intracellular domain (AICD),

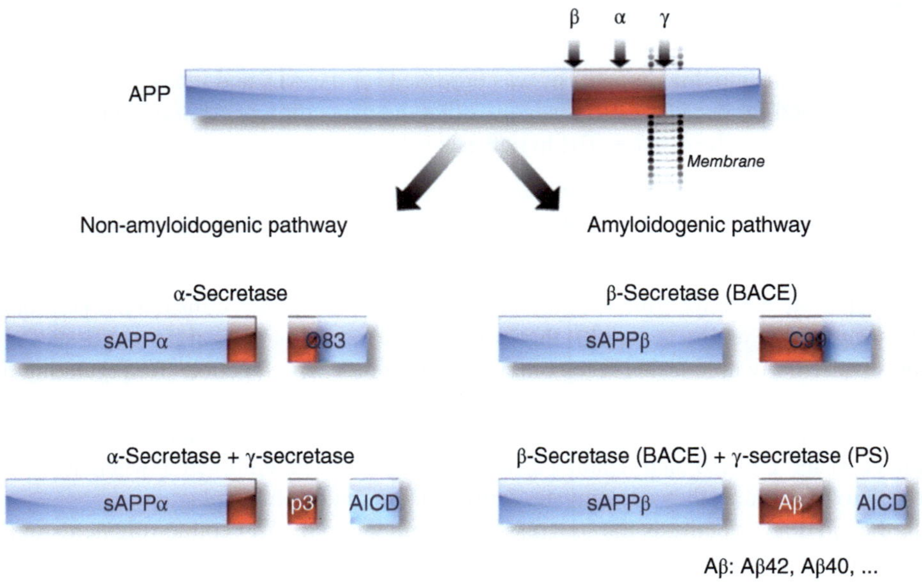

Figure 11.2 *Amyloidogenic and non-amyloidogenic pathways of β amyloid precursor protein (βAPP) processing. Source: [34] / Springer Nature / CC BY 2.0.*

involved in the control of gene transcription, and Aβ peptide fragments, both at the interface with the cytoplasm and the extracellular space, of which the most abundant Aβ40 and Aβ42 are secreted and aggregate to give rise to protofibrils, fibrils, and finally senile plaques [37]. In the cascade of events of the amyloid hypothesis, approximately 5%–10% of Aβ peptide fragments end with residue 42 and appear to be of greater importance in the pathogenesis of AD. Senile plaques formed following the accumulation of Aβ induce the activation of microglia and astrocytes, which activate the response of the immune system that leads to the inflammatory state. Inflammation in turn activates oxidative stress, which would result in the phenomenon of hyperphosphorylation of the tau protein and the formation of NFTs, causing progressive synaptic degeneration and subsequent alteration of neuronal homeostasis [38, 39]. Cell death resulting from these damage mechanisms is responsible for the neurodegeneration of widespread areas of the hippocampus and cerebral cortex, which gives rise to the clinical symptoms of AD [40]. It has been proposed that inflammation is the third major pathological hallmark of AD. In AD, Aβ in brain amyloid deposits and tau in NFTs are thought to have prion-like self-replicating properties [41]. In prion diseases, a normal brain prion protein is misfolded into a pathogenic form called a prion. The prion then causes other prion proteins to misfold, resulting in a significant increase in the abnormal proteins, causing brain damage.

The Aβ protein hypothesis, proposed by Hardy and Higgins in 1992 [42], claims that the overproduction and extracellular accumulation of the Aβ peptide with consequent formation of senile plaques is also due to the presence of mutations in the genes encoding the precursor protein APP and PSEN (a class of heterodimeric peptides composed of *N*- and *C*-terminal fragments [43], resulting from the endoproteolysis of presenilin). According to the amyloid hypothesis, these genetic mutations would be implicated in the pathogenesis of the disease and would represent the main causes of neurotoxicity in AD. APP gene mutations increase the deposits of Aβ1–40 fragments. PSEN gene mutations, in addition to increasing the production of Aβ1–42 fragments, act by determining an increase in the activity of presenilins, which results in an enhancement of the catalytic activity of γ-secretase. These latter mutations represent the most common cause of autosomal AD and in particular, PSEN1 mutations cause the most aggressive form of AD [44, 45]. All these mutations have the same final effect of increasing the formation of Aβ deposits, as APP protein mutations favor proteolysis through the amyloidogenic pathway, while presenilin mutations increase the activity of γ-secretase, and both result in an increase in the production of the Aβ peptide. Although some aspects are still to be clarified, the amyloidogenic hypothesis appears to have greater value among researchers for AD. In this context, potential therapies aimed at blocking the formation of Aβ amyloidogenesis can be used to slow down or even prevent further neurodegeneration.

11.3 Therapeutic Strategies for Alzheimer's Disease

Although no definitive cure for the AD currently exists, several therapeutic approaches have been proposed to clinically slow down disease progression. These strategies primarily aim to pharmacologically modulate key pathological mechanisms causing AD. Recently, anti-amyloid monoclonal antibodies targeting Aβ plaques or oligomers as first disease-modifying agents have been approved [46]. Also, the integration of psychosocial, cognitive, and behavioral interventions indicated to exert beneficial effects can synergistically enhance the efficacy of pharmacological treatments, contributing to the slowing of symptom progression and improving the quality of life for both patients and caregivers.

One of the earliest pharmacological strategies was based on the cholinergic hypothesis, which suggests that cognitive decline in AD is associated with decreased acetylcholine levels. Direct administration of acetylcholine is not possible due to its instability and short half-life. Similarly, cholinergic agonists have several systemic effects and significant side effects, thus unsuitable for clinical use. Instead, treatment efforts have focused on acetylcholinesterase inhibitors, which prevent the enzymatic degradation of acetylcholine, thus increasing its concentration in the synapse. Reversible acetylcholinesterase inhibitors with low peripheral affinity and adequate lipophilicity have been developed, thus allowing them to cross the BBB and to act preferentially on the central nervous system. Early examples include tacrine, donepezil, physostigmine, galantamine, and neostigmine. However, current clinical interest mainly focuses on rivastigmine and galantamine. Rivastigmine is significant for its limited pharmacological interactions, while galantamine is preferred for its high bioavailability and short half-life, thus reducing the possibility of side effects. Tacrine has been discontinued due to hepatotoxicity. Donepezil, a noncompetitive acetylcholinesterase inhibitor with a long half-life, allows once-daily dosing and has demonstrated higher clinical efficacy. However, it is also associated with cholinergic side effects such as insomnia, bradycardia, arrhythmias, nausea, and diarrhea, whereas for galantamine and rivastigmine such effects occur to a lesser extent. A recent pharmacological development involves targeting the glutamatergic system with memantine, an *N*-methyl-D-aspartate (NMDA) receptor antagonist that showed moderate but positive therapeutic effects in reducing cognitive deterioration in patients with moderate-to-severe AD.

In June 2021, the US FDA approved aducanumab, a monoclonal antibody (mAb) targeting Aβ plaques for early-stage AD as a first-in-class disease-modifying agent [47]. Despite its approval, the decision was met with criticism due to inconclusive clinical trial data regarding its efficacy. After aducanumab, other anti-amyloid mAbs were approved as disease-modifying agents for the treatment of early AD, such as lecanemab and donanemab.

11.3.1 Aggregation Inhibitors and β-Sheet Breakers for Anti-β-Amyloid Therapies

Aβ aggregation proceeds from soluble oligomers to insoluble fibrils, contributing to neurotoxicity. Although several peptide-based aggregation inhibitors have been reported [48], a definitive treatment remains elusive. Indeed, peptides are gaining importance in the discovery of therapeutic treatments for AD. Ongoing research is focused on improving the stability and efficacy of peptides that target key pathological features of AD, including Aβ, tau proteins, and metal ions. Accordingly, several peptides and peptidomimetics are under study for their therapeutic potential [49].

Peptides designed to bind specific motifs in Aβ or tau proteins can inhibit the formation and aggregation of toxic oligomers and fibrils, thus intervening at critical stages of AD progression [50]. In 2008, two peptides based on the KLVFF sequence and additional water-soluble amino acid residues, namely OR1 (RGKLVFFGR) and OR2 (RGKLVFFGR-amide), were reported by Austen et al., showing inhibition of Aβ fibril formation [51].Then, Sun et al. developed LK7 (Ac-LVFFARK-NH_2), designed to disrupt $A\beta_{42}$ fibrillization through the incorporation of positively charged residues [52]. A modified version, LK7-HH, demonstrated enhanced biocompatibility, stronger Aβ inhibition, and Cu^{2+}-chelating properties, reducing ROS formation. Further optimization led to cLK7, a cyclic analogue with improved stability, reduced cytotoxicity, and higher $A\beta_{40}$ binding affinity.

Beyond aggregation inhibition, some peptides also possess the unique capacity to disassemble preformed fibrils. Such disassembly is thought to occur *via* a strain-relief mechanism, as peptides induce torsional stress on the fibril backbone, thus weakening inter-strand hydrogen bonds and promoting solvent access to the fibril core [53]. The resultant loss of fibrillar integrity contributes to the breakdown of toxic aggregates. This dual functionality, consisting of blocking aggregation and promoting fibril disassembly, positions therapeutic peptides as particularly promising candidates for the development of disease-modifying treatments for AD. In this context, β-sheet breaker peptides function by targeting hydrophobic regions in Aβ to disrupt fibril formation, reduce neurotoxicity, and improve cognitive outcomes. Strategies like proline substitution, *N*-methylation, and cyclization enhance their therapeutic potential, although BBB penetration and bioavailability are still challenging. Interestingly, β-sheet macrocycles, such as amyloid β-sheet mimics, have shown promise by inhibiting aggregation through β-sheet interactions and intramolecular hydrogen bonding.

Stapled peptides: Following the successful application of stapled peptides for targeting complex interactions, their potential as modulators of amyloidosis in AD was explored [54]. In particular, tail-to-side-chain peptide stapling was performed using various dicarboxylic acids at the *N*-terminus, resulting in effective inhibition of $A\beta_{1-40}$ aggregation and disruption of preformed fibrils under physiological conditions to produce nontoxic species. Comparative analyses revealed that stapled peptides significantly outperformed linear analogues in both inhibition and disruption, with *N*-terminal glutaric acid as the best chemical moiety to address stapling of key GLVFFKR epitope at the Lys residue for the tail-to-side-chain stapling with an amide bond. The conformational rigidity of stapled peptides likely contributed to their efficacy, suggesting the application of stapled macrocycles as promising candidates for anti-amyloid therapeutics.

Peptoids: These peptidomimetic molecules have emerged as promising therapeutic candidates for neurodegenerative amyloidoses, such as AD, Huntington's disease (HD), and Parkinson's disease, which are all characterized by the aggregation of misfolded proteins including Aβ, huntingtin, and α-synuclein, respectively. In AD, amyloid aggregation is also linked to amylin, as observed in the brain's vasculature of late-onset AD patients [55]. As reported by Bezprozvanny et al., the screening of large peptoid libraries has yielded $A\beta_{1-42}$-selective inhibitors, such as inhibitor of amyloid 1 (IAM1) (Figure 11.3) that block both early oligomerization and mature fibril formation in a concentration-dependent manner, with some dimers demonstrating strong neuroprotective effects [56]. Rationally designed peptoids mimicking the hydrophobic core of Aβ, particularly those incorporating aromatic residues, effectively attenuate $A\beta_{1\text{-}40}$ fibrillization and exhibit sensitivity to side-chain placement and chirality that influence aggregate morphology and toxicity.

As another entry in this approach, peptide-peptoid hybrids, designed to mimic Aβ motifs while enhancing structural stability, have outperformed traditional peptides in aggregation inhibition and toxicity attenuation [57]. Specifically, based on the key KLVFF sequence that is known to bind Aβ aggregates, the *N*-terminal position was functionalized with thymine and barbituric acid derivative capable to function as donor–acceptor moieties (Figure 11.4) and *N*-methylation at specific positions of peptoid moieties to provide stability and block the growth of β-sheets, through inhibition of hydrogen-bond-driven fibrillar aggregation. The synthesized peptidomimetic hybrids proved to display activity in both inhibition and dissolution of Aβ aggregates, as shown by thioflavin assay and through a yeast cell model showing Aβ toxicity.

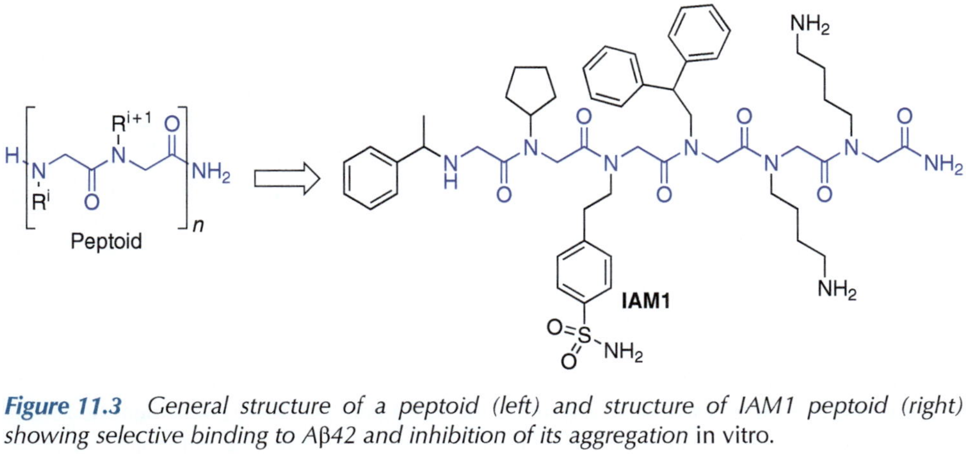

Figure 11.3 *General structure of a peptoid (left) and structure of IAM1 peptoid (right) showing selective binding to Aβ42 and inhibition of its aggregation* in vitro.

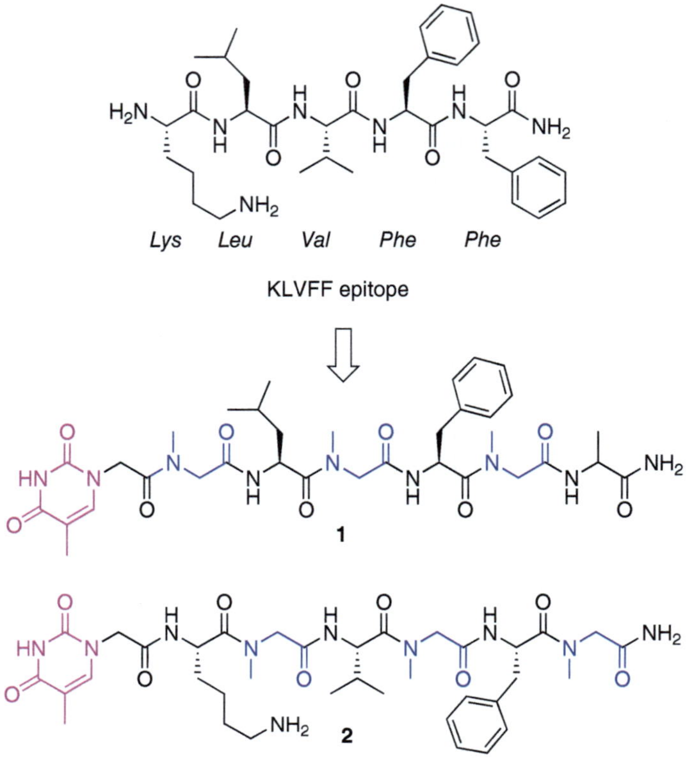

*Figure 11.4 Structure of reference KLVFF peptide and corresponding peptidomimetics **1** and **2** showing peptoid moieties (blue) and thymine (magenta) at the N-terminal position.*

Foldamers: These molecules proved powerful for addressing complex protein–protein interactions, particularly arranged with helical conformations, and providing added stability to the parent peptide epitope. Recent studies have shown that sulfonyl-γ-peptide foldamers can interfere with Aβ aggregation (Figure 11.5, compound **3**) by binding to its central domain and promoting an alternative α-helical conformation, thus inhibiting Aβ from following its normal aggregation pathway [58]. Comparable results have also been observed with other γ-peptide and β-sheet foldamer analogues, which suppress Aβ aggregation through distinct mechanistic routes [59]. Specifically, Ongeri et al. reported the first demonstration of helical foldamers functioning as dual inhibitors of both human islet amyloid polypeptide (hIAPP), relevant for type-2 diabetes, and $A\beta_{1-42}$ aggregation while simultaneously stabilizing the monomeric forms of these amyloidogenic peptides. Particularly, the non-peptide foldamer **4** (Figure 11.5) constructed from 4-amino-(or methyl)-1,3-thiazole-5-carboxylic acid scaffolds, which organizes into a 9-helix architecture mimicking the 3_{10} helix, exhibited remarkable activity, as shown by combining studies of biophysical techniques, including thioflavin-T fluorescence assays, transmission electron microscopy, capillary electrophoresis, and mass spectrometry. In another work on non-peptide-based foldamers, SK-131 was identified from a library of oligoquinoline-based foldamers as a highly effective antagonist of $A\beta_{42}$

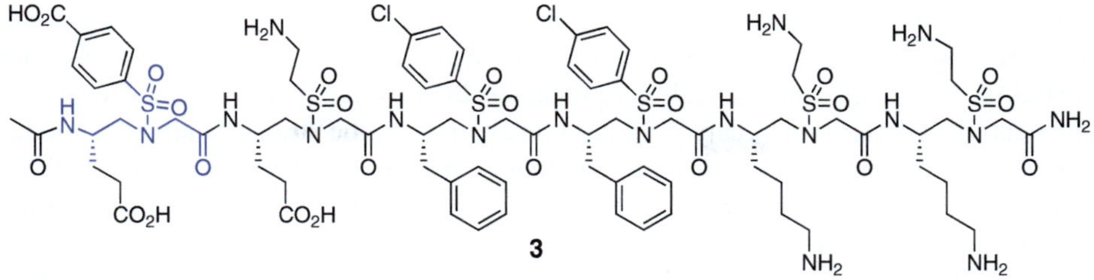

3

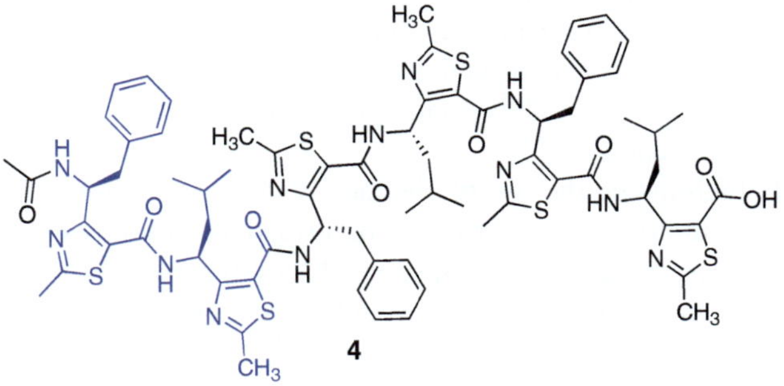

4

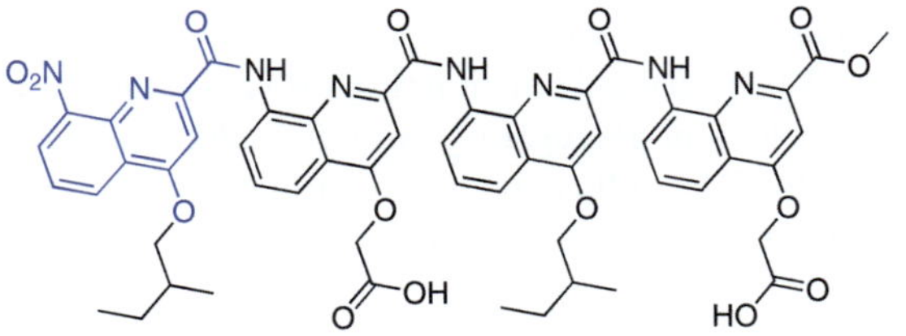

SK-131

Figure 11.5 *Examples of foldamer peptidomimetics targeting Aβ aggregation; the monomeric unit is highlighted in blue.*

aggregation [60], promoting an α-helical conformation in monomeric Aβ to prevent its aggregation. Also, it was demonstrated that SK-131 reduced Aβ-associated pathological features in both AD cell models and multiple *Caenorhabditis elegans* AD models. Specifically, it proved to suppress intracellular $A\beta_{42}$ aggregation and reduce ROS. Also, efficient BBB penetration and favorable pharmacological characteristics were ascertained for SK-131, suggesting that this molecule is a brain-permeable foldamer that rescues AD phenotypes *in vivo*. Of relevance with respect to other approaches, this molecule provided structural remodeling of Aβ42 to achieve therapeutic benefit.

11.3.2 Anti-tau Peptidomimetics

Inhibition of tau protein aggregation has been studied as another approach toward therapeutic strategies for AD, applying the potential of Aβ-derived peptides and peptidomimetics accordingly [61]. Tau is a microtubule-associated protein intrinsically disordered, playing a key role in regulating microtubule stability and dynamics [62]. Pathological modifications of tau in AD reduce its binding affinity to microtubules, thus leading to their destabilization and subsequent disassembly, and impairing axonal structure and function, ultimately contributing to neuronal degeneration. While protein–protein interactions are essential for maintaining normal cellular processes, their dysregulation is characteristic of many neurodegenerative disorders [63]. As a result, therapeutic strategies aimed at modulating or disrupting these aberrant interactions have emerged [64]. Peptides such as KLVFF and LPFFD, known as β-sheet disruptors, have been investigated along with modified peptidomimetics for their capacity to simultaneously inhibit Aβ and tau aggregation. In this view, such peptides were engineered to interfere with hydrophobic and hydrogen-bonding interactions critical to amyloid pathology, incorporating structural modifications such as thymine and sarcosine to improve activity. Among the tested sequences, KLVFF, two peptidomimetics significantly reduced tau aggregation, as evinced by ThS fluorescence, TEM, and CD spectroscopy.

Another study highlighted the use of chemically designed systems, inspired by pathological and non-pathological tau protein conformations, to investigate and modulate tau misfolding and aggregation [65]. Specifically, by incorporating a novel non-natural isooxazoline-$\beta^{2,2}$-amino acid, researchers stabilized either β-hairpin or extended conformations in peptides containing ^{306}VQIVYK^{311} and ^{275}VQIINK^{280} sequences (Figure 11.6, structure **6**). The extended conformation, mimicking a tau monomer, induced aggregation without the need for additives like heparin, thus producing fibrils similar to those observed in AD. In contrast, the β-hairpin conformation (Figure 11.6, structure **5**) proved to maintain the tau protein in a soluble, non-aggregating form. Peptidomimetic foldamers also proved effective in live neuron models, influencing tau-microtubule interactions, thus highlighting the capability of peptidomimetics to modulate tau aggregation both *in vitro* and in living neuron models.

11.3.3 BACE1 Protease

BACE1 or β-secretase [66, 67], is a typical aspartyl protease, monomeric and similar to pepsin, that plays a key role in the production of Aβ, implicated in AD. β-secretase activity can be found in several cells of the body; nevertheless, BACE1 shows its maximum activity in neuronal cells [68]. Its activity appears to be optimal at acidic pH, and the enzyme is preferentially active within endosomes. For this reason, APP must be transported from the surface to the cell interior. The BACE1 product can then be transferred to the cell surface, where it becomes a substrate for β-secretase, thus leading to the formation of Aβ. BACE1

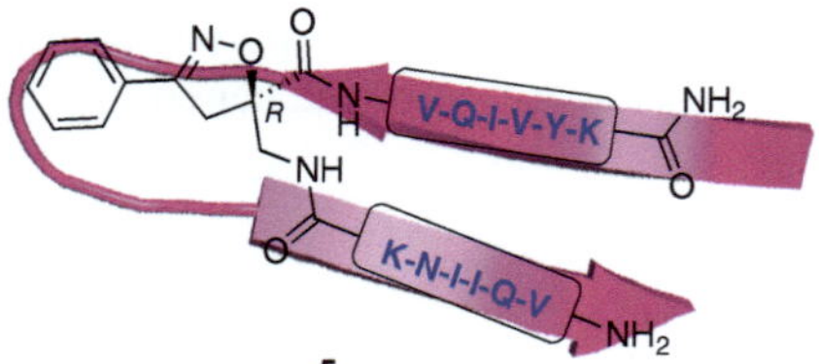

5
β-Hairpin conformation

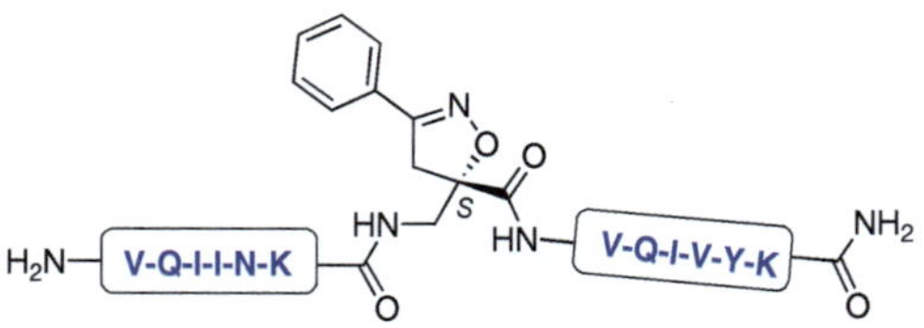

6
Extended conformation

Figure 11.6 Structure of enantiomeric isoxazoline-$\beta^{2,2}$-amino acids stabilizing either β-hairpin or extended conformations of key ^{306}VQIVYK^{311} and ^{275}VQIINK^{280} peptide sequences as a function of scaffold chirality.

is found in the endoplasmic reticulum and in the Golgi apparatus, where it cleaves the terminal portion of some proteins that are important in neural function. BACE1 is formed in the endoplasmic reticulum as a glycosylated pro-peptide, pro-BACE1. Then, pro-BACE1 is transformed into mature BACE1 in the Golgi apparatus. The pro-peptide has two configurations:

- a closed configuration, in which the pro-domain is folded inside the catalytic site;
- an open configuration, in which the bonds at the catalytic site with the pro-domain are cleaved.

The cleavage of the pro-domain, present in pro-BACE1, allows BACE1 to have full enzymatic activity due to the catalytic active site being completely accessible to the substrate [69]. Similar to pro-BACE1, mature BACE1 has two main conformations:

- *Open flap* conformation where BACE1 is free from the substrate and is energetically stable due to the multiple hydrogen bonds present in the flap region of the enzyme (Figure 11.7).
- *Closed flap* conformation characterized by BACE1 bound to the substrate. To achieve this conformation, the hydrogen bonds between multiple amino acid residues are broken. The destabilization caused by the breaking of these bonds is due to the interaction of the enzyme with the substrate.

The catalytic domain of the BACE1 protease contains eight pockets consisting of different catalytic residues, among which the aspartic acid residues Asp32 and Asp228 play a central role in peptide bond cleavage (Table 11.2).

Two water molecules are present in BACE1. The first one is held in position between the Asp32 and Asp228 pair by hydrogen bonds at the active site, which is essential for

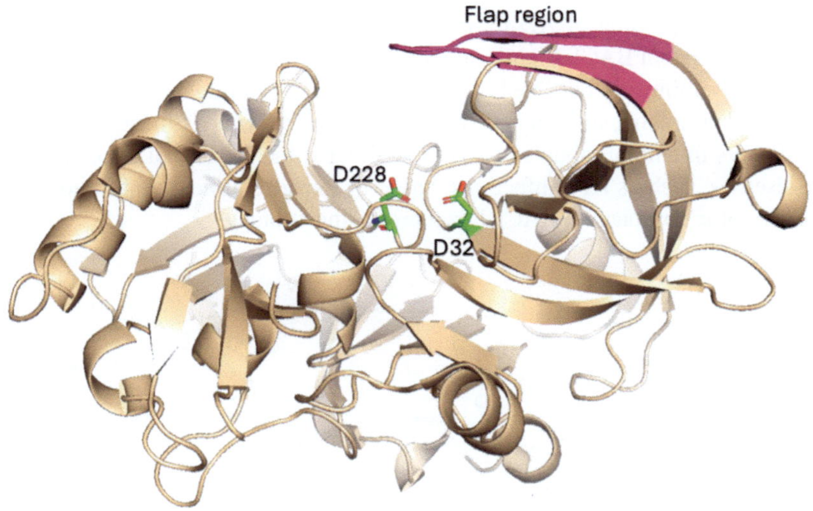

Figure 11.7 *Conformation of mature BACE1 in open flap conformation (PDB: 1SGZ). Catalytic Asp32 (D32) and Asp228 (D228) are highlighted in green; the flap region is shown in magenta.*

Table 11.2 Enzyme pockets and corresponding amino acid residues in the catalytic domain of BACE1.

Pocket	Amino acids	Pocket	Amino acids
S1	Leu30, Asp32, Tyr71, Leu119, Gln73, Phe108	*S1′*	Val131, Tyr71, Thr72, Asp228
S2	Asn233, Arg235, Ser325	*S2′*	Ser35, Val69, Pro70, Tyr198
S3	Leu133, Ile110, Ser229	*S3′*	Arg128, Tyr198
S4	Gln73, Thr232, Arg307	*S4′*	Glu125, Ile126, Tyr197, Tyr198

the formation of the enzyme–substrate bond. The second water molecule is stabilized by hydrogen bonds with a Tyr residue in the flap region of the enzyme and also participates in a hydrogen bond network established with Ser35, Asp32, and Asp228. Therefore, both water molecules assist in the catalytic reaction, influencing the activity of BACE1 [70, 71]. It has been proposed that BACE1 catalyzes hydrolysis of the peptide bond belonging to the substrate through an acid–base catalysis mechanism mediated by a water molecule and the two aspartic acid residues, Asp32 and Asp228. Specifically, one of the aspartic acids catalyzes the nucleophilic attack of the water molecule on the carbonyl carbon, whereas the other residue activates the carbonyl of the peptide bond, forming a tetrahedral intermediate stabilized by hydrogen bonds with the carboxyl group of Asp. The regeneration of the carbonyl leads to the cleavage of the substrate amide bond, releasing the two *N*- and *C*-terminal fragments of the substrate (Figure 11.8) [72].

BACE1 has been identified as a key target for the treatment of AD, as its inhibition is associated with the blockade of Aβ formation at the onset of APP modification. Accordingly, it has been demonstrated that excess Aβ production can be harmful to cognitive function even before neuronal death by limiting synaptic function [73], thus corroborating the strategy to slow down the progression of AD by inhibiting Aβ formation at an early stage. Animal studies have shown that BACE1 inhibition does not lead to major side effects; therefore, its inhibition may be clinically feasible [74]. The therapeutic potential of BACE1 has been further considered promising since genetic inhibition of the enzyme has been shown to prevent memory deficits in animal models of AD [75]. Finally, the success of drugs for treatment of other diseases, such as HIV infections, based on aspartic protease inhibition suggested applying the same approach to BACE1 as a promising strategy.

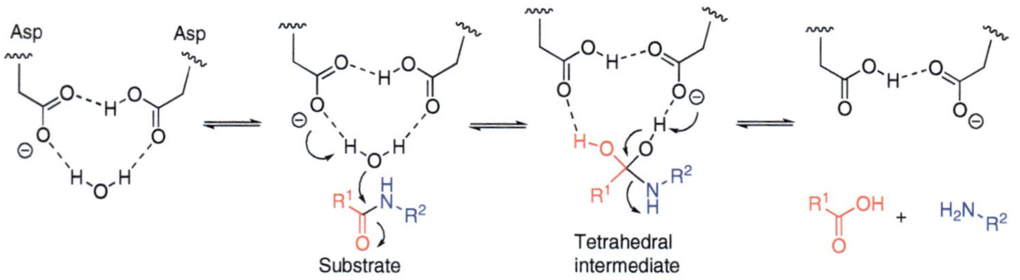

Figure 11.8 Catalytic mechanism of peptide cleavage at the active site of BACE1.

11.4 BACE1 Inhibitors

Since 2000, several small molecules with different structures have been synthesized for the inhibition of BACE1. Examples comprise pseudo-peptide inhibitors and a wide range of peptidomimetic inhibitors, also including non-peptide heterocyclic inhibitors. The development of OM99-2, an inhibitor based on the APP cleavage site sequence and on the transition-state (TS) concept utilizing a non-hydrolyzable isostere, demonstrated the druggability of BACE1 [76]. Then, the elucidation of the enzyme–ligand complex crystal structure gave further insight into the binding mode of inhibitors to BACE1 (Figure 11.9) [77].

The catalytic Asp32 and Asp228 residues are found in the catalytic site of BACE1, establishing hydrogen bonds within the active site. In the complex with OM99-2, four hydrogen bonds are formed between these catalytic residues and the hydroxyl isostere that mimics the transition state, and additional hydrogen bonds are formed between the active site and the flap region. Also, secondary binding subsites are evident, as S2 and S4 subsites display mainly hydrophilic character, whereas S1 and S3 exhibit hydrophobic properties. The P1 leucine and P3 valine residues of OM99-2 occupy the hydrophobic S1 and S3 pockets, with the P1 leucine specifically engaging Tyr71 and Phe108 of BACE1, and the P2 asparagine of the inhibitor establishes a hydrogen bond with both the P4 glutamate of OM99-2 and Arg235 of the enzyme.

BACE1 inhibitors can be classified into two main categories, peptide and non-peptide peptidomimetic inhibitors, and further classifications can be made based on the structure of the inhibitor, as reported by Ghosh in a comprehensive account of peptidomimetic BACE1 inhibitors [78].

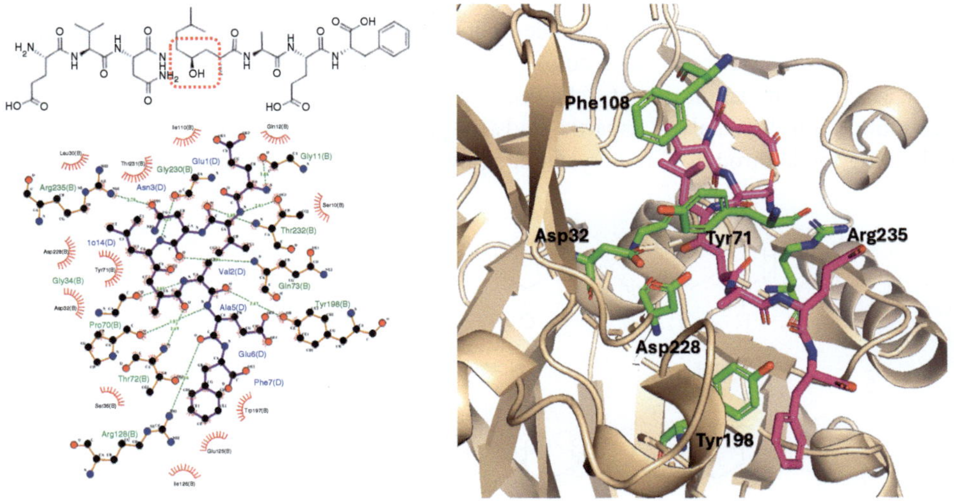

Figure 11.9 *Left: Structure and 2D-binding mode of OM99-2 and BACE1 with LigPlot+; right: close-up of the X-ray crystallographic structure of OM99-2 in the BACE1 active site; key residues are highlighted in green (PDB: 2ZHR).*

11.4.1 Peptide-Based Inhibitors

As anticipated, peptidomimetic compounds have long been a focus of research in this field, and first BACE1 inhibitors developed showed high *in vitro* potency and generally conceived to interfere with the catalytic mechanism of BACE1 by means of transition-state isostere, where a hydroxyl group displaces the water molecule at one of the two aspartic acid residues and forms hydrogen bonds with both Asp32 and Asp228, thus blocking the enzymatic activity. The most commonly used transition-state isostere for the development of peptidomimetic BACE1 inhibitors is reported in Figure 11.10, which enabled the generation of several classes of BACE1 peptidomimetics.

Statines have been used as transition-state inhibitors because of the leucine side chain at the P1 position, thus mimicking the moiety present in the Swedish mutant of APP (M671L) that increases beta-cleavage of APP by the enzyme BACE1, leading to higher levels of Aβ peptides in the brain.

Accordingly, compound **7** (Figure 11.11) was identified from a small library of statine-based peptidomimetics, showing inhibitory activity in the micromolar range in a BACE1 assay (Figure 11.11). Binding mode studies suggested complementary interactions between P1–P3 and the corresponding enzyme's subsites S1–S3.

Phenylnorstatine was included in an octapeptide structure (**8**) to enable inhibition through strong interaction of the hydroxymethyl carbonyl moiety with catalytic aspartate residues and complementary interactions between side-chain amino acids and the corresponding subsites, resulting in a sub-micromolar inhibitor [79]. Subsequent optimization of the structure to reduce the peptide character failed in maintaining enzyme inhibitory activity.

Hydroxyethylene isostere inhibitors were among the first examples of BACE1 inhibitors. These inhibitors were based on a leucine–alanine dipeptide sequence at the cleaving site in analogy with the above-cited "Swedish mutation" of APP. OM99-2 was developed using this transition-state isostere, and all this kind of peptide-based inhibitors were further elaborated to reduce the peptide character and improve pharmacokinetic properties [80]. Nevertheless, low selectivity and reduced potency were found in follow-up compounds when reducing the size of the molecule.

The hydroxyethylamine moiety showed similar behavior as those already discussed, interacting with the catalytic aspartic acids. In particular, compound **10** was studied for its binding mode through X-ray crystallography, showing the hydroxyl group interacting

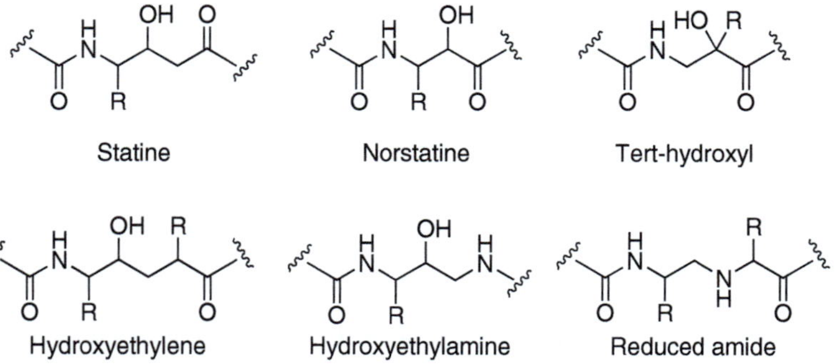

Figure 11.10 Transition-state isosteres used for developing peptidomimetic BACE1 inhibitors.

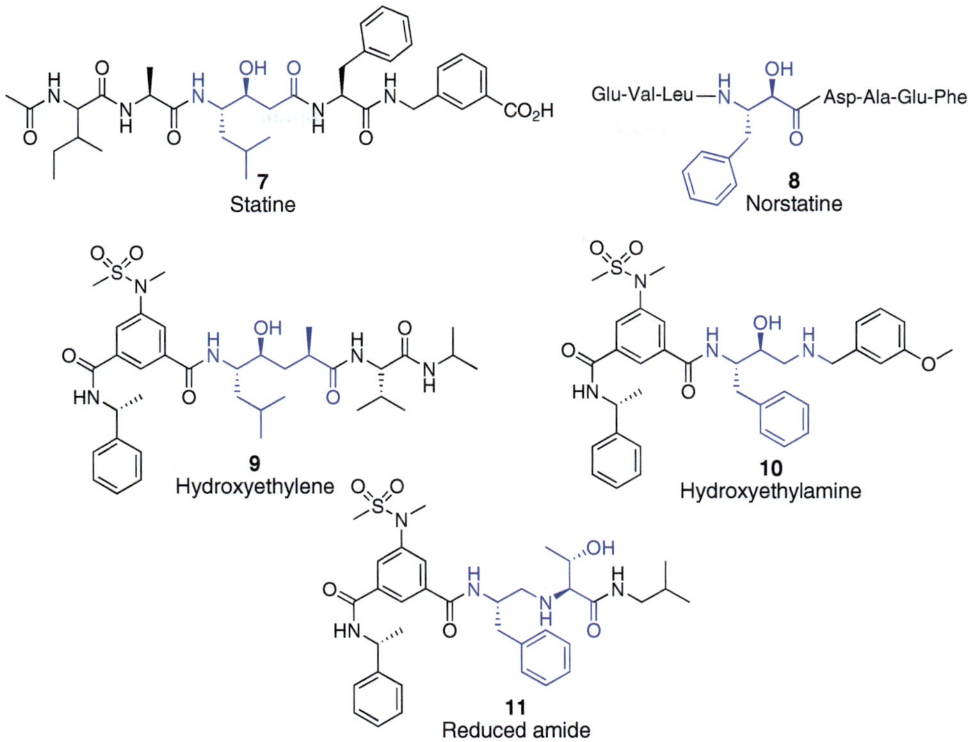

Figure 11.11 *Representative peptidomimetic BACE1 inhibitors; the transition-state isostere is highlighted in blue.*

with Asp32 and the α-amino group with Asp228. These inhibitors are characterized by the presence of conformationally constrained cyclic amines that have been incorporated to interact with the catalytic aspartates. In this class of inhibitors, the nitrogen of the reduced amide appears to bind to the catalytic aspartic acids, while other interactions are useful in promoting further binding of the inhibitor into the active site. An X-ray crystal structure of the complex between **11** and BACE1 showed the reduced amide interacting with Asp228 and Asp32, establishing a series of hydrogen bonds with neighboring residues in the catalytic site, such as Gly230, Gly34, and Ser35. This class of peptidomimetic inhibitors, although capable of strong inhibition of BACE1, proved problematic for their unfavorable pharmacokinetics, and in particular for being unstable to proteolysis and unable to cross the BBB. This prompted pharmaceutical companies to invest in the development of non-peptide inhibitors as drug candidates for the inhibition of BACE1.

11.4.2 Non-peptide Inhibitors

The identification of small molecule BACE1 inhibitors not possessing any peptide character has been based on generating compounds that are resistant to metabolic degradation. Compared with peptide-derived inhibitors, such molecules were conceived to display improved penetration of the BBB and reduced susceptibility to P-glycoprotein (Pgp) efflux.

To this end, heterocyclic scaffolds have been applied as valuable molecular frameworks in the design of non-peptide peptidomimetics. Nevertheless, significant challenges have been faced for developing non-peptide BACE1 inhibitors, as these molecules must be sufficiently decorated in functional groups to achieve high binding affinity at the active site of the enzyme, while remaining small enough to ensure acceptable pharmacokinetic behavior. Moreover, their lipophilic properties must be carefully balanced to promote BBB permeability without compromising solubility or safety. Early efforts to minimize molecular size often resulted in a marked loss of inhibitory potency. Nevertheless, subsequent optimization has yielded small molecule inhibitors with favorable CNS penetration, potent suppression of brain Aβ levels, and encouraging safety profiles in preclinical models.

Progress in this field has been facilitated by fragment-based design strategies and virtual screening approaches, both supported by high-resolution crystallographic data. The combination of such methodologies provided insight into the understanding of drug–target interactions and guided rational modifications to improve efficacy. Importantly, the use of physicochemical parameters in drug development, such as molecular weight, polar surface area, lipophilicity, hydrogen bond donor count, and conformational flexibility, has enabled fine-tuning of compounds to achieve a balance between potency, drug-like properties, and brain availability. Representative examples of this series are reported in Figure 11.12, highlighting the basic scaffold employed to interact with key aspartic acid residues in the catalytic site. Compound **12**, possessing an acyl-guanidine moiety allowed for establishing a network of hydrogen bonds with catalytic aspartic acid residues, as evinced from the crystal structure of **12** and BACE1 [81]. Similar binding mode was exhibited by compounds **13–15**, all interacting with catalytic Asp32 and Asp228 through hydrogen bonds and salt-bridge interactions taking advantage of the basicity of the central scaffold. Interestingly, the amino-hydantoin scaffold of **14** has been recently exploited for the development of highly selective BACE1 inhibitors (see par. 11.3.3).

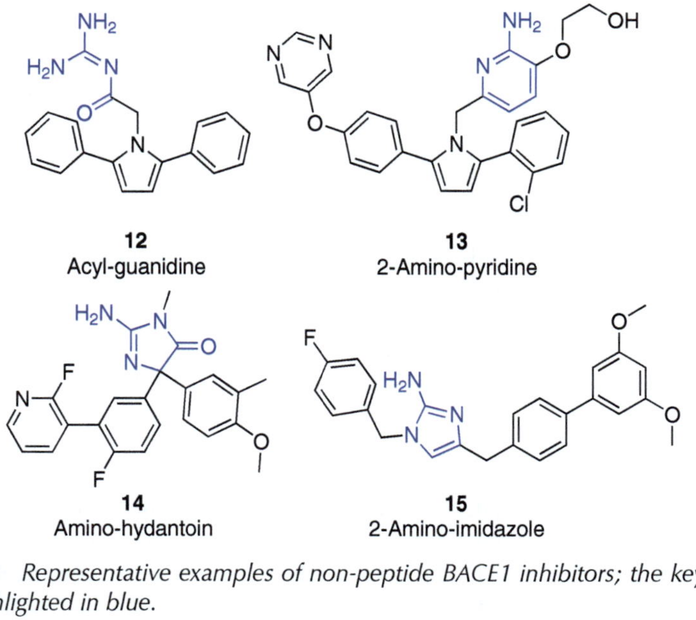

Figure 11.12 *Representative examples of non-peptide BACE1 inhibitors; the key basic scaffolds are highlighted in blue.*

11.5 BACE1 Inhibitors in Clinical Trials

Over the past decade, the therapeutic strategy of targeting BACE1 for inhibition in AD has been widely explored, and small molecule optimization has led to a wide range of preclinical candidates [82, 83], some of which entered clinical trials. Although promising in preclinical studies, they all failed under clinical evaluation in populations ranging from patients with mild-to-moderate AD, highlighting limitations in the clinical utility of BACE1 inhibitors. Four of the most important molecules that have been developed are the following (Figure 11.13):

- MK-8931 (Verubecestat), developed by Merck Research Laboratories, is based on a 3-amino-1,2,4-thidiazine scaffold, and possessing nanomolar activity against BACE1. After accomplishing the clinical phase IIa/b in 2016, the late-stage trial was stopped in February 2017, because it was found to be ineffective in the clinical treatment of patients with moderate AD [84]. The study on Verubecestat was definitively concluded in 2018 after demonstrating that the drug was unlikely to show a positive benefit/risk ratio.
- AZD3839, developed by Astra Zeneca, contains an iminopyrrolidine scaffold.
- LY-2886721, developed by Eli Lilly, is characterized by having a tetrahydro-4*H*-furo [3,4-*d*]thiazin-2-amino scaffold.
- LY3314814/AZD3293 (Lanabecestat), developed jointly by Eli Lilly and Astra Zeneca as amino-2*H*-imidazole compound, entered clinical trials in 2014 for the treatment of patients with early-onset AD. Phase III trials were discontinued in June 2018 as lanabecestat did not demonstrate sufficient efficacy in slowing cognitive or functional decline in patients with early or mild AD.

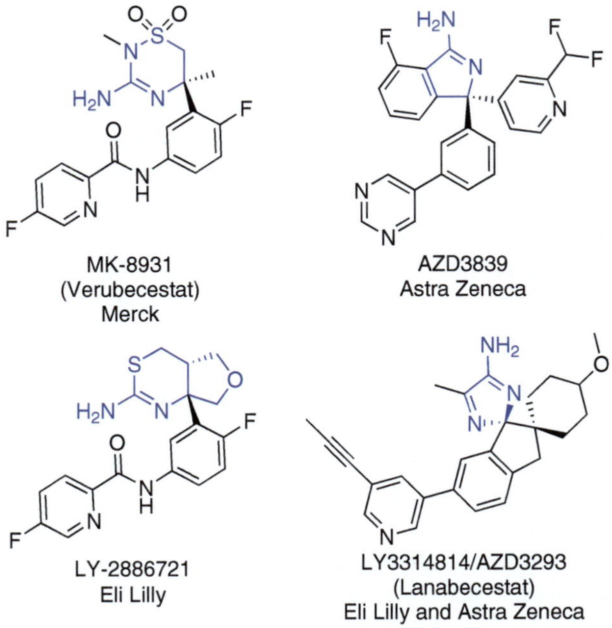

Figure 11.13 *Representative BACE1 inhibitors that entered clinical trials; the core scaffold is highlighted in blue.*

From the analysis of the molecular structure, it can be observed that they are all characterized by the presence of a basic group, such as amidine, guanidine, or an iminopyrrolidine molecular moiety. In fact, although the presence of a highly basic group within the molecular structure of the pharmacophore is necessary to establish strong hydrogen-bonding interactions with the catalytic residues Asp32 and Asp228 of BACE1, the modulation of the intrinsic basicity of this group is essential to allow the inhibitors to efficiently cross the BBB and escape the efflux mediated by P-glycoprotein [85]. While BACE1 inhibitors are capable of substantially reducing Aβ concentrations in the brain, failure of clinical trials in patients with mild-to-moderate AD raised concerns about the validity of the approach and on the design of prior studies. Nevertheless, a growing body of evidence now indicates that the timing of intervention may be a key factor, with earlier administration of BACE1 inhibitors potentially offering greater therapeutic benefit. Also, animal studies suggested that gradual BACE1 deletion can reverse amyloid pathology, indicating potential benefits if inhibition is carefully modulated [86]. Furthermore, longitudinal imaging studies in AD mouse models show that pharmacological BACE1 inhibition slows early Aβ plaque formation but is less effective against established plaques [87]. These findings emphasize the importance of timing in clinical interventions and the need to balance therapeutic efficacy with the essential cognitive functions mediated by BACE1. Therefore, experimental and clinical evidence continues to support the opinion that BACE1 inhibitors possess disease-modifying potential, and further refinement of their design and application may yet translate into benefits for patients with AD.

11.6 Toward Selective BACE1 Inhibitors

As recently reported by Ghosh in recent editorial paper [88], clinical BACE inhibitors have generally exhibited limited selectivity between BACE1 and BACE2 isoforms, which may explain in part why some adverse effects were observed during trials. In this view, developing structurally diverse and highly selective BACE1 inhibitors could clarify both the therapeutic potential of BACE1 targeting and the off-target effects associated with specific compound moieties. In a recent paper, Varghese John et al. described the discovery of an APP-selective BACE1 inhibitor that demonstrated specificity for APP as the substrate and BACE1 as the catalytic target [89]. Specifically, FAH65 based on a fluoro-aminohydantoin (FAH) scaffold was identified from a compound library screen using a maltose-binding protein-conjugated APPC125 substrate and evaluated for inhibition through IC_{50} determination with the P5–P5′ activity assay (Figure 11.14). This is a particularly useful test to determine target selectivity, as it uses a synthetic peptide substrate that contains 10 residues total, 5 upstream and 5 downstream of the cleavage site to mimic the natural context of the APP around the β-secretase cleavage site. As the assay measures how well BACE1 cleaves a sequence similar to APP under controlled conditions, it allows for giving a precise measure of the inhibitors' potency.

Lead compound FAH65 was studied for its selectivity toward other physiological substrates, showing minimal activity against neuregulin-1 (NRG1) and P-selectin glycoprotein ligand-1 (PSGL1). Also, FAH65 exhibited poor inhibition of the related aspartyl proteases cathepsin D and BACE2. Subsequent cell-based and *in vivo* assays were used for measuring the function of the lead inhibitor, showing in both assays the reduction of

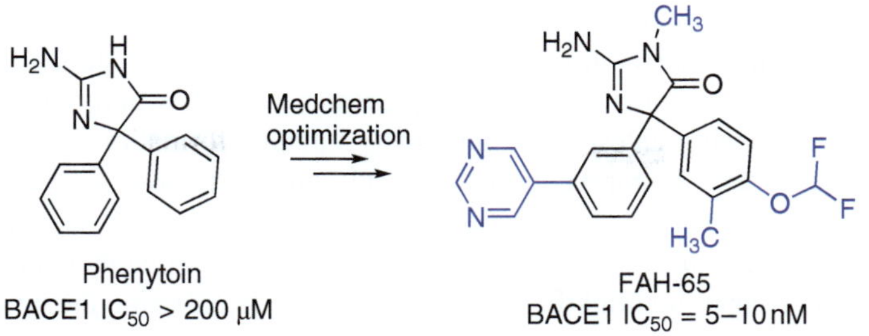

Figure 11.14 Structure of APP-selective BACE1 inhibitor FAH-65 and parent scaffold phenytoin, highlighting in blue the side-chain functional groups targeting key subsites of BACE1 active site.

the levels of BACE1-derived cleavage products, including soluble APPβ, the β *C*-terminal fragment, and Aβ species 1–40 and 1–42. Finally, preclinical assays in a murine AD model showed FAH65 improving recognition memory, confirming this molecule as a promising first-in-class oral APP-selective BACE1 inhibitor with the potential to lower Aβ burden and circumventing the adverse effects that limited the clinical success of non-selective BACE1 inhibitors.

11.7 Conclusion

AD remains a major challenge for drug discovery, as developing effective therapies requires targeting complex pathogenic pathways while ensuring safety and clinical efficacy. Over the past two decades, significant efforts have focused on identifying disease-modifying treatments for AD, particularly on the development of BACE1 inhibitors as one of the most studied approaches for AD therapy, applying peptidomimetic concepts in the design of several classes of inhibitors. Considerable advances in medicinal chemistry yielded compounds with strong potency and favorable pharmacological behavior, advancing several candidates into phase I–III trials over last decade. Despite initial promise, these inhibitors largely failed in clinical trials, and the lack of success with BACE1 and γ-secretase inhibitors has raised questions about the feasibility of targeting Aβ production, challenging aspects of the amyloid cascade hypothesis.

Nevertheless, a key limitation appears to have been the modest selectivity of most inhibitors, which showed little distinction between BACE1 and related isoforms, potentially contributing to side effects. Furthermore, the structural similarity among many of the tested compounds suggests that the chemical diversity of clinical candidates has been limited. However, the discontinuation of these trials does not eliminate the therapeutic potential of this target, as there is evidence that BACE1 inhibitors can act as disease-modifying agents by preventing the generation Aβ peptides, due to the role of BACE1 inhibition in blocking directly a central process that drives synaptic dysfunction, neurotoxicity, and disease progression. The recent approval of lecanemab, which slows cognitive decline in early-stage AD patients, reinforces the relevance of the amyloid hypothesis and highlights the possibility that BACE1-directed therapies could still serve as complementary strategies in future drug development.

In this context, recent advances in diagnostic technologies are making it increasingly feasible to identify individuals at risk of developing AD at earlier stages. This progress is expected to facilitate clinical trials at the preclinical phase, where BACE1 inhibitors are expected to exert the greatest effect on amyloid plaque formation. Growing evidence indicates that the timing of intervention is critical, with early administration of BACE1 inhibitors offering a higher likelihood of therapeutic success. Indeed, the lack of efficacy observed in past clinical trials may be partly attributable to treatments being initiated too late in the disease course, highlighting the importance of determining the optimal window for intervention.

Ongoing research also explores tau-directed therapies, natural products, neurotransmitter modulators, neurotrophic factors, metabolic interventions, and other strategies aiming to develop effective disease-modifying treatments for AD.

References

1. WHO. *Dementia*. Geneva, Switzerland: WHO, 2025. https://www.who.int/news-room/fact-sheets/detail/dementia.
2. Duthey B. *Background Paper – 6.11*. Geneva, Switzerland: World Health Organization, 2013. pp. 1–77.
3. Castellani RJ, Rolston RK, Smith MA. Alzheimer disease. *Dis Mon*. 2010;**56**:484–546.
4. Knopman DS, Amieva H, Petersen RC, Chételat G, Holtzman DM, Hyman BT, Nixon RA, Jones DT. Alzheimer disease. *Nat Rev Dis Primers*. 2021;**7**:33.
5. Small GW, Rabins PV, Barry PP, Buckholtz NS, DeKosky ST, Ferris SH, Finkel SI, Gwyther LP, Khachaturian ZS, Lebowitz BD, McRae TD, Morris JC, Oakley F, Schneider LS, Streim JE, Sunderland T, Teri LA, Tune LE. Diagnosis and treatment of Alzheimer disease and related disorders. Consensus statement of the American Association for Geriatric Psychiatry, the Alzheimer's Association, and the American Geriatrics Society. *JAMA*. 1997;**278**:1363–1371.
6. Hebert LE, Scherr PA, Beckett LA, Albert MS, Pilgrim DM, Chown MJ, Funkenstein HH, Evans DA. Age-specific incidence of Alzheimer's disease in a community population. *JAMA*. 1995;**273**:1354–1359.
7. Zhu CW, Sano M. Economic considerations in the management of Alzheimer's disease. *Clin Interv Aging*. 2006;**1**:143–154.
8. Moutinho S. Women twice as likely to develop Alzheimer's disease as men – but scientists do not know why. *Nat Med*. 2025;**31**:704–707.
9. Lane CA, Hardy J, Schott JM. Alzheimer's disease. *Eur J Neurol*. 2018;**25**:59–70.
10. Wimo A, Seeher K, Cataldi R, Cyhlarova E, Dielemann JL, Frisell O, Guerchet M, Jönsson L, Malaha AK, Nichols E, Pedroza P, Prince M, Knapp M, Dua T. The worldwide costs of dementia in 2019. *Alzheimers Dement*. 2023;**19**:2865–2873.
11. Bertram L, Tanzi RE. Thirty years of Alzheimer's disease genetics: the implications of systematic meta-analyses. *Nat Rev Neurosci*. 2008;**9**:768–778.
12. Giri M, Zhang M, Lü Y. Genes associated with Alzheimer's disease: an overview and current status. *Clin Interv Aging*. 2016;**11**:665–681.
13. Liu CC, Liu CC, Kanekiyo T, Xu H, Bu G. Apolipoprotein E and Alzheimer disease: risk, mechanisms and therapy. *Nat Rev Neurol*. 2013;**9**:106–118.
14. Alzheimer's Association. Alzheimer's disease facts and figures. *Alzheimers Dement*. 2020;**2020**(16):39–460.

15. Brookmeyer R, Johnson E, Ziegler-Graham K, Arrighi HM. Forecasting the global burden of Alzheimer's disease. *Alzheimers Dement.* 2007;**3**:186–191.
16. Peters R, Ee N, Peters J, Booth A, Mudway I, Anstey KJ. Air pollution and dementia: a systematic review. *J Alzheimer's Dis.* 2019;**70**:S145–S163.
17. Oudin A, Forsberg B, Adolfsson AN, Lind N, Modig L, Nordin M, Nordin S, Adolfsson R, Nilsson LG. Traffic-related air pollution and dementia incidence in Northern Sweden: a longitudinal study. *Environ Health Perspect.* 2016;**124**:306–312.
18. Sweeney MD, Kisler K, Montagne A, Toga AW, Zlokovic BV. The role of brain vasculature in neurodegenerative disorders. *Nat Neurosci.* 2018;**21**:1318–1331.
19. Kivipelto M, Ngandu T, Fratiglioni L, Viitanen M, Kåreholt I, Winblad B, Helkala EL, Tuomilehto J, Soininen H, Nissinen A. Obesity and vascular risk factors at midlife and the risk of dementia and Alzheimer disease. *Arch Neurol.* 2005;**62**:1556–1560.
20. Livingston G, Sommerlad A, Orgeta V, Costafreda SG, Huntley J, Ames D, Ballard C, Banerjee S, Burns A, Cohen-Mansfield J, Cooper C, Fox N, Gitlin LN, Howard R, Kales HC, Larson EB, Ritchie K, Rockwood K, Sampson EL, Samus Q, Schneider LS, Selbæk G, Teri L, Mukadam N. Dementia prevention, intervention, and care. *Lancet.* 2017;**390**:2673–2734.
21. Scarmeas N, Luchsinger JA, Schupf N, Brickman AM, Cosentino S, Tang MX, Stern Y. Physical activity, diet, and risk of Alzheimer disease. *JAMA.* 2009;**302**:627–637.
22. Barnes DE, Yaffe K. The projected effect of risk factor reduction on Alzheimer's disease prevalence. *Lancet Neurol.* 2011;**10**:819–828.
23. Wilson RS, Boyle PA, Yu L, Barnes LL, Schneider JA, Bennett DA. Life-span cognitive activity, neuropathologic burden, and cognitive aging. *Neurology.* 2013;**81**:314–321.
24. McKhann G, Drachman D, Folstein M, Katzman R, Price D, Stadlan EM. Clinical diagnosis of Alzheimer's disease: report of the NINCDS-ADRDA Work Group under the auspices of Department of Health and Human Services Task Force on Alzheimer's Disease. *Neurology.* 1984;**34**:939–944.
25. Livingston G, Huntley J, Liu KY, Costafreda SG, Selbæk G, Alladi S, Ames D, Banerjee S, Burns A, Brayne C, Fox NC, Ferri CP, Gitlin LN, Howard R, Kales HC, Kivimäki M, Larson EB, Nakasujja N, Rockwood K, Samus Q, Shirai K, Singh-Manoux A, Schneider LS, Walsh S, Yao Y, Sommerlad A, Mukadam N. Dementia prevention, intervention, and care: 2024 report of the Lancet standing Commission. *Lancet.* 2024;**404**:572–628.
26. Mach RH. New targets for the development of PET tracers for imaging neurodegeneration in Alzheimer disease. *J Nucl Med.* 2014;**55**:1221–1224.
27. Temussi PA, Masino L, Pastore A. From Alzheimer to Huntington: why is a structural understanding so difficult? *EMBO J.* 2003;**22**:355–361.
28. De Loof A, Schoofs L. Alzheimer's disease: is a dysfunctional mevalonate biosynthetic pathway the master-inducer of deleterious changes in cell physiology? *OBM Neurobiol.* 2019;**3**:046.
29. Masters CL, Simms G, Weinman NA, Multhaup G, McDonald BL, Beyreuther K. Amyloid plaque core protein in Alzheimer disease and Down syndrome. *Proc Natl Acad Sci USA.* 1985;**82**:4245–4249.
30. Kulandaivelu KS, Thinakaran G. Amyloidogenic processing of beta-amyloid precursor protein in intracellular compartments. *Neurology.* 2006;**66**:S69–S73.
31. Selkoe DJ. Amyloid beta-protein precursor: new clues to the genesis of Alzheimer's disease. *Curr Opin Neurobiol.* 1994;**4**:708–716.

32. Walter J, Capell A, Hung AY, Langen H, Schnölzer M, Thinakaran G, Sisodia SS, Selkoe DJ, Haass C. Ectodomain phosphorylation of beta-amyloid precursor protein at two distinct cellular locations. *J Biol Chem*. 1997;**272**:1896–1903.
33. Oltersdorf T, Ward PJ, Henriksson T, Beattie EC, Neve R, Lieberburg I, Fritz LC. The Alzheimer amyloid precursor protein. Identification of a stable intermediate in the biosynthetic/degradative pathway. *J Biol Chem*. 1990;**265**:4492–4497.
34. Del Prete D, Checler F, Chami M. Ryanodine receptors: physiological function and deregulation in Alzheimer disease. *Mol Neurodegener*. 2014;**9**:21.
35. Querfurth HW, LaFerla FM. Alzheimer's disease. *New Engl J Med*. 2010;**362**: 329–344.
36. Selkoe D. Alzheimer's disease: genes, proteins, and therapy. *Physiol Rev*. 2001;**81**:741–766.
37. Kirkitadze MD, Kowalska A. Molecular mechanisms initiating amyloid beta-fibril formation in Alzheimer's disease. *Acta Biochim Pol*. 2005;**52**:417–423.
38. Heneka MT, O'Banion MK. Inflammatory processes in Alzheimer's disease. *J Neuroimmunol*. 2007;**184**:69–91.
39. Tuppo EE, Arias HR. The role of inflammation in Alzheimer's disease. *Int J Biochem Cell Biol*. 2005;**37**:289–305.
40. Hardy J, Selkoe DJ. The amyloid hypothesis of Alzheimer's disease: progress and problems on the road to therapeutics. *Science*. 2002;**297**:353–356.
41. Kinney JW, Bemiller SM, Murtishaw AS, Leisgang AM, Salazar AM, Lamb BT. Inflammation as a central mechanism in Alzheimer's disease. *Alzheimers Dement (N Y)*. 2018;**4**:575–590.
42. Hardy JA, Higgins GA. Alzheimer's disease: the amyloid cascade hypothesis. *Science*. 1992;**256**:184–185.
43. Podlisny MB, Citron M, Amarante P, Sherrington R, Xia W, Zhang J, Diehl T, Levesque G, Fraser P, Haass C, Koo EH, Seubert P, St George-Hyslop P, Teplow DB, Selkoe DJ. Presenilin proteins undergo heterogeneous endoproteolysis between Thr291 and Ala299 and occur as stable *N*- and *C*-terminal fragments in normal and Alzheimer brain tissue. *Neurobiol Dis*. 1997;**3**:325–337.
44. Scheuner D, Eckman C, Jensen M, Song X, Citron M, Suzuki N, Bird TD, Hardy J, Hutton M, Kukull W, Larson E, Levy-Lahad E, Viitanen M, Peskind E, Poorkaj P, Schellenberg G, Tanzi R, Wasco W, Lannfelt L, Selkoe D, Younkin S. Secreted amyloid beta-protein similar to that in the senile plaques of Alzheimer's disease is increased in vivo by the presenilin 1 and 2 and APP mutations linked to familial Alzheimer's disease. *Nat Med*. 1996;**2**:864–870.
45. Borchelt DR, Thinakaran G, Eckman CB, Lee MK, Davenport F, Ratovitsky T, Prada CM, Kim G, Seekins S, Yager D, Slunt HH, Wang R, Seeger M, Levey AI, Gandy SE, Copeland NG, Jenkins NA, Price DL, Younkin SG, Sisodia SS. Familial Alzheimer's disease-linked presenilin 1 variants elevate Abeta1-42/1-40 ratio in vitro and in vivo. *Neuron*. 1996;**17**:1005–1013.
46. Chhabra A, Solanki S, Saravanabawan P, Venkiteswaran A, Nimmathota N, Modi NM. A systematic review of the efficacy and safety of anti-amyloid beta monoclonal antibodies in treatment of Alzheimer's disease. *Expert Opin Biol Ther*. 2024;**24**:1261–1269.

47. Yang P, Sun F. Aducanumab: the first targeted Alzheimer's therapy. *Drug Discov Ther*. 2021;**15**:166–168.
48. Sehra N, Parmar R, Jain R. Peptide-based amyloid-beta aggregation inhibitors. *RSC Med Chem*. 2025;**16**:1083–1104.
49. Goyal D, Shuaib S, Mann S, Goyal B. Rationally designed peptides and peptidomimetics as inhibitors of amyloid-β (Aβ) aggregation: potential therapeutics of Alzheimer's disease. *ACS Comb Sci*. 2017;**19**:55–80.
50. Mojarad-Jabali S, Roh KH. Peptide-based inhibitors and nanoparticles: emerging therapeutics for Alzheimer's disease. *Int J Pharm*. 2025;**669**:125055.
51. Austen BM, Paleologou KE, Ali SA, Qureshi MM, Allsop D, El-Agnaf OM. Designing peptide inhibitors for oligomerization and toxicity of Alzheimer's beta-amyloid peptide. *Biochemistry*. 2008;**47**:1984–1992.
52. Zhang H, Zhang C, Dong XY, Zheng J, Sun Y. Design of nonapeptide LVFFARKHH: A bifunctional agent against Cu^{2+} -mediated amyloid β-protein aggregation and cytotoxicity. *J Mol Recognit*. 2018;**31**:e2697.
53. Hou K, Ge P, Sawaya MR, Dolinsky JL, Yang Y, Jiang YX, Lutter L, Boyer DR, Cheng X, Pi J, Zhang J, Lu J, Yang S, Yu Z, Feigon J, Eisenberg DS. How short peptides can disassemble ultra-stable tau fibrils extracted from Alzheimer's disease brain by a strain-relief mechanism. *bioRxiv*. 2024;586668.
54. Kalita S, Kalita S, Paul A, Sarkar A, Mandal B. Peptidomimetics prepared by tail-to-side chain one component peptide stapling inhibit Alzheimer's amyloid-β fibrillogenesis. *Chem Sci*. 2020;**11**:4171–4179.
55. Wolf LM, Servoss SL, Moss MA. Peptoids: emerging therapeutics for neurodegeneration. *J Neurol Neuromed*. 2017;**2**:1–5.
56. Luo Y, Vali S, Sun S, Chen X, Liang X, Drozhzhina T, Popugaeva E, Bezprozvanny I. Aβ42-binding peptoids as amyloid aggregation inhibitors and detection ligands. *ACS Chem Neurosci*. 2013;**4**:952–962.
57. Rajasekhar K, Suresh SN, Manjithaya R, Govindaraju T. Rationally designed peptidomimetic modulators of aβ toxicity in Alzheimer's disease. *Sci Rep*. 2015;**5**:8139.
58. Liu H, Cui Y, Zhao X, Wei L, Wang X, Shen N, Odom T, Li X, Lawless W, Karunarathne K, Muschol M, Guida W, Cao C, Ye L, Cai J. Helical sulfonyl-γ-AApeptides modulating Aβ oligomerization and cytotoxicity by recognizing Aβ helix. *Proc Natl Acad Sci USA*. 2024;**121**:e2311733121.
59. Kaffy J, Berardet C, Mathieu L, Legrand B, Taverna M, Halgand F, Van Der Rest G, Maillard LT, Ongeri S. Helical γ-peptide foldamers as dual inhibitors of amyloid-β peptide and islet amyloid polypeptide oligomerization and fibrillization. *Chemistry*. 2020;**26**:14612–14622.
60. Baysah CZ, Dohoney RA, Palanikumar L, Stillman NH, Penney AL, Sola AD, Paredes DA, Magzoub M, Kumar S. A brain-penetrating foldamer rescues Aβ aggregation-associated alzheimer's disease phenotypes in in vivo models. *ACS Chem Neurosci*. 2025;**16**:1309–1322.
61. Gorantla NV, Sunny LP, Rajasekhar K, Nagaraju PG, Cg PP, Govindaraju T, Chinnathambi S. Amyloid-β-derived peptidomimetics inhibits tau aggregation. *ACS Omega*. 2021;**6**:11131–11138.
62. Gorantla NV, Chinnathambi S. Tau protein squired by molecular chaperones during Alzheimer's disease. *J Mol Neurosci*. 2018;**66**:356–368.

63. Spires-Jones TL, Attems J, Thal DR. Interactions of pathological proteins in neurodegenerative diseases. *Acta Neuropathol.* 2017;**134**:187–205.
64. Ballatore C, Brunden KR, Trojanowski JQ, Lee VM, Smith AB 3rd, Huryn DM. Modulation of protein-protein interactions as a therapeutic strategy for the treatment of neurodegenerative tauopathies. *Curr Top Med Chem.* 2011;**11**:317–330.
65. di Lorenzo D, Bisi N, Bucci R, Ennen I, Lo Presti L, Dodero V, Brandt R, Ongeri S, Gelmi ML, Tonali N. Application of modular isoxazoline-β2,2-amino acid-based peptidomimetics as chemical model systems for studying the tau misfolding. *iScience.* 2025;**28**:112272.
66. Vassar R. BACE1: the beta-secretase enzyme in Alzheimer's disease. *J Mol Neurosci.* 2004;**23**:105–114.
67. Sinha S, Anderson JP, Barbour R, Basi GS, Caccavello R, Davis D, Doan M, Dovey HF, Frigon N, Hong J, Jacobson-Croak K, Jewett N, Keim P, Knops J, Lieberburg I, Power M, Tan H, Tatsuno G, Tung J, Schenk D, Seubert P, Suomensaari SM, Wang S, Walker D, Zhao J, McConlogue L, John V. Purification and cloning of amyloid precursor protein beta-secretase from human brain. *Nature.* 1999;**402**:537–540.
68. Seubert P, Oltersdorf T, Lee MG, Barbour R, Blomquist C, Davis DL, Bryant K, Fritz LC, Galasko D, Thal LJ et al. Secretion of beta-amyloid precursor protein cleaved at the amino terminus of the beta-amyloid peptide. *Nature.* 1993;**361**:260–263.
69. Vassar R, Kovacs DM, Yan R, Wong PC. The beta-secretase enzyme BACE in health and Alzheimer's disease: regulation, cell biology, function, and therapeutic potential. *J Neurosci.* 2009;**29**:12787–12794.
70. Andreeva NS, Rumsh LD. Analysis of crystal structures of aspartic proteinases: on the role of amino acid residues adjacent to the catalytic site of pepsin-like enzymes. *Protein Sci.* 2001;**10**:2439–2450.
71. Gorfe AA, Caflisch A. Functional plasticity in the substrate binding site of beta-secretase. *Structure.* 2005;**13**:1487–1498.
72. Toulokhonova L, Metzler WJ, Witmer MR, Copeland RA, Marcinkeviciene J. Kinetic studies on beta-site amyloid precursor protein-cleaving enzyme (BACE). Confirmation of an iso mechanism. *J Biol Chem.* 2003;**278**:4582–4589.
73. Kamenetz F, Tomita T, Hsieh H, Seabrook G, Borchelt D, Iwatsubo T, Sisodia S, Malinow R. APP processing and synaptic function. *Neuron.* 2003;**37**:925–937.
74. Luo Y, Bolon B, Kahn S, Bennett BD, Babu-Khan S, Denis P, Fan W, Kha H, Zhang J, Gong Y, Martin L, Louis JC, Yan Q, Richards WG, Citron M, Vassar R. Mice deficient in BACE1, the Alzheimer's beta-secretase, have normal phenotype and abolished beta-amyloid generation. *Nat Neurosci.* 2001;**4**:231–232.
75. Ohno M, Sametsky EA, Younkin LH, Oakley H, Younkin SG, Citron M, Vassar R, Disterhoft JF. BACE1 deficiency rescues memory deficits and cholinergic dysfunction in a mouse model of Alzheimer's disease. *Neuron.* 2004;**41**:27–33.
76. Hong L, Koelsch G, Lin X, Wu S, Terzyan S, Ghosh AK, Zhang XC, Tang J. Structure of the protease domain of memapsin 2 (beta-secretase) complexed with inhibitor. *Science.* 2000;**290**:150–153.
77. Shimizu H, Tosaki A, Kaneko K, Hisano T, Sakurai T, Nukina N. Crystal structure of an active form of BACE1, an enzyme responsible for amyloid beta protein production. *Mol Cell Biol.* 2008;**28**:3663–3671.

78. Ghosh AK, Osswald HL. BACE1 (β-secretase) inhibitors for the treatment of Alzheimer's disease. *Chem Soc Rev*. 2014;**43**:6765–6813.
79. Shuto D, Kasai S, Kimura T, Liu P, Hidaka K, Hamada T, Shibakawa S, Hayashi Y, Hattori C, Szabo B, Ishiura S, Kiso Y. KMI-008, a novel beta-secretase inhibitor containing a hydroxymethylcarbonyl isostere as a transition-state mimic: design and synthesis of substrate-based octapeptides. *Bioorg Med Chem Lett*. 2003;**13**:4273–4276.
80. Tang J, Hong L, Ghosh AK. The discovery of β-secretase and development toward a clinical inhibitor for AD: an exciting academic collaboration. In: Ghosh AK, editor. *Aspartic Acid Proteases as Therapeutic Targets*, vol. **45**. Weinheim, Germany: Wiley-VCH, 2010. pp. 413–440.
81. Cole DC, Manas ES, Stock JR, Condon JS, Jennings LD, Aulabaugh A, Chopra R, Cowling R, Ellingboe JW, Fan KY, Harrison BL, Hu Y, Jacobsen S, Jin G, Lin L, Lovering FE, Malamas MS, Stahl ML, Strand J, Sukhdeo MN, Svenson K, Turner MJ, Wagner E, Wu J, Zhou P, Bard J. Acylguanidines as small-molecule beta-secretase inhibitors. *J Med Chem*. 2006;**49**:6158–6161.
82. Hsiao CC, Rombouts F, Gijsen HJM. New evolutions in the BACE1 inhibitor field from 2014 to 2018. *Bioorg Med Chem Lett*. 2019;**29**:761–777.
83. Coimbra JRM, Marques DFF, Baptista SJ, Pereira CMF, Moreira PI, Dinis TCP, Santos AE, Salvador JAR. Highlights in BACE1 inhibitors for Alzheimer's disease treatment. *Front Chem*. 2018;**6**:178.
84. Kennedy ME, Stamford AW, Chen X, Cox K, Cumming JN, Dockendorf MF, Egan M, Ereshefsky L, Hodgson RA, Hyde LA, Jhee S, Kleijn HJ, Kuvelkar R, Li W, Mattson BA, Mei H, Palcza J, Scott JD, Tanen M, Troyer MD, Tseng JL, Stone JA, Parker EM, Forman MS. The BACE1 inhibitor verubecestat (MK-8931) reduces CNS β-amyloid in animal models and in Alzheimer's disease patients. *Sci Transl Med*. 2016;**8**:363ra150.
85. Scarpini E, Scheltens P, Feldman H. Treatment of Alzheimer's disease: current status and new perspectives. *Lancet Neurol*. 2003;**2**:539–547.
86. Hu X, Das B, Hou H, He W, Yan R. BACE1 deletion in the adult mouse reverses preformed amyloid deposition and improves cognitive functions. *J Exp Med*. 2018;**215**: 927–940.
87. Peters F, Salihoglu H, Rodrigues E, Herzog E, Blume T, Filser S, Dorostkar M, Shimshek DR, Brose N, Neumann U, Herms J. BACE1 inhibition more effectively suppresses initiation than progression of β-amyloid pathology. *Acta Neuropathol*. 2018; **135**:695–710.
88. Ghosh AK. BACE1 inhibitor drugs for the treatment of Alzheimer's disease: lessons learned, challenges to overcome, and future prospects. *Glob Health Med*. 2024;**6**:164–168.
89. Campagna J, Jagodzinska B, Wi D, Zhu C, Lee J, Cohn W, Jun M, Elias C, Padder S, Descamps O, Peters-Libeu C, Zhang Q, Gorostiza O, Poksay K, Spilman P, Bredesen D, John V. Discovery of an APP-selective BACE1 inhibitor for Alzheimer's disease. *Neurotherapeutics*. 2025;e00610.

12

Peptidomimetic Drugs

12.1 Introduction

Herein, a series of peptidomimetic drugs that entered the clinic is summarized, with most of the reported molecules having achieved final approval as diagnostic or therapeutic drugs. The discussed peptidomimetics have been developed for a wide array of medical indications, encompassing multiple areas of medicine, thus demonstrating the wide applicability of the peptidomimetic approach in drug discovery. Specifically, the first group accounts for experimental and approved treatments toward infectious diseases, whereas drugs described in the second group deal with cancer therapeutics. Finally, the third group addresses other pathologies such as Rett syndrome and adult growth hormone dysfunction.

The peptidomimetic drugs herein discussed have been selected among those developed over the last two decades, thus not including other well-known peptidomimetic drugs that are undoubtedly considered as milestones in the development of peptidomimetic drugs, such as inhibitors of HIV-protease, ACE inhibitors, bortezomib, first-generation thrombin inhibitors, and first-generation hepatitis C virus (HCV) drugs as protease inhibitors. The compound data sheets contain information about main identification IDs and molecular descriptors such as molecular weight, chemical formula and structure, SMILES code, and data about medical indication and approval date/clinical stages of investigation. Data have been retrieved from DrugBank, FDA Orange Book, Wikipedia, and the Protein Data Bank (PDB). A short description of the key features of the molecule, such as the binding mode, drug mechanism, and information about clinical studies, concludes the characterization of each peptidomimetic drug. The chemical structures of the discussed drugs show a different degree of peptide character. For example, murepavadin, trofinetide, tavilermide, difelikefalin, and macimorelin retain most of the peptide character, whereas venetoclax, brilacidin, and voxilaprevir show a marked non-peptide chemical structure. Nirmatrelvir, ixazomib, and vipivotide (in the warhead portion of the molecule) result in a balanced contribution of peptide structure and local modifications in the overall molecular framework (Figure 12.1).

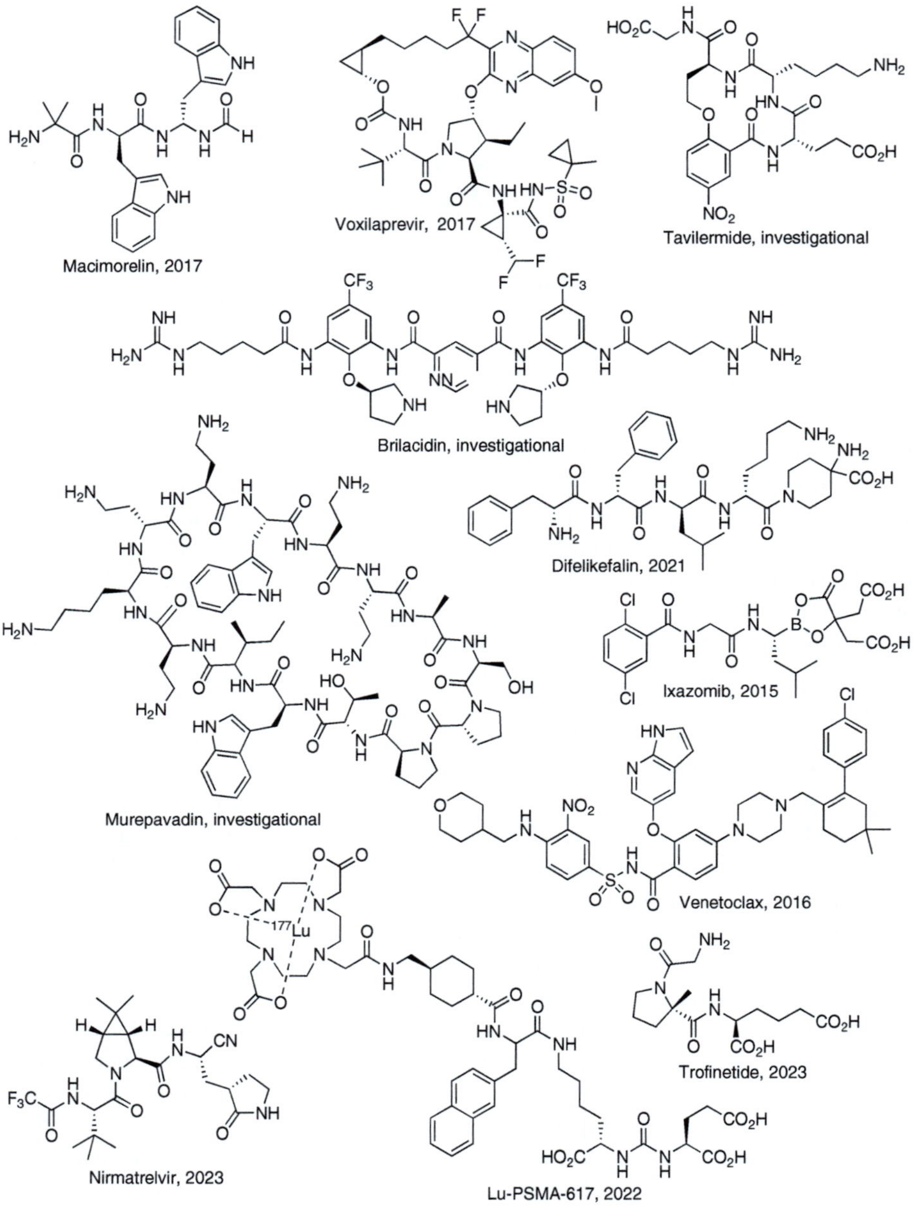

Figure 12.1 *Chemical structure of approved and investigational peptidomimetic drugs herein discussed.*

12.2 Nirmatrelvir

Peptidomimetic inhibitor of the SARS-CoV-2 main protease (Mpro) is shown in Figure 12.2.

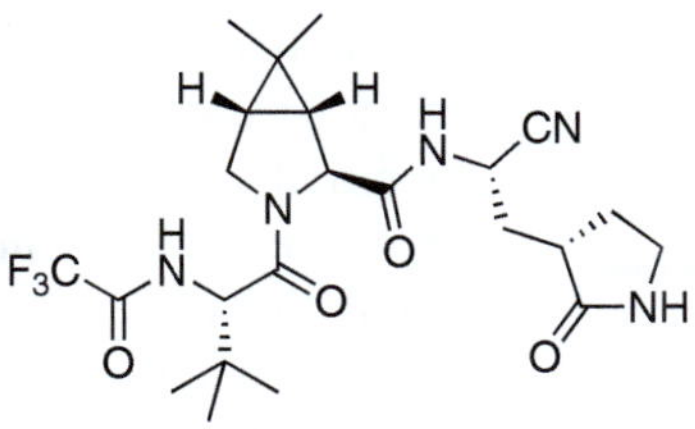

Figure 12.2 *Chemical structure of nirmatrelvir.*

Generic name (INN)	Nirmatrelvir
Brand name	PAXLOVID
Company	Pfizer Inc.
Chemical formula	$C_{23}H_{32}F_3N_5O_4$
Molecular weight	499.535 $g\,mol^{-1}$
SMILES code	CC1(C2C1C(N(C2)C(=O)C(C(C)(C)C)NC(=O)C(F)(F)F)C(=O)NC(CC3CCNC3=O)C#N)C
DrugBank ID	DBDB16691
PubChem ID	155903259
PDB ligand	ZGW
Date of approval/clinical stage	December 22, 2021 (emergency use authorization) May 25, 2023
Medical indication (FDA label) [1]	PAXLOVID, which includes nirmatrelvir, a severe acute respiratory syndrome coronavirus 2 (SARS-CoV-2) main protease (Mpro: also referred to as 3CLpro or nsp5 protease) inhibitor, and ritonavir, an HIV-1 protease inhibitor and CYP3A inhibitor, is indicated for the treatment of mild-to-moderate coronavirus disease 2019 (COVID-19) in adults who are at high risk for progression to severe COVID-19, including hospitalization or death.
Description and mechanism of action (FDA label)	PAXLOVID is nirmatrelvir tablets co-packaged with ritonavir tablets. Nirmatrelvir is a SARS-CoV-2 main protease (Mpro) inhibitor, and ritonavir is an HIV-1 protease inhibitor and CYP3A inhibitor. Nirmatrelvir is a severe acute respiratory syndrome coronavirus 2 (SARS-CoV-2) antiviral drug. Ritonavir is an HIV-1 protease inhibitor but is not active against SARS-CoV-2 Mpro. Ritonavir inhibits the CYP3A-mediated metabolism of nirmatrelvir, resulting in increased plasma concentrations of nirmatrelvir.

Nirmatrelvir is a peptidomimetic inhibitor of the SARS-CoV-2 main protease (Mpro), also referred to as 3C-like protease (3CLpro) or nonstructural protein 5 (nsp5) protease. Inhibition of SARS-CoV-2 Mpro renders it incapable of processing the viral polyproteins pp1a and pp1ab, preventing viral replication. Nirmatrelvir inhibited the activity of recombinant SARS-CoV-2 Mpro in a biochemical assay with a Ki value of 3.1 nM and an IC_{50} value of 19.2 nM. Nirmatrelvir was found to bind directly to the SARS-CoV-2 Mpro active site by X-ray crystallography.

As discussed in Chapter 8, the 3-chymotrypsin-like protease (3CLpro), also known as the main protease (Mpro), is a cysteine protease that plays an essential role in the replication of coronaviruses by processing viral polyproteins into functional nonstructural proteins [2]. Its inhibition is crucial in impairing viral replication, thus representing one of the most validated antiviral drug targets. Unlike other cysteine proteases such as papain or cathepsins, Mpro requires a glutamine residue at the P1 position for substrate recognition, which is not present in any human cysteine proteases [3], thus providing a valuable feature for achieving selective inhibition of this enzyme and minimizing off-target effects in human cells. Mpro is active as a homodimer, and it is characterized by possessing a catalytic activity operated by a dyad formed by Cys145 and His41. The specificity of substrate–enzyme molecular recognition is addressed through six subsites (S1′–S5), which interact with corresponding substrate residues, and show small amino acids such as glycine, alanine, or serine at the key P1′ position. Advances in X-ray crystallography and cryo-electron microscopy have elucidated the active site of Mpro and substrate-binding pockets, facilitating the rational design of inhibitors. Based on its mechanism, peptidomimetic inhibitors have been developed following two major approaches, namely covalent and non-covalent. These inhibitors consist of warhead groups that interact directly with the catalytic cysteine and form reversible or irreversible covalent bonds depending on their electronic and steric properties. The development of covalent inhibitors targeting Cys145 has been achieved using aldehyde, α-ketoamide, ketone, or Michael acceptor warheads, which have been studied since the SARS-CoV-1 outbreak (2002–2003), paving the way for rapid structural optimization during the COVID-19 pandemic. Among the early candidates, Pfizer's PF-00835231 (lufotrelvir), incorporating a hydroxymethylketone warhead, proved to act as a potent and reversible covalent inhibitor of Mpro with an IC_{50} of 0.27 nM [4]. Structural studies confirmed its high-affinity binding to the protease active site through interaction with Cys145. However, its poor oral bioavailability limited clinical development to intravenous administration. Thus, a phosphate prodrug form was created to enhance the solubility for continuous infusion in hospitalized patients. Insights gained from lufotrelvir informed subsequent structure–activity relationship (SAR) optimizations aimed at improving potency and pharmacokinetic properties. Through systematic modifications consisting of the removal of hydrogen-bond donors and reduction of the peptide character,

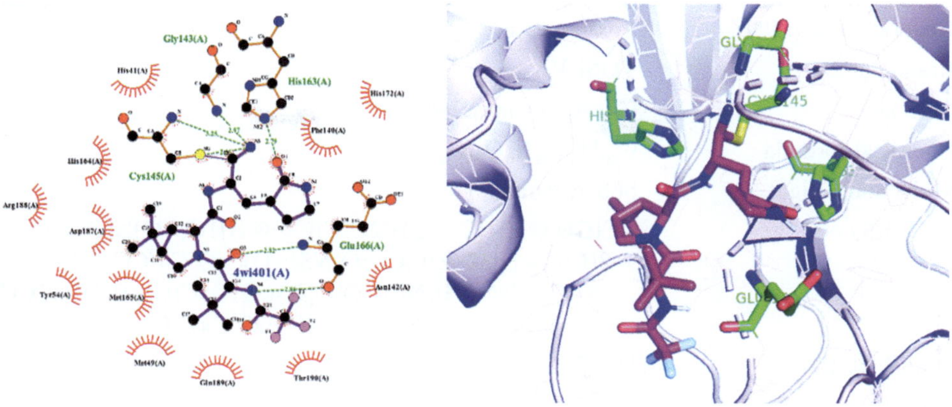

Figure 12.3 *Left: binding mode of nirmatrelvir in the active site of 3CLpro protease (LigPlot+); right: X-ray crystallographic structure of nirmatrelvir in the protease active site (PDB: 7RFW).*

Pfizer scientists retained the P1 γ-lactam as a Gln mimetic while experimenting with new warheads such as nitrile and benzothiazol-2-yl ketone. The nitrile warhead, combined with optimized P2–P4 residues, including a proline derivative and a CF_3-substituted tert-leucine (Figure 12.3), yielded nirmatrelvir [5]. Co-administered with ritonavir to inhibit CYP3A4 metabolism, as formulated in Paxlovid, this combination received Emergency Use Authorization in December 2021, representing a major milestone in structure-based coronavirus drug design [6].

12.3 Voxilaprevir

Non-covalent, reversible peptidomimetic inhibitor of HCV NS3/4A protease is shown in Figure 12.4.

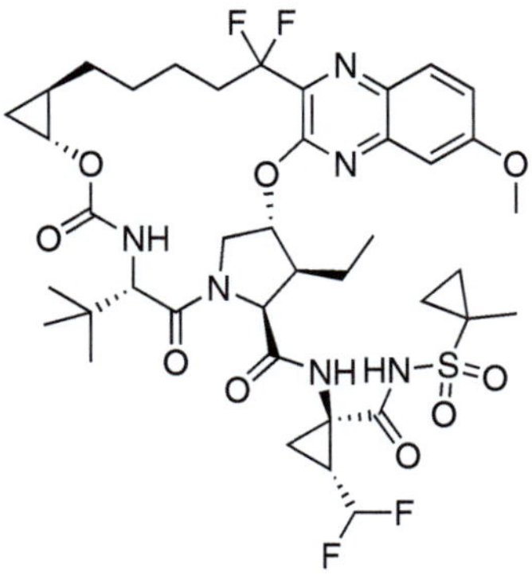

Figure 12.4 *Chemical structure of voxilaprevir.*

Generic name (INN)	Voxilaprevir
Brand name	VOSEVI
Company	Gilead Sciences Inc.
Chemical formula	$C_{40}H_{52}F_4N_6O_9S$
Molecular weight	868.94 g mol^{-1}
SMILES code	CC[C@@H]1[C@@H]2CN([C@@H]1C(=O)N[C@@]3(C[C@H]3C(F)F)C(=O)NS(=O)(=O)C4(CC4)C)C(=O)[C@@H](NC(=O)O[C@@H]5C[C@H]5CCCCC(c6c(nc7cc(ccc7n6)OC)O2)(F)F)C(C)(C)C
DrugBank ID	DB12026
PubChem ID	89921642
PDB ligand	L9P
Date of approval/clinical stage	July 18, 2017
Medical indication (FDA label) [7]	VOSEVI is a fixed-dose combination of sofosbuvir, a hepatitis C virus (HCV) nucleotide analog NS5B polymerase inhibitor, velpatasvir, an HCV NS5A inhibitor, and voxilaprevir, an HCV NS3/4A protease inhibitor and is indicated for the treatment of adult patients with chronic HCV infection without cirrhosis or with compensated cirrhosis (Child-Pugh A) who have (1, 2.2, 14): • genotype 1, 2, 3, 4, 5, or 6 infection and have previously been treated with an HCV regimen containing an NS5A inhibitor. • genotype 1a or 3 infection and have previously been treated with an HCV regimen containing sofosbuvir without an NS5A inhibitor. • Additional benefit of VOSEVI over sofosbuvir/velpatasvir was not shown in adults with genotype 1b, 2, 4, 5, or 6 infection previously treated with sofosbuvir without an NS5A inhibitor.
Description and mechanism of action (FDA label)	VOSEVI is a fixed-dose combination of sofosbuvir, velpatasvir, and voxilaprevir which are direct-acting antiviral (DAA) agents against the hepatitis C virus. Voxilaprevir is a non-covalent, reversible inhibitor of the NS3/4A protease, which is necessary for the proteolytic cleavage of the HCV-encoded polyprotein (into mature forms of the NS3, NS4A, NS4B, NS5A, and NS5B proteins) and is essential for viral replication. In a biochemical inhibition assay, voxilaprevir inhibited the proteolytic activity of recombinant NS3/4A enzymes from clinical isolates of HCV genotypes 1b and 3a with Ki values of 38 and 66 pM, respectively.

Voxilaprevir is an antiviral drug used in the treatment of chronic HCV infection. It is a direct-acting antiviral (DAA) and works as an NS3/4A protease inhibitor, thus targeting a key step of the HCV replication process. The introduction of direct-acting antivirals has revolutionized the management of HCV infection, offering sustained virologic response rates exceeding 95% with short treatment protocols and favorable tolerability profiles. However, the emergence of viral resistance, particularly because of NS5A inhibitor-containing regimens, raised a significant issue for achieving wide-ranging therapeutic efficacy. Thus, voxilaprevir was designed as a next-generation protease inhibitor with enhanced potency and a broad resistance profile to overcome viral escape mechanisms.

Hepatitis C virus is a member of the Flaviviridae family and represents one of the most prevalent viral infections worldwide, with an estimated 71 million people chronically infected worldwide [8, 9]. This infection is strongly associated with a spectrum of hepatic disorders, including chronic hepatitis, liver cirrhosis, and hepatocellular carcinoma, contributing significantly to liver-related mortality. HCV replication consists of a highly error-prone RNA-dependent RNA polymerase, which enables rapid viral evolution and genetic diversity. The viral genome encodes two proteases, namely the NS2 autoprotease and the NS3 serine protease, that mediate the cleavage of the viral polyprotein to generate three structural and seven nonstructural proteins.

The nucleophilic property of the catalytic serine residue within the HCV NS3/4A protease has been effectively exploited in antiviral drug development, giving rise to the first generation of covalent inhibitors, including telaprevir and boceprevir. Subsequent second-generation inhibitors did not apply the classical ketoamide warhead structure but contained large aromatic substituents linked to the P2 position through ether bridges. To further improve the binding affinity through conformational pre-organization, macrocyclic scaffolds were exploited at P1–P3 (e.g., simeprevir) and P2–P4 (e.g., grazoprevir) molecular portions. Third-generation inhibitors, exemplified by voxilaprevir and glecaprevir, retained the P2–P4 macrocyclic framework but were optimized to overcome therapeutic limitations in patients infected with genotype 3 HCV. The structure of these compounds closely resembles grazoprevir but includes fluorine atoms that confer broad and pan-genotypic antiviral activity. Among the 12 NS3/4A protease inhibitors approved by the FDA to date, only five continue to be used clinically, largely due to the persistence of viral resistance mechanisms. Macrocyclization has increasingly emerged as a valuable approach for the synthesis of protease inhibitors. This strategy proved effective to address relatively shallow enzyme surfaces that mediate essential protein–protein interactions (PPIs) while simultaneously imposing conformational constraints to minimize entropic penalties upon target binding, thus improving the inhibitory potency [10].

Voxilaprevir was approved as a component of the pan-genotypic STR Vosevi® (sofosbuvir, velpatasvir, and voxilaprevir) [9]. This combination drug provides a treatment option for the most difficult-to-cure patients who have failed prior therapies with other direct-acting antivirals. Key interactions of this peptidomimetic drug within the active site of the enzyme are summarized as follows (Figure 12.5):

- binding to the catalytic triad, particularly hydrogen bonds are established between the quinoxaline moiety of voxilaprevir with His57 and Asp81; the interaction of the P1 group of the inhibitor with Ser139 facilitates the formation of a covalent bond

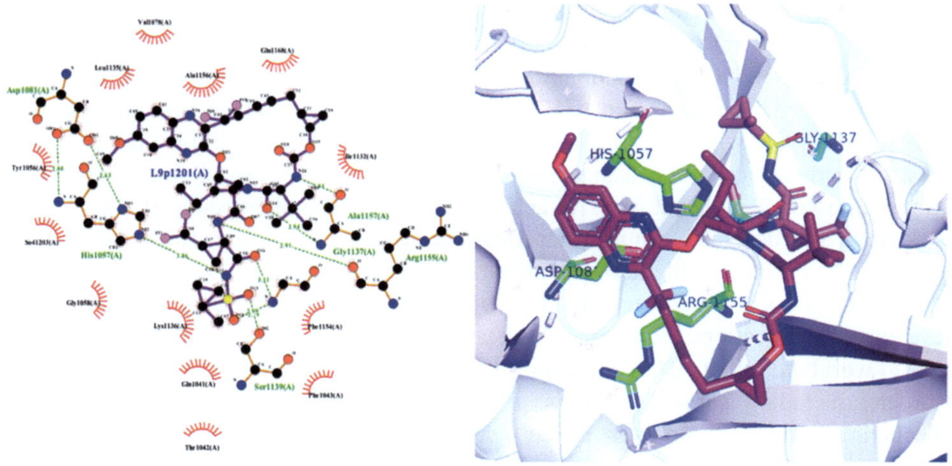

Figure 12.5 Left: binding mode of voxilaprevir in the active site of HCV NS3/4A protease (LigPlot+); right: X-ray crystallographic structure of voxilaprevir in the protease active site (PDB: 6NZT).

- electrophilic interactions established by the difluoromethylene moiety of voxilaprevir and particularly fluorine-specific interactions, including hydrogen-bonding with the backbone carbonyl oxygen of the protease and the carbonyl oxygen of the P1 group, allowing for pre-organizing the inhibitor for the interaction and improving its potency against proteases from all genotypes
- hydrophobic interactions displayed by the cyclopropyl group at the P4 position of voxilaprevir occupying the S4 pocket of the protease and contributing to the overall binding affinity of the inhibitor.

12.4 Brilacidin

A defensin mimetic acting on the bacterial cell membrane, also referred to as an antimicrobial peptidomimetic, is shown in Figure 12.6.

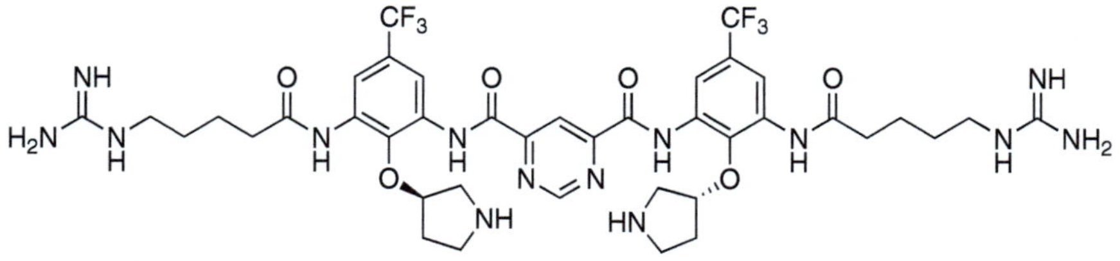

Figure 12.6 Chemical structure of Brilacidin.

Generic name (INN)	Brilacidin
Brand name	N/A
Company	Innovation Pharmaceuticals Inc.
Chemical formula	$C_{40}H_{50}F_6N_{14}O_6$
Molecular weight	936.922 g mol^{-1}
SMILES code	C1CNC[C@@H]1OC2=C(C=C(C=C2NC(=O)C3=CC(=NC=N3)C(=O)NC4=CC(=CC(=C4O[C@@H]5CCNC5)NC(=O)CCCCNC(=N)N)C(F)(F)F)C(F)(F)F)NC(=O)CCCCNC(=N)N
DrugBank ID	DB12997
PubChem ID	25023695
PDB ligand	N/A
Date of approval/clinical stage	Investigational; Phase 2 completed
Medical indication	Brilacidin is under investigation for the supportive care of Mucositis, Stomatitis, Mouth Diseases, and Head and Neck Neoplasms.
Description (FDA label)	N/A

Antibacterial peptides constitute a class of innate antimicrobial agents naturally present in the tissues and cells of humans and animals, particularly within leukocytes. These peptides play a crucial role in the host's innate immune defense by protecting against bacterial infections. The first peptide to advance into clinical development was magainin [11] that was originally isolated from the skin of amphibians. Nevertheless, its development was discontinued due to challenges related to large-scale production, molecular stability, and lack of antimicrobial potency. Since the identification of magainin, numerous antibacterial peptides have been characterized, thus expanding their structural diversity and improving the knowledge of biological functions.

Brilacidin is a synthetic, non-peptide small-molecule peptidomimetic designed to mimic the structure and function of host defense peptides (HDPs), particularly defensins [12]. Defensins are a family of antimicrobial peptides with a cationic and amphipathic profile and capable of disrupting microbial cell membranes and inhibiting pathogen growth. Brilacidin was developed through *de novo* design to replicate the amphiphilic and cationic properties of HDPs, thus enabling the disruption of microbial membranes and the modulation of immune responses. The development of brilacidin was achieved by William F. DeGrado through the application of computational modeling to design small-molecules able to mimic the structural features of HDPs [13–15]. By focusing on the amphiphilic nature and positive charge distribution of defensins, arylamide foldamers capable to maintain the ability to interact with and disrupt microbial membranes were developed, offering improved stability and bioavailability as compared to natural peptides. The development of brilacidin highlights the importance of peptidomimetic drugs in addressing the growing threat of antibiotic resistance. By mimicking natural antimicrobial peptides of the human immune system, brilacidin proved as an innovative approach to fight infections resistant to traditional antibiotics. While some studies have shown promising results, brilacidin has not yet progressed to Phase 3 trials or received regulatory approval. It remains an investigational drug under clinical development by Innovation Pharmaceuticals.

12.5 Murepavadin

A 14-amino-acid macrocyclic peptide antibiotic specifically targeting *Pseudomonas aeruginosa* is shown in Figure 12.7.

Generic name (INN)	Murepavadin
Brand name	N/A
Company	Polyphor Ltd.
Chemical formula	$C_{73}H_{112}N_{22}O_{16}$
Molecular weight	1,553.837 g mol^{-1}
SMILES code	CC[C@H](C)[C@H]1C(=O)N[C@H](C(=O)N[C@H](C(=O)N[C@@H](C(=O)N[C@H](C(=O)N[C@H](C(=O)N[C@H](C(=O)N[C@H](C(=O)N[C@H](C(=O)N[C@H](C(=O)N2CCC[C@@H]2C(=O)N3CCC[C@H]3C(=O)N[C@H](C(=O)N[C@H](C(=O)N1)Cc4c[nH]c5c4cccc5)[C@@H](C)O)CO)C)CCN)CCN)Cc6c[nH]c7c6cccc7)CCN)CCN)CCCN)CCN
DrugBank ID	DB14777
PubChem ID	91824766
PDB ligand	N/A
Date of approval/clinical stage	Investigational; Phase 3
Medical indication	Murepavadin (also known as POL7080) is a peptidomimetic antibiotic developed to treat serious infections caused by *Pseudomonas aeruginosa.*
Description (FDA label)	N/A

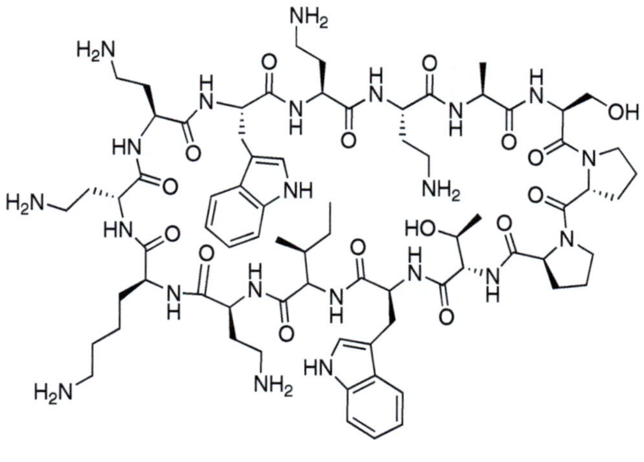

Figure 12.7 Chemical structure of murepavadin.

Murepavadin (POL7080) is a *Pseudomonas*-specific peptidomimetic antibiotic and the first representative of a new class of outer membrane protein-targeting agents. It is a synthetic cyclic β-hairpin peptidomimetic derived from the host defense peptide protegrin I (PG-1) [16]. Unlike conventional broad-spectrum antibiotics, murepavadin acts by targeting lipopolysaccharide transport protein D (LptD), which is a key component in the biogenesis of the outer membrane in *Pseudomonas aeruginosa*. This unique mechanism disrupts lipopolysaccharide transport, causing alterations in the bacterial outer membrane and resulting in cell death.

PG-1, the reference epitope for murepavadin, displays potent antimicrobial activity through membrane disruption, but also shows toxicity and hemolytic properties that limit its therapeutic use. Thus, researchers designed a 14-residue epitope mimetic library using a D-proline/L-proline scaffold to mimic PG-1's β-hairpin conformation. Through iterative optimization, analogues such as L8-1 demonstrated strong antibacterial activity with minimal hemolysis. Follow-up structure–activity relationship studies highlighted the importance of specific residues for antimicrobial potency, such as tryptophan at positions 2 and 8. Conformational studies using nuclear magnetic resonance confirmed that a stable β-hairpin structure is an essential requirement for the biological activity. These investigations resulted in the discovery of murepavadin, which retained potent *Pseudomonas*-specific efficacy, high plasma stability, as well as nonhemolytic behavior at therapeutic concentrations.

Murepavadin proved to exhibit nanomolar-level activity against *P. aeruginosa*, including carbapenemase-producing and colistin-resistant strains, with minimal inhibitory concentrations (MIC_{90}) of approximately $0.12\,mg\,L^{-1}$ across over 1,200 global clinical isolates. The compound showed no cross-resistance with existing antibiotic classes and negligible activity against other Gram-negative or Gram-positive bacteria, thus demonstrating its selectivity through a novel mechanism. In preclinical models, murepavadin achieved exceptional efficacy, with effective doses ranging from 0.25 to 0.55 mg/kg. Pharmacokinetic studies demonstrated strong lung tissue penetration, supporting its development for pneumonia treatment. Clinical evaluations further confirmed the potential of murepavadin. In fact, Phase I trials showed good tolerance of the compound with no serious adverse events at plasma concentrations consistent with therapeutic efficacy. Such favorable pharmacological and toxicological findings enabled murepavadin to advance to Phase II trials, demonstrating efficacy in hospital-acquired *Pseudomonas* infections. It subsequently progressed to Phase III clinical development for the treatment of hospital-acquired and ventilator-associated bacterial pneumonia caused by *P. aeruginosa* [17]. Originally administered intravenously, the Phase III trial for intravenous use in ventilator-associated bacterial pneumonia was terminated due to safety concerns, particularly connected to renal toxicity, in the patient cohort. Subsequently, development has shifted to an inhaled formulation of murepavadin, intended for lung infections, including chronic and resistant *P. aeruginosa* infections in cystic fibrosis and non-cystic fibrosis bronchiectasis.

12.6 Ixazomib

Peptidomimetic bortezomib-derived boronate as a proteasome inhibitor is shown in Figure 12.8.

Generic name (INN)	Ixazomib citrate
Brand name	NINLARO
Company	Takeda Pharmaceuticals USA Inc.
Chemical formula	$C_{14}H_{19}BCl_2N_2O_4$
Molecular weight	361.03 g mol^{-1}
SMILES code	CC(C)C[C@H](NC(=O)CNC(=O)C1=CC(Cl)=CC=C1Cl)B(O)O
DrugBank ID	DB09570
PubChem ID	25183872
PDB ligand	N/A
Date of approval/clinical stage	Nov 20, 2015
Medical indication (FDA label) [18]	NINLARO is a proteasome inhibitor indicated in combination with lenalidomide and dexamethasone for the treatment of patients with multiple myeloma who have received at least one prior therapy.
Description and mechanism of action (FDA label)	NINLARO (ixazomib) is an antineoplastic agent. Ixazomib citrate, a prodrug, rapidly hydrolyzes under physiological conditions to its biologically active form, ixazomib. Ixazomib is a reversible proteasome inhibitor. Ixazomib preferentially binds and inhibits the chymotrypsin-like activity of the beta 5 subunit of the 20S proteasome. Ixazomib induced apoptosis of multiple myeloma cell lines *in vitro*. Ixazomib demonstrated *in vitro* cytotoxicity against myeloma cells from patients who had relapsed after multiple prior therapies, including bortezomib, lenalidomide, and dexamethasone. The combination of ixazomib and lenalidomide demonstrated synergistic cytotoxic effects in multiple myeloma cell lines. *in vivo*, ixazomib demonstrated antitumor activity in a mouse multiple myeloma tumor xenograft model.

The development of proteasome inhibitors has significantly advanced cancer therapeutics. Bortezomib, discovered through optimization of an aldehyde-containing proteasome substrate peptide and approved in 2003 as the first-in-class peptidomimetic boronate proteasome inhibitor [19] demonstrated potent anticancer activity and significantly

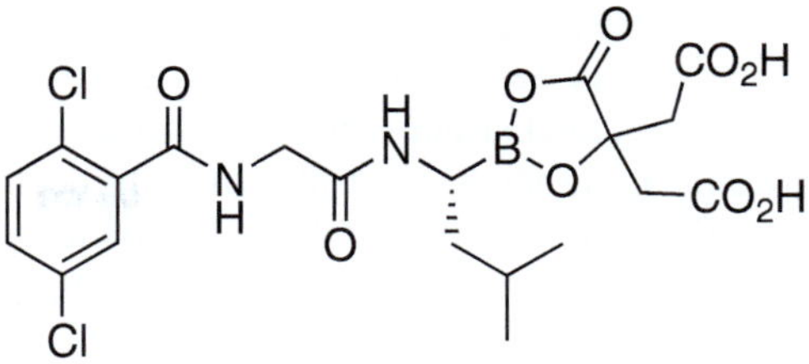

Figure 12.8 Chemical structure of ixazomib.

improved the overall survival and quality of life of multiple myeloma patients, although requiring parenteral administration. Subsequent research led to the discovery and approval of ixazomib as the first orally bioavailable proteasome inhibitor developed for multiple myeloma therapy as a second-generation analog of bortezomib. The ubiquitin-proteasome system plays a pivotal role in maintaining cellular homeostasis by degrading damaged or misfolded proteins. Dysregulation of this system is implicated in the pathogenesis of multiple myeloma and other malignancies, where enhanced proteasomal activity promotes tumor cell survival [20]. Thus, targeting this pathway with proteasome inhibitors has become a key approach for the therapy of hematologic cancers [21, 22]. Nevertheless, the side-effect profile limited its long-term use, thus motivating the development of orally bioavailable analogues.

Ixazomib is a boronic acid-containing peptide analog structurally related to bortezomib [23], that reversibly interacts with the catalytic threonine residue of the proteasome's 20S subunit. Specifically, the chemical structure of ixazomib contains a derivative of the amino acid leucine in which the carboxylic acid group is replaced by a boronic acid. The compound was identified through large-scale screening of boron-containing molecules and optimized for oral bioavailability and metabolic stability. Structure–activity relationship studies and *in vivo* profiling allowed to improve its molecular structure, culminating to the clinically approved form, ixazomib citrate, that is a prodrug that rapidly hydrolyzes in plasma to yield active ixazomib [24].

The compound was developed by Takeda Pharmaceutical Company Limited and represented the culmination of efforts to design next-generation proteasome inhibitors based on the structure of bortezomib. Ixazomib specifically targets the β5 subunit of the 20S proteasome complex, with a dissociation half-life of approximately 18 minutes. In contrast, bortezomib exhibits a dissociation half-life of about 110 minutes, whereas carfilzomib binds irreversibly. The inhibition of the proteasome by ixazomib leads to the accumulation of misfolded and ubiquitinated proteins, resulting in endoplasmic reticulum stress and induction of apoptosis [25], that is particularly critical for cancer cells, that are more dependent on proteasomal degradation and therefore selectively sensitive to this effect. Thus, ixazomib disrupts key processes including cell cycle regulation, DNA repair, and cellular stress responses. Ixazomib received orphan drug designation from the FDA and the EMA for multiple myeloma in 2011 and for AL amyloidosis in 2012. The pivotal TOURMALINE-MM1 Phase III trial evaluated ixazomib in combination with lenalidomide and dexamethasone in patients with relapsed or refractory multiple myeloma. The trial demonstrated significant improvement in progression-free survival compared to placebo, establishing the efficacy of the oral proteasome inhibitor regimen. Based on these findings, the FDA granted priority review in September 2015 and approved ixazomib (in combination with lenalidomide and dexamethasone) on November 20, 2015 for the second-line treatment of multiple myeloma. Although the EMA initially declined marketing authorization in May 2016 due to insufficient efficacy data, it subsequently approved ixazomib on November 21, 2016 after re-evaluation, conditional upon further clinical studies.

12.7 Venetoclax

A non-peptide α-helical mimetic of a BH3 peptide is shown in Figure 12.9.

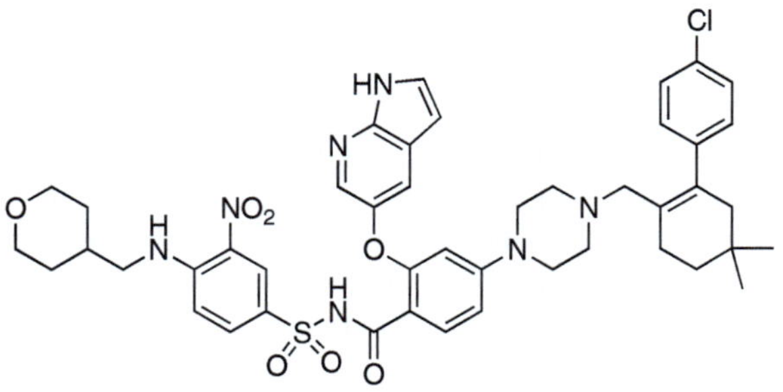

Figure 12.9 Chemical structure of venetoclax.

Generic name (INN)	Venetoclax, ABT-199
Brand name	VENCLEXTA, VENCLYXTO
Company	Abbvie Inc.
Chemical formula	$C_{45}H_{50}ClN_7O_7S$
Molecular weight	868.45 g mol^{-1}
SMILES code	CC1(CCC(=C(C1)c2ccc(cc2)Cl)CN3CCN(CC3)c4ccc(c(c4)Oc5cc6cc[nH]c6nc5)C(=O)NS(=O)(=O)c7ccc(c(c7)[N+](=O)[O−])NCC8CCOCC8)C
DrugBank ID	DB11581
PubChem ID	49846579
PDB ligand	LBM
Date of approval/clinical stage	Apr 11, 2016
Medical indication (FDA label) [26]	VENCLEXTA is a BCL-2 inhibitor indicated: • for the treatment of adult patients with chronic lymphocytic leukemia (CLL) or small lymphocytic lymphoma (SLL); • in combination with azacitidine or decitabine or low-dose cytarabine for the treatment of newly-diagnosed acute myeloid leukemia (AML) in adults who are age 75 years or older, or who have comorbidities that preclude use of intensive induction chemotherapy.

	This indication is approved under accelerated approval based on response rates. Continued approval for this indication may be contingent upon verification and description of clinical benefit in confirmatory trials
Description and mechanism of action (FDA label)	Venetoclax is a selective inhibitor of the BCL-2 protein. It is a light yellow to dark yellow solid with the empirical formula $C_{45}H_{50}ClN_7O_7S$ and a molecular weight of 868.44. Venetoclax has very low aqueous solubility. Venetoclax is a selective and orally bioavailable small-molecule inhibitor of BCL-2, an antiapoptotic protein. Overexpression of BCL-2 has been demonstrated in CLL and AML cells, where it mediates tumor cell survival and has been associated with resistance to chemotherapeutics. Venetoclax helps restore the process of apoptosis by binding directly to the BCL-2 protein, displacing pro-apoptotic proteins like BIM, triggering mitochondrial outer membrane permeabilization, and activating caspases. In nonclinical studies, venetoclax has demonstrated cytotoxic activity in tumor cells that overexpress BCL-2.

Venetoclax, marketed as Venclexta in the United States and Venclyxto in the European Union, is an orally bioavailable, selective inhibitor of BCL-2 developed for the treatment of hematologic malignancies including chronic lymphocytic leukemia (CLL), small lymphocytic lymphoma (SLL), and acute myeloid leukemia (AML) [27, 28]. BCL-2 is an antiapoptotic protein overexpressed in several lymphoid malignancies, where it promotes survival and confers resistance to cytotoxic chemotherapy [29]. Venetoclax binds with high affinity to the hydrophobic groove of BCL-2, antagonizing its antiapoptotic function and restoring programmed cell death in malignant cells, thus contributing with this mechanism to directly addressing a major resistance pathway in hematologic cancers.

It belongs to a novel class of small-molecule BH3-mimetics that target PPIs within the BCL-2 family [30]. BCL-2 normally binds to pro-apoptotic BH3-only proteins, thus inhibiting the mitochondrial outer membrane permeabilization and blocking apoptosis. The BH3 domain of these pro-apoptotic proteins forms an α-helix that engages a hydrophobic groove on BCL-2. By competitively binding to BCL-2, venetoclax displaces pro-apoptotic BH3-only proteins, leading to activation of BAX/BAK, mitochondrial permeabilization, cytochrome c release, and subsequent caspase-mediated apoptosis [31]. This mechanism selectively targets malignant BCL-2-dependent cells, with minimal off-target inhibition of related antiapoptotic proteins such as BCL-XL [32].

Venetoclax was discovered through collaborative efforts between Abbott Laboratories (now AbbVie) and Genentech, utilizing fragment-based and structure-guided design starting from previous work on navitoclax (ABT-263), an orally bioavailable inhibitor of both BCL-2 and BCL-XL. Modifications to enhance selectivity for BCL-2 and to reduce platelet toxicity led to the synthesis of venetoclax [33]. The discovery started with BH3-derived peptides that were subsequently optimized to produce a non-peptide scaffold reproducing the spatial arrangement of pharmacophoric side chains occupying the h1–h4 hydrophobic pockets of BCL-2. Thus, venetoclax can be considered a functional peptidomimetic mimicking the bioactive conformation of the parent peptide epitope without containing a peptide character.

Preclinical studies demonstrated potent and selective cytotoxicity against CLL cells and synergy with anti-CD20 monoclonal antibodies. In 2015, the FDA granted venetoclax a breakthrough therapy designation for patients with relapsed or refractory CLL or SLL. Subsequently, in April 2016, the FDA-approved venetoclax for patients with CLL with 17p deletion who had received at least one prior therapy. Authorization was granted under accelerated approval based on overall response rate, due to its demonstrated benefit in high-risk populations. Then, the European Medicines Agency approved venetoclax in December 2016 for similar indications. Ongoing clinical trials continue to evaluate venetoclax in other hematologic malignancies, including non-Hodgkin lymphoma, multiple myeloma, diffuse large B-cell lymphoma, and follicular lymphoma.

Venetoclax represents the first selective BCL-2 inhibitor to achieve clinical approval, thus validating a new therapeutic paradigm of targeting intracellular PPIs [34]. Its success has encouraged the development of next-generation BH3 mimetics targeting other members of the BCL-2 family, including MCL-1 and BCL-XL. Thus, the discovery and approval of venetoclax mark a milestone in modern oncology and medicinal chemistry as a potent, selective, and orally bioavailable BCL-2 inhibitor for treatment strategies for CLL, SLL, and AML.

12.8 Lu-PSMA-617

^{177}Lu-PSMA-617 is the first radioligand therapy for use in patients with metastatic castration-resistant prostate cancer (mCRPC) is shown in Figure 12.10.

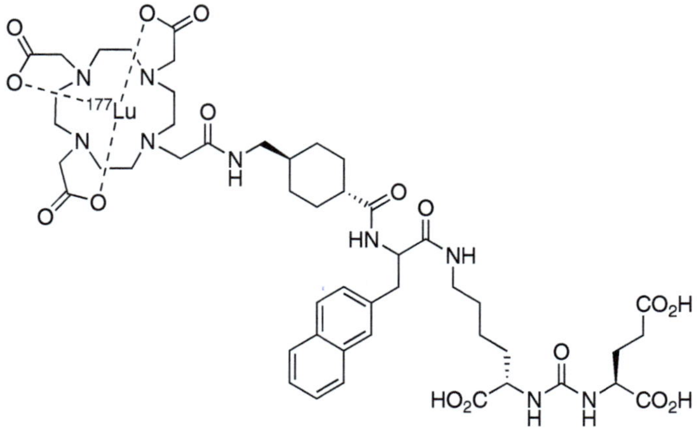

Figure 12.10 Chemical structure of radioligand ^{177}Lu-PSMA-617.

Generic name (INN)	Lutetium Lu-177 Vipivotide Tetraxetan, Lu-PSMA-617
Brand name	PLUVICTO
Company	Novartis Pharmaceuticals Corp.
Chemical formula	$C_{49}H_{68}{}^{177}LuN_9O_{16}$
Molecular weight	1,216.07 g mol^{-1}
SMILES code	[177Lu+3].OC(=O)CC[C@H](NC(=O)N[C@@H](CCCCNC(=O)[C@H](Cc1ccc2ccccc2c1)NC(=O)[C@@H]3CC[C@@H](CNC(=O)CN4CCN(CC(=O)[O−])CCN(CC(=O)[O−])CCN(CC(=O)[O−])CC4)CC3)C(=O)O)C(=O)O
DrugBank ID	DB16778
PubChem ID	122706785
PDB ligand	N/A
Date of approval/clinical stage	Mar 23, 2022
Medical indication (FDA label) [35]	PLUVICTO is a radioligand therapeutic agent indicated for the treatment of adult patients with prostate-specific membrane antigen (PSMA)-positive metastatic castration-resistant prostate cancer (mCRPC) who have been treated with androgen receptor (AR) pathway inhibition and taxane-based chemotherapy.
Description and mechanism of action (FDA label)	PLUVICTO (lutetium Lu 177 vipivotide tetraxetan) is a radioligand therapeutic agent. Lutetium Lu 177 vipivotide tetraxetan is a PSMA-binding ligand bound to a DOTA chelator radiolabeled with lutetium-177. Lutetium Lu 177 vipivotide tetraxetan is a radioligand therapeutic agent. The active moiety of lutetium Lu 177 vipivotide tetraxetan is the radionuclide lutetium-177, which is linked to a moiety that binds to PSMA, a transmembrane protein that is expressed in prostate cancer, including mCRPC. Upon binding of lutetium Lu 177 vipivotide tetraxetan to PSMA-expressing cells, the beta-minus emission from lutetium-177 delivers radiation to PSMA-expressing cells, as well as to surrounding cells, and induces DNA damage, which can lead to cell death.

Prostate cancer ranks as the third most frequently diagnosed malignancy and remains among the principal contributors to cancer-related mortality on a global scale [36]. It is estimated that approximately ten million men are currently affected, with annual deaths exceeding 400,000 [37]. Management of early-stage prostate cancer typically involves surgical intervention and/or radiotherapy. Nevertheless, advanced disease requires pharmacological treatments, including luteinizing hormone-releasing hormone (LHRH) analogs, androgen receptor (AR) signaling inhibitors, poly(ADP-ribose) polymerase (PARP) inhibitors, and taxane-based chemotherapeutic agents. Despite initial success, resistance to these therapies developed, ultimately leading to treatment failure [38]. Pluvicto (lutetium-177 vipivotide tetraxetan) represents an integration of peptidomimetic chemistry with targeted radioligand therapy, demonstrating the translational potential of rational peptide design in precision oncology. Specifically, the therapeutic strategy underlying Pluvicto is based on the recognition of prostate-specific membrane antigen (PSMA), a transmembrane glycoprotein that is overexpressed in metastatic castration-resistant prostate cancer cells and minimally expressed in most normal tissues. The high selectivity of PSMA for malignant cells provides a molecular target for the development of safe cytotoxics following the design rationale for peptidomimetic drugs [39, 40]. In Pluvicto, the PSMA-targeting ligand is a peptidomimetic analogue of the naturally occurring glutamate-urea-lysine motif recognized by PSMA. This small-molecule peptidomimetic was optimized to achieve high-affinity binding and favorable pharmacokinetics. Also, the successful conjugation of this ligand to lutetium-177 *via* the chelator tetraxetan enabled the selective delivery of beta radiation to tumor cells. In Pluvicto, the PSMA-targeting ligand is conjugated to the chelator through a *trans*-4-(aminomethyl)cyclohexanecarboxylic acid (AMCA) linker, with the ε-amino function of Lys side-chain being the anchoring site to the AMCA linker. In this molecular construct, the peptidomimetic component is the molecular warhead system for the cytotoxic payload to PSMA-expressing cells [41].

Upon intravenous administration, Pluvicto binds with high specificity to PSMA on prostate cancer cells. The ligand–receptor complex undergoes internalization, allowing the lutetium-177 radionuclide to emit beta particles, thus inducing localized DNA double-strand breaks and cell apoptosis. The peptidomimetic nature of the ligand, as well as guaranteeing the receptor affinity, allowed for achieving *in vivo* stability and efficient tumor uptake, thus solving those limitations that are often found in peptides [42]. Also, the gamma emissions of the conjugate allow for simultaneous molecular imaging in the context of theranostic applications. Pluvicto has demonstrated clinically efficacy in patients with PSMA-positive mCRPC, improving overall survival and radiographic progression-free survival in the Phase III VISION trial [42]. In this context, the incorporation of a peptidomimetic ligand exemplified the successful application of peptidomimetic chemistry to targeted radiotherapy.

12.9 Trofinetide

First FDA-approved treatment for Rett syndrome, Gly-Pro-Glu peptidomimetic is shown in Figure 12.11.

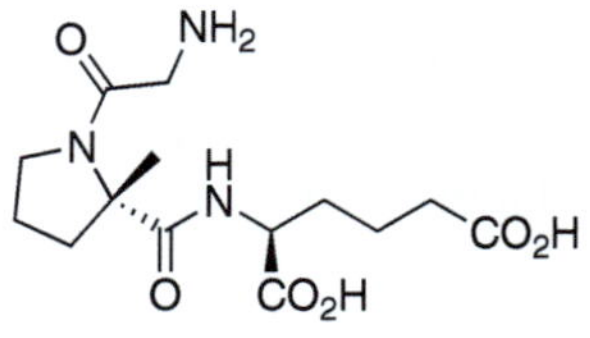

Figure 12.11 Chemical structure of trofinetide.

Generic name (INN)	Trofinetide
Brand name	Daybue
Company	Acadia Pharmaceuticals Inc
Chemical formula	$C_{13}H_{21}N_3O_6$
Molecular weight	315.326 g mol^{-1}
SMILES code	C[C@]1(CCCN1C(=O)CN)C(=O)N[C@@H](CCC(=O)O)C(=O)O
DrugBank ID	DB06045
PubChem ID	11318905
PDB ligand	N/A
Date of approval/clinical stage	Mar 10, 2023
Medical indication (FDA label) [43]	DAYBUE is indicated for the treatment of Rett syndrome in adults and pediatric patients 2 years of age and older.
Description and mechanism of action (FDA label)	Trofinetide is designated chemically as (2*S*)-2-{[(2*S*)-1-(2-aminoacetyl)-2-methylpyrrolidine-2-carbonyl]amino}pentanedioic acid (IUPAC). Its empirical formula is $C_{13}H_{21}N_3O_6$ and its molecular weight is 315.33 g mol^{-1}. The mechanism by which trofinetide exerts therapeutic effects in patients with Rett syndrome is unknown.

Trofinetide (Daybue) represents a milestone in the therapeutic management of Rett syndrome, a rare and debilitating neurodevelopmental disorder occurring almost exclusively in females [44, 45]. Trofinetide was approved by the FDA in March 2023 as the first medication specifically indicated for Rett syndrome in patients aged two years and older. The recent approval represents the successful conclusion of extensive research dedicated to elucidating the molecular mechanisms and to developing therapeutic strategies that specifically address this pathology. Rett syndrome is an X-linked dominant neurodevelopmental disorder primarily attributed to mutations in the MECP2 gene, which encodes methyl-CpG-binding protein 2 as a critical regulator of neuronal gene transcription. Deficiency or malfunction of MECP2 leads to significant neurological impairments, including regression of language and motor skills, seizures, and autonomic dysfunction. Rett syndrome affects about one in every 10,000–15,000 female births worldwide, and historically it has been managed through symptomatic care, as no disease-modifying treatments were available prior to the authorization of this therapeutic agent. Trofinetide emerged from investigations into glycine-proline-glutamate (GPE), a neuroprotective tripeptide derived from insulin-like growth factor 1 (IGF-1) [46]. Due to the short biological half-life of this tripeptide, a stable analog was synthesized to achieve trofinetide, specifically incorporating an α-methylproline to enhance resistance to enzymatic degradation and to improve bioavailability. This chemical variation enabled sustained neuroprotective and anti-inflammatory effects within the central nervous system (CNS). Concerning the biological mechanism of action, trofinetide is supposed to restore synaptic function and promote neuronal survival, thus reducing the synaptic deficits characteristic of Rett syndrome. Although its precise molecular pathways remain under investigation, preclinical and clinical studies have demonstrated functional improvements. The expedited development of trofinetide, thanks to FDA designations including Fast Track, Orphan Drug, and Rare Pediatric Disease status, culminated in the FDA approval following the successful LAVENDER trial, thus marking a key advancement as the first disease-specific pharmacological treatment for Rett syndrome.

12.10 Tavilermide

Selective cyclic tripeptide possessing tyrosine kinase TrkA receptor agonist activity is shown in Figure 12.12.

Generic name (INN)	Tavilermide, MIM-D3 (development code)
Brand name	N/A
Company	Mimetogen Pharmaceuticals USA, Inc.
Chemical formula	$C_{24}H_{32}N_6O_{11}$
Molecular weight	580.551 g mol^{-1}
SMILES code	NCCCC[C@@H]1NC(=O)[C@H](CCC(=O)O)NC(=O)c2cc([N+](=O)[O−])ccc2OCC[C@@H](C(=O)NCC(=O)O)NC1=O
DrugBank ID	DB12441
PubChem ID	9808372
PDB ligand	N/A
Date of approval/clinical stage	Investigational; Phase 3
Medical indication	Tavilermide has been investigated for the treatment of Dry Eye Syndromes.
Description (FDA label)	N/A

Tavilermide is a selective, cyclic tripeptide that acts as a partial agonist of the TrkA receptor [47]. This class of compounds is commonly described as nerve growth factor (NGF) mimetics. Originally synthesized by Burgess and colleagues, tavilermide was designed to function as a TrkA receptor agonist [48, 49]. The compound is being developed by Mimetogen Pharmaceuticals as an ophthalmic formulation for the treatment of dry eye disease [50, 51]. It is currently undergoing evaluation in two multicenter Phase III clinical trials in the United States for this indication.

As a key regulator of neuronal growth and repair, NGF plays a pivotal role in the maintenance, survival, and differentiation of neurons. This dimeric neurotrophin mediates

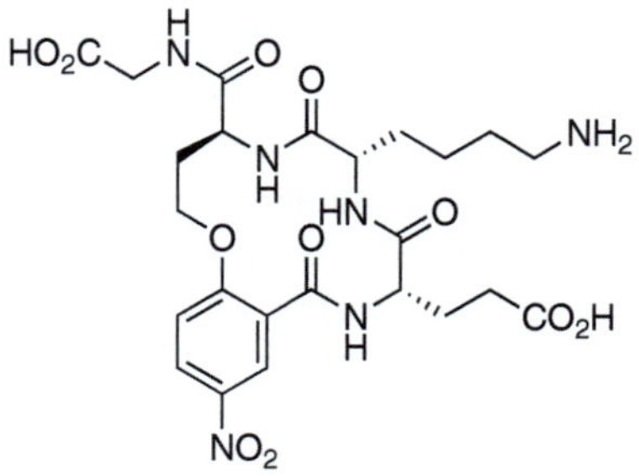

Figure 12.12 Chemical structure of tavilermide.

its effects through two transmembrane receptors, TrkA and p75NTR [52]. The detection of NGF and TrkA in normal and diseased ocular tissues indicates their involvement in corneal homeostasis and regeneration. Numerous studies have shown that NGF enhances corneal epithelial healing and nerve function, and clinical trials have confirmed its therapeutic benefit in neurotrophic keratitis and corneal ulcers [53]. However, the clinical use of topical NGF is limited by ocular pain, poor bioavailability, and high production costs. To overcome these limitations, small-molecule neurotrophin mimetics have been developed [54]. Among them, tavilermide is a stable cyclic peptidomimetic that reproduces a β-turn region of NGF and acts as a partial TrkA agonist, providing neuroprotective effects without engaging p75NTR [55].

12.11 Difelikefalin

Analgesic opioid peptide for the treatment of moderate-to-severe pruritus associated with chronic kidney disease is shown in Figure 12.13.

Generic name (INN)	Difelikefalin
Brand name	KORSUVA
Company	Vifor (International) Inc.
Chemical formula	$C_{36}H_{53}N_7O_6$
Molecular weight	679.863 g mol^{-1}
SMILES code	CC(C)C[C@H](C(=O)N[C@H](CCCCN)C(=O)N1CCC(CC1)(C(=O)O)N)NC(=O)[C@@H](CC2=CC=CC=C2)NC(=O)[C@@H](CC3=CC=CC=C3)N
DrugBank ID	DB11938
PubChem ID	24794466
PDB ligand	N/A
Date of approval/clinical stage	Aug 23, 2021
Medical indication (FDA label) [56]	KORSUVA is a kappa opioid receptor agonist indicated for the treatment of moderate-to-severe pruritus associated with chronic kidney disease in adults undergoing hemodialysis.
Description and mechanism of action (FDA label)	KORSUVA (difelikefalin) is a kappa opioid receptor agonist. Difelikefalin is a synthetic peptide with a single stereoisomer and is present as an acetate salt. KORSUVA is a kappa opioid receptor (KOR) agonist. The relevance of KOR activation to therapeutic effectiveness is not known.

Difelikefalin [57] is a synthetic, peripherally restricted kappa opioid receptor (KOR) agonist designed to alleviate moderate-to-severe pruritus in adult patients with chronic kidney disease undergoing hemodialysis [58]. Its molecular structure is a hydrophilic small peptide composed of D-amino acids, which limits its penetration of the blood–brain barrier and thereby reduces the CNS side-effect profile typical of classical opioid agonists. The mechanism of action of difelikefalin consists of selectively binding to KORs on peripheral sensory neurons and immune cells, triggering downstream pathways that

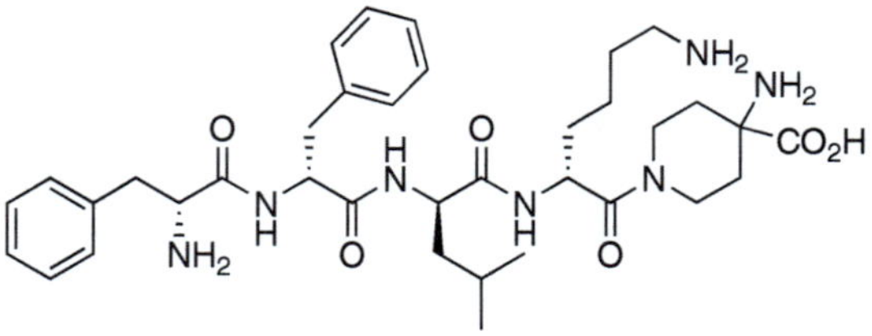

Figure 12.13 Chemical structure of difelikefalin.

modulate itch sensation and inflammatory signaling. As difelikefalin is more than 30k-fold selective over μ and δ opioid receptors and avoids many adverse effects associated with non-selective opioids, such as respiratory depression, euphoria, or dependence.

In clinical trials, difelikefalin demonstrated rapid onset and prolonged efficacy, with significantly more patients achieving significant reductions in itch intensity compared to placebo, resulting in approval by regulatory authorities. Specifically, the FDA granted approval in August 2021, and the EMA adopted a positive opinion in February 2022 for the indication of uremic pruritus [59].

12.12 Macimorelin

Orally active ghrelin mimetic that stimulates growth hormone (GH) secretion is shown in Figure 12.14.

Generic name (INN)	Macimorelin acetate
Brand name	MACRILEN
Company	Novo Nordisk Inc.
Chemical formula	$C_{26}H_{30}N_6O_3$
Molecular weight	474.565 g mol^{-1}
SMILES code	CC(C)(C(=O)N[C@H](CC1=CNC2=CC=CC=C21)C(=O)N[C@H](CC3=CNC4=CC=CC=C43)NC=O)N
DrugBank ID	DB13074
PubChem ID	9804938
PDB ligand	N/A
Date of approval/clinical stage	Dec 20, 2017
Medical indication (FDA label) [60]	MACRILEN is a growth hormone (GH) secretagogue receptor agonist indicated for the diagnosis of adult growth hormone deficiency.
Description and mechanism of action (FDA label)	MACRILEN for oral solution is macimorelin acetate, a synthetic growth hormone secretagogue receptor agonist. Macimorelin acetate is described chemically as D-Tryptophanamide, 2-methylalanyl-*N*-[(1*R*)-1-(formylamino)-2-(1*H*-indol-3-yl)ethyl]-acetate. Macimorelin stimulates GH release by activating growth hormone secretagogue receptors present in the pituitary and hypothalamus.

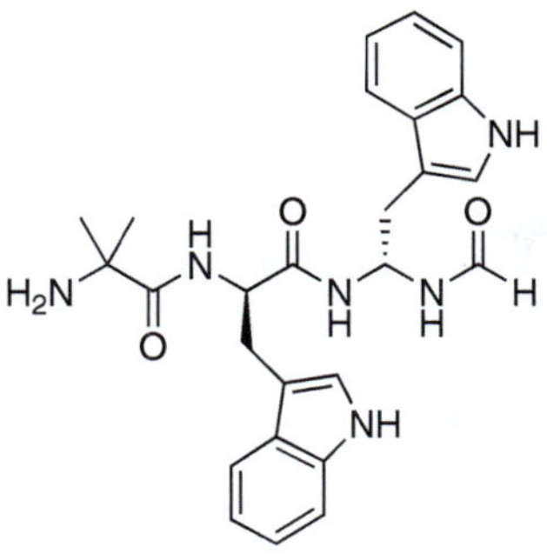

Figure 12.14 *Chemical structure of macimorelin.*

Macimorelin was discovered at the University of Montpellier [61] and approved in 2017 for the diagnosis of adult growth hormone deficiency [62]. It is a growth hormone secretagogue (GHS) pseudotripeptide analog consisting of three residues, namely an aminoisobutyric residue (Aib) at the *N*-terminal position, a D-tryptophan at the central position, and a mimetic of a D-Trp possessing a formylated gem diamino moiety. Macimorelin acetate is a representative entry of the successful application of peptidomimetic design translated into a clinically approved drug for the diagnosis of adult growth hormone deficiency [63]. It is an orally active small-molecule GHS acting as a ghrelin receptor agonist, activating the GHS receptor type 1a [64, 65]. In the context of GHSs, natural ligands like ghrelin are limited by their rapid enzymatic inactivation, thus restricting their clinical utility. Peptidomimetic analogues such as pralmorelin, macimorelin, and related compounds were developed to overcome these challenges, resulting in the achievement of orally bioavailable and stable compounds capable of stimulating GH release. Unlike ghrelin, macimorelin does not require peptide bonds or acylation to activate GHSR-1a, as its molecular structure reproduces key hydrophobic and hydrogen-bonding interactions of the ghrelin *N*-terminal region to enable selective and potent receptor activation [63]. Specifically, the biological activity of macimorelin consists of binding to GHSR-1a with an IC_{50} of 123 nM for the human receptor in the pituitary and hypothalamus, enabling a specific endocrine effect, with minimal off-target activity on other GPCRs. The oral bioavailability of this drug makes macimorelin markedly different from earlier peptide analogues that required intravenous administration.

Macimorelin has been approved by the FDA and EMA for the diagnosis of adult growth hormone deficiency, replacing more invasive stimulation tests such as the insulin tolerance test. In clinical trials, oral administration of macimorelin stimulated rapid and reproducible GH responses, demonstrating superior performance as compared to traditional methods.

References

1. PAXLOVID (Pfizer Inc). *Approved Drug Products with Therapeutic Equivalence Evaluations*. Silver Spring, Maryland: FDA, 2025.
2. Jin Z, Du X, Xu Y, Deng Y, Liu M, Zhao Y, Zhang B, Li X, Zhang L, Peng C, Duan Y, Yu J, Wang L, Yang K, Liu F, Jiang R, Yang X, You T, Liu X, Yang X, Bai F, Liu H, Liu X, Guddat LW, Xu W, Xiao G, Qin C, Shi Z, Jiang H, Rao Z, Yang H. Structure of Mpro from SARS-CoV-2 and discovery of its inhibitors. *Nature*. 2020;**582**:289–293.

3. Rut W, Groborz K, Zhang L, Sun X, Zmudzinski M, Pawlik B, Wang X, Jochmans D, Neyts J, Młynarski W, Hilgenfeld R, Drag M. SARS-CoV-2 Mpro inhibitors and activity-based probes for patient-sample imaging. *Nat Chem Biol*. 2021;**17**:222–228.
4. Hoffman RL, Kania RS, Brothers MA, Davies JF, Ferre RA, Gajiwala KS, He M, Hogan RJ, Kozminski K, Li LY, Lockner JW, Lou J, Marra MT, Mitchell LJ Jr, Murray BW, Nieman JA, Noell S, Planken SP, Rowe T, Ryan K, Smith GJ 3rd, Solowiej JE, Steppan CM, Taggart B. Discovery of ketone-based covalent inhibitors of coronavirus 3CL proteases for the potential therapeutic treatment of COVID-19. *J Med Chem*. 2020;**63**:12725–12747.
5. Owen DR, Allerton CMN, Anderson AS, Aschenbrenner L, Avery M, Berritt S, Boras B, Cardin RD, Carlo A, Coffman KJ, Dantonio A, Di L, Eng H, Ferre R, Gajiwala KS, Gibson SA, Greasley SE, Hurst BL, Kadar EP, Kalgutkar AS, Lee JC, Lee J, Liu W, Mason SW, Noell S, Novak JJ, Obach RS, Ogilvie K, Patel NC, Pettersson M, Rai DK, Reese MR, Sammons MF, Sathish JG, Singh RSP, Steppan CM, Stewart AE, Tuttle JB, Updyke L, Verhoest PR, Wei L, Yang Q, Zhu Y. An oral SARS-CoV-2 Mpro inhibitor clinical candidate for the treatment of COVID-19. *Science*. 2021;**374**:1586–1593.
6. Hammond J, Leister-Tebbe H, Gardner A, Abreu P, Bao W, Wisemandle W, Baniecki M, Hendrick VM, Damle B, Simón-Campos A, Pypstra R, Rusnak JM, Investigators EPIC-HR. Oral nirmatrelvir for high-risk, nonhospitalized adults with Covid-19. *N Engl J Med*. 2022;**386**:1397–1408.
7. VOSEVI (Gilead Sciences Inc). *Approved Drug Products with Therapeutic Equivalence Evaluations*. Silver Spring, Maryland: FDA, 2025.
8. Messina JP, Humphreys I, Flaxman A, Brown A, Cooke GS, Pybus OG, Barnes E. Global distribution and prevalence of hepatitis C virus genotypes. *Hepatology*. 2015;**61**(1):77–87.
9. Taylor JG, Zipfel S, Ramey K, Vivian R, Schrier A, Karki KK, Katana A, Kato D, Kobayashi T, Martinez R, Sangi M, Siegel D, Tran CV, Yang ZY, Zablocki J, Yang CY, Wang Y, Wang K, Chan K, Barauskas O, Cheng G, Jin D, Schultz BE, Appleby T, Villaseñor AG, Link JO. Discovery of the pan-genotypic hepatitis C virus NS3/4A protease inhibitor voxilaprevir (GS-9857): a component of Vosevi®. *Bioorg Med Chem Lett*. 2019;**29**:2428–2436.
10. Vinogradov AA, Yin Y, Suga H. Macrocyclic peptides as drug candidates: recent progress and remaining challenges. *J Am Chem Soc*. 2019;**141**:4167–4181.
11. Zasloff M. Magainins, a class of antimicrobial peptides from Xenopus skin: isolation, characterization of two active forms, and partial cDNA sequence of a precursor. *Proc Natl Acad Sci USA*. 1987;**84**:5449–5453.
12. Mensa B, Howell GL, Scott R, DeGrado WF. Comparative mechanistic studies of brilacidin, daptomycin, and the antimicrobial peptide LL16. *Antimicrob Agents Chemother*. 2014;**58**:5136–5145.
13. Liu D, Choi S, Chen B, Doerksen RJ, Clements DJ, Winkler JD, Klein ML, DeGrado WF. Nontoxic membrane-active antimicrobial arylamide oligomers. *Angew Chem Int Ed*. 2004;**43**:1158–1162.
14. Choi S, Isaacs A, Clements D, Liu D, Kim H, Scott RW, Winkler JD, DeGrado WF. De novo design and in vivo activity of conformationally restrained antimicrobial arylamide foldamers. *Proc Natl Acad Sci USA*. 2009;**106**:6968–6973.

15. Tew GN, Scott RW, Klein ML, Degrado WF. De novo design of antimicrobial polymers, foldamers, and small molecules: from discovery to practical applications. *Acc Chem Res*. 2010;**43**:30–39.
16. Robinson JA, Shankaramma SC, Jetter P, Kienzl U, Schwendener RA, Vrijbloed JW, Obrecht D. Properties and structure-activity studies of cyclic beta-hairpin peptidomimetics based on the cationic antimicrobial peptide protegrin I. *Bioorg Med Chem*. 2005;**13**(6):2055–2064.
17. Martin-Loeches I, Dale GE, Torres A. Murepavadin: a new antibiotic class in the pipeline. *Expert Rev Anti-Infect Ther*. 2018;**16**:259–268.
18. NINLARO (Takeda Pharmaceuticals USA Inc). *Approved Drug Products with Therapeutic Equivalence Evaluations*. Silver Spring, Maryland: FDA, 2025.
19. Bross PF, Kane R, Farrell AT, Abraham S, Benson K, Brower ME, Bradley S, Gobburu JV, Goheer A, Lee SL, Leighton J, Liang CY, Lostritto RT, McGuinn WD, Morse DE, Rahman A, Rosario LA, Verbois SL, Williams G, Wang YC, Pazdur R. Approval summary for bortezomib for injection in the treatment of multiple myeloma. *Clin Cancer Res*. 2004;**10**:3954–3964.
20. Adams J. The proteasome: a suitable antineoplastic target. *Nat Rev Cancer*. 2004; **4**:349–360.
21. Sogbein O, Paul P, Umar M, Chaari A, Batuman V, Upadhyay R. Bortezomib in cancer therapy: mechanisms, side effects, and future proteasome inhibitors. *Life Sci*. 2024;**358**:123125.
22. Brayer J, Baz R. The potential of ixazomib, a second-generation proteasome inhibitor, in the treatment of multiple myeloma. *Ther Adv Hematol*. 2017;**8**:209–220.
23. Gupta N, Hanley MJ, Xia C, Labotka R, Harvey RD, Venkatakrishnan K. Clinical pharmacology of ixazomib: the first oral proteasome inhibitor. *Clin Pharmacokinet*. 2019;**58**:431–449.
24. Kisselev AF, Goldberg AL. Proteasome inhibitors: from research tools to drug candidates. *Chem Biol*. 2001;**8**:739–758.
25. Moreau P, Masszi T, Grzasko N, Bahlis NJ, Hansson M, Pour L, Sandhu I, Ganly P, Baker BW, Jackson SR, Stoppa AM, Simpson DR, Gimsing P, Palumbo A, Garderet L, Cavo M, Kumar S, Touzeau C, Buadi FK, Laubach JP, Berg DT, Lin J, Di Bacco A, Hui AM, van de Velde H, Richardson PG, TOURMALINE-MM1 Study Group. Oral ixazomib, lenalidomide, and dexamethasone for multiple myeloma. *N Engl J Med*. 2016;**374**:1621–1634.
26. VENCLEXTA (Abbvie Inc). *Approved Drug Products with Therapeutic Equivalence Evaluations*. Silver Spring, Maryland: FDA, 2025.
27. Souers AJ, Leverson JD, Boghaert ER, Ackler SL, Catron ND, Chen J, Dayton BD, Ding H, Enschede SH, Fairbrother WJ, Huang DC, Hymowitz SG, Jin S, Khaw SL, Kovar PJ, Lam LT, Lee J, Maecker HL, Marsh KC, Mason KD, Mitten MJ, Nimmer PM, Oleksijew A, Park CH, Park CM, Phillips DC, Roberts AW, Sampath D, Seymour JF, Smith ML, Sullivan GM, Tahir SK, Tse C, Wendt MD, Xiao Y, Xue JC, Zhang H, Humerickhouse RA, Rosenberg SH, Elmore SW. ABT-199, a potent and selective BCL-2 inhibitor, achieves antitumor activity while sparing platelets. *Nat Med*. 2013;**19**:202–208.
28. Stilgenbauer S, Eichhorst B, Schetelig J, Coutre S, Seymour JF, Munir T, Puvvada SD, Wendtner CM, Roberts AW, Jurczak W, Mulligan SP, Böttcher S, Mobasher M,

Zhu M, Desai M, Chyla B, Verdugo M, Enschede SH, Cerri E, Humerickhouse R, Gordon G, Hallek M, Wierda WG. Venetoclax in relapsed or refractory chronic lymphocytic leukaemia with 17p deletion: a multicentre, open-label, phase 2 study. *Lancet Oncol*. 2016;**17**:768–778.
29. Cory S, Adams JM. The Bcl2 family: regulators of the cellular life-or-death switch. *Nat Rev Cancer*. 2002;**2**:647–656.
30. Lessene G, Czabotar PE, Colman PM. BCL-2 family antagonists for cancer therapy. *Nat Rev Drug Discov*. 2008;**7**:989–1000.
31. Adams JM, Cory S. The BCL-2 arbiters of apoptosis and their growing role as cancer targets. *Cell Death Differ*. 2018;**25**:27–36.
32. Wilson WH, O'Connor OA, Czuczman MS, LaCasce AS, Gerecitano JF, Leonard JP, Tulpule A, Dunleavy K, Xiong H, Chiu YL, Cui Y, Busman T, Elmore SW, Rosenberg SH, Krivoshik AP, Enschede SH, Humerickhouse RA. Navitoclax, a targeted high-affinity inhibitor of BCL-2, in lymphoid malignancies: a phase 1 dose-escalation study of safety, pharmacokinetics, pharmacodynamics, and antitumour activity. *Lancet Oncol*. 2010;**11**:1149–1159.
33. Leverson JD, Phillips DC, Mitten MJ, Boghaert ER, Diaz D, Tahir SK, Belmont LD, Nimmer P, Xiao Y, Ma XM, Lowes KN, Kovar P, Chen J, Jin S, Smith M, Xue J, Zhang H, Oleksijew A, Magoc TJ, Vaidya KS, Albert DH, Tarrant JM, La N, Wang L, Tao ZF, Wendt MD, Sampath D, Rosenberg SH, Tse C, Huang DC, Fairbrother WJ, Elmore SW, Souers AJ. Exploiting selective BCL-2 family inhibitors to dissect cell survival dependencies and define improved strategies for cancer therapy. *Sci Transl Med*. 2015;**7**:279ra40.
34. Ashkenazi A, Fairbrother WJ, Leverson JD, Souers AJ. From basic apoptosis discoveries to advanced selective BCL-2 family inhibitors. *Nat Rev Drug Discov*. 2017;**16**:273–284.
35. PLUVICTO (Novartis Pharmaceuticals Corp.). *Approved Drug Products with Therapeutic Equivalence Evaluations*. Silver Spring, Maryland: FDA, 2025.
36. Chia CSB. Novel PSMA-targeting radionuclide peptidomimetics for treating prostate cancer. *ACS Med Chem Lett*. 2022;**14**:3–4.
37. Sung H, Ferlay J, Siegel RL, Laversanne M, Soerjomataram I, Jemal A, Bray F. Global cancer statistics 2020: GLOBOCAN estimates of incidence and mortality worldwide for 36 cancers in 185 countries. *CA Cancer J Clin*. 2021;**71**:209–249.
38. Sandhu S, Moore CM, Chiong E, Beltran H, Bristow RG, Williams SG. Prostate cancer. *Lancet*. 2021;**398**:1075–1090.
39. Lawhn-Heath C, Salavati A, Behr SC, Rowe SP, Calais J, Fendler WP, Pienta KJ, Paller CJ, Pomper MG, Gorin MA, Turkbey B, Hope TA. Prostate-specific membrane antigen PET in prostate cancer: current status and future directions. *Radiology*. 2021;**299**:248–260.
40. Silver DA, Pellicer I, Fair WR, Heston WD, Cordon-Cardo C. Prostate-specific membrane antigen expression in normal and malignant human tissues. *Clin Cancer Res*. 1997;**3**:81–85.
41. Seifert R, Alberts IL, Afshar-Oromieh A, Rahbar K. Prostate cancer theranostics: PSMA targeted therapy. *PET Clin*. 2021;**16**:391–396.
42. Sartor O, de Bono J, Chi KN, Fizazi K, Herrmann K, Rahbar K, Tagawa ST, Nordquist LT, Vaishampayan N, El-Haddad G, Park CH, Beer TM, Armour A, Pérez-Contreras

WJ, DeSilvio M, Kpamegan E, Gericke G, Messmann RA, Morris MJ, Krause BJ, VISION Investigators. Lutetium-177-PSMA-617 for metastatic castration-resistant prostate cancer. *N Engl J Med*. 2021;**385**:1091–1103.

43. DAYBUE (Novartis Pharmaceuticals Corp.). *Approved Drug Products with Therapeutic Equivalence Evaluations*. Silver Spring, Maryland: FDA, 2025.
44. Kennedy M, Glass L, Glaze DG, Kaminsky S, Percy AK, Neul JL, Jones NE, Tropea D, Horrigan JP, Nues P, Bishop KM, Youakim JM. Development of trofinetide for the treatment of Rett syndrome: from bench to bedside. *Front Pharmacol*. 2024;**14**:1341746.
45. Parent H, Ferranti A, Niswender C. Trofinetide: a pioneering treatment for Rett syndrome. *Trends Pharmacol Sci*. 2023;**44**:740–741.
46. Alonso De Diego SA, Gutiérrez-Rodríguez M, Pérez de Vega MJ, Casabona D, Cativiela C, González-Muñiz R, Herranz R, Cenarruzabeitia E, Frechilla D, Del Río J, Jimeno ML, García-López MT. New Gly-Pro-Glu (GPE) analogues: expedite solid-phase synthesis and biological activity. *Bioorg Med Chem Lett*. 2006;**16**:1392–1396.
47. Jain P, Li R, Lama T, Saragovi HU, Cumberlidge G, Meerovitch K. An NGF mimetic, MIM-D3, stimulates conjunctival cell glycoconjugate secretion and demonstrates therapeutic efficacy in a rat model of dry eye. *Exp Eye Res*. 2011;**93**:503–512.
48. Feng Y, Wang Z, Jin S, Burgess K. SNAr cyclizations to form cyclic peptidomimetics of β-turns. *J Am Chem Soc*. 1998;**120**:10768–10769.
49. Feng Y, Burgess K. Solid-phase SNAr macrocyclizations to give turn–extended-turn peptidomimetics. *Chem Eur J*. 1999;**5**:3261–3272.
50. Meerovitch K, Torkildsen G, Lonsdale J, Goldfarb H, Lama T, Cumberlidge G, Ousler GW 3rd. Safety and efficacy of MIM-D3 ophthalmic solutions in a randomized, placebo-controlled phase 2 clinical trial in patients with dry eye. *Clin Ophthalmol*. 2013;**7**:1275–1285.
51. Patil S, Sawale G, Ghuge S, Sathaye S. Quintessence of currently approved and upcoming treatments for dry eye disease. *Graefes Arch Clin Exp Ophthalmol*. 2025;**263**:269–278.
52. Patapoutian A, Reichardt LF. Trk receptors: mediators of neurotrophin action. *Curr Opin Neurobiol*. 2001;**11**:272–280.
53. Lambiase A, Rama P, Bonini S, Caprioglio G, Aloe L. Topical treatment with nerve growth factor for corneal neurotrophic ulcers. *N Engl J Med*. 1998;**338**:1174–1180.
54. Saragovi HU, Zaccaro MC. Small molecule peptidomimetic ligands of neurotrophin receptors, identifying binding sites, activation sites and regulatory sites. *Curr Pharm Des*. 2002;**8**:2201–2216.
55. Bruno MA, Clarke PB, Seltzer A, Quirion R, Burgess K, Cuello AC, Saragovi HU. Long-lasting rescue of age-associated deficits in cognition and the CNS cholinergic phenotype by a partial agonist peptidomimetic ligand of TrkA. *J Neurosci*. 2004;**24**:8009–8018.
56. KORSUVA (Vifor International Inc). *Approved Drug Products with Therapeutic Equivalence Evaluations*. Silver Spring, Maryland: FDA, 2025.
57. Fugal J, Serpa SM. Difelikefalin: a new κ-opioid receptor agonist for the treatment of hemodialysis-dependent chronic kidney disease-associated pruritus. *Ann Pharmacother*. 2023;**57**:480–488.

58. Anecka A, Perlikowska R, Gach K, Wyrebska A, Fichna J. Development of opioid peptide analogs for pain relief. *Curr Pharm Des*. 2010;**16**:1126–1135.
59. Deeks ED. Difelikefalin: first approval. *Drugs*. 2021;**81**:1937–1944.
60. MACRILEN (Novo Nordisk Inc). *Approved Drug Products with Therapeutic Equivalence Evaluations*. Silver Spring, Maryland: FDA, 2025.
61. Guerlavais V, Boeglin D, Mousseaux D, Oiry C, Heitz A, Deghenghi R, Locatelli V, Torsello A, Ghé C, Catapano F, Muccioli G, Galleyrand JC, Fehrentz JA, Martinez J. New active series of growth hormone secretagogues. *J Med Chem*. 2003;**46**:1191–1203.
62. Li Petri G, Di Martino S, De Rosa M. Peptidomimetics: an overview of recent medicinal chemistry efforts toward the discovery of novel small molecule inhibitors. *J Med Chem*. 2022;**65**:7438–7475.
63. Smith RG, Van der Ploeg LH, Howard AD, Feighner SD, Cheng K, Hickey GJ, Wyvratt MJ Jr, Fisher MH, Nargund RP, Patchett AA. Peptidomimetic regulation of growth hormone secretion. *Endocr Rev*. 1997;**18**:621–645.
64. Koch L. Growth hormone in health and disease: Novel ghrelin mimetic is safe and effective as a GH stimulation test. *Nat Rev Endocrinol*. 2013;**9**:315.
65. Garcia JM, Swerdloff R, Wang C, Kyle M, Kipnes M, Biller BM, Cook D, Yuen KC, Bonert V, Dobs A, Molitch ME, Merriam GR. Macimorelin (AEZS-130)-stimulated growth hormone (GH) test: validation of a novel oral stimulation test for the diagnosis of adult GH deficiency. *J Clin Endocrinol Metab*. 2013;**98**:2422–2429.

Index